Computer Communications and Networks

The **Computer Communications and Networks** series is a range of textbooks, monographs and handbooks. It sets out to provide students, researchers, and non-specialists alike with a sure grounding in current knowledge, together with comprehensible access to the latest developments in computer communications and networking.

Emphasis is placed on clear and explanatory styles that support a tutorial approach, so that even the most complex of topics is presented in a lucid and intelligible manner.

More information about this series at http://www.springer.com/series/4198

Yulei Wu • Sukhdeep Singh • Tarik Taleb
Abhishek Roy • Harpreet S. Dhillon
Madhan Raj Kanagarathinam • Aloknath De
Editors

6G Mobile Wireless Networks

 Springer

Editors
Yulei Wu
University of Exeter
Exeter, UK

Tarik Taleb
Aalto University
Aalto, Finland

Harpreet S. Dhillon
Virginia Tech
Blacksburg, VA, USA

Aloknath De
Samsung R&D India
Bangalore, Karnataka, India

Sukhdeep Singh
Samsung R&D India
Bangalore, Karnataka, India

Abhishek Roy
MediaTek Inc
San Jose, CA, USA

Madhan Raj Kanagarathinam
Samsung R&D India
Bangalore, Karnataka, India

ISSN 1617-7975   ISSN 2197-8433   (electronic)
Computer Communications and Networks
ISBN 978-3-030-72779-6   ISBN 978-3-030-72777-2   (eBook)
https://doi.org/10.1007/978-3-030-72777-2

This Springer imprint is published by the registered company Springer Nature Switzerland AG
The registered company address is: Gewerbestrasse 11, 6330 Cham, Switzerland

Preface

3GPP recently completed its Release 16 in June 2020 which is a key milestone for the specification of an initial full 5G system. It has also started the discussion for Release 17 and the planning for Release 18. As a new generation of cellular technology typically appears every 8–10 years, 6G can be expected around 2030. A few organizations, companies, universities, and countries have already started the initial research of 6G. The FCC has taken the first step of opening up terahertz wave spectrum (frequencies between 95 GHz and 3 THz), citing that it will "expedite the deployment of new services in the spectrum above 95 GHz." In early 2018, the University of Oulu in Finland announced the funding of their 6G Flagship program to research materials, antennas, software, and more that could play a role in the launch of 6G. The idea is to explore use cases for which 5G may not be sufficient and then develop algorithms, protocols, and hardware to enable such use cases. 6G could offer high-fidelity holograms, multisensory communications, Terahertz (THz) communications, and pervasive artificial intelligence. The overarching aim of this book is to explore the evolution from 5G towards 6G from a service, air interface, and network perspective, as well as to take a crystal ball perspective on what the future with 6G may look like.

This book lays out a vision for 6G and discusses key research directions towards sustainable development goals and societal challenges while creating true productivity through radically new technological enablers of 6G mobile wireless networks. This book not only shows the potential 6G use cases, requirements, metrics, and enabling technologies but also discusses the emerging technologies and topics such as 6G PHY technologies, reconfigurable intelligent surface, millimeter-wave/THz links, visible light communications, transport layer for Tbit/s communications, high-capacity backhaul connectivity, cloud nativeness, machine-type communications, edge intelligence and pervasive AI, network security and blockchain, and the role of open-source platform in 6G. The state-of-the-art research results, methods, and information on these emerging technologies are of particular interest. How emerging technologies can contribute to making the vision of 6G a reality is also an important aspect to be discussed in this book.

This book provides a more systematic view on these emerging technologies towards 6G in support of a wider variety of verticals. In addition, it presents the expected applications with the requirements and the possible technologies for 6G communications. This book also outlines the possible challenges and research directions to facilitate the future research and development of 6G mobile wireless networks.

This is the world's first book on 6G Mobile Wireless Networks that aims to cover a comprehensive understanding of key drivers, use cases, research requirements, challenges, and open issues for 6G from researchers across the globe. In this book, we have invited world experts in this area, including company representatives, academic researchers, and other builders and members of communication society to contribute the chapters. This book covers the following topics:

- 6G Use Cases, Requirements, Metrics, and Enabling Technologies
- PHY Technologies for 6G Wireless
- Reconfigurable Intelligent Surface for 6G Wireless Networks
- Millimeter-Wave/THz Links for 6G Wireless
- Challenges in Transport Layer for Tbit/s Communications
- High-Capacity Backhaul Connectivity for 6G Wireless
- Cloud Nativeness for 6G Wireless Networks
- Machine-Type Communications in 6G
- Edge Intelligence and Pervasive AI in 6G
- Blockchain: Foundations and Role in 6G
- Role of Open-source Platforms in 6G
- Quantum Computing and 6G Wireless

Targeted audiences: This book targets both academia and industry readers. Graduate students can select promising research topics from this book that are suitable for their thesis or dissertation research. Researchers will have a deep understanding of the challenging issues and opportunities of 6G mobile wireless networks and can thus easily find an unsolved research problem to pursue. Industry engineers from IT companies, service providers, content providers, network operators, and equipment manufacturers can get to know the engineering design issues and corresponding solutions after reading some practical schemes described in the chapters.

We have required all chapter authors to provide as many technical details as possible. Each chapter also includes references for readers' further studies and investigations. If you have any comments or questions on certain chapters, please contact the editors for more information.

Thank you for reading this book. We wish that this book can help you with the scientific research and practical problems of 6G mobile wireless networks.

Disclaimer

The ideas and proposals discussed in the chapters are individual thoughts of the authors and do not reflect any organization's thoughts or information.

Exeter, UK   Yulei Wu
Bangalore, Karnataka, India   Sukhdeep Singh
Aalto, Finland   Tarik Taleb
San Jose, CA, USA   Abhishek Roy
Blacksburg, VA, USA   Harpreet S. Dhillon
Bangalore, Karnataka, India   Madhan Raj Kanagarathinam
Bangalore, Karnataka, India   Aloknath De

The original version of this book was revised: Revised book has been uploaded to Springerlink. The correction to this book is available at https://doi.org/10.1007/978-3-030-72777-2_22

Contents

Chapter 1
Introduction

Sukhdeep Singh, Yulei Wu, and Tarik Taleb

The fifth generation mobile wireless communications steered services across a wide range of sectors such as transport, healthcare, retail, finance, factories, etc. Deployment of ultra-low latency and high bandwidth 5G networks has ushered the way to digitalization and automation across several business domains that lead to ginormous growth of data traffic and number of smart devices. In the next decade, the societal needs will continue to increase prolifically which will give rise to diverse use cases that cannot be served under the umbrella of 5G networks like holographic teleportation, XR, remote surgery, UAVs, etc. that would entail micro-second factor latency and Tbps level bandwidth. Apart from this, shift from Industry 4.0 to Industry X.0 in the next decade will shepherd the way for surge in connectivity density beyond the limits of what 5G networks promise. This also calls for the refinement of energy efficiency practices beyond 5G. As a result the research community is drifting its focus towards the next frontier in mobile communications, i.e., 6G wireless systems which are poised to accommodate use cases and linked technologies to overcome limitations and bottlenecks of current 5G networks. Along with the deployment of 5G networks taking place around the world, 3GPP Rel. 17 and Rel. 18 have started complaisanting the features that would be part of beyond 5G networks. Several R&D centres, universities, industries and standardization bodies across the globe have already led down its vision for 6G wireless and subsequently kicked off the research in the direction of their foresight. Since 6G is anticipated to be a decade away, every organization has made their own proposition of 6G wireless.

S. Singh (✉)
Samsung R&D Institute India-Bangalore, Bangalore, Karnataka, India
e-mail: sukh.sandhu@samsung.com

Y. Wu
Department of Computer Science, University of Exeter, Exeter, UK

T. Taleb
Department of Communication and Network, Aalto University, Espoo, Finland

© The Author(s), under exclusive license to Springer Nature Switzerland AG 2021
Y. Wu et al. (eds.), *6G Mobile Wireless Networks*, Computer Communications
and Networks, https://doi.org/10.1007/978-3-030-72777-2_1

Use cases
Requirements
Metrics
Virtualization
Enabling technologies
Network slicing
PHY technologies
NG 6G RAN
Resource allocation
Federated learning
Blockchain
Transport layer
Quantum Communications
Open source
Edge computing and intelligence
Machine Type Communications
Reconfigurable Intelligent surfaces
Cloudification (AI enabled cloud/Cloud fog architecture)

Fig. 1.1 End to end perspective of 6G mobile wireless networks

However, in reality the 6G wireless use cases, requirements and technologies might or might not go beyond the speculations made as of today. This book aims to bring different prepositions being made from end to end perspective (as shown in Fig. 1.1) which can serve as a pre-standard research for various organizations working in a wide range of topics on 6G wireless. It aims to cover the end to end perspective of 6G in the following 20 chapters.

Chapter 2 covers a variety of unparalleled emerging applications that will prove to be the primary force behind the research for 6G technology. Next, it describes the new requirements generated by these novel applications which will set the standards for 6G technology. Subsequently, it describes the major performance metrics for 6G.

Chapter 3 reviews the use cases and relative key enabling technologies that will characterize the future 6G framework. It summarizes the key challenges, potentials, and use cases of each listed technology. It identifies three critical areas of 6G wireless: (i) communications at Terahertz and optical frequencies for ultra-high-speed broadband access, (ii) cell-less architectures to enable ubiquitous 3D coverage, and (iii) intelligent networks to simplify the management of complex networks and reduce costs. After summarizing envisioned use cases and corresponding Key Performance Indicators (KPIs) in the 6G ecosystem, it provides the review of the characteristics of these innovations and speculate about whether and how they will satisfy the most stringent 6G network demands in a holistic fashion, in view of the foreseen economic, social, technological, and environmental context of the 2030 era.

Chapter 4 presents the overview of a few pressing PHY layer challenges in 6G whose investigation will cross-leverage expertise in signal processing, information theory, electromagnetics and physical implementation. A demanding dialog of the research challenges and possible solutions that must be addressed from applications point of view, to design of the next generation 6G PHY have been discussed in this chapter. The road to overcome the challenges given in this chapter is full of obstacles, and it is able to provide enough insights to begin research towards the many promising open directions.

Chapter 5 provides an overview of several important challenges in the physical layer design on Reconfigurable intelligent surfaces (RIS)-aided 6G wireless networks. The RIS channel estimation problem is investigated and two Bayesian-inference based channel estimation approaches are presented for the RIS-aided single-user MIMO and multi-user MIMO systems. Besides, this chapter presents an analytical framework for evaluating the asymptotic estimation performance for the latter approach. Finally, the high accuracy and efficiency of the presented approaches are demonstrated by extensive numerical simulations.

Chapter 6 provides an in-depth study of millimeter wave and terahertz communication systems and their relevance to the emerging 6G wireless systems. Particular emphasis is given on educating the readers about the propagation characteristics of these bands, their implications on the design considerations, key implementation and deployment challenges, as well as potential applications that are driving innovation in these bands. The chapter concludes with a brief discussion on the current standardization activities related to the use of these bands for commercial communications.

Chapter 7 explores the challenges in the design of next-generation transport layer protocols (NGTP) in 6G Terahertz communication-based networks. Some of the challenges due to user mobility, high-speed and high-bitrate communications, and other issues are discussed. The impact of these issues and potential approaches to mitigate these challenges are also presented. It provides a notion on how a transport layer with a centralized cross-layer (add-on approach) intelligence can help the end-to-end network operate more efficiently.

Chapter 8 proposes the Mode Hopping (MH) scheme which is expected to be a new technique for anti-jamming in 6G wireless communications. Furthermore, it puts forward Mode Frequency Hopping (MFH) scheme to further enhance the anti-jamming performance for wireless communications. The proposed MH and MFH schemes using binary Differential Phase Sifting Keying (DPSK) modulation can achieve better anti-jamming results than those using binary Frequency Shift Keying (FSK) modulation.

Chapter 9 scrutinizes the challenges of hybrid lightwave/RF networking. Furthermore, the interplay between hybrid lightwave/RF networks and enabling technologies of the next generations of wireless access is discussed. Finally, the concept of cross-band network design is presented as an enabling technology for the sixth generation of wireless networks (6G).

Chapter 10 provides resource allocation perspective for 6G optical wireless systems. This chapter introduces and discusses the optimization of resource allocation in an optical wireless communication (OWC) system, specifically, a visible light communication (VLC) system. It introduces an indoor OWC system that provides multiple access using wavelength division multiple access (WDMA). The optimization of resource allocation in terms wavelengths and access points to different is considered in this work by maximizing the sum over all users of Signal to Interference-plus-Noise Ratio (SINRs). A Mixed Integer Linear Programme (MILP) model has been developed to optimize the resource allocation.

Chapter 11 provides a survey of the current works to develop the idea of a future machine-centric 6G network that does not impede on Human-Type Communications (HTC), from the physical to the application layer. It summarizes the devices and applications that will participate in the 6G MTC landscape. It then discuss the fundamental techniques enabling MTC at physical and access layers. Furthermore, it moves on to the network as well as transport layers and discusses the technologies and strategies that 6G should develop to achieve the requirements. Finally, it puts forward the application-specific requirements and discusses how a cross-layer approach can bring convergence between these requirements.

Chapter 12 gives an overview on edge intelligence and how it is a key enabler for 6G technologies. Specifically, it provides an insight into the wireless requirements and transformations seen in this new breed of services that necessitate an intelligent backbone supporting their network interactively far from the conventional role of edge computing in 5G. Subsequently, it looks into the necessary changes that need to occur on the AI mechanism governing the edge, in order to successfully deliver complex 6G applications with their promised performance.

Chapter 13 covers in-depth study of a distributed, autonomous, and federated AI-enabled cloud and edge computing for 6G wireless. It provides an in-depth study on the envisage use cases, network architectures, deployment scenarios, and technology-driven paradigm shifts for the 6G networks. Besides, it presents AI as an efficient platform for realizing high levels of automation that are essential for the optimization and management of the current and future complex networks. Moreover, key technical areas and technology advances such as cognitive spectrum sharings, photonics-based cognitive radio, innovative architecture models, terahertz communications, holographic radio, and advanced modulation schemes are presented for the 5G and beyond networks.

Chapter 14 evaluates a cloud fog architecture for future 6G networks paying special attention to energy efficiency and latency. It proposes a Mixed Integer Linear Programming (MILP) model that is generic and independent of technology and application. It later proves the effectiveness of cloud fog/cloud DC with respect to power savings, processing capacity and efficiency.

Chapter 15 aims to provide a comprehensive overview on key concepts towards the RAN of the 6G system. It provides a study of various legacy RAN implementations, reported in literature, from different perspectives, and the motivation of their reconstruction and redesign with respect to the requirements of their forthcoming generations. Furthermore, it presents an exclusive review of the NG-RAN with special emphasize on the cloudification and virtualization of resources and services, and on the management and orchestration of RAN slices. It identifies the key drivers that will leverage advanced mMTC, uRLLC, and eMBB services and applications; and finally addresses key challenges and future research arising from the deployment of NG 6G RAN.

Chapter 16 addresses the preliminaries and key challenges of combining Federated Learning (FL) techniques with 6G wireless networks. Moreover, it investigates the delay minimization problem of FL over wireless communication networks. The tradeoff between computation delay and transmission delay is determined by the learning accuracy. To solve this problem, first it proves that the total delay is a convex function of the learning accuracy. Then, the optimal solution is obtained by using the bisection method. Simulation results show the various properties of the proposed solution.

Chapter 17 discusses the need of open source in the 6G wireless. Subsequently, it puts forward two major backbone of 6G i.e. Intelligence and Automation. It discusses the role of open source in providing the swiftness and interoperability to the key enablers of Intelligence and Automation. Finally, it investigates the open issues and challenges that can be faced while realizing the openness.

Chapter 18 provides an overview of the potential benefits blockchain has to offer for 6G networks and services, and the interplay of blockchain with key enabling technologies for 6G. Furthermore, it provides an overview of blockchain technology and related concepts, and gives insight into blockchain-based applications leveraging mobile networks. Finally, it provides an overview of the challenges related to blockchain-based applications.

Chapter 19 explores various tenets of quantum communication and describes the challenges related to the same. Furthermore, it describes the role quantum communication will play in 6G and beyond communications followed by some practical Considerations and future of quantum communications.

Chapter 20 provides an overview of the state-of-the-art post-quantum secure public-key primitives for key establishment, encryption and digital signatures selected into the third round of the NIST PQC competition for post-quantum secure cryptography standardization. It also discusses their properties and the effect on the performance of the future 6G networks.

Chapter 21 covers the open issues and provides the concluding remarks from the perspective of following seven dimensions of 6G research covered in the book by various authors across the globe:

1. Base layer design for Tera-hertz communication.
2. High-precision network, dynamic topology and open source.
3. Machine-communication in IoT with thrust on energy efficiency.
4. All-pervasive AI with learnability and explainability.
5. Edge and fog computing with split processing.
6. Security, privacy and trustworthiness of system.
7. Rich-media use cases and services.

Chapter 2
6G Use Cases, Requirements, and Metrics

Navrati Saxena, Eshita Rastogi, and Ayush Rastogi

Abstract 5G wireless network, commercially launched in recent years, promises to improve network performance and ensure ubiquitous connectivity. However, numerous emerging applications like extended reality, telepresence, telesurgery, autonomous driving etc. require very high data rate while simultaneously maintaining the related reliability and latency constraints. Coupled with a massive surge in number of smart devices and Internet of Things (IoT), these requirements are likely to saturate the 5G network in future and have, therefore, motivated the research community to think beyond 5G wireless technology. The discussions on 6G wireless technology have started taking shape with the primary objective to serve future demands in 2030. The 6G cellular wireless network will intend to satisfy the huge data rate, latency, and reliability requirements across a myriad of smart devices. Through this chapter, we explain the interesting and novel applications of 6G along with its key requirements and performance metrics.

1 Introduction

A gradual and steady evolution of wireless networks from second to fifth generations has been witnessed over the last three decades. The penetration of smartphones, IoT devices, and emerging new multimedia applications has significantly contributed towards an unprecedented increase in mobile data traffic. With smart devices

N. Saxena (✉)
Department of Computer Science, San Jose State University, San Jose, CA, USA
e-mail: navrati.saxena@sjsu.edu

E. Rastogi
Department of Electrical and Computer Engineering, Sungkyunkwan University, Suwon, South Korea
e-mail: eshita@skku.edu

A. Rastogi
System Design Lab, Networks Division, Samsung Electronics Corporation, Suwon, South Korea
e-mail: ra.ayush@samsung.com

Y. Wu et al. (eds.), *6G Mobile Wireless Networks*, Computer Communications and Networks, https://doi.org/10.1007/978-3-030-72777-2_2

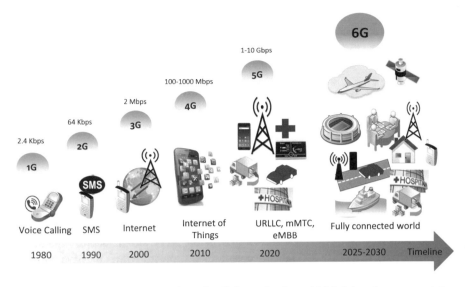

Fig. 2.1 Road map of the evolution of cellular technology, highlighting the representative applications, from 1G to 6G

becoming popular by each passing day, all-IP based 4G and 5G networks are regularly used by millions of people. This has resulted in the emergence of a set of novel applications like video conferencing, augmented/virtual reality (XR) applications, streaming of video in real time, telemedicine, and online gaming. Other such important applications include smart cities, inter-connected vehicles for autonomous driving, connected smart healthcare systems, etc.. Along with satisfying users' requirements, these applications also enable to wireless operators to boost their revenues by opening up new business use-cases.

Figure 2.1 delineates the continual increase in demand and myriad of use-cases which have come up from the initial 1G cellular wireless networks technology to their evolution towards 6G technology. The demands for higher data rate and further increase in network capacity is the primary reason for the continuous evolution of cellular wireless networks where, in 2018, 3GPP-Release 15 had standardized the 5G of cellular networks [1]. 5G technology is a game-changer and connects an massive number of devices and machines. 5G New Radio (NR) has a potential to cover up to 65% total population round the globe by 2025. This will generate up to 45% of all the mobile data traffic in the world.[1] 5G NR offers flexible network design to deal with a large variety of applications related to enhanced Mobile Broadband (eMBB), Ultra-Reliable Low Latency Communications (URLLC), massive Machine Type Communications (mMTC), and enhanced

Vehicle to Everything (eV2X). Considering the current trends, 5G will lead to massive site densification causing a large increase in network capacity. 3GPP has also included the applicability of NR over Unlicensed band (NR-U) for both uplink and downlink operations [2]. NR-U has the potential to enhance the performance of several applications such as industrial IoT, private networks, etc. by extending the bandwidth resources on unlicensed band.

With the advances in communication and other technologies such as Artificial Intelligence (AI) and Big Data, some new and interesting applications like Extended Reality (XR) and holographic display are emerging. These applications require extremely high data rate and are likely to exhaust the existing wireless capacity. For instance, 16K Virtual Reality (VR) requires a throughput of 0.9 Gbps to deliver sufficient user experience. The existing 5G technology cannot provide this high data rate required for seamless streaming. Similarly, extremely high data rate transmission is required to provide holographic display. The hologram display over a normal sized mobile phone requires at least 0.58 Tbps of data rate, which is much higher than the peak data rate offered by 5G technology.[2] Thus, with the perpetual introduction of new services, wireless vendors and operators predict that 5G will reach its limits by 2030 as the chase for data, capacity, and coverage will continue unabated.

Sixth Generation (6G): A Step Beyond 5G
While the 5G cellular system is in its deployment phase, a discussion of 6G is slowly but surely gaining momentum and 6G technology expected to emerge by 2030. A plethora of interesting data-intensive applications ranging from XR applications, e-healthcare, and Brain-Computer Interaction (BCI) to flying vehicles and connected autonomous driving might become a reality in the future due to 6G, which aims to revolutionize the evolution of cellular wireless from *"connected things"* to *"connected intelligence"* with more stringent Key Performance Indicators (KPIs) [3]. The vision of 6G can be summarized by the phrase *"ubiquitous wireless intelligence"* [4], where ubiquitous refers to omnipresent coverage and seamless accessibility to services. The users form an inter-connected network over the wireless link making the wireless connectivity crucial. 6G wireless technology envisions to make the network and the smart devices cognizant and content-aware for all the users by exploiting the concepts of AI, Machine Learning (ML), and Big Data techniques, which will facilitate autonomous decision making in smart devices [4]. ML and Big Data techniques will also help the network components to handle and evaluate a currently-intractable amount of real-time data. This will enable the network to automatically modify itself based on users' experiences; for example, by allowing fast spectrum allocation and re-allocation.

Wireless networks must deliver massive data rate with high reliability and extremely low latency to successfully handle these services simultaneously over

---

[2]https://news.samsung.com/global/samsungs-6g-white-paper-lays-out-the-companys-vision-for-the-next-generation-of-communications-technology.

both uplink and downlink. This fundamental rate-latency-reliability trade-off will make it mandatory to exploit frequencies beyond GHz-range and transform the wireless network. The emergence of 6G will hopefully provide multi-fold improvement in data rate, latency, user capacity, and 3-dimensional coverage. It will integrate several verticals such as sensing, communication, computing, positioning, navigation, and imaging. Quantum computers and networks will solve previously hard-to-solve problems and will also help in ensuring privacy and security of personal data.

The aforementioned applications and use-cases will give rise to new set of requirements. Although this is an early stage to formally discuss and describe future 6G technology, yet it has no doubt attracted researchers and academicians across the globe and started taking shape. Currently, there are already several initiatives, such as *6G Flagship*[3] research program of University of Oulu, Finland and *Terabit Bidirectional Multi-user Optical Wireless System (TOWS) for 6G LiFi*[4] by the government of U.K., working towards developing 6G technologies. Global telecommunication organizations such as SK Telecom, Ericsson, Nokia, Samsung, etc. and academic researchers have also started working towards building 6G network. Compared to 5G wireless system, 6G is a paradigm shift which includes Terahertz Communications (THz), Super-Massive Multi-Input Multi-Output (SM-MIMO), Large Intelligent Service (LIS), Holographic Beamforming (HBF), Orbital Angular Momentum Multiplexing (OAM), Laser Communication, Visible-Light Communication (VLC), blockchain-based spectrum sharing, and Quantum Computing [5, 6]. We briefly describe the key enablers of 6G technology in Fig. 2.2.

Organization of the Chapter In the rest of the chapter, first, we cover a variety of unparalleled emerging applications that will prove to be the primary force behind the research for 6G technology. Next, we describe the new requirements generated by these novel applications which will set the standards for 6G technology. Subsequently, we describe the major performance metrics for 6G.

2 Emerging Application of 6G Wireless Network

The major services offered by 5G can be categorised as eMBB, URLLC, and mMTC. The applications supported by these services continue to improve as the time progresses and have motivated the researchers to think beyond 5G technology. Some of the applications include holographic telepresence, XR, ultra-smart cities, remote surgery, HD imaging, distance education, autonomous driving, etc., as shown in Fig. 2.3. These applications set different requirement criteria and

---

[3] A research program lead by University of Oulu. It primarily focuses of the adoption of 5G ecosystem and innovation of 6G.

[4] A research organization which focuses on the increase in the wireless capacity by exploring new spectrum.

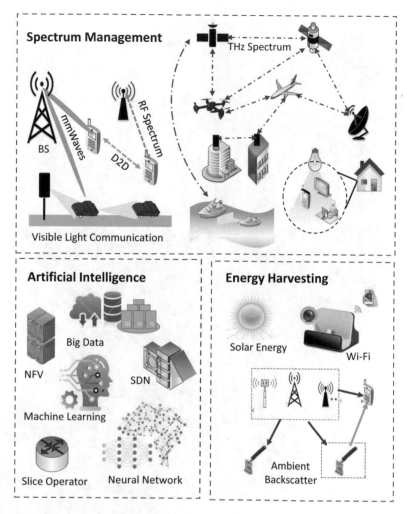

Fig. 2.2 Enabling technologies of 6G wireless networks

standards, such as an enormous number of connections, multifold increase in data rate, and stringent reliability and latency requirements, that need to be explored while developing 6G technology. Some of aforementioned applications are already handled by 5G technology but 6G, in addition, envisions to deal with the excess burden which the resource-constrained 5G is unable to handle. Interestingly, the 6G network will expand the communication environment to cover not only ground and sky but also space and under-water. The 6G technology is still in its infancy and it is hard to forecast the complete horizon of applications of 6G. In this section, we enumerate some of the plausible uses-cases.

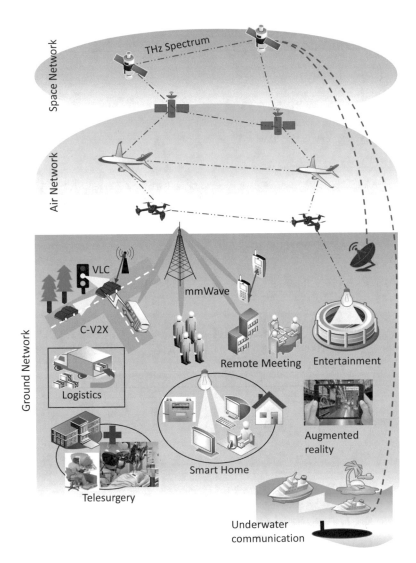

Fig. 2.3 Emerging applications of next-generation wireless 6G network

2.1 Virtual, Augmented, and Mixed Reality

Virtual Reality (VR) services are a simulated experience which will enable the user
to experience a virtual and immersive environment with a first-person view. VR
technology has a potential to allow geographically-separated people to communicate effectively in groups. They can make eye contact and can manipulate common
virtual objects. It will require the real-time movement of extremely high resolution

electro-magentic signals to geographically-distant places to relay various thoughts and emotions. XR with high definition imaging and high resolution of 4K/8K will be suitably used in innovative applications such as entertainment services (including video games and 3D cameras), education and training, meetings with physical and social experience, and work-space communication, etc. [7]. These new applications are likely to saturate the existing 5G spectrum because they will require data rates above 1 Tbps. Additionally, real-time user interaction in an immersive environment is bound to require minimum latency and ultra-high reliability.

2.2   Holographic Telepresence

With the on-going developments in technology, humans have become heavily dependent on the innovations supported by technology. One such innovation is holographic telepresence, which can make persons or objects present in different locations appear right in front of another person. It has the potential to bring a tremendous change in the way we communicate and is slowly becoming a part of the mainstream communication systems. Some interesting applications of telepresence include enhancing movie and television viewing experience, gaming, robot control, remote surgery, etc. The human tendency to connect remotely might experience a transition from a conventional video conference to a virtual in-person meeting, thus reducing the necessity of travel for business. To capture the physical presence, a three-dimensional image coupled with stereo voice is required. The three-dimensional holographic display will require a massive data rate of approximately 4.32 Tbps over ultra-reliable communication network [8]. Similarly, the latency requirements will be in the order of sub-ms, which will help in synchronizing many viewing angles. This will obviously pose severe communication constraints on the existing 5G network.

2.3   Automation: The Future of Factories

Over the past few years, advances in robotic, augmented reality, industrial IoT, AI, and ML have the potential to radically revamp the manufacturing industries. While the concept of Industry 4.0 emerged with 5G, future 6G along with the AI tools will substantially transform the industries and manufacturing processes. The potential of 6G will provide a foundation of Industry 5.0. The use of an automatic control system and communication technology will most likely reduce the need for human intervention in the industrial processes. Additionally, these industries will demand high standards in terms of reliability, transmission latency, and security to ensure effective management, maintenance, and operation of real factories. This industrial setup is expected to be dependent on mobile robots and drones to realize automation and transportation within warehouses. Since robotic communication will suffer from severe instabilities due to delays or malicious action, the future 6G network is expected to precisely consider the requirements like reliability, delay-sensitivity, security etc..

2.4 Smart Lifestyle with Integrated Massive IoT

Future 6G technology will contribute towards building a smart city environment. This environment will connect millions of applications including utilities (electricity, water, and waste management), smart transportation, smart grid, residential surroundings, telemedicine, happy shopping with assured security, etc. Major cities will further see the advent of flying taxis, which is expected to have stringent requirements of connectivity. The quality of life of people will substantially ameliorate owing to the seamless and ubiquitous connectivity of the smart devices.

Smart homes are an important part of smart lifestyle. Initially, the concept of smart home mainly progressed via the development of utility meters and smart devices but now, this trend is revamping because IoT is in a mature state and is capable of providing ubiquitous connectivity at all times by facilitating close integration among the home appliances. Naturally, this requires high data rates and extremely high security of users' personal data. 6G system envisions to fulfill these stringent requirements of data rate, latency, and security by providing the necessary infrastructure and by fully integrating the devices with AI for autonomous decision-making at home.

2.5 Autonomous Driving and Connected Vehicles

Transport industry is likely to experience a massive transformation in the upcoming years. Self-driving, remote-driving, platooning, and intelligent roads, etc. are expected to take a new shape leveraging the benefits of 6G. Along with improved reliability, extremely-low latency, and high precision positioning, V2X communication also needs to securely exchange massive amounts of data related to driving and ambient environment. Continuous updates of live traffic and real-time hazard information on the road are essential for ensuring the safety of humans. Also, the vehicles move at a very high speed due to which the round trip time for communications should be extremely low. 6G networks provide efficient mobility management and handover in vehicular communication while maintaining the reliability and latency requirements. Additionally, AI will undoubtedly enhance the V2X communication scenario by processing autonomous commands to control transport network. Leveraging the benefits of 6G networks and AI, the traffic control system can proactively explore and evaluate the road traffic load, topology, accidents, etc., thus enhancing public safety.

2.6   Healthcare

The healthcare sector has undergone a colossal change because of the benefits of 5G cellular network and will continue to be revolutionized with the help of upcoming 6G technology. Aging populations impose a huge burden on the existing healthcare system. Because of the proliferation of e-Health services, hospitals are coming up with ubiquitous health monitoring systems that can monitor various human health indicators such as temperature, blood pressure, blood sugar, etc. and report the medical data to the concerned department. Besides regular monitoring, health industry is working towards remote surgeries which, again, will have stringent QoS requirements. Multiple requirements such as extremely low latency, network coverage, ultra-high data rate, high reliability, etc. are the prerequisites of a successful tele-surgery. Both industry and academia are extremely interested in the research in communication and networking domain related to accessing remote healthcare.

2.7   Non-terrestrial Communication

Connectivity outage during natural disasters causes immense damage to human life, property, and business. 6G technology will research Non-Terrestrial Communication (NTC) to support ubiquitous coverage and massive-capacity global connectivity. To overcome the coverage limitations of 5G, 6G technology endeavours to explore Non-Terrestrial Networks (NTN) to support global, ubiquitous, and continuous connectivity. NTN can help in dynamically offloading traffic from terrestrial components and also, in reaching unserved areas. Thus, non-terrestrial stations such as Unmanned Aerial Vehicles (UAVs), High Altitude Platform Stations (HAPSs), drones, and satellites are likely to complement terrestrial networks [9]. This will impart multiple benefits like cost-effective coverage in congested locations, support high-speed mobility, and high throughput services. NTNs can be considered to support applications which include meteorology, surveillance, broadcasting information, remote sensing, and navigation [10]. They can complement the terrestrial towers by serving as an alternative path in case the terrestrial tower is out of service. Similarly, NTC can help in broadcasting warning and entertainment content to a large static as well as mobile audience. Long distance and inter-satellite transmission can be achieved by laser communication.

2.8   Under-Water Communication

The under-water communication environment is crucial to ensure worldwide connectivity as water covers the majority of the Earth's surface. Under-water networks will be employed to provide connectivity as well as observing and monitoring various oceanic and deep-sea activities. Unlike land, water exhibits different propagation characteristics and therefore, bi-directional under-water communication requires more under-water hubs and also needs to use acoustic and laser communications to accomplish data transmission with a high speed [11]. Underwater networks establish the connection between submerged base stations and communication nodes like submarines, sensors, divers, etc. Moreover, this underwater communication network can also coordinate with terrestrial networks.

2.9 Disaster Management

The primary focus of the upcoming 6G technology is to provide rate-latency-reliability, deep coverage, and localization accuracy, as mentioned above. Leveraging the benefits of these stringent requirements of 6G, it can be assumed that upcoming technology might prove to be more efficient while dealing with natural disasters and pandemics such as COVID-19. With the deeper penetration of internet access, it will become easier to propagate safety messages and track people with higher accuracy. This will lead to people in disaster zones being well-informed of the situation as it unfolds and being connected to the rest of the world in times of crisis. Drones and flying taxis will help in dealing with the crises by providing food and medicines without much human intervention, thus ensuring human safety. The inclusion of NTN and UAVs are expected to help in reaching out to the under-served and the un-served areas in case of any natural calamity.

2.10 Environment

6G wireless network is expected to provide massive coverage with the help of UAVs, NTN, satellites, etc. and aims to cover major remote and inaccessible areas with the help of these technologies. This is likely to help in remotely monitoring environmental conditions, wildlife, pollution, agriculture etc. Due to the capability of flying autonomously in a pre-defined path, UAVs can monitor active hazardous locations in real-time. Similarly, UAVs can be equipped with sensors that can sense pollution level in air, emissions in industrial zones etc. [12]. The condition of forests in remote areas and the wildlife there can also be traced in real-time.

3 Requirements and KPI Targets of 6G

The aforementioned innovative use-cases call for the necessity of re-defining the requirements of upcoming 6G technology. 5G technology has abundant potential

Table 2.1 Comparative study of KPIs of 5G and 6G

Parameters	5G	6G
Data rate: downlink	20 Gb/s	> 1 Tb/s
Data rate: uplink	10 Gb/s	1 Tb/s
Traffic capacity	10 Mb/s/m^2	1–10 Gb/s/m^3
Latency	1 ms	10–100 μs
Reliability	Upto 99.999%	Upto 99.99999%
Mobility	Upto 500 km/hr	Upto 1000 km/hr
Connectivity density	10^6 devices/Km^2	10^7 devices/Km^2
Security and privacy	Medium	Very high

but it cannot satisfy the stringent rate-reliability-latency requirements of the new applications. The requirements and KPIs of 6G technology will be stricter and more diverse. For example, while the 5G network is already operated in the very high frequency mm-waves region, 6G could require even higher frequencies for operation. The 6G technology will focus on achieving higher peak data rate, seamless ubiquitous connectivity, non-existent latency, high reliability, and strong security and privacy for providing ultimate user experience. Table 2.1 describes the comparative study of the KPIs of both 5G and 6G. In this section, we formally describe the major requirements of 6G technology.

3.1 High Data Rate

The major concern of 6G wireless system design will be to provide extremely high data rate as compared to 5G. It aims to deliver a peak data rate of more than 1 Tbps per user and an average data rate of 1 Gbps per user by exploiting new spectrum. In current wireless networks, while the data rate can be increased by improving spectral efficiency using MIMO, combining multiple RATs, etc. but it still cannot go up to 1 Tbps. Such high data rate can be achieved by using very high bandwidths in the THz band and visible light spectrum and by exploring OAM multiplexing, laser communication, and SM-MIMO with more than 1000 antenna elements. Quantum communication also has a potential to contribute considerably towards increasing the data rate. This massive increase in data rate is expected to enable a new horizon of applications.

3.2 Extremely Low Latency

6G will have stringent requirement of sub-ms or even non-existent latency to support the emerging new applications. The highly interactive real-time applications such as tactile internet and VR targeted for 6G require ultra-low latency. The end-to-end (E2E) delay and air latency is targeted to be 1 ms and 100 µs, respectively. Inclusion of the technologies like THz communication, SM-MIMO, and quantum computing are expected to meet the rigorous latency requirements. Additionally, AI, ML, and Big Data will be involved in assisting the 6G network in determining the optimum way to exchange data among the users and the base station. This is expected to reduce the latency to some extent [13].

3.3 Low Power Consumption

In the 4G and 5G era, the smart phones and other devices need regular charging. High computation power required for processing AI algorithms will lead to fast depletion of battery of a device. 6G technology aims to give sufficient attention to reducing power consumption and ensuring longer battery life of smart devices [14]. To lower the energy consumption, users can off-load their computing tasks to a smart base station with reliable power supply. A decrease in the transmit power of the device can observed by exploring cooperative relay communication and network densification. Several energy harvesting techniques will be applied in 6G to alleviate the need of battery replacement and achieving long battery life. The energy can be harvested from electro-magnetic signals in the ambient environment, sunlight, and micro-vibrations, which will turn into vital power sources for low-power applications [15]. Wireless battery charging technique from a long distance might also help in extending the battery life of the devices.

3.4 High Frequency Bands

Data rate has increased by tenfolds for every subsequent cellular wireless network generation. This massive increase cannot be achieved merely by improving the spectral efficiency of the network. Currently, wider bandwidth at higher frequencies are exploited to accomplish the data rate requirements. Researcher are investigating sub-THz and THz bands as a suitable candidate for 6G. THz communication band is defined from 0.1–10 THz and offers rich spectrum resources for both electromagnetic and light waves. These frequencies will provide multi-Tbps data rate for transmitting data both indoor and outdoor scenarios. THz frequencies, being directional in nature, can significantly mitigate inter-cell interference. Additionally, at higher frequencies, the size of the antenna and the associated radio frequency circuitry will become smaller in size but it will also make the fabrication on chip difficult. These major challenges need meticulous attention of the research community in the coming years [5].

3.5 Ultra-Reliability

6G intends to improve the reliability up to 99.99999% which is 100 times higher as compared to the existing 5G technology. Majority of the applications like industrial automation, remote surgery, etc. require extremely high reliability. Similarly, the performance of services like holography, XR, and tactile internet heavily depends on unparalleled levels of reliability requirements. This *five-nines reliability* is likely to guarantee the QoS requirements for a wide variety of aforementioned applications.

3.6 Security and Privacy

Security and privacy are crucial performance parameters in 6G which are overlooked in both 4G and 5G technologies. 6G technology is likely to satiate security and privacy requirements for a wide variety of supported applications [14]. Quantum Computing is a heavily-researched technology that can help in achieving ultra-secure networks by applying quantum key distribution. However, requirements of supporting a massive data rate and a wide variety of applications are likely to impose certain challenges for security in wireless computing in 6G. Block-chain technology, which is used to encrypt data so that it is difficult to modify and manipulate, will help in securing and authenticating future communication systems.

3.7 Massive Connection Density

One of the major drivers of the development in wireless technology is the perpetual increase of connected users. 10^7 connected devices per square kilometer are expected to emerge by end of this decade. Super-smart and fully-connected environment and lifestyle with millions of inter-connected devices will result in enormous number of inter-connected devices. There will be a plethora of autonomous services giving rise to the concept of Internet of Everything (IoE). These devices will include smart phones, IoT devices, vehicles, drones, and many more. Considering the massive connection density expected to emerge, researchers are exploring new avenues like THz communication, VLC, and blockchain-based spectrum sharing to satisfy it.

3.8 Extreme Coverage Extension

The existing 5G technology mostly concentrates on metropolitan areas and not so much on the developing and the rural areas. Thus, a large swath of population still cannot access good internet services. 6G technology envisions to provide wide area coverage with deep indoor penetration, thus guaranteeing sufficient fairness to rural areas. To accomplish the goal of worldwide connectivity, 6G will provide an inter-connected communication network extending from ground and sky to under-water environment and satellite communication. Since propagation characteristics of water are different, therefore, electromagnetic signals used for conventional wireless communication are not suitable for under-water communication. Laser-based communication with ultra-high bandwidth can achieve high data rates for both free space and under-water communication. Similarly, VLC can be used to provide coverage for indoor hotspots [5].

3.9 Mobility

6G networks are expected to be highly dynamic and will support high user speeds depending on the progression in transportation system. While 5G supports user speeds of up to 500 km/hr, 6G intends to support speeds of approximately 1000 km/hr. However, this will lead to the need of frequent handovers. Additionally, other service requirements like high data rate, high reliability, and low latency are likely to make efficient handover difficult [16]. For example, these issues occur while providing communication services on aeroplanes. The concepts of AI have the potential to support complex decision-making and optimize handover strategies in real-time while still maintaining the latency and reliability requirements.

4   Performance Metrics

6G promises to deliver high Quality of Service (QoS), Quality of Experience (QoE), and Quality of Physical Experience (QoPE) [17]. QoS deals with the network performance while QoE focuses on the performance perceived by the users. QoPE is a newly-introduced metric which also takes human perception into account. Emerging applications like XR, holographic telepresence, autonomous driving, eHealth, under-water communication, etc. are being adopted quickly. To ensure the best possible user experience, network operators are emphasizing more on E2E service quality and therefore, while migrating to 6G networks, it is essential that the advancements in architecture maintain these quality requirements. Hence, a 6G cellular network is expected to deliver QoS, QoE, and QoPE along with reliability and security. The transition from Self-Organizing Networks (SON) to Self-Sustaining Networks (SSN) is likely to help 6G in achieving highly-automated networks which can maintain QoS, QoE, and QoPE on their own [17]. Although measuring the performance metrics is subjective at this moment because of the fact that 6G is still in its infancy, yet, this section highlights the fundamental advancements in QoS, QoE, QoPE, and SSN, which can be effective in future 6G communications.

4.1   Improving Quality of Service

Applications such as XR, holographic telepresence, autonomous driving, e-Health etc. are not a distant future. These applications challenge our ability to meet stringent QoS requirements i.e. massive data rate of 1 Tbps, < 1 ms E2E round trip latency, 99.99999% reliability, ground to aerial coverage, and high spectral and energy efficiency. Hence, the major attributes of QoS include network availability, performance reliability, uninterrupted coverage, security and privacy, and system integrity. Abundance of THz spectrum and super-massive MIMO are likely to reduce resource constraints while guaranteeing higher QoS. AI and ML will also facilitate in efficient utilization of user information which will help in improving the quality of service. However, the emerging 6G applications impose new QoS challenges, which cannot sufficiently addressed by conventional QoS models and parameters. Therefore, exploring new QoS metrics and models, and their inter-dependencies in wireless links under various constraints are necessary to strengthen 6G mobile wireless networks.

4.2   Refining Quality of Experience

QoE is considered as an important performance metric for the 6G era because of the rate-reliability-latency sensitive applications like telepresence, telesurgery, etc. The concept of QoE is difficult to measure because of its subjective nature [18]. 5G QoS metrics, which primarily comprise of rate of packet loss, E2E latency, and round-trip-times, are now considered less effective for the emerging applications of 6G wireless networks. QoE emphasizes on user's perceived satisfaction, aesthetic experience, understanding, and application manageability. Hence, along with the QoS, usage outcome is also an essential parameter to guarantee QoE, as shown in Fig. 2.4. For further improvement in QoE, network and service management requires a significant transformation by further investigating the advances in automation, cognitive operation, AI, and ML.

4.3   Quality of Physical Experience

The upcoming 6G will witness a rise in human-centric services that are tightly coupled with human users. The emerging concept of wireless Brain Computer Interaction (BCI) introduces a new emotion-driven use-case that requires 6G

Fig. 2.4 Relation between QoS, QoE and QoPE; where QoS deals with network performance, QoE focuses on user-perceived performance, and QoPE takes human perception into account

connectivity [17]. With the use of wireless BCI technology, users can not only interact with their surroundings and other people but also the devices embedded in the world around them. Using such technology might help people to control and monitor their environment using gestures and communicate through haptic messages. Performance metrics that will support wireless BCI services will be different from that of 5G technology. Along with rate-latency-reliability requirements, BCI also deals with physical perception and actions of humans, thus, giving rise to a new set of QoPE metrics which will integrate the factors from human physiology with QoS and QoE metrics as shown in Fig. 2.4.

4.4   Self-Sustaining Networks

5G technology initiated the concept of intelligent communication network by introducing AI and ML-based operations. User experience and network automation is likely to improve by decreasing the chances of human interference and automating functionalities like wireless network configuration, optimization, and healing [18]. 6G will require a transition from classical self-organizing networks into an AI-based intelligent, cognitive, and self-sustaining network. This will have the potential to maintain long-term KPIs under highly dynamic and complicated scenarios originating from 6G applications. AI and ML will empower 6G to enable operations like self-aggregation, self-learning, self-optimization, and self-adaptiveness without any human intervention, as shown in Fig. 2.5 [19].

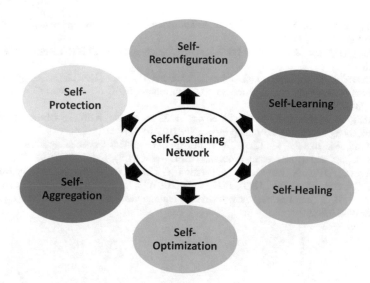

Fig. 2.5 Concept of self-sustaining network (SSN)

References

1. 3GPP TS 38.300, Technical specification group radio access network; NR; NR and NG-RAN overall description. Rel. 16, V. 16.2.0 (2020)
2. 3GPP TR 38.889, Technical specification group radio access network; study on NR-based access to unlicensed spectrum, Rel. 16, V. 16.0.0 (2018)
3. K.B. Letaief, W. Chen, Y. Shi, J. Zhang, Y.J. Zhang, The roadmap to 6G: AI empowered wireless networks. IEEE Commun. Mag. **57**(8), 84–90 (2019)
4. M. Latva-Aho, K. Leppänen, Key drivers and research challenges for 6G ubiquitous wireless intelligence (white paper). Oulu, 6G Flagship (2019)
5. Z. Zhang, Y. Xiao, Z. Ma, M. Xiao, Z. Ding, X. Lei, G.K. Karagiannidis, P. Fan, 6G wireless networks: vision, requirements, architecture, and key technologies. IEEE Veh. Technol. Mag. **14**(3), 28–41 (2019)
6. M. Giordani, M. Polese, M. Mezzavilla, S. Rangan, M. Zorzi, Toward 6g networks: use cases and technologies. IEEE Commun. Mag. **58**(3), 55–61 (2020)
7. E. Bastug, M. Bennis, M. Médard, M. Debbah, Toward interconnected virtual reality: opportunities, challenges, and enablers. IEEE Commun. Mag. **55**(6), 110–117 (2017)
8. X. Xu, Y. Pan, P.P. Lwin, X. Liang, 3D holographic display and its data transmission requirement, in *2011 International Conference on Information Photonics and Optical Communications* (IEEE, Piscataway, 2011), pp. 1–4
9. M. Giordani, M. Zorzi, Satellite communication at millimeter waves: a key enabler of the 6G era, in *2020 International Conference on Computing, Networking and Communications (ICNC)*. (IEEE, Piscataway, 2020), pp. 383–388
10. M. Giordani, M. Zorzi, Non-terrestrial communication in the 6G era: challenges and opportunities (2019, preprint). arXiv:1912.10226
11. Z. Zeng, S. Fu, H. Zhang, Y. Dong, J. Cheng, A survey of underwater optical wireless communications. IEEE Commun. Surv. Tuts. **19**(1), 204–238 (2016)
12. Z. Zeng, S. Fu, H. Zhang, Y. Dong, J. Cheng, A survey of underwater optical wireless communications. IEEE Commun. Surv. Tuts. **19**(1), 204–38 (2016)

13. P. Yang, Y. Xiao, M. Xiao, S. Li, 6G wireless communications: vision and potential techniques. IEEE Netw. **33**(4), 70–5 (2019)
14. S. Dang, O. Amin, B. Shihada, M.S. Alouini, What should 6G be? Nat. Electron. **3**(1), 20–29 (2020)
15. S. Ulukus, A. Yener, E. Erkip, O. Simeone, M. Zorzi, P. Grover, K. Huang, Energy harvesting wireless communications: a review of recent advances. IEEE J. Sel. Areas Commun. **33**(3), 360–381 (2015)
16. H. Yang, A. Alphones, Z. Xiong, D. Niyato, J. Zhao, K. Wu, Artificial intelligence-enabled intelligent 6g networks (2019, preprint). arXiv:1912.05744
17. W. Saad, M. Bennis, M. Chen, A vision of 6G wireless systems: applications, trends, technologies, and open research problems. IEEE Netw. **34**(3), 134–142 (2019)
18. M. Agiwal, A. Roy, N. Saxena, Next generation 5G wireless networks: a comprehensive survey. IEEE Commun. Surv. Tuts. **18**(3), 1617–1655 (2016)
19. M. Piran, D. Y. Suh, Learning-driven wireless communications, towards 6G (2019, preprint). arXiv:1908.07335

Chapter 3
6G Enabling Technologies

Michele Polese, Marco Giordani, Marco Mezzavilla, Sundeep Rangan,
and Michele Zorzi

Abstract While network operators have already started deploying commercial 5th
generation (5G) networks, existing cellular technologies may lack the level of
reliability, availability, and responsiveness requested by future wireless applications.
For this reason, the research community at large is already defining the most
promising technologies that can enable 6th generation (6G) wireless systems. We
have identified three critical innovations: (1) communications at Terahertz and
optical frequencies for ultra-high-speed broadband access, (2) cell-less architectures
to enable ubiquitous 3D coverage, and (3) intelligent networks to simplify the management of complex networks and reduce costs. In this chapter, after summarizing
envisioned use cases and corresponding Key Performance Indicators (KPIs) in the
6G ecosystem, we will review the characteristics of these innovations and speculate
about whether and how they will satisfy the most stringent 6G network demands
in a holistic fashion, in view of the foreseen economic, social, technological, and
environmental context of the 2030 era.

1 Introduction

The advent of a multitude of new data-hungry, delay-sensitive mobile services will
likely introduce severe challenges to current 5G systems. Figure 3.1 illustrates how
the evolution of mobile applications has introduced over the years an exponential
increase of mobile data consumption. In the next decade, the radical automation

M. Polese
Institute for the Wireless Internet of Things, Northeastern University, Boston, MA, USA
e-mail: m.polese@northeastern.edu

M. Giordani · M. Zorzi
Department of Information Engineering, University of Padova, Padova, Italy
e-mail: giordani@dei.unipd.it; zorzi@dei.unipd.it

M. Mezzavilla (⊠) · S. Rangan
NYU Tandon School of Engineering, Brooklyn, NY, USA
e-mail: mezzavilla@nyu.edu; srangan@nyu.edu

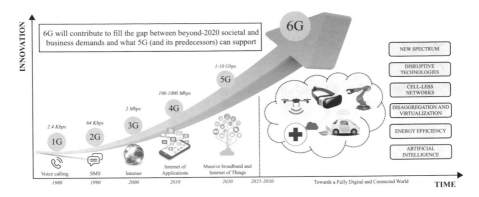

Fig. 3.1 Cellular networks generations (from 1G to 6G) and representative applications. Adapted from [1]

of industrial manufacturing, with concepts such as digital twin and Industry 4.0, the widespread proliferation of unmanned systems, both terrestrial and aerial, along with the millions of sensors that will be embedded into our cities, will call for a fundamental redesign of current cellular networks.

Mobile networks will provide a fundamental support to such smart environments, constituting their nervous system, as we discuss in [1]. Wireless links will transfer much greater amounts of data, at gigabit-per-second rates. Moreover, 6G connection will continue the trend toward a ubiquitous connectivity, not only for human communications, but, following the Internet of Things (IoT) paradigm, also to network together autonomous vehicles, sensors, wearable and medical devices, distributed computing resources and robots [2].

5G networks have already enabled impressive performance advancements toward a fully connected wireless fabric, by (1) expanding mobile cellular networks into new frequency bands (e.g., with millimeter wave (mmWave) communications), (2) introducing advanced spectrum usage and management, and (3) completely redesigning the core of the cellular network. However, the expected requirements of 6G applications will exceed the capabilities of the current 5G specifications. Wireless networks will need to support data rates in the order of terabits per second, latency values below the millisecond, and ten million connected devices per km^2.

This has recently sparked the interest of the research community toward the definition of the requirements and the technologies of a new generation of mobile networks, i.e., 6G systems, that will meet the connectivity demands of future intelligent and autonomous digital ecosystems. Following our discussion in [1], this chapter aims at illustrating what is the set of technologies that we envision as a likely candidate for more advanced and vertical-specific wireless networking solutions with respect to the state of the art of communication and networking technologies. In particular, we first analyze several potential use cases for future connected systems, and then map them to their key requirements in terms of latency, throughput, coverage, reliability, and other factors. Along these lines, this book chapter describes some scenarios in which the 5G networks being deployed today would be able to satisfy such performance demands.

Based on this analysis, the second part of the chapter highlights possible technological enablers for 6G, which include radically new communication technologies, network architectures, and deployment models to satisfy the relevant KPIs in each connected scenario. In particular, we foresee the development of:

- *New technologies at the physical layer, and the exploitation of frequencies above 100 GHz:* the vastly available spectrum in the Terahertz and optical bands will unlock unprecedented wireless capacity [3–5]. However, to unleash the true potential of these portions of the wireless spectrum, new disruptive communication technologies need to be introduced.
- *Multi-dimensional network architectures:* more complex networks will emerge as a result of the more advanced 6G use cases. To this end, we expect that 6G networks will provide a 3D coverage [6, 7], i.e., to support aerial platforms, the aggregation of heterogeneous technologies for access and backhaul, and the fully virtualized radio access and core network elements [8].
- *Prediction-based network optimization:* the multi-dimensional architecture described in the previous point will dramatically increase the complexity of the networks. To address this issue, 6G will have to increasingly rely on automated and intelligent techniques, deployed at each layer and node of such multi-dimensional networks [9]. In particular, distributed and unsupervised learning, and knowledge sharing will constitute key enablers for real-time decisions, that will be critical to maintaining and operating these complex networks.

Prior work has discussed possible technological advancements for 6G networks (e.g., [10, 11]). With respect to these contributions, in this book chapter and in [1] we outline a system-level vision, which starts by identifying the future use cases under development today for 6G systems and their performance requirements, and then highlights the challenges and opportunities associated to 6G technological enablers from a full-stack, end-to-end perspective. Most importantly, we select with a critical approach a subset of the various solutions that have been identified in the wireless research literature as the most promising candidates for an actual deployment ten to 15 years from today. Some of these technologies are incremental with respect to 5G, while other will represent a major breakthrough. The combination of these two approaches will clearly define a new generation of mobile networks, with solutions that have not been thoroughly addressed or cannot be properly included in current 5G standards developments.

We expect that this book chapter will help promote research efforts toward the identification of new communication and networking paradigms to meet the boldest requirements of 6G scenarios.

2   6G Use Cases

5G technologies are associated with trade-offs on power consumption, latency, cost of deployment and operations, hardware complexity, end-to-end reliability, throughput, and communication resilience. 6G innovations, on the contrary, will be developed in such a way that stringent network demands (in terms of ultra-high reliability, capacity, energy efficiency, and low latency) are jointly met in a holistic fashion.

In this section, we review the proprieties, characteristics and foreseen requirements of applications that, for their complementarity and generality, can be considered as good representatives of next-generation 6G services. Although some of these applications have already been discussed in 5G, we believe that they will likely not be part of future 5G deployments either due to technological limitations or because the market will not be mature enough to support them (especially within the very short timeframe in which 5G is supposed to be released). Figure 3.2 illustrates the KPIs of the use cases we will describe later in this section.

Augmented Reality (AR) and Virtual Reality (VR)

Current mobile networks have paved the way toward wireless video streaming, which represents one of the applications that contribute to the largest portion of mobile data traffic. The increasing use of such streaming and multimedia services has justified the introduction of new frequency bands (i.e., mmWaves) in 5G networks, to increase the capacity of the network. Nonetheless, following the multi-Gbps opportunity introduced by 5G mmWave communications, the multimedia ecosystem is developing new technologies and applications (i.e., augmented reality (AR) and virtual reality (VR)) which are more data-hungry, and extend the two-dimensional video screen to 3D application. Then, just like wireless video streaming

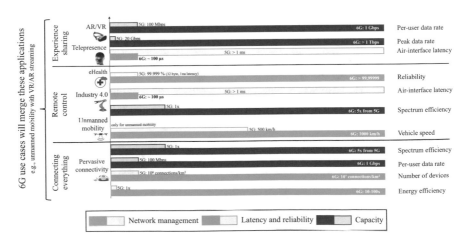

Fig. 3.2 KPIs of future 6G use cases, together with the improvements with respect to 5G networks, using data from [2, 10–17]. Adapted from [1]

has saturated 4G deployments, the spreading of AR and VR will exhaust the bandwidth available in the 5G spectrum, and will eventually call for wireless system capacity above 1 Tbps, which exceeds the 20 Gbps peak throughput objective outlined for 5G [2]. Furthermore, real-time interaction among different users of the same AR or VR setup introduces tight constraints on the latency that the network needs to provide. Therefore, it is not possible to heavily encode and compress AR/VR content, as coding and decoding may take too long. Consequently, the data rate to be allocated to each user needs to be in the order of Gpbs, while 5G only foresees a 100 Mbps target.

Holographic Telepresence (Teleportation)
Humans have always pursued the dream of teleportation as a means to deliver life-sized three-dimensional digital representation of all human senses in real time. Among several other benefits, this technology can allow virtual interaction of people during business events and meetings, and remove geographical time and distance barriers. This innovation, however, raises several challenges in 6G networks from a communication point of view. Some literature works, e.g., [12], make the case that the throughput requirements for the transmission of a 3D raw hologram, without compression, would be in the order of several Tbps, depending on the sensor's resolution and frame rate. The latency requirement will then go below the ms threshold to make the holographic experience smoother and more immersive. Furthermore, teleportation will require the processing of a very large number of streams generated from sensors at different view angles, as opposed to the few required for VR/AR, thereby raising stringent synchronization requirements.

eHealth
6G will revolutionize the health-care sector promoting workflow optimization, remote/virtual patient monitoring, and robotic telesurgery, while guaranteeing more efficient and affordable patient assistance. In particular, 6G will accomplish a transition towards a "care outside hospital" paradigm, where health-care services can be offered directly in the homes of the patients, thus reducing management and administrative costs for health facilities as well as making health-care assistance more accessible to the most unprivileged countries in the world. Besides the high cost, the current major issue for the development of eHealth services is the lack of real-time tactile feedback [13]. Furthermore, the proliferation of telemedicine applications will require the maintenance of continuous connectivity with very high reliability (>99.99999%, due to the potentially fatal consequences of communication failures), ultra-low latency (sub-ms), and support for mobility. These ambitious KPIs will be satisfied with increased spectrum availability, together with the evolution of current artificial intelligence paradigms, as expected in 6G networks [2].

Pervasive Connectivity
Although 5G networks support more than 1,000,000 connections per km^2, mobile traffic will grow threefold from 2016 to 2021, thereby pushing the number of connected nodes to the extreme, with at least an order of magnitude increase in

the number of devices per km^2 [2]. 6G networks will indeed connect smartphones, personal devices and wearables, IoT sensors in smart city deployments, robots, Unmanned Aerial Vehicles (UAVs), vehicles, and so on. This will further increase the load on already stressed deployments, which will not be capable of supporting the connectivity requests of each and every user with the performance requirements illustrated in Fig. 3.2. Moreover, while 80% of the wireless data traffic is consumed by indoor users, cellular networks never really targeted indoor coverage. For example, 5G infrastructures, which may be operating in the mmWave spectrum, will hardly provide indoor connectivity as high-frequency radio signals cannot easily penetrate solid material. Furthermore, 5G densification presents scalability issues and high deployment and management costs for operators. 6G networks will instead provide seamless and pervasive connectivity in different scenarios, satisfying demanding Quality of Service (QoS) requirements in both outdoor and indoor scenarios with a resilient and low-cost infrastructure. Additionally, 6G deployments will need to be more energy efficient with respect to 5G (with a 10–100x improvement in efficiency), otherwise the overall energy consumption would prevent scalable deployments with low environmental impact.

Industry 4.0 and Robotics
6G will foster and further develop the manufacturing revolution known as Industry 4.0, already started with the support of 5G networks. Industry 4.0 foresees a digital conversion of the manufacturing process toward a full deployment of cyber physical systems on the production lines, enabling connected services such as predictive diagnostics, maintenance, and flexible, cost-effective and efficient machine-to-machine interactions [14]. Moreover, digital twins will support remote inspection and development of industrial products, through a highly reliable, high fidelity digital representation of the real systems. Fully automated processes, however, introduce an additional set of requirements in terms of reliable and isochronous data transfer [15], which 6G needs to address through the disruptive set of technologies we will describe later in this book chapter. For example, the control of industrial actuators needs real-time communications with a delay jitter that should be in the order of 10^{-6} s, while digital twin and AR/VR industrial use cases require Gbps peak data rates.

Unmanned Mobility
The transition towards smart and connected transportation offers safer traveling, improved traffic management, automated driving, and support for infotainment for drivers and passengers, with an estimated market of more than 7 trillion USD [16]. Connected vehicles require high volumes of data to be exchanged among cars and clouds for high-resolution dynamic mapping, sensor sharing, and computational offloading: the data rate requirements will reach the Terabytes per driving hour [17], far beyond current network capabilities. Furthermore, automated driving requires unprecedented levels of reliability and low latency (>99.99999% and <1 ms, respectively), especially in high mobility scenarios (up to 1000 km/h). Another very important requirement will be ensuring accurate positioning for moving objects (up

to 10 cm, depending on the target use case) [18]. Besides cars, flying vehicles (e.g., drones) also have a great potential for 6G. In this perspective, advances in hardware, software, and spectrum solutions will pave the way towards more efficient and flexible deployment and management of groups of vehicles of a different nature, as we will discuss in Sect. 3.

This wide diversity of use cases is a unique characteristic of the 6G paradigm, whose potential will be fully unleashed only through breakthrough technological advancements and novel network designs, as described in the next section.

3 6G Enabling Technologies

In the following paragraphs, we will review the technologies we identified as enablers of the applications introduced in Sect. 2, following also the discussion in [1]. Notably, we focus on the technologies that were deliberately left out from current 5G standards specifications (i.e., from 3GPP NR Releases 15 and 16), and on solutions that are part of today's research but not yet ready for a commercial deployment. We will consider wireless technologies that exploit new portions of the spectrum in Sect. 3.1, new multi-dimensional architectural breakthroughs in Sect. 3.2, and finally disruptive applications of machine learning and artificial intelligence for predictive network optimization in Sect. 3.3. Table 3.1 summarizes the main technological innovations that could be introduced in 6G networks, considering their potential, the associated challenges, and the use cases introduced in Sect. 2 they empower.

3.1 Novel Wireless Paradigms and Frequencies Above 100 GHz

It is possible to characterize a new generation of wireless networks by identifying a set of new communication techniques that extend the capabilities of mobile devices beyond what was possible before, for example, in terms of latency and data rates. To this end, connected services in 5G networks are enabled by massive Multiple Input, Multiple Output (MIMO) and the mmWave spectrum. Along this line, to meet the KPIs described in Sect. 2, 6G deployments will combine traditional frequency bands (i.e., mmWaves and sub-6 GHz) and portions of the spectrum that have not been included in any cellular standard so far, i.e., Visible Light Communications (VLC) and the Terahertz band. Figure 3.3 depicts the path loss associated with each of these bands, for deployment scenarios that are representative of each technology, to illustrate how the different portions of the wireless spectrum provide distinct challenges and opportunities. In the following paragraphs, we will focus on the two novel spectrum bands that will be most likely considered for use in 6G.

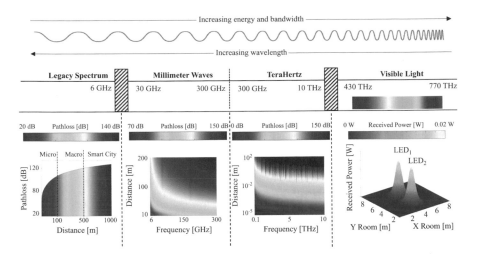

Fig. 3.3 Pathloss for sub-6 GHz, mmWave and Terahertz bands, and received power for VLC. The sub-6 GHz and mmWave pathloss follows the 3GPP models considering both LOS and NLOS conditions, while LOS-only is considered for Terahertz [19] and VLC [20]. This figure is taken from [1]

- **Terahertz communications** operate between 100 GHz and 10 THz [19] and, compared to mmWaves, bring to the extreme the potential of high-frequency connectivity, enabling data rates in the order of hundreds of Gbps, in line with the boldest 6G requirements. The THz bands would be combined with highly directional antenna arrays for massive spatial multiplexing. Also, due to the small wavelength, even high-rank line-of-sight MIMO links [21] may be possible at moderate distances to support multiple spatial paths in the absence of scattering. The small wavelengths may also enable new kinds of ultra-small-scale electronic packaging solutions for the RF and antenna circuitry, particularly in short-range applications. There are, however, significant challenges: THz RF devices are still in the early stages of development and currently significantly less power efficient (see, e.g. [22]). Due to the wide bandwidths and large number of elements to support, the digital baseband processing power can also be significant [23, 24]. Also, similar to the mmWave bands, THz signals are extremely susceptible to blockage. For very long links (e.g. >1 km), molecular absorption can also be problematic, particularly in heavy rain and fog (See [19] and Fig. 3.3). Nevertheless, some early experiments have demonstrated the potential for medium-range (e.g. 100 m) NLOS paths in micro-cellular type settings [25]. Further channel modeling is required [26], as well as a deeper understanding of the challenges associated to the full-stack, end-to-end design of terahertz networks [5].

- **VLC** can complement Radio Frequency (RF) communications, by exploiting on the wide adoption of cheap Light Emitting Diode (LED) luminaries. LED lamps can indeed modulate signals through quick variations in light intensity, which are not visible to the human eye [27]. The research on VLC is more mature than that on Terahertz communications, partly because experimental platforms at these frequencies are more affordable and have been available for some time. A standard for VLC (i.e., IEEE 802.15.7) has also been defined; however, this technology has never been considered by the 3GPP for inclusion in a cellular network standard. VLC are non-coherent, i.e., the transmitter and receiver do not exploit the knowledge of the channel, thus the path loss is proportional to the distance raised to the power of four. Therefore, as shown in Fig. 3.3, VLC exhibit a limited coverage range. Moreover, this technology requires an illumination source (i.e., it cannot be used in the dark), and suffers from shot noise of other light sources (e.g., the sun). For these reasons, it can be mostly used indoors [27]. Moreover, this technology relies on RF for the uplink. Nonetheless, VLC could be used to introduce cellular coverage in indoor scenarios, as discussed in Sect. 2, as this is a use case that has not been fully addressed by current cellular standards. In indoor scenarios, VLC can exploit a very large unlicensed band, and be deployed without cross-interference among different rooms and with relatively cheap hardware.

Although standardization bodies are studying Terahertz and VLC solutions for future wireless systems (with the IEEE 802.15.3d and 802.15.7 standard specifications, respectively), these technologies have not yet been included in a cellular network deployments. Therefore, additional research is needed to allow 6G mobile users to transmit and receive signals in the THz and VLC frequency bands, with advances to be developed in hardware and algorithms for flexible multi-beam acquisition and tracking in NLOS environments.

Besides the addition of new frequency bands, 6G will also leverage disruptive innovation at the physical layer, and at the circuits level. The following will be key enablers for 6G:

- **Full-duplex communication stack.** A careful design of self-interference suppression circuits [28] can enable concurrent transmission and reception of wireless signals. Future cellular implementations will enable simultaneous downlink and uplink transmissions to increase the multiplexing capabilities and the overall system throughput without using additional bandwidth. 6G networks will require attentive planning for the realization of full-duplex procedures in order to avoid interference, in particular from a scheduling standpoint [28].
- **Novel channel estimation techniques.** Given the directional nature of millimeter- and Terahertz-wave communications, 6G systems will need to investigate new channel estimation techniques in these new conditions. On one side, it has been demonstrated that out-of-band information can be used to estimate the angle of arrival of the signal, thus improving the timeliness and accuracy of beam management. This can be achieved by overlaying the omnidirectional propagation profile of sub-6 GHz signals with channel

estimation at mmWaves [29]. At the same time, the sparsity of the mmWave and Terahertz channels can be leveraged to design compressive sensing techniques that estimate the channel characteristics with a reduced number of samples.

• **Sensing and network-based localization.** Even though the literature on localization and mapping is quite widespread, the use of RF signals to improve positioning has never been applied in practice to cellular networks. 6G networks will exploit a unified interface for localization and communications which can be leveraged to optimize procedures such as beamforming and handovers, or even enable new user services in the context, for example, of vehicular communications and telemedicine.

3.2 Multi-Dimensional Network Architectures

Structural network modifications will be critical to support, for example, the multi-Gbps data rates enabled by Terahertz communications. As a consequence, additional points of access to fiber, along with expanded backhaul capacity, will be necessary. Moreover, the integration of different and heterogeneous communication technologies in the same network will pose increased challenges to the management of the overall system. In this section we describe some of the main architectural innovations that will be introduced in the 6G ecosystem, as presented in Fig. 3.4 and summarized in the following paragraphs.

• **Heterogeneous access.** 6G networks will support multiple radio technologies. This feature permits multi-connectivity solutions to be applied beyond the boundaries of a cell, with the users seen as connected to the network as a whole, and not to a specific cell. This concept, which is usually referred to as "cell-less" paradigm, guarantees seamless mobility support and near-zero latency due to minimal handover overhead. The devices will then be able to automatically

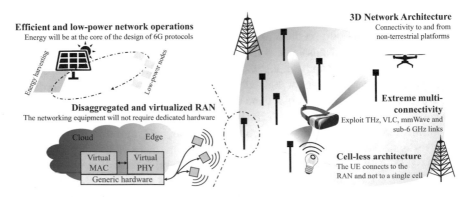

Fig. 3.4 Architectural innovations introduced in 6G networks, as illustrated in [1]

transition across heterogeneous access technologies, thereby exploiting the complementary characteristics of the different network interfaces. For example, the inherent robust nature of the sub-6 GHz layer can be leveraged for control operations, while a multi-Gbps data plane can be enabled by mmWave, Terahertz, and/or VLC links, as described in Sect. 3.1.

- **3D networks.** We envision future 6G network architectures to provide three-dimensional (3D) coverage, thereby overlaying terrestrial infrastructures with aerial/space platforms, from drones to balloons, up to satellites. Not only can these elements provide extra connectivity in the most crowded areas, e.g., during events or when terrestrial base stations are overloaded, but they can also guarantee seamless service continuity and reliability in those rural areas where fixed infrastructures are not even deployed [7]. Despite such promising opportunities, however, there are several issues to be addressed before non-terrestrial platforms can be practically deployed in wireless networks, like accurate air-to-ground and air-to-air channel modeling, topology optimization of satellite constellations and drone swarms, resource management, and energy efficiency [30].

- **Core network virtualization.** 5G networks have started the process of disaggregating once-monolithic network equipment: for example, 5G base stations can be deployed with distributed units with the lower layers of the protocol stack, and centralized units in data centers at the edge. However, the 3GPP has not yet specified how to implement virtualization, nor have existing 5G studies discussed the potential vulnerabilities associated with virtual network functions, which may indeed be subject to cyber-attacks. 6G networks will push disaggregation to its limits by virtualizing the whole protocol stack, including the Medium Access Control (MAC) and Physical (PHY) layers requiring, today, dedicated hardware implementations. This approach will allow the realization of low-cost distributed platforms with minimal baseband processing for the RF components, and will decrease the costs of networking equipment, making a massively dense deployment economically sustainable.

- **Advanced access-backhaul techniques.** For the support of terabit-per-second data rates, as envisioned in 6G, it will be fundamental to massively expand the backhaul capacity, in particular as long as Terahertz and VLC deployments, that are typically associated which a very large access point density, are concerned. The huge capacity available from 6G technologies can thus be exploited to realize self-backhauling solutions in which the radios in the base stations can offer both access and backhaul connectivity [31]. While such an approach is already under the umbrella of current 5G studies, the larger scale of 6G networks will pose new challenges and opportunities: the networks will need higher autonomous configuration capabilities, but the increase in access capacity will not need to be matched by an increase in fiber points of presence.

- **Low-power consumption.** 6G devices will be deployed in a pervasive manner to satisfy the future connectivity requirements. User terminals and networking equipment will then need to be powered with energy sources and, given the scale expected in 6G networks, it is imperative that the systems are designed to be more efficient and less energy consuming than current networks. The main challenge associated with incorporating energy-harvesting mechanisms into 5G is the efficiency loss that takes place when converting harvested signals into electric current. One solution is to implement circuits that allow devices to be self-powered, a critical pre-requisite to enable off-grid operations or long-lived IoT sensors which are often in stand-by mode.

3.3   Predictive Models for Network Operations

The complexity introduced in the 6G architecture will likely prevent the realization of closed-form optimizations. In particular, 6G deployments will be much denser, more heterogeneous, and characterized by stricter performance requirements compared to the 5G baseline. As a consequence, while integrating intelligence in cellular networks is already under discussion within the research community, it is expected that intelligence will play a more prominent role in future 6G networks. It should be noticed that, even though the standard may not directly indicate which techniques and learning strategies should be implemented in networks, a data-driven approach still represents a promising tool that network operators and telecommunication vendors should use to meet the 6G requirements [32]. In particular, 6G research will be oriented towards the following aspects.

- **Data selection and feature extraction.** The large amount of data that will be disseminated to and from future connected devices (e.g., vehicles in a fully-autonomous framework) will overwhelm already congested communication networks. It is therefore fundamental that end terminals can discriminate the *value of information* to use their (limited) network resources for the transmission of the data contents that are considered more critical for potential receiver(s) [33, 34]. In this context, machine learning (ML) solutions can be used to compute the temporal and spatial correlation between consecutive observations, as well as to extract features from the sensors' acquisitions and predict the a-posteriori probability of a sequence given its previous history. In 6G, labeling the data for supervised learning approaches may be infeasible. Unsupervised learning, on the other hand, does not need labeling, and can be used to autonomously build representations of the complex network to perform general optimizations, going beyond the capabilities of a supervised approach. Moreover, by coupling the unsupervised representation with reinforcement learning methods it is possible to let the network fully operate in an autonomous fashion.

- **Inter-user inter-operator knowledge sharing.** With learning-based systems, mobile operators and users can share not only spectrum and infrastructures, as in traditional networks, but also learned representations of different network deployments and/or use cases, thus providing the system with improved multiplexing capabilities. Examples of applications include speeding up the network installation in new markets, or better adapting to new unexpected operational scenarios. For the development of those systems, 6G research will need to study the trade-offs associated to latency, energy consumption, and system overhead, together with the cost of on-board vs. edge-cloud-assisted processing for the data.
- **User-centric network architectures.** ML-driven networks are still in their infancy, but will represent a key component of future 6G systems. Specifically, we envision a distributed artificial intelligence paradigm aimed at realizing a user-centric network architecture. This way, it will be possible for end terminals to make autonomous network decisions depending on the results of previous operations, thus removing the overhead introduced when communicating with centralized controllers. Distributed methods can process ML algorithms with a sub-ms latency in a quasi real-time manner, thereby yielding more responsive network management.

4 Conclusions

In this chapter, we reviewed the use cases and relative key enabling technologies that we believe will characterize the future 6G framework. Table 3.1 summarizes the key challenges, potentials, and use cases of each listed technology. We make the case that 6G research can evolve the traditional wireless networking paradigms of 5G and previous generations, introducing, among other innovations, the support for Terahertz and visible light bands, cell-less and non-terrestrial architectures, and massively distributed intelligence. However, research is still needed before these technologies can be market-ready for the unforeseen digital use cases of the society of 2030 and beyond.

Table 3.1 Comparison of 6G enabling technologies and relevant use cases

Enabling technology	Potential	Challenges	Use cases
Novel wireless paradigms and frequencies above 100 GHz			
Terahertz	High data rate, small antenna size, focused beams	Circuit design, propagation loss	Pervasive connectivity, industry 4.0, teleportation
VLC	Low-cost hardware, limited interference, unlicensed spectrum	Limited coverage, need for RF uplink	Pervasive connectivity, eHealth
Full duplex	Relaying and simultaneous TX/RX	Interference management and scheduling	Pervasive connectivity, industry 4.0
Out-of-band channel estimation	Flexible multi-spectrum communications	Need for reliable frequency mapping	Pervasive connectivity, teleporting
Sensing and localization	Novel services and context-based control	Efficient multiplexing of communication and localization	eHealth, unmanned mobility, industry 4.0
Multi-dimensional network architectures			
Multi-connectivity and cell-less architecture	Seamless mobility and integration of different kinds of links	Scheduling, need for new network design	Pervasive connectivity, unmanned mobility, teleporting, eHealth
3D network architecture	Ubiquitous 3D coverage, seamless service	Modeling, topology optimization and energy efficiency	Pervasive connectivity, eHealth, unmanned mobility
Disaggregation and virtualization	Lower costs for operators for massivelydense deployments	High performance for PHY and MAC processing	Pervasive connectivity, teleporting, industry 4.0, unmanned mobility
Advanced access-backhaul integration	Flexible deployment options, outdoorto- indoor relaying	Scalability, scheduling and interference	Pervasive connectivity, eHealth
Energy-harvesting and low-power operations	Energy-efficient network operations, resiliency	Need to integrate energy source characteristics in protocols	Pervasive connectivity, eHealth

Enabling technology	Potential	Challenges	Use cases
Predictive models for network operations			
Intelligence in the network			
Learning for value of information assessment	Intelligent and autonomous selection of the information to transmit	Complexity, unsupervised learning	Pervasive connectivity, eHealth, teleporting, industry 4.0, unmanned mobility
Knowledge sharing	Speed up learning in new scenarios	Need to design novel sharing mechanisms	Pervasive connectivity, unmanned mobility
User-centric network architecture	Distributed intelligence to the endpoints of the network	Real-time and energy-efficient processing	Pervasive connectivity, eHealth, industry 4.0
Not considered in 5G		With new features/capabilities in 6G	

Acknowledgments This work was partially supported by NIST through Award No. 70NANB17H166, by the U.S. ARO under Grant no. W911NF1910232, by MIUR (Italian Ministry for Education and Research) under the initiative "Departments of Excellence" (Law 232/2016), by NSF grants 1302336, 1564142, and 1547332, the SRC and the industrial affiliates of NYU WIRELESS.

References

1. M. Giordani, M. Polese, M. Mezzavilla, S. Rangan, M. Zorzi, Toward 6G networks: use cases and technologies. IEEE Commun. Mag. **58**(3), 55–61 (2020)
2. Z. Zhang, Y. Xiao, Z. Ma, M. Xiao, Z. Ding, X. Lei, G.K. Karagiannidis, P. Fan, 6G wireless networks: vision, requirements, architecture, and key technologies. IEEE Vehic. Technol. Mag. **14**(3), 28–41 (2019)
3. T.S. Rappaport, et al., Wireless communications and applications above 100 GHz: opportunities and challenges for 6G and beyond. IEEE Access **7**, 78729–78757 (2019)
4. I.F. Akyildiz, J.M. Jornet, C. Han, Terahertz band: next frontier for wireless communications. Phys. Commun. **12**, 16–32 (2014)
5. M. Polese, J. Jornet, T. Melodia, M. Zorzi, Toward toward end-to-end, full-stack 6G Terahertz networks. IEEE Commun. Mag. **58**, 48–54 (2020). https://arxiv.org/abs/2005.07989
6. M. Boschiero, M. Giordani, M. Polese, M. Zorzi, Coverage analysis of UAVs in Millimeter wave networks: A stochastic geometry approach, in *Proceedings of the 16th Intl Wireless Communications and Mobile Computing Conference (IWCMC 2020)*, Limassol, Cyprus (2020). https://arxiv.org/pdf/2003.01391.pdf
7. M. Giordani, M. Zorzi, Satellite Communication at Millimeter waves: A Key Enabler of the 6G Era, *IEEE International Conference on Computing, Networking and Communications (ICNC)* (2020)
8. L. Bonati, M. Polese, S. D'Oro, S. Basagni, T. Melodia, Open, Programmable, and Virtualized 5G Networks: State-of-the-Art and the Road Ahead (2020). arXiv:2005.10027 [cs.NI]
9. M. Polese, R. Jana, V. Kounev, K. Zhang, S. Deb, M. Zorzi, Machine learning at the edge: A data-driven architecture with applications to 5G cellular networks. IEEE Trans. Mobile Comput. Early Access (2020)
10. W. Saad, M. Bennis, M. Chen, A vision of 6G wireless systems: applications, trends, technologies, and open research problems. IEEE Netw. **34**(3), 134–142 (2020)
11. E. Calvanese Strinati, S. Barbarossa, J.L. Gonzalez-Jimenez, D. Ktenas, N. Cassiau, L. Maret, C. Dehos, 6G: the next frontier. IEEE Vehic. Technol. Mag. **14**(3), 42–50 (2019)
12. X. Xu, Y. Pan, P.P.M.Y. Lwin, X. Liang, 3D holographic display and its data transmission requirement, in *International Conference on Information Photonics and Optical Communications* (2011), pp. 1–4
13. Q. Zhang, J. Liu, G. Zhao, Towards 5G enabled tactile robotic telesurgery (2018). Preprint arXiv:1803.03586
14. J. Lee, B. Bagheri, H.-A. Kao, A cyber-physical systems architecture for industry 4.0-based manufacturing systems. Manuf. Lett. **3**, 18–23 (2015)
15. M. Wollschlaeger, T. Sauter, J. Jasperneite, The future of industrial communication: automation networks in the era of the internet of things and industry 4.0. IEEE Ind. Electron. Mag. **11**(1), 17–27 (2017)
16. N. Lu, N. Cheng, N. Zhang, X. Shen, J.W. Mark, Connected vehicles: solutions and challenges. IEEE Int. Things J. **1**(4), 289–299 (2014)
17. J. Choi, V. Va, N. Gonzalez-Prelcic, R. Daniels, C.R. Bhat, R.W. Heath, Millimeter-wave vehicular communication to support massive automotive sensing. IEEE Commun. Mag. **54**(12), 160–167 (2016)

18. F. Mason, M. Giordani, F. Chiariotti, A. Zanella, M. Zorzi, An adaptive broadcasting strategy for efficient dynamic mapping in vehicular networks. IEEE Trans. Wirel. Commun. **19**(8), 5605–5620 (2020)

19. J.M. Jornet, I.F. Akyildiz, Channel modeling and capacity analysis for electromagnetic wireless nanonetworks in the Terahertz band. IEEE Trans. Wireless Commun. **10**(10), 3211–3221 (2011)

20. T. Komine, M. Nakagawa, Fundamental analysis for visible-light communication system using LED lights. IEEE Trans. Consum. Electron. **50**(1), 100–107 (2004)

21. F. Bohagen, P. Orten, G.E. Oien, Construction and capacity analysis of high-rank line-of-sight MIMO channels, in *Proceedings of the IEEE Wireless Communications and Networking Conference*, vol. 1 (2005), pp. 432–437

22. A. Simsek, S.-K. Kim, M.J.W. Rodwell, A 140 GHz MIMO transceiver in 45 nm SOI CMOS, in *Proceedings of the IEEE BiCMOS and Compound Semiconductor Integrated Circuits and Technology Symposium (BCICTS)* (2018)

23. P. Skrimponis, S. Dutta, M. Mezzavilla, S. Rangan, S.H. Mirfarshbafan, C. Studer, J. Buckwalter, M.J.W. Rodwell, Power consumption analysis for mobile mmwave and sub-THz receivers, in *Proceedings of the IEEE 6G Wireless Summit (6G SUMMIT)* (2020)

24. S.H. Mirfarshbafan, A. Gallyas-Sanhueza, R. Ghods, C. Studer, Beamspace Channel Estimation for Massive MIMO mmWave Systems: Algorithm and VLSI Design (2019). Preprint arXiv:1910.00756

25. N.A. Abbasi, H. Arjun, A.M. Nair, A.S. Almaiman, F.B. Rottenberg, A.E. Willner, A.F. Molisch, Double directional channel measurements for THz communications in an urban environment (2019). Preprint arXiv:1910.01381

26. Y. Xing, T.S. Rappaport, Propagation measurement system and approach at 140 GHz-moving to 6G and above 100 GHz, in *Proceedings of the IEEE Global Communications Conference (GLOBECOM)* (2018)

27. P.H. Pathak, X. Feng, P. Hu, P. Mohapatra, Visible light communication, networking, and sensing: a survey, potential and challenges. IEEE Commun. Surveys Tuts. **17**(4), 2047–2077 (2015) Fourth Quarter

28. S. Goyal, P. Liu, S.S. Panwar, R.A. Difazio, R. Yang, E. Bala, Full duplex cellular systems: will doubling interference prevent doubling capacity? IEEE Commun. Mag. **53**(5), 121–127 (2015)

29. A. Ali, N. González-Prelcic, R.W. Heath, Millimeter wave beam-selection using out-of-band spatial information. IEEE Trans. Wireless Commun. **17**(2), 1038–1052 (2018)

30. M. Giordani, M. Zorzi, Non-Terrestrial networks in the 6G Era: challenges and opportunities. IEEE Netw. **35**, 244–251 (2020)

31. M. Polese, M. Giordani, T. Zugno, A. Roy, S. Goyal, D. Castor, M. Zorzi, Integrated access and backhaul in 5G mmWave networks: potentials and challenges. IEEE Commun. Mag. **58**(3), 62–68 (2020)

32. M. Wang, Y. Cui, X. Wang, S. Xiao, J. Jiang, Machine learning for networking: workflow, advances and opportunities. IEEE Netw. **32**(2), 92–99 (2018)

33. M. Giordani, A. Zanella, T. Higuchi, O. Altintas, M. Zorzi, Investigating value of information in future vehicular communications, in *2nd IEEE Connected and Automated Vehicles Symposium (CAVS)* (2019)

34. T. Higuchi, M. Giordani, A. Zanella, M. Zorzi, O. Altintas, Value-anticipating V2V communications for cooperative perception, in *30th IEEE Intelligent Vehicles Symposium (IV)* (2019)

Chapter 4
Physical Layer Design Challenges for 6G Wireless

Satinder Paul Singh

Abstract 6G networks will be based on a combination of 5G and LTE with other known technologies that are not established enough for being included in 5G. 6G networks will be based on various new mechanisms that were never encountered when drafting and creating 5G specifications in combination with vast enrichments of technologies that were already present in the previous group of wireless cellular networks. PHY layer is the key to achieving reliable and high-speed data transmission over wireless channels. At the transmitter, the bitstreams are processed by such modules as channel coding, channel, and signal modulation, precoding MIMO, and Orthogonal frequency-division multiplexing (OFDM) modulation; the converse procedure is achieved at the receiver to recover the desired bits. At the protocol/algorithmic level, the key parameters include enhanced coding, modulation, and waveforms to achieve inferior latency, advanced reliability, and concentrated complexity. Different decisions will be needed to optimally support various possible use cases. The source productivity can be further improved by using various amalgamations of full-duplex radios, interference management based on rate splitting, machine-learning based optimization, coded caching, and distribution. This chapter paper describes the various design issues at the ASIC/SoC level for PHY to achieve 6G connectivity with enormously high data rates, ranging to Tbps/THz, 6g PHY modeling, Scalability, and Implementation issues at the chip level. We discuss digital/analog signal processing challenges followed by directions to focus on next-generation AI-based physical layer for 6G networks and security level issues in PHY Design at the System on a Chip (SoC) level.

S. P. Singh (✉)
Geography Head - Europe, Cogknit Semantics Pvt. Ltd., Bengaluru, Karnataka, India
e-mail: satinder.singh@cogknit.com

© The Author(s), under exclusive license to Springer Nature Switzerland AG 2021
Y. Wu et al. (eds.), *6G Mobile Wireless Networks*, Computer Communications
and Networks, https://doi.org/10.1007/978-3-030-72777-2_4

1   Introduction

The Physical Layer (PHY) is the key to achieving reliable and high-speed data transmission over wireless channels. At the transmitter, the bit streams are processed by such modules as channel coding, channel and signal modulation, precoding MIMO, and Orthogonal frequency-division multiplexing (OFDM) modulation; the converse procedure is achieved at the receiver to recover the desired bits.

At the protocol/algorithmic level, the key parameters include enhanced coding, modulation, and waveforms to achieve inferior latency, advanced reliability, and concentrated complexity. Different decisions will be needed to optimally support different use cases. The source productivity can be further improved by using various amalgamations of full-duplex radios, interference management based on rate splitting, machine-learning based optimization, coded caching, and distribution.

In terms of technologies, especially at PHY layer, former generations of mobile communications are often hallmarked by convinced way of multiple access, such as FDMA (Frequency Division Multiple Access), TDMA (Time Division Multiple Access), CDMA (Code Division Multiple Access), OFDM (Orthogonal Frequency Division multiple Access) for straightforwardness [1]. This just accentuates the importance of PHY level technology advancements which are not only about the air-interface designs, but also various breakthrough in electronic/photonic materials, microelectronic fabrication, device manufacturing, and so on. The comprehension of wireless communications at THz frequencies requires the development of precise channel models to capture the influence of both channel individualities including the high atmospheric attenuation and molecular absorption rates at various broadcast windows, as well as the propagation effects including reflection, scattering, and diffraction, with reverence to different resources. Based on various studies, surely 6G PHY systems be like:

- Make an immense use of "heavily circulated digital signal processing" and a lot of cache memories, e.g., in the form of cloud-Radio Access Network technology.
- Influence network slicing and multi-access edge computing to enable new services with highly influenced performance requirements and to give the needed resources real support for vertical markets
- Spectator an increasing combination of terrestrial and satellite wireless networks, unmanned aerial vehicles (UAVs) and low-Earth orbit (LEO) micro satellites, and to fill exposure holes and offload the network in heavy-load circumstances.
- Influence Artificial intelligence (AI) and Machine Learning (ML) methodologies to enhance the efficiency of traditional DSP model-based algorithms and FPGA neural network-based accelerators for signal processing and resource allocation.

Figure 4.1 shows different 6G targets that represents a radical increase in KPIs This chapter paper describes the physical-layer (PHY) methodologies for achieving 6G connectivity with very high data rates, up to the Tbps range. Theoretical 6g PHY modeling encounters, ASIC hardware implementation issues and scalability.

Fig. 4.1 6G targets

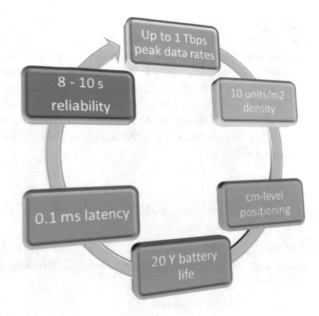

1. Signal processing challenges
2. AI ML in 6G PHY
3. Security level issues in PHY Design

2   The ROle of PHY Layer in 6G

With the introduction of new use cases such as Ultra-Reliable Low Latency Communication (URLLC) and Massive Machine Type Communications (MMTC), one of the principal motivation is to use Artificial Intelligence (AI) techniques especially Machine Learning (ML) to advance adaptive, dynamic and self-learning systems for the PHY layer in B5G and 6G. Research community always say that the PHY layer in 6G will have a secondary role. And there is a common briefing that "the PHY layer will be departed" [2]. The apprehension then was that physical layer (PHY) research would slowly decline and limit to sustaining short-term and derivative needs of the Industry [3]. Those apprehensions were short-lived, as proven by the authoritative revival propelled by ground-breaking innovations such as nonorthogonal multiple access (NOMA), full duplex (FD) radio and hybrid mm Wave RF technologies.

PHY research will not become vanished but will renovate and solve future encounters along the road. Concretely, we believe that 6G PHY will be split in two components: The Primary PHY and the Secondary-PHY.

Primary PHY will devoted only to solving complex DSP (Digital Signal processing) problems close to real time Application Specific Integrated Circuit (ASIC) with Software Defined Radio (SDR) and including the handling of waveforms, beamforming & signal interference supervision, all taking into account the hardware impairments imposed by the compacting of massive MIMO antennas, full-duplex operation and mm Wave RF, as well as the imperfections of channel state information (CSI), caused by all the dynamics of 6G protocol and mechanism.

Secondary PHY is architected more of high application software driver of the Primary PHY and focus on the interface with the Artificial Intelligence (AI) core, developing the code-domain technologies required to control and interact with the Primary PHY.

3   6G PHY at Terahertz (THz) Band

The Terahertz (THz) band (0.1–10 THz) is revealing its potential as a key wireless technology to fulfill the future demands for 6G wireless systems, neighboring to the mm-wave spectrum,

1. GHz bandwidth resource
2. Pico-second level symbol duration
3. Incorporation of millions of submillimeter long antennas,
4. Frail interference deprived of full legacy guideline.

THz band has been one of the smallest amounts explored frequency bands in the "Electromagnetic (EM) spectrum", for absence of efficient THz transceivers and antennas. For an efficient THz communication system, very much detailed and accurate yet tractable applicable for both Indoor and Outdoor environment THz multipath channel models need to be developed System on Chip scientists need to discourse below challenges:

- Conveying the signal within the joined system and to the antenna with low loss
- Packaging of the integrated system without substantial loss while maintaining cooling of system
- Lowering the mixer phase-noise
- Low power multi-Giga-samples-per-second analog-to-digital converters (ADCs) and digital-to-analog convertors (DACs); and lastly
- Low power digital input/output (IO) to DACs and ADCs to transfer data at Tbps data rate within power consumed and energy dissipated within required targets

3.1   Signal Processing Challenges

In snowballing processing power, there are always complications such as the flexibility of a processing unit, data pipes that transition data from processing resource to processing resource, and the elasticity of the configuration. With complex developments in MAC layer functionality, added focus on SDNs, and more robust and complex schedulers, the need for time-accurate processing and parallel computing becomes superior [4]. For SDN case extremely multifaceted requirements from application software, The concern for today's experienced leaders in 6G is not whether they should use ASICs (or FPGAs), but how best to combine this technology with traditional CPUs and GPUs in the multi-cloud environment and how best to manage costs through the software development and production life cycle implementation.

Simple raw data can't be processed like current scenarios manager type 3rd party applications EDA software's, data can come from channel, base station, or a mobile phone or human body, extreme low latencies between different modules of System on Chip (SoC) are required. Any bid that assimilates heterogenous processor architectures (SoC, ASIC, ASIP, RISC-V, FPGA etc.) requires that the data be transferred over high-speed serial boundaries with low latency and extensive data bandwidth buses.

Accurate estimation of channel and robust and selective filtering are two signal processing challenges:

3.1.1 Challenge 1: Accurate Estimation of Channel

Supporting high data rates (Gbps) in high mobility scenarios a prime concern of operators will require dealing with much shorter channel coherence times and Ultra-low latency requirements will see transmission intervals significantly condensed, and the number of limitations to estimate will be enormously large as a consequence of the scaling (not only of antennas/access points, but also in the numbers of users/devices).

3.1.2 Challenge 2: Robust and Selective Filtering

Traditional Filters (e.g., SCM) rely on the adequate readiness of samples the number of samples should be far-off greater than the number of signals and on the postulation that these samples are valid. Given these conditions led to a contradiction between true and estimated covariance leads to highly inaccurate filters with severe performance losses, in terms of connectivity, reliability, and data rates. A major task is then to develop Signal processing filtering results which are strong to the effects of finite-sampling, temporal correlation, and corrupted (outlying) samples. Appropriate tools and hopeful directions can be leveraged from the fields of robust statistics, RMT, and high-dimensional covariance approximation

4 Artificial Intelligence (AI)/Machine Learning (ML) in 6G PHY

Requirements from augmented reality (AR) and virtual reality (VR) over wireless are considered in current research as these technologies are becoming major application drivers for various verticals including health and entertainment. Other important drivers are autonomous vehicles and Industry 4.0. These application domains feature large data sets and heavy computation with real-time or near real-time responsivity—a combination requiring distributed solutions and AI to achieve the targeted performance [9].

The most ambitious hope is to build an AI-based end-to-end PHY architecture, which would potentially transform and replace those aspects that require expert knowledge and modules. Autoencoders from deep learning can be used to build the end-to-end PHY as three modules, including one transmitter, channel, and receiver. However, because of the complexity, the application of AI to independently design and enhance one or more PHY functions rather than the end-to-end PHY is more readily possible.

Deep learning is an important technique for designing and enhancing the PHY, where convolutional neural networks (CNNs) could be used as signal classification and channel decoding, deep neural networks (DNNs) are potential technologies for channel estimation and signal detection, and complex CNNs (CCNNs) can be used to build OFDM receivers.

Conventional PHY designs always target average channel conditions and use system of measurement such as ergodic capacity and average error rate. These approaches improve average QoS, which suits large, long transmissions such as those associated with classical mobile broadband, but they ignore the specifics of individual data-transmission sessions and, therefore, not successful to characterize the real QoS of short transmission sessions, which are connected with low range IoT.

ML has started to breach all walks of life, as well as the field of wireless communication [5]. The physical layer is generally architected based on extreme statistically proven mathematical models, and several major modules are modeled and optimized discretely. This design method can adjust to the fast time-varying physiognomies of the physical layer, but often some nonlinear influences in the PHY are not straight forward to model. To proceed further and embark a significant process therefore it is mandatory to assimilate advanced Machine Learning (ML) features and algorithms into the physical layer of 6G wireless systems [10].

Hardware issues at PHY and ASIC level, many optimization problems are not in convex nature, e.g. maximizing throughput by means of power control, multi-user spectrum optimization in multi-carrier systems, optimization of spectrum sensing for cognitive radios, and optimal beamforming formulated as a sum-rate maximization problem under a total power budgeting constraint. Such problems may be solved using dual decomposition techniques that require iterative algorithms, which in turn often cannot be computed in real time due to a high computational load.

To alleviate the high computational complexity and resulting latency associated with existing iterative algorithms, empirical solutions have been proposed for some physical layer problems such as beamforming design [2]. Although experiential solutions can be attained with low computational delay, this benefit comes at the expense of performance loss. Deep learning techniques have great probable to find solutions to these problems in real time, while continuing good performance and plummeting computational delay. Deep learning is a very powerful methodology for scheming, enhancing, and enlightening one and may be several occupations in the physical layer for 6G. This comprises CNNs for signal classification, and DNNs for channel estimation and signal detection.

ML-driven can be very useful in PHY layer in these areas:

1. Security
2. Channel Coding
3. Synchronization
4. Channel estimation
5. Beamforming

4.1 Hardware Architectures for AI in Radio

Different hardware architecture for 6g PHY enables:

- Dense, organized systems in the field using AI to regulate actionable intelligence in real time that leverage the typical CR architecture of an FPGA and general-purpose processor (GPP), sometimes with the addition of a graphics processing unit (GPU) module
- Segmental, scalable, more compute-intensive systems typically consisting of CRs attached to high-end servers with controlling GPUs that conduct offline processing

For systems with low size, weight, and power, coupling the hardware-processing efficiency and low-latency performance of the FPGA with the programmability of a GPP makes a lot of sense. Though the FPGA may be harder to program, it is the key enabler to achieving low Power, Performance and timing in real-time systems. Major factors to simulate in ASIC simulations are

1. Signal-to-noise ratio (SNR)
2. Phase noise
3. Linear and nonlinear impairments
4. Waveform selection

Greater, compute-intensive structures need a SoC System Architecture that scales and heterogeneously controls best-in-class processors. These designs typically comprise FPGAs for baseband processing, GPPs for control, and ASIP/GPUs for AI processing. GPUs offer a fine mixture of programming ease and the ability to process enormous amounts of data.

5 Security Issues at 6G PHY "enabler of confidentiality in 6G connectivity"

No doubt practically moving outside of 5G, new technology 6G will certainly increase the key performance indicators of old generation technologies, enabling the definition of additional challenging applications, ranging from AR (Augmented Reality) and holographic projection to extremely ultra-sensitive applications. Keeping in mind this reference, in this context, a complete approach to security and based solutions are required to handle the overabundance of different systems and platforms (Fig. 4.2).

Physical layer security, changing the security strategy on the SoC hardware level, might be one of the confidentiality enablers in 6G connectivity. Its features combined with advances in Artificial intelligence (AI) algorithms and the trend toward distributed computing architectures can be subjugated either to enhance classical cryptographic techniques or to meet security requirements when dealing with simple but sensitive devices which are unable to implement cryptographic methods. These include devices and nano-devices of the Internet of Things, and the Internet of Bio-Nano-Things where in-body nano-devices will become nodes of the future Internet [3].

Physical layer security addresses one of the most important applications of 6G: human-centric mobile communications [6]. In this outline, an increasing interest of scientific research has been oriented to wireless body area networks and in particular to on-body and in-body nano-devices, including biochemical communications and signal processing. In the future, the entire human body will be part of the entire network architecture, it will be seen as a node of the network or a set of nodes (wearable devices, implantable sensors, nano-devices, etc.) that collects extremely sensitive information to be exchanged for multiple purposes (e.g. health, statistics, safety, etc.). By managing with the high security and privacy requirements and the energy and miniaturization constraints of the new communication terminals physical

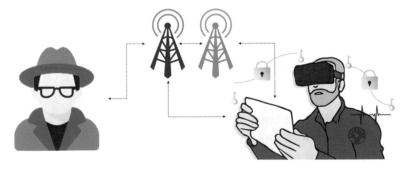

Fig. 4.2 6G Phy security

layer security techniques could represent well-organized solutions for safeguarding the most life-threatening and less investigated network segments which are the those between the body sensors and a sink or a pivot node.

Two very fascinating potential application scenarios for physical layer security in the 6G context are human bond communication and molecular communication. The former requires a secure transmission of all the five human senses for replicating human biological features, allowing disease diagnosis, emotion detection, as well gathering biological characteristics and handling human body remote interactions. The latter is based on shifting information theory concepts into the biochemical domain (communications between various biological cells inside the human body) and it requires advanced, low-complexity and reliable secured mechanisms for securing intra-body signal processing communications and to make sure enabling trustworthy sensing and actuation in the human body, which is a very challenging environment (e.g. secure Internet of bio-nano things) [7]. The ETSI group is working on the tuning of security and privacy for future body area networks, and physical layer security as one of potential candidate to handle the confidentiality of in- and on-body network devices with the typically low available resources. This is an extremely significant when 6G will include in- or on-body nodes as part of the network.

State-of-the-art encryption itself is considered unassailable when it comes to data confidentiality and integrity, however, there exist doubts concerning the traditional authentication and key distribution design for the future. PHY-based key generation solutions distinguish themselves from traditional key exchange solutions by being completely decentralized and not relying on any fixed parameters designed by a particular entity, but rather on the distributed entropy source which is the wireless channel.

In the future, newly developed or extended communication protocol implementations should see PHY-layer attributes (channel state information (CSI), received signal strength indicator (RSSI), carrier frequency offset (CFO), etc.) of all PHY-exchanges easily available to higher layers, allowing for a much deeper level of integration, control, and interchangeability of security modules. Such attribute granularity and unprecedented network visibility at the PHY-level will encourage the development of security and authentication solutions which leverage these previously unused characteristics.

If comprehensive security is not embedded into these services, new attack vectors will arise at the PHY-layer in the form of range extension/reduction attacks that can spoof distances between devices. With an availability of secure communication link, the available PHY-layer attributes can then also be used for PHY-threat detection [8]. These can be implemented using classical signal processing techniques that highlight and discover anomalies in the PHY-attributes of the particular received signal or through the detection of abnormalities in the packet exchanges

6 Conclusion

Given the 6G era has miles to travel across and is a decade away, it is very well timed to understand the key challenges for communications engineers that require immediate research attention. This chapter has identified few pressing challenges whose investigation will cross-leverage expertise in signal processing, information theory, electromagnetics and physical implementation. A demanding dialog of the research challenges and possible solutions that must be addressed from applications point of view, to design of the next generation PHY. The road to overcome the challenges given in this chapter is full of obstacles, and we are able to provide enough insights to begin research towards the many promising open directions. As a result, it will definitely serve as a motivation for research approaching the next decade.

References

1. M. Latva-Aho, K. Leppänen (eds.), Key drivers and research challenges for 6G ubiquitous wireless intelligence [White Paper]. University of Oulu (2019). http://urn.fi/urn:isbn: 9789526223544
2. R. Mitra, S. Jain, V. Bhatia, Least minimum symbol error rate based post-distortion for VLC using random fourier features. IEEE Commun. Lett. **24**(4), 830–834 (2020)
3. A. Zappone, M.D. Renzo, M. Debbah, Wireless networks design in the era of deep learning: Model-based, AI-based, or both? IEEE Commun. Mag. (2020)
4. Z. Ghahramani, Probabilistic machine learning and artificial intelligence. Nature **521**(7553), 452–459 (2015)
5. R. Mitra, F. Miramirkhani, V. Bhatia, M. Uysal, Mixture-kernel based post-distortion in RKHS for time-varying VLC channels. IEEE Trans. Veh. Technol. **68**(2), 1564–1577 (2019)
6. M. Chen, H.V. Poor, W. Saad, S. Cui, Convergence time optimization for federated learning over wireless networks. Preprint (2020). arXiv:2001.07845
7. A. Ferdowsi, W. Saad, Generative adversarial networks for distributed intrusion detection in the Internet of Things. Preprint (2019). arXiv:1906.00567
8. A. Taleb Zadeh Kasgari, W. Saad, M. Mozaffari, H.V. Poor, Experienced deep reinforcement learning with generative adversarial networks (GANs) for model-free ultra reliable low latency communication. Preprint (2019). arXiv–1911
9. M. Chen, U. Challita, W. Saad, C. Yin, M. Debbah, Artificial neural networks-based machine learning for wireless networks: A tutorial. IEEE Commun. Surv. Tutorials **21**(4), 3039–3071 (2019)
10. M. Chen, Z. Yang, W. Saad, C. Yin, H.V. Poor, S. Cui, A joint learning and communications framework for federated learning over wireless networks. Preprint (2019). arXiv:1909.07972

Chapter 5
PHY-Layer Design Challenges in Reconfigurable Intelligent Surface Aided 6G Wireless Networks

Hang Liu, Xiaojun Yuan, and Ying-Jun Angela Zhang

Abstract Reconfigurable intelligent surfaces (RISs), made of nearly-passive, low-cost, and reconfigurable meta-materials, can artificially customize the propagation environment by introducing controllable and independent phase shifts on electro-magnetic waves. Integrating RISs into 6G wireless networks leads to unprecedented energy focusing at receivers, and hence significantly enhances the primary communications. This chapter first overviews the new PHY-layer design challenges accompanying RIS-aided 6G communications. Particularly, channel estimation in RIS-aided systems is more challenging than that in traditional communication systems, since the nearly-passive RISs have very limited capability of receiving, processing, and transmitting incident signals. This chapter further illustrates two state-of-the-art approaches for estimating the cascaded transmitter-RIS and RIS-receiver channels by exploiting channel structural features in RIS-aided systems. Specifically, the first approach artificially introduces signal sparsity by controlling the on/off states of the RIS elements, and estimates the cascaded channels by sparse matrix factorization and matrix completion. The second approach directly factorizes the cascaded channels by exploiting the slow-varying information of the RIS-receiver channel and the inherent sparsity of the transmitter-RIS channel. Finally, this chapter discusses open issues in RIS channel estimation and concludes with future research directions.

H. Liu · Y.-J. Angela Zhang
Department of Information Engineering, The Chinese University of Hong Kong, Hong Kong
SAR, China
e-mail: lh117@ie.cuhk.edu.hk; yjzhang@ie.cuhk.edu.hk

X. Yuan (✉)
Center for Intelligent Networking and Communications, The University of Electronic Science
and Technology of China, Chengdu, China
e-mail: xjyuan@uestc.edu.cn

© The Author(s), under exclusive license to Springer Nature Switzerland AG 2021
Y. Wu et al. (eds.), *6G Mobile Wireless Networks*, Computer Communications
and Networks, https://doi.org/10.1007/978-3-030-72777-2_5

53

1   Introduction

With the launch of standardisation and commercialisation of fifth-generation (5G) communications, sixth-generation (6G) research effort has been kicked off from both academic and industrial perspectives. The 6G wireless networks are expected to provide a paradigm shift for future communications from "connected things" to "connected intelligence", where the later combines mobile devices as a distributed intelligent platform integrating the capabilities of sensing, communications, and computations. It is believed that the existing 5G technologies, such as massive multiple-input multiple-output (MIMO), millimeter-wave (mmWave) communications, and small cells, are not sufficient to meet the requirements of ultra-low latency, ultra-high reliability, and ultra-massive connectivity in the future networks. This inherent insufficiency of 5G wireless networks is largely due to the postulation that the propagation environment of wireless signals is pre-determined and uncontrollable in the system design. However, smart radio environments, as an emerging concept for 6G wireless communications, treat wireless channels as controllable and programmable entities that can be jointly configured with the transmitters and receivers [1].

Reconfigurable intelligent surfaces (RISs, a.k.a. large intelligent metasurfaces and passive holographic MIMO surfaces) are envisioned to be the key enabler of the smart radio environments [2]. A RIS is a thin sheet made of nearly passive and inexpensive meta-materials, and is controlled by using external stimuli to manipulate the incident electromagnetic (EM) waves. RISs usually have the following features:

- The RIS elements are capable of introducing independent reconfigurations (e.g., phase shifts) on the incident EM waves, and flexibly adapting their responses in real-time.
- RISs are made nearly passive by using low-cost and low-power-consumption electronics without radio frequency chain.
- RISs can be easily placed upon ordinary objects such as walls, ceilings, or facades of buildings.

Thanks to these advantages, RISs can be integrated into the wireless networks to aid communications with a low cost. Specifically, a RIS acts like a large reflecting antenna array, and its elements induce controllable and independent phase shifts on the incident signals. Through optimizing the phase shifts, the RIS is able to combat the unfavorable wireless channel conditions and achieve desired channel responses at the receivers by coherently enhancing useful signals and suppressing interference. To exploit the full benefits of the RIS, new challenges in the physical layer design need to be addressed, as listed as follows.

Channel Estimation Channel estimation in a RIS-aided system, compared with that in a conventional communication system without RIS, is quite different and more challenging. The reason is two-fold: First, in addition to the transmitter-receiver direct link, the existence of the RIS brings an extra transmitter-RIS-receiver channel link to be estimated; second, since a nearly passive RIS has very limited signal processing capability, it is the responsibility of the receiver to seperate the transmitter-RIS and RIS-receiver channels by observing a cascade of them. Therefore, new algorithms are necessary for RIS channel estimation.

Passive Beamforming Adjusting the RIS phase shifts to enhance the end-to-end communications is referred to as passive beamforming (a.k.a. reflect beamforming). Passive beamforming is usually optimized together with the active beamforming/precoding at the transmitter. In practice, the RIS is implemented with a finite number of phase shift levels, and hence have a quantized set of available angles. Therefore, joint beamforming optimization problems are typically non-convex with discrete feasible sets.

Resource Allocation and System Optimization To fully exploit the advantages of RISs, resource allocation and system optimization on transmit power, the RIS size and placement, and phase shifter quantization levels, etc., should be carefully addressed. Besides, new transceiver strategies are needed to accommodate the new system design.

It is clear that the solutions to the last two challenges critically depend on the knowledge of channel state information (CSI). Therefore, efficient CSI acquisition algorithms are of paramount importance for the RIS system design. In this chapter, we focus on tackling the channel estimation problem in RIS-aided wireless networks. In Sects. 3 and 4, we present two state-of-the-art approaches from [3, 4] on estimating the cascaded transmitter-RIS and RIS-receiver channels in two different setups. Furthermore, we discuss open challenges in RIS channel estimation in Sect. 5. For more discussions on other channel estimation approaches and the solutions to passive beamforming and resource allocation, we refer interested readers to the literature summarized in Sect. 6.

Notation Throughout this chapter, we use $\mathbb{R}$ and $\mathbb{C}$ to denote the real and complex number sets, respectively. Regular letters, bold small letters, and bold capital letters are used to denote scalars, vectors, and matrices, respectively. We use $j \triangleq \sqrt{-1}$ to denote the imaginary unit. We use $(\cdot)^\star$, $(\cdot)^T$, $(\cdot)^{-1}$, and $(\cdot)^H$ to denote the conjugate, the transpose, the inverse, and the conjugate transpose, respectively. We use x_{ij} to denote the (i, j)-th entry of $\mathbf{X}$. We use $\mathcal{N}(\mathbf{x}; \boldsymbol{\mu}, \boldsymbol{\Sigma})$ and $\mathcal{CN}(\mathbf{x}; \boldsymbol{\mu}, \boldsymbol{\Sigma})$ to denote that $\mathbf{x}$ follows the real normal and the circularly-symmetric normal distributions with mean $\boldsymbol{\mu}$ and covariance $\boldsymbol{\Sigma}$, respectively. We use $\mathbf{I}$ to denote the identity matrix with an appropriate size, $\text{diag}(\mathbf{x})$ to denote a diagonal matrix with the diagonal entries specified by $\mathbf{x}$, $\|\cdot\|_p$ to denote the ℓ_p norm, $\|\cdot\|_F$ to denote the Frobenius norm, $\delta(\cdot)$ to denote the Dirac delta function, $\propto$ to denote equality up to a constant multiplicative factor, and $\mathbb{E}[\cdot]$ to denote the expectation operator.

2 System Model

We consider a single-cell RIS-aided MIMO system, as shown in Fig. 5.1. An M-antenna BS serves K users, where each user is equipped with N antennas. A RIS comprising L passive phase-shift elements is deployed to assist the communication between the users and the BS. We assume a quasi-static block-fading channel model, where the channel coefficients remain invariant within the coherence time. The k-th-user-RIS channel coefficients and the RIS-BS channel coefficients are denoted by $\mathbf{H}_{UR,k} \in \mathbb{C}^{L \times N}$ and $\mathbf{H}_{RB} \in \mathbb{C}^{M \times L}$, respectively.

Throughout this chapter, we assume that the direct channels between the BS and the users are accurately estimated (and cancelled from the model) via conventional MIMO channel estimation methods by turning off the RIS reflecting elements. Therefore, the channel estimation approaches introduced in this chapter focus on the estimation of the user-RIS and RIS-BS channels at the BS from a noisy observation of a cascade of them. We refer to the resultant RIS channel estimation problem as the *cascaded* channel estimation problem.

The RIS elements induce independent phase shifts on the incident signals. We denote the RIS phase-shift vector at time t by

$$\boldsymbol{\psi}(t) \triangleq [\varpi_1(t)e^{j\psi_1(t)}, \varpi_2(t)e^{j\psi_2(t)}, \cdots, \varpi_L(t)e^{j\psi_L(t)}]^T, \tag{5.1}$$

where $\varpi_l(t) \in \{0, 1\}$ represents the on/off state of the l-th RIS element, and $\psi_l(t) \in [0, 2\pi)$ represents the phase shift of the l-th RIS element.

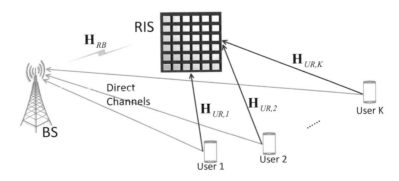

Fig. 5.1 The considered RIS-aided MIMO system [4]

In the subsequent two sections, we present two cascaded channel estimation approaches for the single-user system (i.e., $K = 1$ and $N \geq 1$) and the multi-user system with single-antenna users (i.e., $K \geq 1$ and $N = 1$).

3 Channel Estimation in RIS-Aided Single-User Systems

This section presents a two-stage approach for cascaded channel estimation in RIS-aided *single-user* MIMO systems (i.e., $K = 1$) based on the work in [3]. For ease of notation, we drop the user index k throughout this section.

3.1 Channel Estimation Protocol

The user transmits a length-T training pilot sequence for cascaded channel estimation. Denote by $\mathbf{x}(t) \in \mathbb{C}^{N \times 1}$ the transmitted pilot at time t. The received signal at time t is given by

$$\mathbf{y}(t) = \mathbf{H}_{RB} \left(\boldsymbol{\psi}(t) \circ (\mathbf{H}_{UR}\mathbf{x}(t)) \right) + \mathbf{n}(t), \tag{5.2}$$

where $\circ$ represents the Hadamard product; and $\mathbf{n}(t)$ represents the additive noise drawn from $\mathcal{CN}(\mathbf{n}(t); \mathbf{0}, \tau_N \mathbf{I})$ at time t. By summarizing all the T samples, the received signal can be recast as

$$\mathbf{Y} = \mathbf{H}_{RB} \left(\boldsymbol{\Psi} \circ (\mathbf{H}_{UR}\mathbf{X}) \right) + \mathbf{N}, \tag{5.3}$$

where $\mathbf{Y} = [\mathbf{y}(1), \cdots, \mathbf{y}(T)]$; $\mathbf{N} = [\mathbf{n}(1), \cdots, \mathbf{n}(T)]$; $\mathbf{X} = [\mathbf{x}(1), \cdots, \mathbf{x}(T)]$; and $\boldsymbol{\Psi} = [\boldsymbol{\psi}(1), \cdots, \boldsymbol{\psi}(T)]$.

Throughout this section, we assume that the user-RIS channel $\mathbf{H}_{UR}$ is rank-deficient. That is, $\text{rank}(\mathbf{H}_{UR}) < \min\{N, L\}$. The rank-deficiency property commonly arises in mmWave MIMO channels under the far-field and limited-scattering assumptions.

Furthermore, we generate the states of the phase shifts $\{\varpi_l(t), \forall l, t\}$ independently from a Bernoulli distribution Bernoulli(λ) with (small) λ being the probability of taking the value of 1. We then generate the phase shifts by $\psi_l(t) = 0, \forall l, t$. This RIS phase shift design ensures the matrix $\boldsymbol{\Psi} \circ (\mathbf{H}_{UR}\mathbf{X})$ to be a sparse matrix, which plays an important role in the following channel estimation design. Finally, The transmitted pilot signal $\mathbf{X}$ is designed to be a full-rank matrix. That is, $\text{rank}(\mathbf{X}) = \min\{T, N\}$.

3.2 Ambiguity Problem in Channel Estimation

We note that the following identity holds for any full-rank diagonal matrix $\mathbf{D} \in \mathbb{C}^{L \times L}$:

$$\mathbf{H}_{RB} \left(\boldsymbol{\Psi} \circ (\mathbf{H}_{UR} \mathbf{X}) \right) = \underbrace{\mathbf{H}_{RB} \mathbf{D}}_{\triangleq \mathbf{H}'_{RB}} \left(\boldsymbol{\Psi} \circ \underbrace{(\mathbf{D}^{-1} \mathbf{H}_{UR}}_{\triangleq \mathbf{H}'_{UR}} \mathbf{X}) \right), \tag{5.4}$$

As a result, we can only estimate $\mathbf{H}'_{UR}$ and $\mathbf{H}'_{RB}$ instead of the ground-truth $\mathbf{H}_{UR}$ and $\mathbf{H}_{RB}$ from the model in (5.3). This is called the ambiguity problem in cascaded channel estimation. However, as explained in the remark below, the CSI acquired in the presence of the diagonal ambiguity does not affect the passive beamforming design.

Remark 1 In passive beamforming, the value of the phase shift vector $\boldsymbol{\psi}(t)$ remains unchanged unless the channel changes. Therefore, $\{\boldsymbol{\psi}(t)\}$ are generally set as a constant vector in the data transmission phase of each coherence block. That is, $\boldsymbol{\psi}(t) = \boldsymbol{\psi}, \forall t$. The signal model in the data transmission phase is thus given by

$$\mathbf{Y} = \mathbf{H}_{RB} \left(\text{diag}(\boldsymbol{\psi}) \circ (\mathbf{H}_{UR} \mathbf{X}) \right) + \mathbf{N}. \tag{5.5}$$

Therefore, we have

$$\mathbf{H}_{RB} \left(\text{diag}(\boldsymbol{\psi}) \circ (\mathbf{H}_{UR} \mathbf{X}) \right) = \mathbf{H}'_{RB} \left(\text{diag}(\boldsymbol{\psi}) \circ (\mathbf{H}'_{UR} \mathbf{X}) \right). \tag{5.6}$$

The optimization over $\boldsymbol{\psi}$ based on $\mathbf{H}_{UR}$ and $\mathbf{H}_{RB}$ is exactly the same as that based on $\mathbf{H}'_{UR}$ and $\mathbf{H}'_{RB}$. In other words, the ambiguity problem in channel estimation does not affect the passive beamforming performance.

3.3 Two-Stage Channel Estimation Approach

Upon receiving $\mathbf{Y}$ in (5.3), we aim to estimate the channel matrices $\mathbf{H}_{UR}$ and $\mathbf{H}_{RB}$ at the BS (subject to some diagonal ambiguity) with the knowledge of $\mathbf{X}$ and $\boldsymbol{\Psi}$. We first rewrite the model in (5.3) as

$$\mathbf{Y} = \mathbf{H}_{RB} \mathbf{Z} + \mathbf{N}, \tag{5.7}$$

where $\mathbf{Z} \triangleq \mathbf{S} \circ (\mathbf{H}_{UR} \mathbf{X})$ is a sparse matrix. This motivates us to consider a two-stage scheme for channel estimation: First estimate $\mathbf{H}_{RB}$ and $\mathbf{Z}$ by sparse matrix factorization, and then estimate $\mathbf{H}_{UR}$ by low-rank matrix completion. The details are describes as follows.

Sparse Matrix Factorization Stage We employ the bilinear approximate message passing (BiG-AMP) algorithm [5] to approximately calculate the minimum mean squared error (MMSE) estimators of $\mathbf{Z}$ and $\mathbf{H}_{RB}$, i.e., the means of the marginal posteriors $\{p(z_{lt}|\mathbf{Y})\}$ and $\{p(h_{RB,ml}|\mathbf{Y})\}$, under the Bayesian framework. Specifically, a factor graph is constructed based on the following factorizable posterior distribution:

$$p(\mathbf{Z}, \mathbf{H}_{RB}|\mathbf{Y}) \propto p(\mathbf{Y}|\mathbf{B})p(\mathbf{Z})p(\mathbf{H}_{RB}), \tag{5.8}$$

where $\mathbf{B} \triangleq \mathbf{H}_{RB}\mathbf{Z}$. In (5.8), the prior distributions are given by

$$p(\mathbf{Y}|\mathbf{B}) = \prod_{m=1}^{M} \prod_{t=T}^{M} \mathcal{CN}(y_{mt}; b_{mt}, \tau_N), \tag{5.9}$$

$$p(\mathbf{Z}) = \prod_{l=1}^{L} \prod_{t=T}^{M} ((1-\lambda)\delta(z_{lt}) + \lambda\mathcal{CN}(z_{lt}; 0, \tau_Z)), \tag{5.10}$$

$$p(\mathbf{H}_{RB}) = \prod_{m=1}^{M} \prod_{l=1}^{L} \mathcal{CN}(h_{RB,ml}; 0, \tau_{H_{RB}}), \tag{5.11}$$

where τ_Z represents the variance of the non-zero entries of $\mathbf{Z}$; and $\tau_{H_{RB}}$ represents the variance of the entries of $\mathbf{H}_{RB}$. Note that we adopt the Bernoulli Gaussian prior distribution to model the sparsity in $\mathbf{Z}$ that is caused by the sparse phase shift matrix $\mathbf{S}$.

With the above probabilistic model, BiG-AMP performs sum-product loopy message passing to iteratively approximate the marginal distributions of $\mathbf{H}_{RB}$ and $\mathbf{Z}$. Furthermore, to achieve computational efficiency, BiG-AMP leverages the central limit theorem (CLT) and the second-order Taylor expansion to approximate the involved messages as Gaussian distributions. As a result, we only need to update the marginal posterior means with tractable expressions during the iteration. Due to space limitation, the detailed updating formulas are omitted here and can be found in [3, Algorithm 1].

Matrix Completion Stage Based on the estimated $\mathbf{Z}$, denoted by $\hat{\mathbf{Z}}$, in the first stage and the sparse phase shift matrix $\mathbf{\Psi}$, the matrix completion stage is to retrieve $\mathbf{H}_{UR}$ by using the Riemannian manifold gradient-based (RGrad) algorithm [6]. Specifically, we first recover the missing entries of $\hat{\mathbf{Z}}$ by solving the following matrix completion problem:

$$\min_{\mathbf{A}} \tfrac{1}{2} \|\mathbf{\Psi} \circ (\mathbf{A} - \hat{\mathbf{Z}})\|_F^2 \text{ subject to } \text{rank}(\mathbf{A}) = r, \tag{5.12}$$

for some predetermined value of r. The RGrad algorithm computes $\mathbf{A}$ iteratively by

$$\mathbf{A}(k+1) = \mathcal{H}_r\left(\mathbf{A}(k) + \alpha_k \mathcal{P}_{\mathcal{S}(k)}(\mathbf{\Psi} \circ (\hat{\mathbf{Z}} - \mathbf{A}(k)))\right), \tag{5.13}$$

where $k = 1, 2, \cdots, K_{max}$ is the iteration index; and we set $\mathbf{A}(0) = \mathbf{0}$. In (5.13), $\mathcal{P}_{\mathcal{S}(k)}(\cdot)$ stands for the projection operation to the left singular vector subspace (denoted by $\mathcal{S}(k)$) of the current estimate $\mathbf{A}(k)$, corresponding to the first r eigenvalues of $\mathbf{A}(k)$; α_k is given by

$$\alpha_k = \frac{\|\mathcal{P}_{\mathcal{S}(k)}(\mathbf{\Psi} \circ (\hat{\mathbf{Z}} - \mathbf{A}(k)))\|_F^2}{\|\mathbf{\Psi} \circ \mathcal{P}_{\mathcal{S}(k)}(\mathbf{\Psi} \circ (\hat{\mathbf{Z}} - \mathbf{A}(k)))\|_F^2}; \qquad (5.14)$$

and $\mathcal{H}_r(\cdot)$ is the hard-thresholding operator for the best rank-r approximation of the associated singular value decomposition (SVD) of the input matrix. In other words, suppose that the SVD of matrix $\mathbf{M}$ is $\mathbf{M} = \mathbf{U}\mathbf{\Sigma}\mathbf{V}^H$ and $\Sigma(i, i)$ represents the (i, i)-th entry of $\mathbf{\Sigma}$. Then, we have

$$\mathcal{H}_r(\mathbf{M}) = \mathbf{U}\mathbf{\Sigma}_r\mathbf{V}^H, \; \Sigma_r(i, i) = \begin{cases} \Sigma(i, i), & i \le r, \\ 0, & i > r. \end{cases} \qquad (5.15)$$

Finally, the estimate of the channel matrix $\mathbf{H}_{UR}$ can be computed as

$$\hat{\mathbf{H}}_{UR} = \hat{\mathbf{A}}\mathbf{X}^\dagger, \qquad (5.16)$$

where $\mathbf{X}^\dagger = (\mathbf{X}\mathbf{X}^H)^{-1}\mathbf{X}$ is the Moore-Penrose inverse and $\hat{\mathbf{A}}$ is the final output of the RGrad algorithm. Here, we assume that the pilot length T is no less than the number of transmit antennas N and rank$(\mathbf{X}) = N$, so as to ensure the existence of $\mathbf{X}^\dagger$.

We summarize the overall joint bilinear factorization and matrix completion (JBF-MC) algorithm in Algorithm 1. As a final remark, we note that the final estimates of $\mathbf{H}_{RB}$ and $\mathbf{H}_{UR}$ have a diagonal ambiguity resulting from the sparse matrix factorization stage. Nevertheless, as discussed in Sect. 3.2, there is no need to eliminate this diagonal ambiguity.

Algorithm 1 The JBF-MC algorithm [3]
___

Input: $\mathbf{Y}$; $\mathbf{X}$; $\mathbf{\Psi}$; r; λ; τ_N; τ_Z; $\tau_{H_{RB}}$.
Use BiG-AMP to compute $\hat{\mathbf{Z}}$ and $\hat{\mathbf{H}}_{RB}$ (cf. [3, Algorithm 1, Lines 1–24]) .
Initialize $\mathbf{A}(0) = \mathbf{0}$.
for $k = 1, 2, \cdots, K_{max}$ **do**
Update α_k by (5.14);
Update $\mathbf{A}(k + 1)$ by (5.13);
end for
$\hat{\mathbf{A}} \leftarrow \mathbf{A}(k + 1)$;
$\hat{\mathbf{H}}_{UR} \leftarrow \hat{\mathbf{A}}\mathbf{X}^\dagger$.
Output: $\hat{\mathbf{H}}_{RB}$ and $\hat{\mathbf{H}}_{UR}$.

Algorithm 1 can be directly extended to the multi-user case by activating users one by one and repeatedly applying the algorithm to estimate each user's channels. However, this leads to an excessively large training overhead, and hence is very inefficient. In the next section, we present an alternative channel estimation algorithm that jointly estimates all users' channels.

3.4 Numerical Results

We now carry out numerical experiments to evaluate the performance of the JBF-MC algorithm. For simplicity, we assume that the antenna elements form half-wavelength uniform linear arrays (ULAs) at the user, the RIS, and the BS. Following the superposition principle of different paths in the prorogation environment, the true user-RIS and RIS-BS channel matrices are generated by the multi-path channel model, where each channel path is determined by the steering vector associated with the ULA and the corresponding angle parameters. Specifically, all the path gain coefficients are independently drawn from $CN(\cdot; 0, 1)$; the angle parameters (i.e., the sine of the angle-of-arrivals and angle-of-departures) are independently and uniformly drawn from $[0, 1)$; the number of paths in the channel matrix $\mathbf{H}_{UR}$ is set to 4 such that it has a low-rank structure to facilitate its estimation in the matrix completion stage; and the number of paths in $\mathbf{H}_{RB}$ is set to $\max\{M, L\}/2$. Finally, we set $L = 70$, $M = N = 64$, and $r = 4$.

The pilot symbols in $\mathbf{X}$ are generated from $CN(\cdot; 0, 1)$; and the signal-to-noise ratio (SNR) is defined as $10\log_{10}(1/\tau_N)$ dB. The estimation performance is evaluated in terms of the normalized mean-square-errors (NMSEs). All the simulation results are obtained by averaging 200 independent trials. Besides, we use the true values of $\mathbf{H}_{RB}$ and $\mathbf{H}_{UR}$ to eliminate the diagonal ambiguity in their estimates. We compare the performance of the JBF-MC algorithm with K-SVD [7] and SPAMS [8] in the sparse matrix factorization stage; and with iterative hard thresholding (IHT) and iterative soft thresholding (IST) [9] in the matrix completion stage.

The NMSEs versus SNR (with $T = 300$) and the number of pilots (with SNR = 10 dB) are depicted in Fig. 5.2 under the sparsity level (sampling rate) $\lambda = 0.2$. It is observed that the proposed JBF-MC algorithm has a significant performance gain over the baseline methods. Figure 5.3 shows the phase transitions of the channel estimation performance versus the sparsity level and the number of pilots. We find that there is a performance tradeoff between the sparse matrix factorization and the matrix completion: The matrix completion fails if λ is too small; and the sparse matrix factorization fails if λ is too large. This is because when a larger number of samples (i.e., larger λ) are available for matrix completion, the performance of BiG-AMP becomes worse as a larger number of non-zero variables are needed to be estimated in the sparse matrix factorization stage.

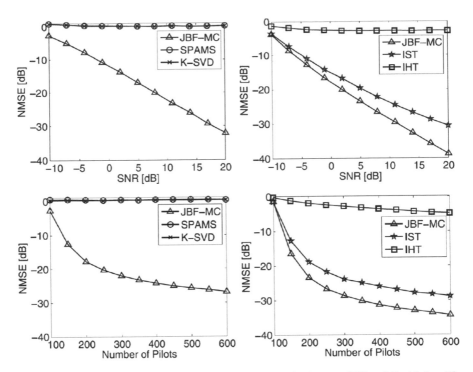

Fig. 5.2 NMSEs of $\mathbf{H}_{RB}$ (left subplots) and $\mathbf{H}_{UR}$ (right subplots) versus SNR and T with $L = 70$, $M = N = 64$, and $\lambda = 0.2$ [3]

4 Channel Estimation in RIS-Aided Multi-User Systems

In this section, we present a matrix-calibration-based cascaded channel estimation algorithm for RIS-aided *multi-user* systems with single-antenna users (i.e., $K \geq 1$ and $N = 1$). The content in this section is based on the work in [4].

4.1 Channel Estimation Protocol

As $N = 1$, we represent the k-th-user-RIS channel coefficients by $\mathbf{h}_{UR,k} \in \mathbb{C}^{L \times 1}$ and define $\mathbf{H}_{UR} \triangleq [\mathbf{h}_{UR,1}, \cdots, \mathbf{h}_{UR,K}]$. Different from Sect. 3, we assume that a ULA is adopted at the BS, and the passive reflecting elements in the RIS are arranged in the form of an $L_1 \times L_2$ uniform rectangular array (URA) with $L_1 L_2 = L$. The steering vector associated with the BS (or RIS) antenna geometry is denoted by $\mathbf{a}_B$ (or $\mathbf{a}_R$) with

$$\mathbf{a}_B(\theta) = \mathbf{f}_M(\sin(\theta)), \tag{5.17a}$$

$$\mathbf{a}_R(\phi, \sigma) = \mathbf{f}_{L_2}(-\cos(\sigma)\cos(\phi)) \otimes \mathbf{f}_{L_1}(\cos(\sigma)\sin(\phi)), \tag{5.17b}$$

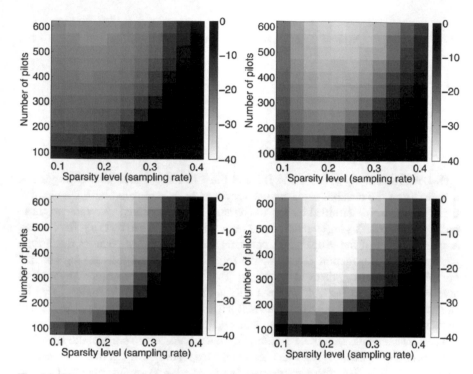

Fig. 5.3 Phase transitions (in terms of NMSEs) of $\mathbf{H}_{RB}$ (left subplots) and $\mathbf{H}_{UR}$ (right subplots) versus the sparsity level and the number of pilots. The upper subplots are with SNR = 10 dB and the lower subplots with SNR = 20 dB [3]

where θ, ϕ, and σ are the corresponding angle parameters; $\otimes$ denotes the Kronecker product; and

$$\mathbf{f}_N(x) \triangleq \frac{1}{\sqrt{N}} \left[1, e^{-j\frac{2\pi}{\varrho}dx}, \cdots, e^{-j\frac{2\pi}{\varrho}d(N-1)x} \right]^T. \tag{5.18}$$

In (5.18), ϱ denotes the carrier wavelength; and d denotes the distance between any two adjacent antennas. Throughout this section, we set $d/\varrho = 1/2$ for simplicity.

To facilitate RIS channel estimation, the users simultaneously transmit training sequences with length T to the BS. Denote by $\mathbf{x}_k = [x_{k1}, \cdots, x_{kT}]^T$ the training sequence of user k, where x_{kt} is the training symbol of user k in time slot t. We assume that the users transmit at constant power τ_X, i.e., $\mathbb{E}[|x_{kt}|^2] = \tau_X, \forall k, t$.

Over the time duration T, all the RIS elements are turned on and are set to have the same phase shift. Without loss of generality, we assume that $\boldsymbol{\psi}(t) = \mathbf{1}, 1 \leq t \leq T$. The received signal at the BS in time slot t is given by

$$\mathbf{y}(t) = \sum_{k=1}^{K} \mathbf{H}_{RB} \mathbf{h}_{UR,k} x_{kt} + \mathbf{n}(t), \, 1 \leq t \leq T, \qquad (5.19)$$

where $\mathbf{n}(t)$ is defined as in (5.2).

4.2 Channel Model

RIS-BS Channel $\mathbf{H}_{RB}$ Since the BS and the RIS rarely move after deployment, a majority of the channel paths evolve very slowly, as compared with the channel coherence time (determined by the movement of mobile ends). We refer to these paths as the slow-varying channel components of $\mathbf{H}_{RB}$, denoted by $\bar{\mathbf{H}}_{RB}$. In contrast, a small portion of the propagation paths may experience fast changes due to the change of the propagation geometry. We refer to them as the fast-varying channel components of $\mathbf{H}_{RB}$, denoted by $\tilde{\mathbf{H}}_{RB}$. To illustrate this, consider an outdoor scenario depicted in Fig. 5.4, where a RIS is mounted on an advertising panel by the road. The propagation paths from static scattering clusters between the BS and the RIS changes very slowly and $\bar{\mathbf{H}}_{RB}$ sums up the paths from all the static scattering clusters. Meanwhile, suppose, for example, that a truck is moving past the RIS. The path through the truck evolves fast, as the corresponding propagation geometry is quickly changing. This fast-varying path is modelled in $\tilde{\mathbf{H}}_{RB}$. To capture $\bar{\mathbf{H}}_{RB}$ and $\tilde{\mathbf{H}}_{RB}$, we model $\mathbf{H}_{RB}$ by the MIMO Rician fading model as [10]

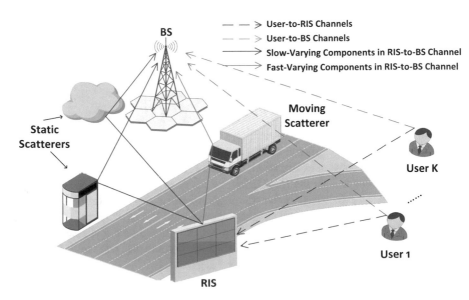

Fig. 5.4 An example of the outdoor propagation geometry. The propagation path through a moving truck belongs to the fast-varying channel components between the BS and the RIS, while the paths from static scatterers belong to the slow-varying channel components [4]

$$\mathbf{H}_{RB} = \sqrt{\frac{\kappa}{\kappa + 1}}\bar{\mathbf{H}}_{RB} + \sqrt{\frac{1}{\kappa + 1}}\tilde{\mathbf{H}}_{RB}, \qquad (5.20)$$

where κ is the Rician factor denoting the power ratio between $\bar{\mathbf{H}}_{RB}$ and $\tilde{\mathbf{H}}_{RB}$. By specifying the paths in the prorogation environment, $\bar{\mathbf{H}}_{RB}$ and $\tilde{\mathbf{H}}_{RB}$ are given by

$$\bar{\mathbf{H}}_{RB} = \sqrt{\beta_0}\sum_{p=1}^{\bar{P}_{RB}}\alpha_p\mathbf{a}_B(\theta_p)\mathbf{a}_R^H(\phi_p, \sigma_p), \qquad (5.21)$$

$$\tilde{\mathbf{H}}_{RB} = \sqrt{\beta_0}\sum_{p=1}^{\tilde{P}_{RB}}\alpha_p\mathbf{a}_B(\theta_p)\mathbf{a}_R^H(\phi_p, \sigma_p), \qquad (5.22)$$

where β_0 is the large-scale path gain encompassing distance-dependent path loss and shadowing; $\bar{P}_{RB}$ (or $\tilde{P}_{RB}$) is the number of the slow-varying (or fast-varying) paths; α_p is the corresponding complex-valued channel coefficient of the p-th path; θ_p is the corresponding azimuth angle-of-arrival (AoA) to the BS; and ϕ_p (or σ_p) is the corresponding azimuth (or elevation) angle-of-departure (AoD) from the RIS.

User-RIS Channels $\{\mathbf{h}_{UR,k}\}$ Similarly to (5.21)–(5.22), we represent $\mathbf{h}_{UR,k}$ as

$$\mathbf{h}_{UR,k} = \sqrt{\beta_k}\sum_{p=1}^{P_k}\alpha_p\mathbf{a}_R(\phi_p, \sigma_p), \qquad (5.23)$$

where β_k is the large-scale path gain for the channel between the k-th user and the RIS; and P_k is the number of paths between the k-th user and the RIS.

Finally, we assume without loss of generality that $\mathbb{E}[\|\bar{\mathbf{H}}_{RB}\|_F^2] = \mathbb{E}[\|\tilde{\mathbf{H}}_{RB}\|_F^2] = \beta_0 ML$ and $\mathbb{E}[\|\mathbf{h}_{UR,k}\|_2^2] = \beta_k L, \forall k$. In other words, β_0 (or β_k) can be regarded as the average power attenuation for each RIS-BS (or user-RIS) antenna pair.

4.3 Virtual Channel Representation

We assume that the slow-varying component matrix $\bar{\mathbf{H}}_{RB}$ in (5.21) keeps static over a time interval much larger than the coherence block length. As a consequence, $\sqrt{\kappa/(\kappa + 1)}\bar{\mathbf{H}}_{RB}$ can be accurately estimated by long-term channel averaging prior to the RIS channel estimation procedure.[1]

---

[1]The assumption of the perfect knowledge of $\sqrt{\kappa/(\kappa + 1)}\bar{\mathbf{H}}_{RB}$ does not lose any generality since any possible error in the channel averaging process can be absorbed into $\sqrt{1/(\kappa + 1)}\tilde{\mathbf{H}}_{RB}$.

Furthermore, as illustrated above, the fast-varying component matrix $\widetilde{\mathbf{H}}_{RB}$ contains a limited number of paths, i.e., $\widetilde{P}_{RB}$ in (5.22) is small. Following [11], we employ a pre-discretized sampling grid $\boldsymbol{\vartheta}$ with length M' ($\geq M$) to discretize $\{\sin(\theta_p)\}_{1 \leq p \leq \widetilde{P}_{RB}}$ over $[0, 1]$. Similarly, we employ two sampling grids $\boldsymbol{\varphi}$ with length L_1' ($\geq L_1$) and $\boldsymbol{\varsigma}$ with length L_2' ($\geq L_2$) to discretize $\{\cos(\sigma_p)\sin(\phi_p)\}_{1 \leq p \leq \widetilde{P}_{RB}}$ and $\{-\cos(\sigma_p)\cos(\phi_p)\}_{1 \leq p \leq \widetilde{P}_{RB}}$, respectively. Then, we represent the fast-varying component matrix $\widetilde{\mathbf{H}}_{RB}$ under the angular bases as [11]

$$\sqrt{\frac{1}{\kappa+1}}\widetilde{\mathbf{H}}_{RB} = \mathbf{A}_B \mathbf{S} \left(\underbrace{\mathbf{A}_{R,v} \otimes \mathbf{A}_{R,h}}_{=\mathbf{A}_R} \right)^H, \qquad (5.24)$$

where $\mathbf{A}_B \triangleq [\mathbf{f}_M(\vartheta_1), \cdots, \mathbf{f}_M(\vartheta_{M'})]$ is an over-complete array response with $\mathbf{f}_M$ defined in (5.18); $\mathbf{A}_{R,h} \triangleq [\mathbf{f}_{L_1}(\varphi_1), \cdots, \mathbf{f}_{L_1}(\varphi_{L_1'})]$ (or $\mathbf{A}_{R,v} \triangleq [\mathbf{f}_{L_2}(\varsigma_1), \cdots, \mathbf{f}_{L_2}(\varsigma_{L_2'})]$) is an over-complete horizontal (or vertical) array response; and $\mathbf{S} \in \mathbb{C}^{M' \times L'}$ is the corresponding channel coefficient matrix in the angular domain with $L' = L_1'L_2'$. Note that the (i, j)-th entry of $\mathbf{S}$ corresponds to the channel coefficient of $\widetilde{\mathbf{H}}_{RB}$ along the path specified by the i-th AoA steering vector in $\mathbf{A}_B$ and the j-th AoD steering vector in $\mathbf{A}_R$. Since the total number of paths $\widetilde{P}_{RB}$ is small, only a few entries of $\mathbf{S}$ are nonzero with each corresponding to a channel path. That is, $\mathbf{S}$ is a sparse matrix.

Similarly to (5.24), $\mathbf{h}_{UR,k}$ can be represented as

$$\mathbf{h}_{UR,k} = \mathbf{A}_R \mathbf{g}_k, \qquad (5.25)$$

where $\mathbf{g}_k \in \mathbb{C}^{L' \times 1}$ represents the channel coefficients of $\mathbf{h}_{UR,k}$ in the angular domain. Experimental studies have shown that the propagation channel often exhibits limited scattering geometry [12]. As a consequence, $\mathbf{g}_k$ is also sparse, where each nonzero value corresponds to a channel path. We note that the sparsity of $\mathbf{S}$ and $\{\mathbf{g}_k\}$ plays an important role in our channel estimation design.

Unlike the setup in the previous section, we do not assume the low-rankness of the user-RIS channels here. In contrast, we exploit the hidden channel sparsity of the user-RIS channels, which arises in propagation channels under the limited-scattering assumption in multi-user communication systems. Note that the sparse channel assumption is looser than the low-rank assumption, since the latter implies the former.

4.4 Problem Formulation

With the above channel representations, we rewrite (5.19) as

$$\mathbf{Y} = \mathbf{H}_{RB}\mathbf{H}_{UR}\mathbf{X} + \mathbf{N}, \tag{5.26}$$

where $\mathbf{X} = [\mathbf{x}_1,\cdots,\mathbf{x}_K]^T$; $\mathbf{N} = [\mathbf{n}(1),\cdots,\mathbf{n}(T)]$, $\mathbf{Y} = [\mathbf{y}(1),\cdots,\mathbf{y}(T)]$. Define $\mathbf{H}_0 \triangleq \sqrt{\kappa/(\kappa+1)}\bar{\mathbf{H}}_{RB}\mathbf{A}_R \in \mathbb{C}^{M\times L'}$; $\mathbf{R} \triangleq \mathbf{A}_R^H\mathbf{A}_R \in \mathbb{C}^{L'\times L'}$; and $\mathbf{G} \triangleq [\mathbf{g}_1,\cdots,\mathbf{g}_K] \in \mathbb{C}^{L'\times K}$. By plugging (5.20), (5.24), and (5.25) into (5.26), we obtain the system model as

$$\mathbf{Y} = \left(\sqrt{\frac{\kappa}{\kappa+1}}\bar{\mathbf{H}}_{RB} + \mathbf{A}_B\mathbf{S}\mathbf{A}_R^H\right)\mathbf{A}_R\mathbf{G}\mathbf{X} + \mathbf{N} = (\mathbf{H}_0 + \mathbf{A}_B\mathbf{S}\mathbf{R})\,\mathbf{G}\mathbf{X} + \mathbf{N}. \tag{5.27}$$

Upon receiving $\mathbf{Y}$ in (5.27), the BS aims to factorize the channel matrices $\mathbf{S}$ and $\mathbf{G}$ with the knowledge of the training signal matrix $\mathbf{X}$. Once the angular bases $\mathbf{A}_B$ and $\mathbf{A}_R$ are predetermined and the slow-varying component matrix $\sqrt{\kappa/(\kappa+1)}\bar{\mathbf{H}}_{RB}$ is given, the sensing matrices $\mathbf{A}_B$, $\mathbf{H}_0$, and $\mathbf{R}$ are also known to the BS. We refer to the above problem as *matrix-calibration* based cascaded channel estimation.

> Different to the algorithm in Sect. 3, the matrix-calibration based cascaded channel estimation problem does not suffer from the ambiguity problem. This is because the knowledge of $\mathbf{H}_0$ eliminates the potential ambiguities in matrix factorization.

4.5 Matrix-Calibration Based Cascaded Channel Estimation Algorithm

In this subsection, we first derive the MMSE estimators of $\mathbf{S}$ and $\mathbf{G}$ given $\mathbf{Y}$ in (5.27) under the Bayesian inference framework. We then resort to sum-product message passing to compute the estimators. To reduce the computational complexity, we impose additional approximations to simplify the message updates in the large-system limit.

4.5.1 Bayesian Inference

Define $\mathbf{W} \triangleq \mathbf{H}_0 + \mathbf{A}_B\mathbf{S}\mathbf{R}$, $\mathbf{Z} \triangleq \mathbf{W}\mathbf{G}$, and $\mathbf{Q} \triangleq \mathbf{Z}\mathbf{X}$. Under the assumption of AWGN, we have

$$p(\mathbf{Y}|\mathbf{Q}) = \prod_{m=1}^{M} \prod_{t=1}^{T} \mathcal{CN}(y_{mt}; q_{mt}, \tau_N). \tag{5.28}$$

Motivated by the sparsity of $\mathbf{S}$ and $\mathbf{G}$, we employ Bernoulli-Gaussian distributions to model their prior distributions as

$$p(\mathbf{S}) = \prod_{m'=1}^{M'} \prod_{l'=1}^{L'} (1 - \lambda_S)\delta(s_{m'l'}) + \lambda_S \mathcal{CN}(s_{m'l'}; 0, \tau_S), \tag{5.29}$$

$$p(\mathbf{G}) = \prod_{l=1}^{L'} \prod_{k=1}^{K} (1 - \lambda_G)\delta(g_{lk}) + \lambda_G \mathcal{CN}(g_{lk}; 0, \tau_G), \tag{5.30}$$

where λ_S (or λ_G) is the corresponding Bernoulli parameter of $\mathbf{S}$ (or $\mathbf{G}$); and τ_S (or τ_G) is the variance of the nonzero entries of $\mathbf{S}$ (or $\mathbf{G}$).

With the prior distributions (5.28)–(5.30), the following proposition states the MMSEs of $\mathbf{S}$ and $\mathbf{G}$.

Proposition 1 *The posterior distribution is given by*

$$p(\mathbf{S}, \mathbf{G}|\mathbf{Y}) = \frac{1}{p(\mathbf{Y})} p(\mathbf{Y}|\mathbf{S}, \mathbf{G}) p(\mathbf{S}) p(\mathbf{G}), \tag{5.31}$$

where $p(\mathbf{Y}) = \int p(\mathbf{Y}|\mathbf{S}, \mathbf{G}) p(\mathbf{S}) p(\mathbf{G}) \mathrm{d}\mathbf{S}\mathrm{d}\mathbf{G}$. Furthermore, the MMSEs of $\mathbf{S}$ and $\mathbf{G}$ are given by

$$MMSE_{\mathbf{S}} = \frac{1}{M'L'} \mathbb{E}\left[\|\mathbf{S} - \hat{\mathbf{S}}\|_F^2\right], MMSE_{\mathbf{G}} = \frac{1}{L'K} \mathbb{E}\left[\|\mathbf{G} - \hat{\mathbf{G}}\|_F^2\right], \tag{5.32}$$

where the expectations are taken over the joint distribution of $\mathbf{S}$, $\mathbf{G}$, and $\mathbf{Y}$; and $\hat{\mathbf{S}} = [\hat{s}_{m'l'}]$ (or $\hat{\mathbf{G}} = [\hat{g}_{lk}]$) is the posterior mean estimator of $\mathbf{S}$ (or $\mathbf{G}$), given by

$$\hat{s}_{m'l'} = \int s_{m'l'} p(s_{m'l'}|\mathbf{Y})\mathrm{d}s_{m'l'}, \hat{g}_{lk} = \int g_{lk} p(g_{lk}|\mathbf{Y})\mathrm{d}g_{lk}. \tag{5.33}$$

In the above, $p(s_{m'l'}|\mathbf{Y}) = \int \int p(\mathbf{S}, \mathbf{G}|\mathbf{Y})\mathrm{d}\mathbf{G}\mathrm{d}(\mathbf{S} \setminus s_{m'l'})$ and $p(g_{lk}|\mathbf{Y}) = \int \int p(\mathbf{S}, \mathbf{G}|\mathbf{Y})\mathrm{d}\mathbf{S}\mathrm{d}(\mathbf{G} \setminus g_{lk})$ are the marginal distributions with respect to $s_{m'l'}$ and g_{lk}, respectively, where $\mathbf{X} \setminus x_{ij}$ means the collection of the elements of matrix $\mathbf{X}$ except for the (i, j)-th one.

Proof See [4, Proposition 2]. □

Exact evaluation of $\hat{\mathbf{S}}$ and $\hat{\mathbf{G}}$ are generally intractable due to the high-dimensional integrations involved in the marginalization. In the following, we provide an approximate solution by following the message passing principle.

4.5.2 Message Passing for Marginal Posterior Computation

Plugging (5.28)–(5.30) into (5.31), we obtain

$$
p(\mathbf{S}, \mathbf{G}|\mathbf{Y}) = \frac{1}{p(\mathbf{Y})} \left(\prod_{m=1}^{M} \prod_{t=1}^{T} p(y_{mt}|q_{mt}) p(q_{mt}|z_{mk}, \forall k) \right)
$$

$$
\times \left(\prod_{m=1}^{M} \prod_{l=1}^{L'} p(w_{ml}|s_{m'l'}, 1 \le m' \le M', 1 \le l' \le L') \right) \left(\prod_{l=1}^{L'} \prod_{k=1}^{K} p(g_{lk}) \right)
$$

$$
\times \left(\prod_{m=1}^{M} \prod_{k=1}^{K} p(z_{mk}|w_{ml}, g_{lk}, 1 \le l \le L') \right) \left(\prod_{m'=1}^{M'} \prod_{l'=1}^{L'} p(s_{m'l'}) \right), \tag{5.34}
$$

where the factorizable distributions are defined in Table 5.1.

We construct a factor graph to represent (5.34) and apply the canonical message passing algorithm to approximately compute the estimators in (5.33). The factor graph is depicted in Fig. 5.5. The variables $\mathbf{S}$, $\mathbf{G}$, $\mathbf{W}$, $\mathbf{Z}$, and $\mathbf{Q}$ are represented by the variable nodes $\{s_{m'l'}\}_{1 \le m' \le M', 1 \le l' \le L'}$, $\{g_{lk}\}_{1 \le l \le L', 1 \le k \le K}$, $\{w_{ml}\}_{1 \le m \le M, 1 \le l \le L'}$, $\{z_{mk}\}_{1 \le m \le M, 1 \le k \le K}$, and $\{q_{mt}\}_{1 \le m \le M, 1 \le t \le T}$, respectively. The factorizable pdfs in

Table 5.1 Notation of factor nodes [4]

Factor	Distribution	Exact form
$p(s_{m'l'})$	$p(s_{m'l'})$	$(1 - \lambda_S)\delta(s_{m'l'}) + \lambda_S \mathcal{CN}(s_{m'l'}; 0, \tau_S)$
$p(g_{lk})$	$p(g_{lk})$	$(1 - \lambda_G)\delta(g_{lk}) + \lambda_G \mathcal{CN}(g_{lk}; 0, \tau_G)$
ws_{ml}	$p(w_{ml}|s_{m'l'}, 1 \le m' \le M', 1 \le l' \le L')$	$\delta(w_{ml} - h_{0,ml} - \sum_{m',l'} a_{B,mm'} s_{m'l'} r_{l'l})$
zwg_{mk}	$p(z_{mk}|w_{ml}, g_{lk}, 1 \le l \le L')$	$\delta(z_{mk} - \sum_{l=1}^{L'} w_{ml} g_{lk})$
qz_{mt}	$p(q_{mt}|z_{mk}, 1 \le k \le K)$	$\delta(q_{mt} - \sum_{k=1}^{K} z_{mk} x_{kt})$
$p(y_{mt}|q_{mt})$	$p(y_{mt}|q_{mt})$	$\mathcal{CN}(y_{mt}; q_{mt}, \tau_N)$

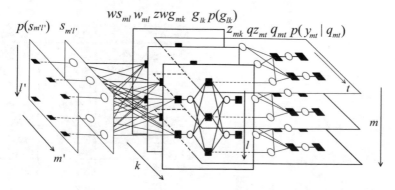

Fig. 5.5 An illustration of the factor graph representation for $M = M' = K = 3$ and $T = L' = 2$, where blank circles and black squares represent variable nodes and factor nodes, respectively [4]

(5.34), represented by factor nodes $\{p(s_{m'l'})\}$, $\{p(g_{lk})\}$, $\{ws_{ml}\}$, $\{zwg_{mk}\}$, $\{qz_{mt}\}$, and $\{p(y_{mt}|q_{mt})\}$, are connected to their associated arguments. We summarize the notation of the factor nodes in Table 5.1. Denote by $\Delta_{a \to b}^i(\cdot)$ the message from node a to b in iteration i, and by $\Delta_c^i(\cdot)$ the marginal message computed at variable node c in iteration i. Applying the sum-product rule, we obtain the expressions of the messages for computing the marginal posterior distributions. Due to space limitation, the detailed formulas are omitted here and can be found in [4, Sect. IV-B].

4.5.3 Approximations for Message Passing

The exact calculations of the messages are computationally intractable in general due to the high-dimensional integrations and normalizations therein. To tackle this, we simplify the calculation of loopy message passing by following the idea of AMP [13] in the large-system limit, i.e., $M, M', K, L, L', T, \tau_N \to \infty$ with the ratios M/K, M'/K, L/K, L'/K, T/K, and τ_N/K^2 fixed. For ease of notation, we define the means and variances of the messages as in Table 5.2. In the following, we sketch the main approximations involved in the algorithm design, while the detailed derivations can be found in [4]. The resultant algorithm is summarized in Algorithm 2.

- By applying second-order Taylor expansions and the CLT argument, we approximate $\prod_{j \neq k} \Delta_{z_{mj} \to qz_{mt}}^i$ as a Gaussian distribution. Then, we characterize both $\Delta_{z_{mt} \to qz_{mk}}^{i+1}$ and $\Delta_{z_{mk}}^{i+1}$ as Gaussian distributions with tractable means and variances.
- To further reduce the computational complexity, we show that $\Delta_{z_{mk} \to qz_{mt}}^{i+1}$ differs from $\Delta_{z_{mk}}^{i+1}$ in only one term, and this term vanishes in the large-system limit. As a consequence, we use the mean and variance of $\Delta_{z_{mk}}^{i+1}$ to approximate those of $\Delta_{z_{mk} \to qz_{mt}}^{i+1}$, and obtain closed-loop updating formulas for $\hat{z}_{mk}$ and v_{mk}^z in Lines 2–8 in Algorithm 2.

Table 5.2 Notations of means and variances for messages [4]

Message	Mean	Variance
$\Delta_{g_{lk} \to zwg_{mk}}^i(g_{lk})$	$\hat{g}_{lk,m}(i)$	$v_{lk,m}^g(i)$
$\Delta_{s_{m'l'} \to ws_{ml}}^i(s_{m'l'})$	$\hat{s}_{m'l',ml}(i)$	$v_{m'l',ml}^s(i)$
$\Delta_{w_{ml} \to zwg_{mk}}^i(w_{ml})$	$\hat{w}_{ml,k}(i)$	$v_{ml,k}^w(i)$
$\Delta_{z_{mk} \to qz_{mt}}^i(z_{mk})$	$\hat{z}_{mk,t}(i)$	$v_{mk,t}^z(i)$
$\Delta_{g_{lk}}^i(g_{lk})$	$\hat{g}_{lk}(i)$	$v_{lk}^g(i)$
$\Delta_{s_{m'l'}}^i(s_{m'l'})$	$\hat{s}_{m'l'}(i)$	$v_{m'l'}^s(i)$
$\Delta_{w_{ml}}^i(w_{ml})$	$\hat{w}_{ml}(i)$	$v_{ml}^w(i)$
$\Delta_{z_{mk}}^i(z_{mk})$	$\hat{z}_{mk}(i)$	$v_{mk}^z(i)$

Algorithm 2 The message passing algorithm [4]

Input: $\mathbf{Y}$; $\mathbf{A}_B$; $\mathbf{H}_0$; $\mathbf{R}$; $\mathbf{X}$; τ_N; λ_S; λ_G; τ_S; τ_G.

Initialization: $\hat{\gamma}_{mt}(0) = \hat{\xi}_{mk}(0) = \hat{\alpha}_{ml}(0) = 0$; $v_{m'l'}^s(1) = v_{lk}^g(1) = v_{ml}^w(1) = v_{mk}^z(1) = 1$; $\hat{g}_{lk}(1) = \hat{z}_{mk}(1) = 0$; $\hat{s}_{m'l'}(1)$ drawn from $p(s_{m'l'})$; $\hat{w}_{ml}(1) = h_{0,ml} + \sum_{m',l'} a_{B,mm'} \hat{s}_{m'l'}(1) r_{l'l}$.

1: **for** $i = 1, 2, \cdots, I_{\max}$

%Update mean and variance for z_{mk}:

2: $\quad \forall m, t$: $v_{mt}^\beta(i) = \sum_{k=1}^K v_{mk}^z(i)|x_{kt}|^2$, $\hat{\beta}_{mt}(i) = \sum_{k=1}^K \hat{z}_{mk}(i)x_{kt} - v_{mt}^\beta(i)\hat{\gamma}_{mt}(i-1)$;

3: $\quad \forall m, t$: $v_{mt}^\gamma(i) = 1/\left(v_{mt}^\beta(i) + \tau_N\right)$, $\hat{\gamma}_{mt}(i) = v_{mt}^\gamma(i)\left(y_{mt} - \hat{\beta}_{mt}(i)\right)$;

4: $\quad \forall m, k$: $\bar{v}_{mk}^p(i) = \sum_{l=1}^{L'}\left(|\hat{w}_{ml}(i)|^2 v_{lk}^g(i) + v_{ml}^w(i)|\hat{g}_{lk}(i)|^2\right)$;

5: $\quad \forall m, k$: $v_{mk}^p(i) = \bar{v}_{mk}^p(i) + \sum_{l=1}^{L'} v_{ml}^w(i)v_{lk}^g(i)$, $v_{mk}^e(i) = 1/\left(\sum_{t=1}^T v_{mt}^\gamma(i)|x_{kt}|^2\right)$;

6: $\quad \forall m, k$: $\hat{e}_{mk}(i) = \hat{z}_{mk}(i) + v_{mk}^e(i)\sum_{t=1}^T x_{kt}^\star \hat{\gamma}_{mt}(i)$;

7: $\quad \forall m, k$: $\hat{p}_{mk}(i) = \sum_{l=1}^{L'} \hat{w}_{ml}(i)\hat{g}_{lk}(i) - \hat{\xi}_{mk}(i-1)\bar{v}_{mk}^p(i)$;

8: $\quad \forall m, k$: $v_{mk}^z(i+1) = \frac{v_{mk}^p(i)v_{mk}^e(i)}{v_{mk}^p(i)+v_{mk}^e(i)}$, $\hat{z}_{mk}(i+1) = \frac{v_{mk}^p(i)\hat{e}_{mk}(i)+\hat{p}_{mk}(i)v_{mk}^e(i)}{v_{mk}^p(i)+v_{mk}^e(i)}$;

%Update mean and variance for g_{lk}:

9: $\quad \forall m, k$: $v_{mk}^\xi(i) = \frac{v_{mk}^p(i)-v_{mk}^z(i)}{(v_{mk}^p(i))^2}$, $\hat{\xi}_{mk}(i) = \frac{\hat{z}_{mt}(i)-\hat{p}_{mt}(i)}{v_{mk}^p(i)}$;

10: $\quad \forall l, k$: $v_{lk}^b(i) = 1/\left(\sum_{m=1}^M |\hat{w}_{ml}(i)|^2 v_{mk}^\xi(i)\right)$;

11: $\quad \forall l, k$: $\hat{b}_{lk}(i) = \left(1 - v_{lk}^b(i)\sum_{m=1}^M v_{ml}^w(i)v_{mk}^\xi(i)\right)\hat{g}_{lk}(i) + v_{lk}^b(i)\sum_m \hat{w}_{ml}^\star(i)\hat{\xi}_{mk}(i)$;

12: $\quad \forall l, k$: $\hat{g}_{lk}(i+1) = \int g_{lk} \Delta_{g_{lk}}^{i+1}(g_{lk})\mathrm{d}g_{lk}$;

13: $\quad \forall l, k$: $v_{lk}^g(i+1) = \int g_{lk}^2 \Delta_{g_{lk}}^{i+1}(g_{lk})\mathrm{d}g_{lk} - |\hat{g}_{lk}(i+1)|^2$;

%Update mean and variance for w_{ml}:

14: $\quad \forall m, l$: $v_{ml}^c(i) = 1/\left(\sum_{k=1}^K |\hat{g}_{lk}(i)|^2 v_{mk}^\xi(i)\right)$;

15: $\quad \forall m, l$: $\hat{c}_{ml}(i) = \left(1 - v_{ml}^c(i)\sum_{k=1}^K v_{lk}^g(i)v_{mk}^\xi(i)\right)\hat{w}_{ml}(i) + v_{ml}^c(i)\sum_{k=1}^K \hat{g}_{lk}^\star(i)\hat{\xi}_{mk}(i)$;

16: $\quad \forall m, l$: $v_{ml}^\mu(i) = \sum_{m'=1}^{M'}\sum_{l'=1}^{L'} |a_{B,mm'}|^2 v_{m'l'}^s(i)|r_{l'l}|^2$;

17: $\quad \forall m, l$: $\hat{\mu}_{ml}(i) = \sum_{m'=1}^{M'}\sum_{l'=1}^{L'} a_{B,mm'}\hat{s}_{m'l'}(i)r_{l'l} - v_{ml}^\mu(i)\hat{\alpha}_{ml}(i-1)$;

18: $\quad \forall m, l$: $v_{ml}^w(i+1) = \frac{v_{ml}^\mu(i)v_{ml}^c(i)}{v_{ml}^\mu(i)+v_{ml}^c(i)}$, $\hat{w}_{ml}(i+1) = \frac{v_{ml}^\mu(i)\hat{c}_{ml}(i)+v_{ml}^c(i)\hat{\mu}_{ml}(i)+v_{ml}^c(i)h_{0,ml}}{v_{ml}^\mu(i)+v_{ml}^c(i)}$;

%Update mean and variance for $s_{m'l'}$:

19: $\quad \forall m, l$: $v_{ml}^\alpha(i) = 1/\left(v_{ml}^\mu(i) + v_{ml}^c(i)\right)$, $\hat{\alpha}_{ml}(i) = v_{ml}^\alpha(i)\left(\hat{c}_{ml}(i) - h_{0,ml} - \hat{\mu}_{ml}(i)\right)$;

20: $\quad \forall m', l'$: $v_{m'l'}^d(i) = 1/\left(\sum_{m=1}^M\sum_{l=1}^{L'} |a_{B,mm'}|^2 v_{ml}^\alpha(i)|r_{l'l}|^2\right)$;

21: $\quad \forall m', l'$: $\hat{d}_{m'l'}(i) = \hat{s}_{m'l'} + v_{m'l'}^d(i)\sum_{m=1}^M\sum_{l=1}^{L'} a_{B,mm'}^\star \hat{\alpha}_{ml}(i)r_{l'l}^\star$;

22: $\quad \forall m', l'$: $\hat{s}_{m'l'}(i+1) = \int s_{m'l'} \Delta_{s_{m'l'}}^{i+1}(s_{m'l'})\mathrm{d}s_{m'l'}$;

23: $\quad \forall m', l'$: $v_{m'l'}^s(i+1) = \int s_{m'l'}^2 \Delta_{s_{m'l'}}^{i+1}(s_{m'l'})\mathrm{d}s_{m'l'} - |\hat{s}_{m'l'}(i+1)|^2$;

24: **if** $\sqrt{\frac{\sum_{m'}\sum_{l'} |\hat{s}_{m'l'}(i+1)-\hat{s}_{m'l'}(i)|^2}{\sum_{m'}\sum_{l'} |\hat{s}_{m'l'}(i)|^2}} \le \epsilon$ **and** $\sqrt{\frac{\sum_l \sum_k |\hat{g}_{lk}(i+1)-\hat{g}_{lk}(i)|^2}{\sum_l \sum_k |\hat{g}_{lk}(i)|^2}} \le \epsilon$, **stop**

25: **end for**

Output: $\hat{s}_{m'l'}$ and $\hat{g}_{lk}$.

- With the tractable form of $\Delta_{z_{mk}}^{i+1}$, we show in a similar way that $\Delta_{w_{ml}}^{i+1}$, $\prod_m \Delta_{zwg_{mk} \to g_{lk}}^i$, and $\prod_{m,l} \Delta_{ws_{ml} \to s_{m'l'}}^i$ can be approximated as Gaussian distributions. As a result, we obtain the closed-loop updating formulas for $\hat{g}_{lk}$ and v_{lk}^g in Lines 9–13; for $\hat{w}_{ml}$ and v_{ml}^w in Lines 14–18; and for $\hat{s}_{m'l'}$ and $v_{m'l'}^w$ in Lines 19–23.

4.6   Asymptotic Mean Square Error (MSE) Analysis

As discussed in Sect. 4.5.1, the MMSEs in (5.32) are difficult to evaluate in general. In the following, we derive an asymptotic performance bound of the MSEs under some mild assumptions by employing the replica method [14]. The analysis here can be regarded as an extension of the replica framework in [14] to the considered matrix-calibration-based cascaded channel estimation problem in (5.27).

We assume that the prior distributions (5.28)–(5.30) are perfectly known. Furthermore, we assume that both the sampling grids φ and ς uniformly cover $[-1, 1]$ with $L_1' = L_1$ and $L_2' = L_2$. As a consequence, the array response matrices $\mathbf{A}_{B,v}$ and $\mathbf{A}_{B,h}$ in (5.24) become two normalized discrete Fourier transform (DFT) matrices and we have $\mathbf{R} = \mathbf{I}$. Discussions on the case of over-complete sampling bases with $L_1' > L_1$ and/or $L_2' > L_2$ can be found in the remark at the end of this subsection. Under these assumptions, the analysis is conducted by evaluating the average free entropy in the large-system limit, i.e., $M, M', K, L, L', T, \tau_N \to \infty$ with the ratios M/K, M'/K, L/K, L'/K, T/K, and τ_N/K^2 fixed. In the sequel, we use $K \to \infty$ to denote this limit for convenience.

To facilitate the computation, we define

$$Q_S = \lambda_S \tau_S, \; Q_G = \lambda_G \tau_G, \tag{5.35a}$$

$$Q_W = \frac{M'}{M} Q_S + \tau_{H_0}, \; Q_Z = L' Q_W Q_G, \tag{5.35b}$$

where $\tau_{H_0} \triangleq \mathbb{E}[|h_{0,ml}|^2]$.

Proposition 2 *As $K \to \infty$, the MMSEs of $\mathbf{S}$ and $\mathbf{G}$ in (5.32) converge to MSE_S and MSE_G, which is given by the solution to the following fixed-point equations:*

$$\tilde{m}_Z = \frac{T \tau_X}{\tau_N + K \tau_X (Q_Z - m_Z)}, \tag{5.36a}$$

$$\tilde{m}_W = \frac{K m_G}{1/\tilde{m}_Z + Q_Z - L' m_W m_G}, \tag{5.36b}$$

$$\tilde{m}_G = \frac{M m_W}{1/\tilde{m}_Z + Q_Z - L' m_W m_G}, \tag{5.36c}$$

$$\widetilde{m}_S = \frac{1}{1/\widetilde{m}_W + Q_W - \tau_{H_0} - M' m_S / M}, \tag{5.36d}$$

$$m_Z = Q_Z - \frac{Q_Z - L' m_W m_G}{1 + \widetilde{m}_Z (Q_Z - L' m_W m_G)} \tag{5.36e}$$

$$m_W = Q_W - \frac{Q_W - \tau_{H_0} - M' m_S / M}{1 + \widetilde{m}_W (Q_W - \tau_{H_0} - M' m_S / M)}, \tag{5.36f}$$

$$m_G = (1 - \lambda_G) \mathbb{E}_{N_G} \left[\left| f_G \left(\frac{N_G}{\sqrt{\widetilde{m}_G}}, \frac{1}{\widetilde{m}_G} \right) \right|^2 \right] + \lambda_G \mathbb{E}_{N_G} \left[\left| f_G \left(\frac{N_G \sqrt{\widetilde{m}_G + 1}}{\sqrt{\widetilde{m}_G}}, \frac{1}{\widetilde{m}_G} \right) \right|^2 \right], \tag{5.36g}$$

$$m_S = (1 - \lambda_S) \mathbb{E}_{N_S} \left[\left| f_S \left(\frac{N_S}{\sqrt{\widetilde{m}_S}}, \frac{1}{\widetilde{m}_S} \right) \right|^2 \right] + \lambda_S \mathbb{E}_{N_S} \left[\left| f_S \left(\frac{N_S \sqrt{\widetilde{m}_S + 1}}{\sqrt{\widetilde{m}_S}}, \frac{1}{\widetilde{m}_S} \right) \right|^2 \right], \tag{5.36h}$$

$$MSE_G = Q_G - m_G, \tag{5.36i}$$

$$MSE_S = Q_S - m_S, \tag{5.36j}$$

In (5.36g)–(5.36h), we define $N_S, N_G \sim CN(\cdot; 0, 1)$; and

$$f_G(x, y) = \frac{\int G p(G) CN(G; x, y) dG}{\int p(G) CN(G; x, y) dG}, f_S(x, y) = \frac{\int S p(S) CN(S; x, y) dS}{\int p(S) CN(S; x, y) dS}, \tag{5.37}$$

where $p(S) \triangleq (1 - \lambda_S) \delta(S) + \lambda_S CN(S; 0, \tau_S)$; and $p(G) \triangleq (1 - \lambda_G) \delta(G) + \lambda_G CN(G; 0, \tau_G)$;

Proof See [4, Sect. V].   □

As a result, we evaluate MSE_S and MSE_G by computing the fixed-point solution of (5.36), which asymptotically describes the MMSEs in (5.32). This can be efficiently implemented by iteratively updating $\{MSE_S, MSE_G, m_o, \widetilde{m}_o : o \in \{Z, W, S, G\}\}$ following (5.36) until convergence.

Remark 2 It is worth noting that the derivation of Proposition 2 adopts the CLT to approximate some intermediate random variables as Gaussian variables. To apply the CLT, we restrict the sampling bases $\mathbf{A}_{B,v}$ and $\mathbf{A}_{B,h}$ to be two normalized DFT matrices. In a more general case where $\mathbf{A}_{B,v}$ and $\mathbf{A}_{B,h}$ are two over-complete bases, the Gaussian approximations may be inaccurate and consequently affect the accuracy of the replica method. In this case, MSE_S and MSE_G in (5.36) may not exactly correspond to the asymptotic MMSEs of $\mathbf{G}$ and $\mathbf{S}$. As a consequence, the performance bound derived in Proposition 2 may become loose, as verified by the numerical results presented in the next section.

4.7   Numerical Results

4.7.1   Simulation Results Under Channel Generation Model (5.28)–(5.30)

We conduct Monte Carlo simulations to verify the analysis in Sect. 4.6. In this sub-
section, we assume that the channel is generated according to the prior distributions
in (5.28)–(5.30). This allows us to calculate the MMSEs of the cascaded channel
estimation problem defined in (5.32) by using the replica method as in Sect. 4.6.
Note that MSE_S and MSE_G in Proposition 2 derived by the replica method will be
used as a benchmark to evaluate the performance of the proposed algorithm (i.e.,
Algorithm 2).[2]
  First, we investigate the performance of the proposed algorithm with normalized
DFT bases $\mathbf{A}_{B,v}$ and $\mathbf{A}_{B,h}$. We set $M = 1.28K$, $M' = 1.6K$, $T = 1.5K$, $L = L' = 0.5K$, $\lambda_G = 0.1$, $\lambda_S = 0.05$, and $\tau_S = \tau_G = \tau_{H_0} = \tau_X = 1$. Figure 5.6
plots the MSE performance versus noise power for $K = 40$ and $K = 100$. The

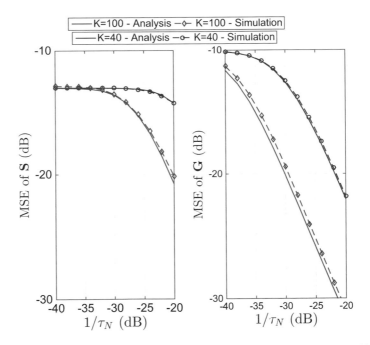

Fig. 5.6 MSEs versus τ_N under normalized DFT bases $\mathbf{A}_{B,v}$ and $\mathbf{A}_{B,h}$. For $K = 40$, we set
$L_1 = L'_1 = 4$ and $L_2 = L'_2 = 5$; For $K = 100$, we set $L_1 = L'_1 = 10$ and $L_2 = L'_2 = 5$ [4]

---

[2]Although the replica analysis in Sect. 4.6 is valid only in the large-system limit, we still use the
derived performance bound as a benchmark even when we conduct simulations with finitely large
systems.

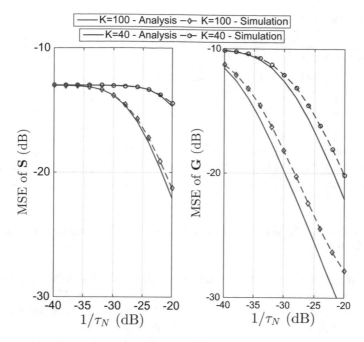

Fig. 5.7 MSEs versus τ_N under over-complete bases $\mathbf{A}_{B,v}$ and $\mathbf{A}_{B,h}$. For $K = 40$, we set $L_1 = L_1' = 4$, $L_2 = 5$, and $L_2' = 7$; For $K = 100$, we set $L_1 = L_1' = 10$, $L_2 = 5$, and $L_2' = 7$ [4]

simulation results are obtained by averaging over 5000 Monte Carlo trials. We see that the performance of the proposed message passing algorithm closely matches the analytical bound derived in Proposition 2.

Next, we simulate on the scenario with an over-complete $\mathbf{A}_{B,v}$. Specifically, we set $L' = 0.7K$ and keep the other parameters unchanged. From Fig. 5.7, we find that the analytical result has a small gap from the simulation result. As discussed in Remark 2, this is because the Gaussian approximations in the derivation of Proposition 2 are less accurate, and hence the performance bound is not as tight as in Fig. 5.6.

Finally, we investigate the required training length T for RIS channel estimation in the noiseless case (i.e., $\tau_N = 0$). In Table 5.3, we list the minimum training length T of the proposed algorithm to achieve near-zero estimation errors (MSE$_S$ < −50 dB and MSE$_G$ < −50 dB), where we set $\lambda_S = 0.05$, $\lambda_G = 0.1$, $M' = 1.25M$, and $L' = L$. The performance bound derived from the replica analysis is included as well. Besides, we compare the proposed algorithm with the methods in [15, 16]. Specifically, when $\tau_N = 0$, the training length T in [15] is $L + \max\{K − 1, \lceil(K − 1)L/M\rceil\}$, where $\lceil\cdot\rceil$ is the ceiling function. Moreover, the method in [16] requires $T = LK$. It can be seen that the proposed algorithm requires a smaller T compared with the two baselines. In particular, for the considered settings, the proposed algorithm only needs about 30% of the training overhead of [15] and about

Table 5.3 Minimum training length T required in different methods [4]

System size	Algorithm 2	Replica method	Method in [15]	Method in [16]
$K = 100, L = 50,$ $M = 128$	44	11	149	5000
$K = 100, L = 100,$ $M = 100$	60	26	199	10000

1% of the overhead of [16]. This is because Algorithm 2 exploits the information on the slowing-varying channel components and the channel sparsity in the angular domain, which significantly reduces the number of variables to be estimated.

4.7.2 Simulation Results Under a More Realistic Channel Generation Model

We now consider a more realistic channel generation model as follows. The large-scale fading component is given by $\beta_i = \beta_{\text{ref}} \cdot d_i^{-\alpha_i}, 0 \leq i \leq K$, where β_{ref} is the reference path loss at the distance 1 m; d_i is the corresponding link distance; and α_i is the corresponding pass loss exponent. We set $\beta_{\text{ref}} = -20$ dB for all the channel links; $\alpha_0 = 2$; $\alpha_k = 2.6, 1 \leq k \leq K$; $d_0 = 50$ m; and d_k uniformly drawn from [10 m, 12 m].

Moreover, we generate $\bar{\mathbf{H}}_{RB}$ by (5.21) with 20 clusters of paths and 10 subpaths per cluster. We draw the central azimuth AoA at the BS of each cluster uniformly over $[-90°, 90°]$; draw the central azimuth (or elevation) AoD at the RIS of each cluster uniformly over $[-180°, 180°]$ (or $[-90°, 90°]$); and draw each subpath with a $10°$ angular spread. The channels $\tilde{\mathbf{H}}_{RB}$ and $\mathbf{h}_{UR,k}$ are generated by (5.22) and (5.23) in a similar way both with a cluster of 10 subpaths. Moreover, every α_p is drawn from $\mathcal{CN}(\alpha_p; 0, 1)$ and is normalized to satisfy $\|\mathbf{H}_{RB}\|_F^2 = \beta_0 ML$ and $\|\mathbf{h}_{UR,k}\|_2^2 = \beta_k L$. We set $K = 20, M = 60, T = 35, L_1 = L_2 = 4$ (i.e., $L = 16$), $\tau_X = 1$, and $\kappa = 9$. For the proposed algorithm, we set $I_{\max} = 2000, \epsilon = 10^{-4}$, and the over-complete bases ϑ, φ and ς in (5.24)–(5.25) to be uniform sampling grids covering $[-1, 1]$. The lengths of the sampling grids are set to have a fixed ratio to the antenna dimensions, i.e., $M'/M = L_1'/L_1 = L_2'/L_2 = 2$, unless otherwise specified. All the results in the sequel are conducted by averaging over 1500 Monte Carlo trials.

Apart from the proposed algorithm, the following baselines are involved for comparisons:

- Concatenate linear regression (LR): By setting aside $\tilde{\mathbf{H}}_{RB}$, we first set the estimate of $\mathbf{H}_{RB}$ as $\hat{\mathbf{H}}_{RB} = \sqrt{\kappa/(\kappa + 1)}\bar{\mathbf{H}}_{RB}$, which is assumed to be deterministic. From (5.27), we obtain the following linear regression problem:

$$\text{vec}(\mathbf{Y}) = (\mathbf{X}^T \otimes \hat{\mathbf{H}}_{RB}\mathbf{A}_R) \text{vec}(\mathbf{G}) + \mathbf{n}', \tag{5.38}$$

where vec(·) represents the vectorization operator, and $\mathbf{n}'$ represents the effective AWGN. We then infer $\hat{\mathbf{G}}$ from (5.38) by employing generalized AMP (GAMP) [17]. Finally, we employ GAMP to estimate $\hat{\mathbf{H}}_{RB}$ with the estimated $\hat{\mathbf{G}}$.

- Oracle bound with $\mathbf{H}_{RB}$ known: Assume that an oracle gives the accurate value of $\mathbf{H}_{RB}$. Similarly to (5.38), we employ GAMP to obtain $\hat{\mathbf{G}}$ with $\hat{\mathbf{H}}_{RB} = \mathbf{H}_{RB}$.

Besides, the following channel estimation methods are included as baselines in the sequel.

- Method in [16]: RIS channels are estimated sequentially. From time slots $(l-1)K+1$ to lK, $1 \le l \le L$, we turn off all the RIS elements but the l-th one. With orthogonal training symbols from the users, the BS computes the linear MMSE estimators of the channel coefficients associated with the l-th RIS element.
- Method in [15]: First, the first user sends an all-one training sequence with length no less than L to the BS, and the BS estimates the channel coefficients of this user (i.e., $\mathbf{H}_{RB} \operatorname{diag}(\mathbf{h}_{UR,1})$). Then, the other users sequentially send a training symbol to the BS, and the BS estimates $\{\mathbf{H}_{RB} \operatorname{diag}(\mathbf{h}_{UR,k})\}_{2 \le k \le K}$ by exploiting the correlations with the channel of the first user.
- Method in [18]: RIS Channel estimation is divided into P phases with $P = \lfloor T/K \rfloor \le L$, where $\lfloor \cdot \rfloor$ is the floor function. In the p-th phase, the RIS phase-shift vector ψ_p is set as the p-th column of an $L \times L$ DFT matrix. The users send orthogonal training sequences with length no less than K to the BS. The BS collects the received signals in P phases and estimates the cascaded channels $\mathbf{H}_{RB}$ and $\{\mathbf{h}_{UR,k}\}$ alternatively by using parallel factor decomposition.

The estimates of all the three baselines suffer from diagonal ambiguities. *For the comparison purpose*, we remove the diagonal ambiguities by assuming the perfect knowledge of $\mathbf{h}_{UR,1}$. Besides, we set the total training length T to be the same for all the algorithms.

We use the normalized MSEs (NMSEs) of $\mathbf{H}_{RB}$ and $\{\mathbf{h}_{UR,k}\}$ to evaluate the performance of the cascaded channel estimation algorithms. Specifically, they are given by

$$\text{NMSE of } \mathbf{H}_{RB} = \frac{\|\hat{\mathbf{H}}_{RB} - \mathbf{H}_{RB}\|_F^2}{\|\mathbf{H}_{RB}\|_F^2}, \tag{5.39a}$$

$$\text{Average NMSE of } \mathbf{h}_{UR,k} = \frac{1}{K} \sum_{k=1}^{K} \frac{\|\hat{\mathbf{h}}_{UR,k} - \mathbf{h}_{UR,k}\|_2^2}{\|\mathbf{h}_{UR,k}\|_2^2}. \tag{5.39b}$$

Figure 5.8 investigates the RIS channel estimation performance as τ_N varies. It can be seen that (1) the proposed algorithm achieves an NMSE of $\mathbf{h}_{UR,k}$ that is very close to the oracle bound, which assumes NMSE of $\mathbf{H}_{RB}$ is zero; (2) Algorithm 2 outperforms the baselines, especially when the noise power is large. This is because the proposed algorithm exploits the information on the slowing-varying channel components and the hidden channel sparsity; (3) the NMSEs of concatenate LR does not decrease when $\tau_N \le -90$ dB. The reason is that the effective noise $\mathbf{n}'$ in (5.38)

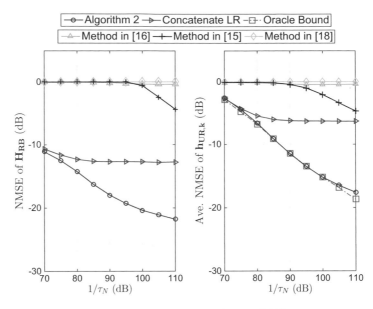

Fig. 5.8 The NMSE performance versus noise power with $T = 35$ [4]

is a combination of the original AWGN noise $\text{vec}(\mathbf{N})$ and the error resulted from the model mismatch, i.e., the ignored term $\left(\mathbf{X}^T \otimes \sqrt{1/(\kappa + 1)}\widetilde{\mathbf{H}}_{RB}\mathbf{A}_R\right)\text{vec}(\mathbf{G})$. Essentially, the latter term strongly correlates with the variable to be estimated in (5.38) (i.e., $\mathbf{G}$). When the model mismatch error dominates the effective noise in the low noise power regime, the correlation issue compromises the convergence of GAMP. On the contrary, the proposed method avoids this problem since the estimate of $\widetilde{\mathbf{H}}_{RB}$, or equivalently $\mathbf{S}$, are updated iteratively during the message passing iteration.

Then, we study the effect of the grid lengths M', L_1', and L_2'. We use η to represent the ratio between the grid length and the number of antennas, i.e., $M'/M = L_1'/L_1 = L_2'/L_2 = \eta$. Figure 5.9 plots the NMSEs of the channel estimation algorithms under various η, where τ_N is fixed to -95 dB. We have the following observations: (1) the NMSEs of the proposed algorithm, concatenate LR and the oracle estimator decrease as η increases, since increasing the sampling grid length leads to a higher angle resolution and hence sparser $\mathbf{S}$ and $\mathbf{G}$; (2) the baselines [15, 16, 18] do not exploit the channel sparsity in the angular domain and its performance is invariant to η; (3) the proposed method substantially improves the estimation performance compared to the other algorithms and closely approaches the oracle bound.

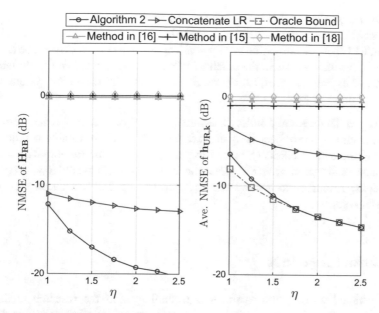

Fig. 5.9 The NMSE performance versus sampling resolution η, where we set $\tau_N = -95$ dB [4]

5 Conclusions

This chapter overviews several important challenges in the physical layer design on RIS-aided 6G wireless networks. The RIS channel estimation problem is investigated and two Bayesian-inference based channel estimation approaches are presented for the RIS-aided single-user MIMO and multi-user MIMO systems. Besides, this chapter presents an analytical framework for evaluating the asymptotic estimation performance for the latter approach. Finally, the high accuracy and efficiency of the presented approaches are demonstrated by extensive numerical simulations.

We note that the present design on RIS channel estimation is far from mature. There are still many interesting open problems related to CSI acquisition in RIS-aided networks, including:

Scalable CSI Acquisition Protocol Many of the existing cascaded channel estimation algorithms in RIS-aided systems have restrictions on the dimension of the RIS channels, no matter it is stated explicitly or not. Specifically, the size of the RIS (L) is required to be no larger than the number of antennas at the BS (M). However, in practice, L is usually in the order of hundreds or even thousands due to the low cost nature of RIS elements, and thus we have $L \gg M$. It is interesting to see how to efficiently and accurately estimate such a large number of RIS channel coefficients in this case.

System Design Under CSI Uncertainty The existing studies on RIS-aided system optimization are mostly based on the knowledge of perfect CSI. As discussed in this chapter, RIS channel estimation is very challenging, and perfect CSI can be costly to acquire in practice. As such, the practical RIS-aided system design should factor in the channel estimation error, calling for research explorations on the system design under CSI uncertainty.

Analysis on Theoretical Limits The fundamental limits of RIS-aided communication systems have not been fully characterized. How much training is required in a given RIS-aided wireless network? How many users can be supported? How does the channel estimation accuracy affect the system performance? How to optimize the training resources to achieve the best performance? These questions call for urgent answers.

6　Further Readings

For a detailed introduction to the RIS technology and the research challenges therein, we refer interested readers to [19]; for more derivation details and more numerical results on the RIS channel estimation approaches presented in this chapter, we refer to [3, 4]; for other state-of-the-art RIS channel estimation algorithms (possibly applicable to different system setups), we refer to [15, 16, 18, 20, 21]; for state-of-the-art passive beamforming designs in RIS-aided wireless networks, we refer to [22–26]; for discussions on resource allocation and system optimization challenges in RIS-aided wireless networks, we refer to [19].

References

1. D. Renzo, et al., Smart radio environments empowered by reconfigurable AI meta-surfaces: an idea whose time has come. EURASIP J. Wireless Commun. Netw. **2019**, 129 (2019)
2. C. Liaskos, S. Nie, A. Tsioliaridou, A. Pitsillides, S. Ioannidis, I. Akyildiz, A new wireless communication paradigm through software-controlled metasurfaces. IEEE Commun. Mag. **56**(9), 162–169 (2018)
3. Z. He, X. Yuan, Cascaded channel estimation for large intelligent metasurface assisted massive MIMO. IEEE Wireless Commun. Lett. **9**(2), 210–214 (2020)
4. H. Liu, X. Yuan, Y.-J.A. Zhang, Matrix-calibration-based cascaded channel estimation for reconfigurable intelligent surface assisted multiuser MIMO. IEEE J. Sel. Areas Commun. **38**(11), 2621–2636 (2020)
5. J.T. Parker, P. Schniter, V. Cevher, Bilinear generalized approximate message passin–Part I: derivation. IEEE Trans. Signal Process. **62**(22), 5839–5853 (2014)
6. T.F.C.K. Wei, J.-F. Cai, S. Leung, Guarantees of riemannian optimization for low rank matrix recovery. SIAM J. Matrix Anal. Appl. **37**(3), 1198–1222 (2016)
7. M. Aharon, M. Elad, A. Bruckstein, K-SVD: an algorithm for designing overcomplete dictionaries for sparse representation. IEEE Trans. Signal Process. **54**(11), 4311–4322 (2006)

8. J. Mairal, F. Bach, J. Ponce, G. Sapiro, Online learning for matrix factorization and sparse coding. J. Mach. Learn. Res. **11**, 19–60 (2010)
9. E.J.C.J.-F. Cai, Z. Shen, A singular value thresholding algorithm for matrix completion. SIAM J. Optim. **20**(4), 1956–1982 (2010)
10. D. Tse, P. Viswanath, *Fundamentals of Wireless Communication* (Cambridge University Press, New York, 2005)
11. X. Li, J. Fang, H. Li, P. Wang, Millimeter wave channel estimation via exploiting joint sparse and low-rank structures. IEEE Trans. Wireless Commun. **17**(2), 1123–1133 (2018)
12. A.F. Molisch, A. Kuchar, J. Laurila, K. Hugl, R. Schmalenberger, Geometry-based directional model for mobile radio channels–principles and implementation. Eur. Trans. Telecommun. **14**(4), 351–359 (2003)
13. D.L. Donoho, A. Maleki, A. Montanari, Message passing algorithms for compressed sensing. Proc. Nat. Acad. Sci. **106**(45), 18914–18919 (2009)
14. Y. Kabashima, F. Krzakala, M. Mézard, A. Sakata, L. Zdeborová, Phase transitions and sample complexity in Bayes-optimal matrix factorization. IEEE Trans. Inf. Theory **62**(7), 4228–4265 (2016)
15. Z. Wang, L. Liu, S. Cui, Channel estimation for intelligent reflecting surface assisted multiuser communications. IEEE Wireless Communications and Networking Conference (WCNC), 1–6 (2020)
16. Q.-U.-A. Nadeem, A. Kammoun, A. Chaaban, M. Debbah, M.-S. Alouini, Intelligent reflecting surface assisted multi-user MISO communication (2019, preprint). preprint arXiv:1906.02360
17. S. Rangan, P. Schniter, E. Riegler, A.K. Fletcher, V. Cevher, Fixed points of generalized approximate message passing with arbitrary matrices. IEEE Trans. Inf. Theory **62**(12), 7464–7474 (2016)
18. L.Wei, C. Huang, G.C. Alexandropoulos, C. Yuen, Parallel factor decomposition channel estimation in RIS-assisted multi-user MISO communication. IEEE Sensor Array and Multichannel Signal Processing Workshop (SAM), 1–5 (2020)
19. X. Yuan, Y.J. Zhang, Y. Shi, W. Yan, H. Liu, Reconfigurable-intelligent-surface empowered wireless communications: challenges and opportunities. IEEE Wireless Commun. Mag., Early Access (2021)
20. A. Taha, M. Alrabeiah, A. Alkhateeb, Enabling large intelligent surfaces with compressive sensing and deep learning. IEEE Access **9**, 44304–44321 (2021)
21. J. Chen, Y.-C. Liang, H.V. Cheng, W. Yu, Channel estimation for reconfigurable intelligent surface aided multi-user MIMO systems (preprint, 2019). arXiv:1912.03619
22. Q. Wu, R. Zhang, Intelligent reflecting surface enhanced wireless network via joint active and passive beamforming. IEEE Trans. Wireless Commun. **18**(11), 5394–5409 (2019)
23. Q. Nadeem, A. Kammoun, A. Chaaban, M. Debbah, M. Alouini, Asymptotic max-min SINR analysis of reconfigurable intelligent surface assisted MISO systems. IEEE Trans. Wireless Commun. **19**, 7748–7764 (2020)
24. C. Huang, A. Zappone, G.C. Alexandropoulos, M. Debbah, C. Yuen, Reconfigurable intelligent surfaces for energy efficiency in wireless communication. IEEE Trans. Wireless Commun. **18**(8), 4157–4170 (2019)
25. Y. Han, W. Tang, S. Jin, C. Wen, X. Ma, Large intelligent surface-assisted wireless communication exploiting statistical CSI. IEEE Trans. Veh. Technol. **68**(8), 8238–8242 (2019)
26. W. Yan, X. Yuan, X. Kuai, Passive beamforming and information transfer via large intelligent surface. IEEE Wireless Commun. Lett. **9**(4), 533–537 (2020)

Chapter 6
Millimeter-Wave and Terahertz Spectrum for 6G Wireless

Shuchi Tripathi, Nithin V. Sabu, Abhishek K. Gupta, and Harpreet S. Dhillon

Abstract With the standardization of 5G, commercial millimeter wave (mmWave) communications has become a reality despite all the concerns about the unfavorable propagation characteristics of these frequencies. Even though the 5G systems are still being rolled out, it is argued that their gigabits per second rates may fall short in supporting many emerging applications, such as 3D gaming and extended reality. Such applications will require several hundreds of gigabits per second to several terabits per second data rates with low latency and high reliability, which are expected to be the design goals of the next generation 6G communications systems. Given the potential of terahertz (THz) communications systems to provide such data rates over short distances, they are widely regarded to be the next frontier for the wireless communications research. The primary goal of this chapter is to equip readers with sufficient background about the mmWave and THz bands so that they are able to both appreciate the necessity of using these bands for commercial communications in the current wireless landscape and to reason the key design considerations for the communications systems operating in these bands. Towards this goal, this chapter provides a unified treatment of these bands with particular emphasis on their propagation characteristics, channel models, design and implementation considerations, and potential applications to 6G wireless. A brief summary of the current standardization activities related to the use of these bands for commercial communications applications is also provided.

S. Tripathi · N. V. Sabu · A. K. Gupta
Indian Institute of Technology Kanpur, Kanpur, Uttar Pradesh, India
e-mail: shuchi@iitk.ac.in; nithinvs@iitk.ac.in; gkrabhi@iitk.ac.in

H. S. Dhillon (✉)
Bradley Department of ECE, Virginia Tech, Blacksburg, VA, USA
e-mail: hdhillon@vt.edu

1 Background and Motivation

The standardization of 5G new radio (NR) was driven by the diverse throughput, reliability, and latency requirements of the ever-evolving ecosystem of applications that need to be supported by modern cellular networks. Within 5G, these applications are categorized as enhanced mobile broadband (eMBB), ultra-reliable low latency communication (URLLC), and massive machine-type communication (mMTC). Right from the onset, it was clear that a one-size-fits-all solution may not work for all the applications because of which the recent generations of cellular systems have explored the use of advanced communications and networking techniques, such as network densification through the use of small cells, smarter scheduling, and multiple antenna systems for improved spectral efficiency, just to name a few. Perhaps the most striking difference of 5G from the previous generations of cellular systems is the acknowledgment that the *classical* sub-6 GHz spectrum is not going to be sufficient to support the requirements of the emerging applications. The millimeter wave (mmWave) spectrum naturally emerged as a potential solution. Although these bands were earlier thought to be unsuitable for the mobile operations due to their unfavorable propagation characteristics, the modern device and antenna technologies made it feasible to use them for commercial wireless applications [1]. As a result, the 5G standards resulted in the birth of commercial mmWave communication.

Now, as we look into the future, it is evident that we are slowly moving towards applications, such as virtual and augmented reality, ultra-HD video conferencing, 3D gaming, and the use of wireless for brain machine interfaces, which will put even more strict constraints on the throughput, reliability, and latency requirements. With the advancement of device fabrication methods, it is also reasonable to expect that the nano-scale communications will see the light of the day soon. With the recent success of mmWave communication, it was quite natural for the researchers to start looking at the other unexplored bands of the radio frequency (RF) spectrum, primarily the terahertz (THz) band that lies above the mmWave band. The THz waves with enormous bandwidth can be used in many applications that require ultra-high data rates. This along with the existing sub-6 GHz and mmWave bands can help us achieve the true potential of many emerging applications. Further, owing to their small wavelength, they can also be used for micro and nano-scale communication. In the past, the use of THz bands was limited to imaging and sensing due to the unavailability of feasible and efficient devices that can work on these frequencies. However, with the recent advancements in THz devices, THz communication is expected to play a pivotal role in the upcoming generations of communication standards [2].

The primary goal of this chapter is to equip readers with sufficient background about the mmWave and THz bands so that they are able to both appreciate the necessity of using these bands for commercial communications in the current wireless landscape and to reason the key design considerations for the communications systems operating in these bands. This is achieved through a systematic treatment of this topic starting with a detailed discussion of the propagation characteristics at these frequencies leading naturally to the discussion on channel models that capture these characteristics. Throughout this discussion, we carefully compare and contrast the propagation characteristics of these new bands with the better known sub-6 GHz cellular bands and explain how the key differences manifest in the channel models. Building on this background, we then explain the implications of these differences on the design considerations for mmWave and THz communications systems and their potential applications to 6G systems. The chapter is concluded with a brief discussion about the current standardization activities related to the use of these bands for commercial communications.

2   Introduction to mmWave and THz Spectrum

Until the 4G cellular standard, the commercial (cellular) communication was limited to the conventional bands up to 6 GHz, which are now referred to as the sub-6 GHz cellular bands. However, there are many bands in the 6–300 GHz range (with enormous bandwidths) that have been used for a variety of non-cellular applications, such as satellite communications, radio astronomy, remote sensing, radars, to name a few. Due to recent advancement in antenna technology, it has now become possible to use this spectrum for mobile communication as well. The frequency band from 30–300 GHz with the wavelengths ranging from 1 to 10 mm is termed the *mmWave band* and offers hundreds of times more bandwidth compared to the sub-6 GHz bands. Although higher penetration and blockage losses are the major drawbacks of mmWave communication systems, researchers have shown that the same effects are helpful in mitigating interference in modern cellular systems, which exhibit dense deployment of small cells. This naturally results in a more aggressive frequency reuse and increased data security due to higher directionality requirement at the mmWave frequencies [3]. The mmWave frequencies from about 24 GHz to about 100 GHz are already being explored as a part of the 5G standard. As we think ahead towards 6G and beyond systems, researchers have also started exploring the 0.1–10 THz band, which is collectively referred to as the *THz band* (with the lower end of this spectrum being obviously of more interest for communications applications).

2.1   Need for the mmWave and THz Bands

It is well-known that the mobile data traffic has been exponentially increasing for more than a decade and this trend is expected to continue for the foreseeable future. With the penetration of the wireless Internet of Things (IoT) devices in new verticals, such as supply chains, health care, transportation, and vehicular communications, this trend is further expected to accentuate. It is estimated that 9.5 billion IoT devices are connected globally in 2019 [4]. The International

Telecommunications Union (ITU) has further estimated that the number of connected IoT devices will rise to 38.6 billion by 2025 and 50 billion by 2030 [5, 6]. Handling this data deluge and the massive number of IoT devices are two of the key design goals for 5G networks [7]. Three possible solutions to meet these demands are to develop better signal processing techniques for an improved spectral efficiency of the channel, the extreme densification of cellular networks, and the use of additional spectrum [8, 9]. Various advanced techniques, such as carrier aggregation, coordinated multi-point processing, multi-antenna communications as well as novel modulation techniques have already been explored in the context of current cellular networks. The chances of getting orders of improvement from these techniques are slim. Likewise, network densification increases interference, which places fundamental limits on the performance gains that can be achieved with the addition of more base stations [10, 11]. The focus of this chapter is on the third solution, which is to use higher frequency bands.

The amount of available spectrum at mmWave frequencies is very large when compared to sub-6 GHz frequencies ($\sim$ 50–100 times). As the bandwidth appears in the pre-log factor of the achievable data-rate, mmWave communication can potentially achieve an order of magnitude higher data rate, which made it attractive for inclusion in the 5G standards. While 5G deployments are still in their infancy, emerging applications such as extended reality may require terabits-per-second (Tbps) links that may not be supported by the 5G systems (since the contiguous available bandwidth is less than 10 GHz). This has created a lot of interest in exploring the THz band to complement the sub-6 GHz and mmWave bands in 6G and beyond systems [12, 13].

2.2 What Can mmWave and THz Frequencies Enable?

Larger bandwidths available in the mmWave spectrum make multi-gigabit wireless communication feasible, thus opening doors for many innovations [3]. For instance, the mmWave frequencies can enable wireless backhaul connections between outdoor base stations (BSs), which will reduce the land-acquisition, installation and maintenance costs of the fiber-optic cables, especially for ultra-dense networks (UDNs). Further, it enables to transform current "wired" data centers to completely wireless data centers with data-servers communicating over mmWave frequencies with the help of highly-directed pencil-beams. Another potential application is the in-boggy vehicle-to-vehicle (V2V) communication in high mobility scenarios including bullet trains and airplanes where mmWave communication systems together with sub-6 GHz systems have the potential of providing better data rates [14].

Further, the THz spectrum consists of bands with available bandwidths of a few tens of GHz, which can support a data rates in the range of Tbps. The communication at THz is further aided by the integration of thousands of sub-millimeter antennas and lower interference due to higher transmission frequencies. It is therefore capable of supporting bandwidth-hungry and low latency applications, such as virtual-reality gaming and ultra-HD video conferencing. Other applications that will benefit from the maturity of THz communications include nano-machine communication, on-chip communications, the internet of nano things (IoNT) [15], and intra-body communication of nano-machines. It can also be combined with bio-compatible and energy-efficient bio-nano-machines communicating using chemical signals (molecules) [16]. Such communication is termed *molecular communication (MC)* [17].

2.3 Available Spectrum

Due to the varying channel propagation characteristics and frequency-specific atmospheric attenuation, researchers have identified specific bands in mmWave/THz spectrum that are particularly conducive for the communications applications. In the world radiocommunication conference (WRC) 2015, ITU released a list of proposed frequency bands in between 24–86 GHz range for global usage [20]. The selection of these bands was done based on a variety of factors, such as channel propagation characteristics, incumbent services, global agreements, and the availability of contiguous bandwidth. WRC-2019 was focused on the conditions for the allocation of high-frequency mmWave bands dedicated to the 5G systems. A total of 17.25 GHz of spectrum had been identified [18]. For the implementation of future THz communication systems, WRC 2019 has also identified a total of 160 GHz spectrum in the THz band ranging between 252 to 450 GHz. A brief description of these mmWave and THz bands are given in Table 6.1.

Although mmWave and THz bands have a huge potential for their usage in communication, there are significant challenges in their commercial deployments. In particular, communication in these bands suffer from poor propagation characteristics, higher penetration, blockage and scattering losses, shorter coverage range, and a need for strong directionality in transmission. These challenges have obstructed the inclusion of mmWave and THz bands in standards and commercial deployments until now. With the advancements in modern antenna and device technologies, it is now becoming feasible to use these bands for communications. However, there are still various design issues that need to be addressed before they can be deployed at a large scale [14, 21]. In this chapter, we will discuss the propagation characteristics of these bands in detail as well as the challenges involved in using them for communications applications.

Table 6.1 Available bands at the mmWave [8, 18] and THz spectrum [19]

Name	Specific bands	Remarks
26 GHz band	26.5–27.5 GHz, 24.25–26.5 GHz	Incumbent services: fixed link services, satellite Earth station services, and short-range devices. Earth exploration satellites and space research expeditions, inter-satellites, backhaul, TV broadcast distribution, fixed satellite Earth-to-space services and high altitude platform station (HAPS) applications
28 GHz band	27.5–29.5 GHz, 26.5–27.5 GHz	Proposed mobile communication. Incumbent services: Local multi-point distribution service (LMDS), Earth-to-space fixed-satellite service and Earth stations in motion (ESIM) application
32 GHz band	31.0–31.3 GHz, 31.8–33.4 GHz	Highlighted as a promising band. Incumbent services: HAPS applications, Inter-satellite service (ISS) allocation
40 GHz lower band	37.0–39.5 GHz, 39.5–40.5 GHz	Incumbent services: Fixed and mobile satellite (space-to-Earth) and Earth exploration and space research satellite (space-to-Earth and Earth-to-space) services, HAPS applications
40 GHz upper band	40.5–43.5 GHz	Incumbent services: Fixed and mobile satellite (space-to-Earth), broadcasting satellite services, mobile services, and radio astronomy
50 GHz	45.5–50.2 GHz, 47.2–47.5 GHz, 47.9–48.2 GHz, 50.4–52.6 GHz	Incumbent services: Fixed non-geostationary satellite and international mobile telecommunication (IMT) services, HAPS applications
60 GHz lower band	57.0–64.0 GHz	Unlicensed operation for personal indoor services, device to device communication via access and backhaul links in the ultra-dense network scenario
60 GHz upper band	64.0–71.0 GHz	Upcoming generations of mobile standards with unlicensed status in UK and USA. Incumbent services: The aeronautical and land mobile services
70/80/90 GHz band	71.0–76.0 GHz, 81.0–86.0 GHz, 92.0–95.0 GHz	Fixed and broadcasting satellite services (space-to-Earth) services. Unlicensed operation for wireless device to device and backhaul communication services in the ultra-dense network scenario in the USA
252–296 GHz band	252–275 GHz, 275–296 GHz	Early proposal for land mobile and fixed service. Suitable for outdoor usage
306–450 GHz band	306–313 GHz, 318–333 GHz, 356–450 GHz	Early proposal for land mobile and fixed service. Suitable for short range indoor communication

3   Propagation at the mmWave and THz Frequencies

3.1   Differences from the Communication in Conventional Bands

The communication at mmWave/THz frequencies differs significantly from the communication at conventional microwave frequencies. This is attributed to the following important factors.

3.1.1   Signal Blockage

The mmWave/THz signals have a much higher susceptibility to blockages compared to the signals at the lower frequencies. The mmWave/THz communication relies heavily on the availability of the line of sight (LOS) links due to very poor propagation characteristics of the non line of sight (NLOS) links [22]. For instance, these signals can be easily blocked by buildings, vehicles, humans, and even foliage. A single blockage can lead to a loss of 20–40 dB. For example, the reflection loss due to glass for the mmWave signal is 3–18 dB while that due to building material like bricks is around 40–80 dB. Even the presence of a single tree amounts to a foliage loss of 17–25 dB for the mmWave signals [9, 14, 23, 24]. Moreover, the mmWave/THz signals also suffer from the self-body blockage caused by the human users which can itself cause an attenuation of around 20–35 dB [23]. These blockages can drastically reduce the signal strength and may even result in a total outage. Therefore, it is of utmost importance to find effective solution to avoid blockages and quick handovers in case a link gets blocked. On the flip side, blockages, including self-body blockages, may also reduce interference, especially from the far off BSs [25]. Therefore, it is crucial to accurately capture the effect of blockages in the analytical and simulation models of mmWave/THz communications systems.

3.1.2   High Directivity

The second important feature of mmWave/THz communication is its high directivity. In order to overcome the severe path loss at these high frequencies, it is necessary to use a large number of antennas at the transmitter and/or receiver side [22]. Fortunately, it is possible to accommodate a large number of antennas in small form factors because antennas at these frequencies are smaller than those at traditional frequencies due to the smaller wavelengths. The use of a large antenna array results in a highly directional communication. High beamforming gain with small beamwidth increases the signal strength of the serving links while reducing the overall interference at the receivers. However, high directionality also introduces the *deafness problem* and thus higher latency. This latency occurs due to the longer beam search process which is a key step to facilitate the directional transmission and reception. This problem is aggravated in the high mobility scenario because both the user as well as the BSs suffer from excessive beam-training overhead. Therefore, new random access protocols and adaptive array processing algorithms are needed such that systems can adapt quickly in the event of blocking and handover due to high mobility at these frequencies [26].

3.1.3 Atmospheric Absorption

Electromagnetic (EM) waves suffer from transmission losses when they travel through the atmosphere due to their absorption by molecules of gaseous atmospheric constituents including oxygen and water. These losses are greater at certain frequencies, coinciding with the mechanical resonant frequencies of the gas molecules [27]. In mmWave and THz bands, the atmospheric loss is mainly due to water and oxygen molecules in the atmosphere, however, there is no prominent effect of atmospheric losses at the microwave frequencies. These attenuations further limit the distance mmWave/THz can travel and reduce their coverage regions. Therefore, it is expected that the systems operating at these frequencies will require much denser BS deployments.

3.2   Channel Measurement Efforts

Many measurement campaigns have been carried out to understand the physical characteristics of the mmWave frequency bands both in the indoor and outdoor settings. These measurement campaigns have focused on the study of path-loss, the spatial, angular, and temporal characteristics, the ray-propagation mechanisms, the material penetration losses and the effect of rain, snow and other attenuation losses associated with different mmWave frequencies. See Table 6.2 for a summary [28].

Likewise, in [45], measurements have been carried out to characterize THz wireless links for both indoor and outdoor environments. In the case of outdoor environments, [45] showed that interference from unintentional NLOS paths could limit the BER performance. The impact of weather on high capacity THz links was discussed in [46]. The frequency ranges which are suitable for THz communication have been studied in [47, 48]. In [49], intra-wagon channel characterization at 60 GHz and 300 GHz are done using measurements, simulations, and modeling.

3.3   Propagation at mmWave and THz Frequencies

We now discuss key propagation characteristics of the mmWave and THz frequencies.

Table 6.2 Mmwave channel measurements efforts for various environment

Scenario/Environment	Measurement efforts
Indoor settings such as office room, office corridors, university laboratory	• Narrowband propagation characteristics of the signal, received power and bit error rate (BER) measurements [29] • Measurements of the fading characteristics/distribution [30] • Effects of frequency diversity on multi-path propagation [31] • RMS delay spread measurement [32] • Effects of transmitter and receiver heights on normalized received power for LOS and NLOS regions [33]
Outdoor settings such as university campus, urban environments, streets, rural areas, natural environments, and grasslands	• Comparison of the propagation mechanisms and fading statistics of the received signals [34] • Channel impulse response, CDFs of received signal envelope and RMS delay spread [35] • Effects of multi-path scattering over foliage attenuation, mean and standard deviation of the path-loss [36] • Impacts of rain attenuation on the link availability and signal depolarization [37] • The path loss exponents and mean RMS delay spread of LOS and NLOS paths [38, 39] • Outdoor measurement over a distance of 5.8 km at 120 GHz [40] • Effects of ground reflections and human shadowing on LOS path-loss measurements [41, 42]
High-speed train (HST) channel propagation measurements in the outdoor scenario	• The reflection and scattering parameters for the materials of the deterministic and random objects present in the HST environment. The verification of channel model in terms of path-loss, shadow-fading, power delay profile and small-scale fading [43]
Outdoor to indoor (O2I) propagation measurements	• Effects of outdoor to indoor penetration losses on the number of multi-path components, RMS delay spread, angular spread and receiver beam-diversity [44]

3.3.1 Atmospheric Attenuation

The atmospheric attenuation is caused by the vibrating nature of gaseous molecules when exposed to the radio signals. Molecules with sizes comparable to the wavelength of EM waves excite when they interact with the waves, and these excited molecules vibrate internally. As a result of this vibration, a part of the propagating wave's energy is converted to kinetic energy. This conversion causes loss in the signal strength [50]. The rate of absorption depends upon the temperature, pressure, altitude and the operating carrier frequency of the signal. At lower frequencies (sub-

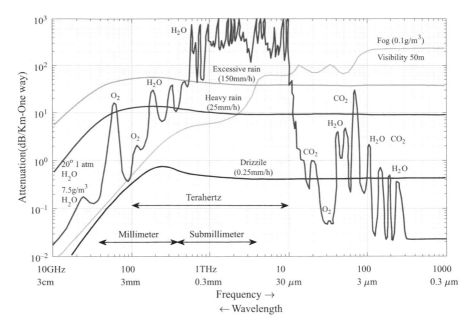

Fig. 6.1 Variation of atmospheric absorption due to various factors with respect to frequencies in the band 10GHz–1000THz. This figure is reproduced using data from [53]

6 GHz), this attenuation is not significant. But, higher frequency waves undergo significant attenuation since their wavelength becomes comparable to the size of dust particles, wind, snow, and gaseous constituents. The two major absorbing gases at mmWave frequencies are oxygen (O_2) and water vapor (H_2O). As seen in Fig. 6.1, the peaks of O_2 absorption losses are observed at 60 GHz and 119 GHz which are associated with a loss of 15 dB/km and 1.4 dB/km, respectively. Similarly, the peaks of H_2O absorption losses are observed at 23 GHz, 183 GHz and 323 GHz, which are associated with a loss of 0.18 dB/km, 28.35 dB/km and 38.6 dB/km, respectively. Similarly, 380 GHz, 450 GHz, 550 GHz, and 760 GHz frequency bands also suffer a higher level of attenuation. However, for short-distance transmission, the combined effects of these atmospheric losses on mmWave signals are not significant [23]. THz communication is even more prone to the atmospheric effects in the outdoor environment. We can see that the spectrum around 600 GHz suffers 100 to 200 dB/km attenuation which is $\approx$10–20 dB over the distance of approximately 100 m [51]. The absorption process can be described with the help of Beer-Lambert's law which states that the amount of radiation of frequency f that is able to propagate from a transmitter to the receiver through the absorbing medium (termed the transmittance of the environment) is defined as [52]

$$\tau(r, f) = \frac{P_{rx}(r, f)}{P_{tx}(f)} = \exp(-\kappa_a(f)r), \qquad (6.1)$$

where $P_{rx}(r, f)$ and $P_{tx}(f)$ are the received and transmitter power, and r is the distance between the transmitter and the receiver. Here, $\kappa_a(f)$ denotes the absorption coefficient of the medium. The $\kappa_a(f)$ is the sum of the individual absorption coefficient of each gas constituent, which depends on its density and type [50].

3.3.2 Rainfall Attenuation

The wavelengths of the mmWave spectrum range between 1 to 10 millimeters, whereas the average size of a typical raindrop is also in the order of a few millimeters. As a result, the mmWave signals are more vulnerable to blockage by raindrops than conventional microwave signals. The light rain (say, 2 mm/hr) imposes a maximum loss of 2.55 dB/km whereas the heavy rain (say, 50 mm/hr) imposes a maximum loss of 20 dB/km. In tropical regions, a monsoon downpour at 150 mm/hr has a maximum attenuation of 42 dB/km at frequencies over 60 GHz. However, in the lower bands of the mmWave spectrum, such as the 28 GHz and 38 GHz bands, lower attenuations of around 7 dB/km are observed during heavy rainfall, which drops to 1.4 dB for the coverage range of up to 200 m. Thus by considering short-range communications and lower bands of mmWave spectrum, the effect of rainfall attenuation can be minimized [23].

3.3.3 Blockage

(i) *Foliage attenuation:* The presence of vegetation can cause further attenuation at mmWave/THz frequencies. The severity of foliage attenuation depends on the carrier frequency and the depth of vegetation. For example, the foliage attenuation loss of 17 dB, 22 dB and 25 dB is observed at 28 GHz, 60 GHz and 90 GHz carrier frequencies, respectively [23].
(ii) *Material penetration losses:* The mmWave and higher frequencies cannot propagate well through obstacles like room furniture, doors and walls. For example, a high penetration losses of 24.4 dB and 45.1 dB was observed at 28 GHz signal when penetrating through two walls and four doors, respectively [23]. The higher penetration losses limit the coverage region of the mmWave transmitter in the indoor-to-outdoor and outdoor-to-indoor scenarios.

A LOS probability model can be used to incorporate the effects of static blockages on the channel. This model assumes that a link of distance d will be LOS with probability $p_L(d)$ and NLOS otherwise. The expressions of $p_L(d)$ are usually obtained empirically for different settings. For example, for the urban macro-cell (UMa) scenario [54]

$$p_L(d) = \min\left(\frac{d_1}{d}, 1\right)\left(1 - e^{-\frac{d}{d_2}}\right) + e^{-\frac{d}{d_2}},$$

where d is the 2D distance in meters and d_1 and d_2 were the fitting parameters equal to 18 m and 63 m, respectively. The same model is also applicable for the urban micro-cell (UMi) scenario, with $d_2 = 36$ m. There are some variations in the LOS probability expressions across different channel measurement campaigns and environments. For example, the LOS probability model developed by NYU [55] is

$$p_L(d) = \left(\min \left(\frac{d_1}{d}, 1 \right) \left(1 - e^{-\frac{d}{d_2}} \right) + e^{-\frac{d}{d_2}} \right)^2.$$

where the fitting parameters d_1 and d_2 were equal to 20 m and 160 m, respectively.

These empirical models can be justified theoretically. In [56], a cellular network with random rectangular blockages was considered where blockages were modeled using the Boolean process and it was shown that LOS probability is given as

$$p_L(d) = e^{-\beta d}, \qquad \text{where } \beta = \frac{2\mu (\mathbb{E}[W] + \mathbb{E}[L])}{\pi},$$

where L and W are the length and width of a typical rectangular blockage and μ is the density of blockages. A different blockage model known as the *LOS ball model* was introduced in [57] which assumes that all links inside a fixed ball of radius R_B are LOS, i.e.

$$p_L(d) = \mathbb{I}(d < R_B), \qquad \text{where } R_B = \frac{\sqrt{2}\mu\mathbb{E}[L]}{\pi},$$

which can also be used in the analysis of mmWave cellular networks.

3.3.4 Human Shadowing and Self Blockage

As discussed earlier, propagation at mmWave/THz frequencies can suffer significant attenuation due to the presence of humans including the self-blockage from the user equipment itself. In [58], human body blockages were modeled using a Boolean model in which humans are modeled as 3D cylinders with centers forming a 2D Poisson point process (PPP). Their heights were assumed to be normally distributed. In indoor environments, human blockages have also been modeled as 2D circles of fixed radius r with centers forming a PPP (with density μ) [59]. The LOS probability for a link of length d, in this case, comes out to be

$$p_L = 1 - e^{-\mu(rd + \pi r^2)}$$

The self-blockage of a user can also be modeled using a 2D cone of angle δ (which is determined by the user equipment width and user to equipment distance), such that all BSs falling in this cone are assumed to be blocked [60].

3.3.5 Reflections and Scattering

Consider an EM wave impinging on a surface. If the surface is smooth and electrically larger than the wavelength of the wave, we see a single reflection in a certain direction. The fraction of the incident field that is reflected in the specular direction is denoted by the reflection coefficient of the smooth surface, termed Γ_s, which also accounts for the penetration loss. The reflected power is thus

$$\overline{P}_R = P\Gamma_s^2,$$

where P is the power of the incident wave. However, if the surface is rough, the wave gets scattered into many directions in addition to a reflected component in the specular direction. This phenomenon is known as *diffuse scattering* [61], which is also exhibited by the mmWave/THz signals. As discussed next in detail, this behavior is attributed to the smaller wavelengths, which are comparable to the size of small structural features of the buildings surfaces.

Most importantly, whether a surface will be perceived smooth or rough depends upon the incident wave's properties. The *Rayleigh criteria* can be used to determine the smoothness or roughness of a surface based on the critical height associated to the wave h_c, which is given as [61]

$$h_c = \frac{\lambda}{8\cos\theta_i},$$

where h_c depends on the incident angle θ_i and wavelength λ. Let the minimum-to-maximum surface protuberance of the given surface be denoted by h_0, while the RMS height of the surface is h_{rms}. Then, if $h_0 < h_c$, the surface can be considered smooth, and if $h_0 > h_c$, the surface can be considered rough for the particular wave with wavelength λ. This implies that as λ decreases, the same surface which was smooth at higher λ, may start becoming rough. Therefore, at lower frequencies, reflection phenomenon is significant, while scattering is negligible as most surfaces are smooth compared to the wave. As a result, reflections are more prominent in the lower mmWave bands while the scattering is moderate. However, as we go higher in frequency to the THz bands, scattering becomes significant since the roughness in the surface of building walls and terrains becomes comparable to the carrier wavelength. As a result, the scattered signal components at THz are more significant compared to the reflected paths.

For rough surfaces, scattering results in additional loss in the reflected wave, if there is one. Therefore, the scattering loss factor (denoted by ρ) has to be considered to obtain the reflection coefficient Γ of a rough surface [61]

$$\Gamma = \rho\Gamma_s, \qquad \text{with } \rho \approx \exp\left[-8\left(\frac{\pi h_{\text{rms}}\cos\theta_i}{\lambda}\right)^2\right].$$

Therefore, the scattered power from this surface is given by

$$\overline{P}_S = P\left(1 - \rho^2\right)\Gamma_s^2,\tag{6.2}$$

and the reflected power is given by

$$\overline{P}_R = P\Gamma^2 = P\rho^2\Gamma_s^2.\tag{6.3}$$

The fraction of the incident wave that is scattered is represented by scattering coefficient S^2. The scattering coefficient S^2 is given by

$$S^2 = \frac{\overline{P}_S}{P} = \left(1 - \rho^2\right)\Gamma_s^2.$$

There are various models to characterize the variation of scattering power with scattering direction. One of the widely used models is the directive scattering (DS) model, which states that the main scattering lobe is steered in the general direction of the specularly reflected wave (θ_r in Fig. 6.2) and the scattered power in a direction θ_s is

$$P_S(\theta_s) \propto \left(\frac{1 + \cos(\theta_s - \theta_r)}{2}\right)^{\alpha_R},$$

where α_R represents the width of the scattering lobe. In [62], DS model is used to model the propagation of a 60GHz wave in the hospital room. The DS model was found to agree with rural and suburban buildings scattering when validated with 1.29 GHz propagation measurements [63].

We can now compute the scattered power at a receiver from a transmitter located at r_i distance away from the surface. From the Friis equation and (6.2), the total scattered power from the surface can be expressed as

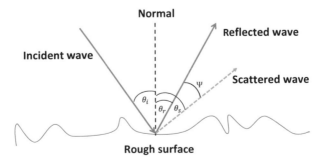

Fig. 6.2 Schematic diagram of a radio wave incident at a surface. θ_i is the incident angle, θ_r is the reflected angle, θ_s is the scattered angle and Ψ is the angle between reflected and scattered waves

$$\overline{P}_S = S^2 A_s \frac{P_t G_t}{4\pi r_i^2},$$

where P_t is the transmitted power, G_t is the transmitter antenna gain, and A_s is the effective aperture of the scattering surface. Now from the DS model, the scattered power P_S at a distance r_s in the direction θ_s, is

$$P_S(\theta_s) = P_{S0} \left(\frac{1 + \cos(\theta_s - \theta_r)}{2} \right)^{\alpha_R},$$

where P_{S0} is the maximum scattered power given as

$$P_{S0} = \frac{\overline{P}_S}{r_s^2 \int \int \left(\frac{1+\cos(\theta_s - \theta_r)}{2} \right)^{\alpha_R} d\theta_s d\phi_s}.$$

If we define $F_\alpha = \int \int \left(\frac{1+\cos(\theta_s - \theta_r)}{2} \right)^{\alpha_R} d\theta_s d\phi_s$, then

$$P_{S0} = \frac{\overline{P}_S}{r_s^2 F_\alpha} = S^2 A_s \frac{P_t G_t}{4\pi r_i^2} \frac{1}{r_s^2 F_\alpha}.$$

Hence, the received power at the receiver located at an angle θ_r and a distance r_s from the surface, is given as

$$p_r = P_S(\theta_s) \times \text{Effective antenna aperture} = P_S \frac{\lambda^2}{4\pi} G_r$$

$$= S^2 A_s \frac{P_t G_t}{4\pi} \frac{1}{r_i^2 r_s^2} \frac{\lambda^2}{4\pi} G_r \frac{1}{F_\alpha} \left(\frac{1 + \cos(\theta_s - \theta_r)}{2} \right)^{\alpha_R}, \tag{6.4}$$

where G_r is the receiver antenna gain. The model can also be extended to consider the backscattered lobe.

3.3.6   Diffraction

Owing to its short wavelength, in mmWave/THz frequencies, diffraction will not be as prominent as it is at microwave frequencies [51]. In these frequencies, NLOS has significantly less power compared to that of the LOS path [64]. However, it may be possible to establish THz links in the shadow of objects with the help of diffraction [65].

3.3.7 Doppler Spread

Since the Doppler spread is directly proportional to the frequency and the speed of users, it is significantly higher at mmWave frequencies than the sub-6 GHz frequencies. For example, the Doppler spread at 30 GHz and 60 GHz is 10 and 20 times higher than at 3 GHz [23].

3.3.8 Absorption Noise

Along with attenuation in the signal power, molecular absorption causes the internal vibration in the molecules which results in the emission of EM radiation at the same frequency as that of the incident waves that provoked this vibration. Due to this, molecular absorption introduces an additional noise known as absorption noise. Since absorption is significant in the THz bands, absorption noise is included in the total noise as an additional term. It is generally modeled using an equivalent noise temperature of the surroundings caused by the molecular absorption [52].

3.3.9 Scintillation Effects

Scintillation refers to the rapid fluctuation in the wave's phase and amplitude due to the fast local variation in the refractive index of the medium through which the wave is travelling. Local variation in temperature, pressure, or humidity causes small refractive index variations across the wavefront of the beam which can destroy the phase front, and the beam cross-section appears as a speckle pattern with a substantial local and temporal intensity variation in the receiver. Infrared (IR) wireless transmission distance is limited by scintillation effects [66]. The result of scintillation on practical THz communication is smaller than the IR beams. The THz waves traveling close to the surface of the earth may be influenced by atmospheric turbulence [67]. However, the extent to which scintillation effects impact the THz bands is still not well understood.

3.4 Beamforming and Antenna Patterns

In multiple inputs and multiple outputs (MIMO) systems, beamforming is used to focus a wireless signal towards a specific receiver (or away from certain directions to avoid interfering with devices in those directions). The gain thus achieved in the signal to noise ratio (SNR) at the intended receiver is called the *beamforming gain*, which is essential in mmWave systems to ensure reliable reception. Traditional MIMO systems were based on the *digital beamforming*, where each element in the antenna array has its separate digital-to-analog (D/A) conversion unit and the RF chain. However, fully digital beamforming is not suitable for mmWave frequencies due to a many-fold increase in the number of antenna elements which not only increases the cost of the overall system but also the substantial power consumption [14]. Further the power consumption generally scales linearly with the sampling rate and exponentially with the number of bits per samples [9, 14, 23, 24].

In order to lower the power consumption, *analog beamforming* has been proposed for mmWave systems where a single RF chain is shared by all antenna elements. Each antenna is fed with the phase shifted version of the same transmit signal where phase shift is determined according to the beamforming direction. However, such transmission is limited to a single stream and single user transmission/reception. To enable multi-user/multi-stream transmission for mmWave networks [9, 14, 23, 24], *hybrid beamforming* has been proposed in which more than one RF chains are used. The hybrid beamforming architectures are broadly classified into two types, the *fully connected hybrid beamforming architecture*, where each RF chain is connected to all antennas and the *partially connected hybrid beamforming architecture*, where each RF chain is connected to a subset of antenna elements. Clearly, hybrid beamforming provides a tradeoff between low-complexity but restrictive analog beamforming and the high-complexity but most flexible fully digital beamforming.

3.4.1 Analog Beamforming Patterns

Due to analog beamforming, the effective gain in the received signal can be computed using the transmitter and receiver antenna patterns which represents the gain in different directions around the antenna array (e.g., see (6.8)). Various antenna patterns have been proposed in the literature to aid the evaluation of mmWave systems. Some examples are discussed below.

Uniform Linear Array (ULA) Model For the antenna element spacing d and signal wavelength λ, the antenna gain of an N-array ULA [68] is

$$G_{\text{act}}(\phi) = \frac{\sin^2(\pi N\phi)}{N^2 \sin^2(\pi\phi)}, \tag{6.5}$$

where $\phi = \frac{d}{\lambda}\cos\theta$ is the cosine direction corresponding to the spatial angle of departure (AoD), θ, of the transmit signal. In order to avoid the grating lobes at mmWave frequencies, the antenna element spacing d is generally kept to be half of the wavelength. Since the spatial angle ϕ depends on d, we can use the approximation $\sin(\pi\phi) \simeq \pi\phi$ in the denominator. Therefore the array gain function in (6.5) can be approximated as a squared sinc-function

$$G_{\text{sinc}}(\phi) \triangleq \frac{\sin^2(\pi N\phi)}{(\pi N\phi)^2}. \tag{6.6}$$

This *sinc antenna pattern* has been widely used for the numerical analysis in antenna theory. Authors in [69] have verified the accuracy of tight lower bound provided by

Fig. 6.3 The sectorized antenna model [14] which provides analytical tractability in the system level evaluations of the mmWave systems

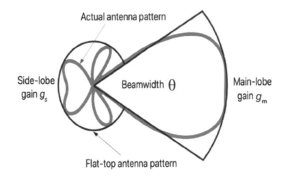

sinc antenna model for the actual antenna pattern that makes it highly suitable for the network performance analysis of the mmWave systems.

Sectorized Antenna Model To maintain the analytical tractability in the network coverage analysis, many researchers approximate the actual antenna pattern with the *flat-top antenna pattern*, also known as the *sectorized antenna model* (see Fig. 6.3). In this model, the array gains within the half-power beam-width (HPBW) θ_{3dB} are approximated to the maximum main-lobe gain g_m while the array gains corresponding to the remaining AoDs are approximated to the first side-lobe gain g_s of the actual antenna pattern [57]. Hence, the gain in the direction θ is given as

$$G_{Flat}(\theta) = \begin{cases} g_m & \text{if } \theta \in [-\theta_{3dB}, \ \theta_{3dB}] \\ g_s & \text{otherwise.} \end{cases}$$

Thus, the flat-top antenna pattern models the continuously varying actual antenna array gains using the fixed main-lobe and side-lobe gains. For highly dense network scenarios, the aggregated interference from the side lobes is significant because of which the term g_s must not be ignored in the analysis. This model is limited in its ability to fit arbitrary antenna patterns and is not suitable to analyze beam misalignments.

Multi-lobe Antenna Model In order to generalize the flat-top antenna pattern, a *multi-lobe antenna model* was proposed in [70] where there are K number of lobes, each with a constant gain. The array gain and the width of each lobe are obtained by minimizing the error function between the multi-lobe pattern and the actual antenna pattern. The limitation of the model includes the lack of the roll-off characteristic of the actual antenna pattern because of which the predicted analytical performance of the network may deviate from the actual network performance [71].

Gaussian Antenna Model The Gaussian antenna model is proposed in order to capture the effects of roll-off in actual antenna pattern which generally occur due to small perturbations and misalignment between the receiver and transmitter [72, 73]. The antenna gain for this model is given as

$$G_{\text{Gaussian}}(\theta) = (g_m - g_s)e^{-\eta\theta^2} + g_s. \tag{6.7}$$

where g_m is the maximum main-lode gain which occurs as $\theta = 0$, g_s is the side-lobe gain and η is a parameter that controls the 3 dB beam-width.

Cosine Antenna Model The antenna pattern for *cosine antenna pattern* is given as [69]

$$G_{\cos}(\theta) = \cos^2\left(\frac{\pi N}{2}\theta\right) \mathbb{1}\left(|\theta| \leq \frac{1}{N}\right).$$

This model can be extended to include multiple lobes [74] to give additional flexibility.

3.4.2 Antenna Patterns for Multi-User/Stream Transmission

The above discussion can be extended to include the hybrid beamforming supporting multi-stream or multi-user transmission. As a *layered* technique, hybrid beamforming can be seen as linear combination of the digital and analog beamforming. Hence the effective antenna pattern in case of multi user or multi-stream transmission will consists of individual analog beam patterns, one for each stream or user.

3.4.3 THz Beamforming

The limited transmission range of THz waves can be extended somewhat via very dense ultra-massive multiple-input multiple-output (UM-MIMO) antenna systems. Since the number of antennas that can fit into the same footprint increases with the square of the wavelength, the THz systems can accommodate even larger number of antenna elements than mmWave systems. This large array of compact antennas results in highly focused beams (pencil beams) of high gain that aids in increasing the transmission distance.

Similar to mmWave communication, the high cost and the high power consumption in digital beamforming makes it unsuitable for THz communication. The analog beamforming at THz waveband can reduce the number of required phase shifters in the RF domain. Nevertheless, it is subject to the additional hardware constraints because the analog phase shifters are digitally controlled and just have quantized phase values, which will significantly restrict analog beamforming performance in practice. On the other hand, the hybrid analog/digital beamforming is again a better trade-off between the analog and digital methods. The hybrid beamforming can have fewer RF chains than antennas and approaches the fully digital performance in sparse channels [75].

The types of antennas that can be used in THz communication are photocon-
ductive antennas, horn antennas, lens antennas, microstrip antennas, and on-chip
antennas. Initially, THz antennas were designed inside the semiconductor using
Indium Phosphide (InP) or Gallium Arsenide (GaAs) in which controlling radiation
pattern was difficult due to high dielectric constant. Therefore lens-based antennas
that were fed by horns were proposed. Other approaches, like stacking different
substrate layers with different dielectric properties, were proposed to improve the
antenna efficiency [76]. In addition to metallic antennas and dielectric antennas,
antennas based on new materials are also possible e.g., carbon-nanotube based
antennas and the planar graphene antennas [77].

3.5 Channel Models

In order to evaluate the performance of the communication system, the very first
step is to construct an accurate channel model. Not surprisingly, researchers have
developed different channel models for mmWave to be used in simulators and
analysis. For example, in 2012, the mobile and wireless communications enablers
for the twenty-twenty information society (METIS) project proposed three-channel
models, namely the stochastic, the map-based and the hybrid model, where the
stochastic model is suitable for frequencies up to 70 GHz, while the map-based
model is applicable for frequencies up to 100 GHz. In 2017, 3GPP 3D channel
model for the sub-100 GHz band was proposed. NYUSIM is another channel
model developed with the help of real-world propagation channel measurements at
mmWave frequencies ranging from 28 GHz to 73 GHz in different outdoor scenarios
[78]. Statistical channel models for UM-MIMO are classified into *matrix-based
models* and *reference-antenna-based models*. Matrix-based models characterize the
properties of the complete channel transfer matrix. On the other hand, the reference-
antenna-based models consider a reference transmitting and receiving antenna first
and analyze the point to point propagation model between them. Then, based on this
model, the complete channel matrix is statistically generated [79].

As discussed above already, THz channels exhibit very different propagation
characteristics compared to the lower frequency bands. Therefore, modeling the
channel and noise is essential for the accurate performance evaluation of the THz
communications systems [80]. Even the free-space scenario is not straightforward
to model in this case because of the significant level of molecular absorption.
Therefore, one needs to be include an additional exponential term along with
the power-law model in the path-loss equation. Overall, the peculiar propagation
characteristics of THz waves already discussed in Sect. 3.3 make their analysis
challenging.

Except for the recent measurements at the sub-terahertz (sub-THz) frequencies
[51], the rest of the THz channel modeling work is driven by ray tracing [81–83] or
statistical channel modeling [49, 84–90]. In particular, a statistical model for THz
channel based on a universal stochastic spatiotemporal model has been introduced

in [85] for indoor channels ranging from 275 GHz to 325 GHz. A 2D geometrical statistical model for device-to-device scatter channels at sub-THz frequencies was proposed in [86, 87]. In addition to the outdoor channel model, the indoor model for intra-wagon channel characterization at 300 GHz was discussed in [49]. At the nano-scale level, the channel models were presented in [88, 89] for intra body THz communication. A hybrid channel model was discussed in [90] for chip-to-chip communication via THz frequencies.

We now describe a simple yet powerful analytically tractable channel model which can be adapted to various propagation scenarios. This is suitable for the system-level performance analysis including using ideas from stochastic geometry.

3.5.1 mmWave Channel

Consider a link between a transmitter and a receiver located at r distance apart of type s where $s \in \{L, N\}$ denoting whether the link is LOS and NLOS [14]. For simplicity, let us assume narrow-band communication and analog beamforming. The received power at the receiver is given as

$$P_r = P_t \ell_s(r) g_R(\theta_R) g_t(\theta_t) H, \tag{6.8}$$

where

1. $\ell_s(r)$ denotes the standard path loss at distance r which is due to spreading loss. It is given by a path-loss function, typically modeled using power-law as

$$\ell_s(r) = c_s r^{-\alpha_s},$$

   where c_s is the near-field gain and α_s is the path-loss exponent.
2. p_t is the transmit power,
3. g_t and g_R are the transmitter and receiver antenna patterns while θ_t and θ_R are the angles denoting beam-direction of the transmitter and receiver. Therefore, $g_t(\theta_t)$ and $g_R(\theta_R)$ are respectively the transmitter and receiver antenna gains.
4. H denotes the small scale fading coefficient. Nakagami fading is often assumed with different parameters μ_L and μ_N for LOS and NLOS link [14]. Therefore H is a Gamma random variable with parameter μ_s.

The above channel model can be extended for different environments and propagation scenarios, for example, to include multiple paths [14], multi-rank channel [91, 92], hybrid beam-forming [92] and massive MIMO. Since the specific bands with high absorption loss are avoided, the effect of molecular absorption can be ignored for mmWave communication.

3.5.2   THz Channel

Since atmospheric attenuation and scattering are prominent at THz frequencies, the THz channel model is expected to be different from the one discussed above for the mmWave communications. Due to the huge difference between LOS and NLOS links, most of the works have considered LOS links only [93, 94]. For simplicity, we will assume narrowband communication. If we consider a LOS link between a transmitter and a receiver located at r distance apart of type s, the received power P_r is given by [45, 66]

$$P_r = P_t \ell(r) g_R(\theta_R) g_t(\theta_t) \tau(r), \qquad (6.9)$$

where $\tau(r)$ is an additional loss term due to molecular absorption defined in (6.1). In LOS links, path-loss can be given by free space path loss *i.e.*

$$\ell(r) = \left(\frac{\lambda^2}{4\pi} \right) \frac{1}{4\pi r^2}.$$

The model can be extended to include scatters/reflectors. If r_1 is the distance between the transmitter and the surface while r_2 is the distance between the surface and the receiver, then the scattered and reflected power are

$$P_{r,S} = P_t g_R(\theta_R) g_t(\theta_t) \ell(r_1) l(r_2) \tau(r_1) \tau(r_2) \Gamma_R$$

and

$$P_{r,R} = P_t g_R(\theta_R) g_t(\theta_t) \ell(r_1 + r_2) \Gamma^2 \tau(r_1 + r_2) \Gamma_S,$$

respectively, where Γ_R and Γ_S are coefficients related to reflection and scattering and may depend on surface orientation and properties. The above channel model can be extended to include other scenarios, for example, multiple paths and wideband communication [81].

4   The mmWave Communications Systems

As discussed above already, the major advantage of using mmWave communications is the availability of abundant spectrum, which is making multi-gigabit-per-second communication possible [95]. However, mmWave signals are more susceptible to blockages and foliage losses, which necessitates highly directional transmission. The combination of high signal attenuation and directional transmission offers several advantages and disadvantages for practical mmWave systems. On the positive side, these make mmWave systems more resilient to interference and hence more likely to operate in the noise-limited regime [1]. Because of this, it is possible for the operators to use a higher frequency reuse factor, thereby resulting in higher network capacity [96, 97]. For the same reasons, mmWave transmissions are inherently more secure compared to the sub-6 GHz transmissions [98–101]. For instance, the high attenuation of susceptibility to blockages make it difficult for the remote eavesdroppers to even overhear mmWave transmissions unless they are located very close to the transmitters. Finally, as will be discussed in detail next, these reasons also make spectrum sharing more feasible at the mmWave frequencies.

On the flip side, with high directivity, the initial cell search becomes a critical issue. Because of the use of directional beams, both the BSs and users need to perform a spatial search over a wide range of angles to align their transmission and reception beams in the correct direction. This adds significant delay and overhead to the communication. The situation degrades further when users are highly mobile due to increased occurrences of handovers. Further, higher susceptibility to blockages can result in outages. One approach to mitigate this is to utilize the concept of macro-diversity [102, 103] and [104], where simultaneous connections with multiple BSs are maintained for each user so that it does not experience any service interruption in the event of blocking of one BS.

After summarizing these key features of mmWave communications, we now discuss a few key implications of these features on the system design. This section will be concluded with a discussion on the potential uses of mmWave communications in future 6G systems.

4.1 Key System Design Implications

Coexistence with Lower Frequency Systems Due to their limited transmission range, a mmWave system may not work effectively in a standalone deployment [105]. In particular, they need to coexist with conventional cellular networks operating on more favorable sub-6 GHz bands such that all the control level management, including load balancing and handovers, is performed over sub-6 GHz microwave transmissions while the data transmissions occur over the mmWave bands. Such networks will provide high capacity and better throughput in comparison to the standalone networks without decreasing reliability [106]. Further, macro-diversity can be utilized, where multiple BSs (some can be sub-6 GHz and some are mmWave) can connect to a user simultaneously to improve LOS probability and link throughput [102].

Spectrum Sharing At lower frequencies, owning an exclusive license of a spectrum band ensures reliability and provides performance guarantees to applications with time-critical operations for an operator. However, mmWave systems often operate in a noise-limited regime because of which exclusive licensing at these frequencies may result in an under-utilization of the spectrum [8]. It has been shown that spectrum sharing at mmWave frequencies does not require sophisticated inter-cell coordination and even uncoordinated spectrum sharing between two or more operators is feasible [107]. This is an attractive option for the *unlicensed spectrum* located at 59–64 GHz and 64–71 GHz bands which will allow multiple users to access the spectrum without any explicit coordination. Such use of unlicensed spectrum increases spectrum utilization and helps minimize the entry barrier for new or small-scale operators. Even at licensed bands, shared use of spectrum can help increase the spectrum utilization and reduce licensing costs. It has also been shown that simple inter-cell interference coordination mechanisms can be used to improve the sharing performance [9, 14, 23, 24, 108]. Furthermore, the bands where mmWave communication coexists with other services (including incumbent services and newly deployed applications) may need to protect each other in the case of dense deployments. For this, spectrum license sharing mechanisms such as uncoordinated, static, and dynamic are the viable options in these bands. Spectrum sharing opportunities at mmWave bands also bring the need to evolve new methods of spectrum licensing which need to be more flexible, opportunistic, dynamic and area specific [8].

Ultra-Dense Networks Ultra-dense networks (UDN) are characterized by very short inter-site distances. They are generally used to provide local coverage in highly populated residential areas, office buildings, university campuses, and city centers. The mmWave frequencies are a natural candidate for UDNs because of the directional transmission and blockage sensitivity which limits interference even at ultra dense deployments. Further, self backhauling provides an inexpensive way to connect these densely deployed APs/BSs to backhaul.

Deep Learning-Based Beamforming The performance of the mmWave systems in a high mobility scenario is severely affected by large training overhead, which occurs due to the frequent updating of large array beamforming vectors. In the last few years, deep learning-based beamforming techniques have attracted considerable interest due to their ability to reduce this training overhead. At the transmitter, pilot signals from the UE are first transmitted to learn the RF signature of the neighboring environment and then this knowledge is used to predict the best beamforming vectors for the transmitted data RF signature. Thus, after successful learning phase, the deep learning models require negligible training overhead which ensures reliable coverage and low latency for the mmWave applications [109].

4.2   Potential Applications of mmWave Communications in 6G

Wireless Access Applications Due to the abundance of bandwidth around 60 GHz band, various technologies are expected to be developed to support unlicensed operations for the wireless local and personal area networks (WLANSs and WPANs) with potential applications in internet access at home, offices, transportation centers, and city hotspots. These technologies are expected to support multi-gigabit data transmission, with examples including IEEE 802.11ad and IEEE 802.11ay [110, 111]. In future, IEEE 802.11ay based mmWave distribution networks (mDNs) may become an alternative-low-cost solution for the fixed optical fiber links. The purpose of mDNs is to provide point-to-point (P2P) and point-to-multi-point (P2MP) mmWave access in indoor as well as outdoor scenario as well as wireless backhaul services to the small cells in an ad-hoc network scenario. The benefits of IEEE 802.11ay based mDN networks are cheaper network infrastructure and the high-speed ubiquitous coverage, while the major challenges include dealing with blockages, interference management, and developing efficient beam-training algorithms [112, 113]. Also, 5G is seen as a significant step in enabling cellular communication over mmWave bands which is expected to mature further in 6G and beyond systems.

Backhaul Infrastructure It is well known that providing fiber backhaul in highly dense small cell deployments is challenging due to increased installation and operational cost [114–116]. Not surprisingly, many researchers have recently invested their efforts to enable wireless backhaul in mmWave bands owing to their directional communication and high LOS throughput. The present 5G cellular backhaul networks are expected to operate on the 60 GHz and 71–86 GHz bands, which are expected to be extended to the 92–114.25 GHz band due to its similar propagation characteristics. Significant efforts have also gone into developing new technologies including cross-polarization interference cancelation (XPIC), bands and carriers aggregation (BCA), LOS MIMO, and orbital angular momentum (OAM) in order to increase the capacity of the current mmWave backhaul solutions [117]. Further work is needed to provide backhaul solutions for the data-hungry future applications of 6G by using higher mmWave bands (above 100 GHz) as well as the THz spectrum [117, 118].

Information Showers Information showers (ISs) are high bandwidth ultra-short range hot spots in which mmWave BSs operating at the unlicensed 60 GHz band are mounted on the ceilings, doorways, entrances of the commercial buildings or pavements, which deliver multi-gigabit data rates over a coverage range of about 10 m [119]. Thus they provide an ideal platform to exchange a huge amount of data between different kind of networks, devices and users over a very short span of time. Unlike conventional small-cell cellular networks, ISs can be used for both offloading as well as pre-fetching of data from the long haul wireless network for applications like instant file transfer and video streaming. ISs also help in improving the energy efficiency and battery life of the mobile terminal due to its ability to download videos and large files within a few seconds. However, installations of ISs requires a very robust architecture that can work seamlessly with the current cellular networks and is still an open area of research [120, 121].

Aerial Communications Many frequency bands in the mmWave spectrum region are already being used to support the high-capacity satellite to ground transmission. However, with the maturity of the drone technology, the future wireless networks is expected to have a much more dynamic aerial component with drones used in a diverse set of applications, such as agriculture, mapping, traffic control, photography, surveillance, package delivery, telemetry, and on-demand handling of higher network loads in large public gatherings like music concerts. Because of higher likelihood of LOS in many of these applications, mmWave communications is expected to play a particularly promising role. Further, quick (on-demand) and easy deployment of drones also make them attractive for many public safety applications, especially when the civil communication infrastructure is compromised or damaged. Naturally, mmWave communications can play a promising role in such applications as well.

Vehicular Communications The ability of vehicles to communicate among themselves as well as the wireless infrastructure not only helps in the navigation of completely autonomous vehicles but is also helpful in the avoidance of accidents in semi-autonomous and manually driven vehicles through timely alerts and route guidance [122]. Because of the high likelihood of LOS and the need to support high data rates, the mmWave (and THz) spectrum is naturally being considered for the vehicular communications systems [38, 123–126]. Further, unified vehicular communications and radar sensing mechanisms are needed for the massive deployments of interconnected smart cars which can easily cope with the rapidly maturing automotive environments, consisting of the networked road signs, connected pedestrians, video surveillance systems, and smart transportation facilities [127].

5   The THz Communications Systems

After discussing mmWave communications systems in detail in the previous section, we now focus on the THz communications systems in this section. Since the THz band is higher in frequency than the mmWave band, the communication at THz band faces almost all the critical challenges that we discussed in the context of mmWave communications. In order to avoid repetition, we will therefore focus on the challenges and implications that are more unique (or at least more pronounced) to the THz communications systems.

Smaller Range Due to high propagation and molecular absorption losses, communication range of THz bands is further limited compared to the mmWave transmission. For instance, in small cells, the THz band may provide coverage up to only about 10 m [128]. Further, the frequency-dependent molecular absorption in the THz bands results in band-splitting and bandwidth reduction [2].

THz Transceiver Design In THz communication, the transceivers need to be wideband, which is a major challenge. The frequency band of the signal to be generated is too high for conventional oscillators, while it is too low for optical photon emitters. This problem is known as the *THz gap*. Another challenge is the design of antennas and amplifiers which support ultra-wideband transmission for THz communication [80]. Currently, the THz waves are generated using either conventional oscillators or optical photon emitters along with frequency multiplier/divider.

THz Beam Tracking Just like mmWave systems, the THz communications systems require beamforming to overcome large propagation losses. However, beamforming requires channel state information, which is challenging to obtain when the array sizes are large, as is the case in THz communications systems. Therefore, it is vital to accurately measure the AoD of transmitters and the angle of arrival (AoA) of receivers using beam tracking techniques. While such beam tracking techniques have been studied extensively for the lower frequencies, it is not so for the THz frequencies. In THz communication, in order to achieve beam alignment, beam switching must be done before beam tracking. However, due to large array sizes, the codebook design for beam switching is computationally complex. On the flip side, these complex codebooks will generate high-resolution beams, which help in accurate angle estimation [75]. This provides a concrete example of the type of challenges and subtle tradeoffs that need to be carefully understood while designing THz communications systems.

The implications of these challenges are similar to the ones we discussed for the mmWave systems in the previous section. For instance, due to the limited coverage area, the THz communication systems are more likely to be deployed for indoor applications [129]. In particular, the indoor links have been found to be robust even in the presence of one or two NLOS reflection components [45]. Likewise, owing to small coverage areas and high directionality, it is expected that the THz systems would efficiently share spectrum without much coordination (similar to the mmWave systems). In bands where passive services like radio astronomy and satellite-based earth monitoring are already present, THz communication systems need to share the spectrum under some protection rules.

5.1   Potential Applications of THz Communications in 6G

THz communication has many applications in macro as well as in micro/nano scale. Some of the applications are discussed in this section.

5.1.1   Macroscale THz Communication

Most of the macroscale use cases of THz communications will be driven by emerging applications requiring Tbps links, which are not possible using the mmWave spectrum. Such applications include ultra HD video conferencing and streaming, 3D gaming, extended reality, high-definition holographic video conferencing, haptic communications, and tactile internet, to name a few. Within conventional cellular network settings, the THz bands are most suitable for small cell indoor applications or high-speed wireless backhaul for small cells [128]. Likewise, in the conventional WLAN applications, the Terabit Wireless Local Area Networks (T-WLAN) can provide seamless interconnection between high-speed wired networks, such as optical fiber links, and personal devices, such as mobile phones, laptops, and smart

TVs. Along similar lines, the Terabit Wireless Personal Area Networks (T-WPAN) can enable ultra-high-speed communication among proximate devices. A special type of WPAN application is *kiosk downloading*, where a fixed kiosk download station is used to transfer multimedia contents, such as large videos, to mobile phones located in its proximity [130]. Other potential applications and advantages of THz communications, such as enhanced security, relevance for aerial and vehicular communications [131, 132], as well as the potential use for providing wireless connections in data centers, can be argued along the same lines as we did already for the mmWave networks in Sect. 4. In order to avoid repetition, we do not discuss this again.

5.1.2 Micro/nanoscale THz Communication

The THz band can also be used for enabling communications between nanomachines [128]. These nanomachines can perform simple tasks, such as computations, data storage, actuation, and sensing. Depending on the application, the transmission distance can vary from a few micrometers to a few meters. Some representative applications of nanomachine communications are discussed below.

(i) *Health monitoring:* Nanosensors or nanomachines deployed inside the human body can measure the level of glucose, cholesterol, the concentration of various ions, biomarkers emitted by the cancer cells, etc. [128]. The measured data can be wirelessly transmitted to a device outside the human body (e.g., mobile phone or a smart band) using THz communication. The external device can process the data and further send it to a medical equipment or to a doctor.

(ii) *Nuclear, biological and chemical defenses:* Nanosensors are capable of sensing harmful chemical and bio-weapon molecules effectively [128]. In contrast to the classical macro-scale chemical sensors, the nanosensors can detect very small concentrations (as small as a single molecule). As a result, the nano devices communicating in the THz bands can be used in defense applications for the detection of harmful chemical, biological and nuclear agents.

(iv) *Internet of nano-things (IoNT) and Internet of bio-nano-things (IoBNT):* The interconnection of nanomachines with the existing communication network is known as IoNT [15]. These interconnected nanodevices via IoNT can serve a variety of purposes ranging from tracking atmospheric conditions and health status to enabling real-time tracking. Also, nano-transceivers and antennas can be embedded in nearly all devices to be connected to the Internet. IoBNT is conceptually the same as IoNT but consists of biological nanomachines as opposed to the silicon-based nanomachines [133]. Biological nanomachines can be made from synthetic biological materials or a modified cell via genetic engineering. IoBNT has many applications in the biomedical field.

(iv) *Wireless network on-chip communication:* THz waves can enable communication among processing cores embedded on-chip with the help of planar nano-antennas of a few micrometers in size [134]. This creates ultra-high-speed inter-core communication for applications where area is a constraint. Graphene-based nano-antennas can be used for the design of scalable and flexible wireless networks on the chips.

5.2 Nanonetworks

While the capabilities of a single nanomachine are limited to simple computations, sensing, and actuation, a network of inter-connected nanomachines can perform much more complex tasks. The nanomachines can communicate with each other or with a central device. Such networks have a wide variety of applications ranging from cancer treatment to environmental monitoring. The two potential carriers of information between nanomachines are EM waves and chemical molecules. Inside the human body, molecular communication has several advantages over EM waves, such as bio-compatibility and energy efficiency.

Integration with Molecular Communication A nano-scale communication network consist of five fundamental components [16]:

 (i) *Message carrier:* Chemical molecules or waves that carry information from the transmitter to the receiver.
 (ii) *Motion component:* Provides force that is needed for the message carrier to move in the communication medium.
(iii) *Field component:* Guides the message carrier in the communication medium. External fields include the EM field, molecular motors, and non-turbulent fluid flow. Internal fields include swarm motion or flocking behavior.
(iv) *Perturbation:* This represents the variation of the message carrier to represent the transmit information. Perturbation is similar to modulation in telecommunication. It can be achieved by varying the concentration or type of molecules based on the transmit information.
 (v) *Specificity:* Reception process of the message at the target. For example, the binding of molecules with the receptor structures present in the target.

   A hybrid communication system that combines molecular and EM paradigms was proposed in [16] (See Fig. 6.4). In this hybrid communication network, MC is used inside the body due to its bio-compatibility, energy efficiency, and the lack of need for communication infrastructure for propagation methods like diffusion-based MC (molecules propagates in the medium based on concentration gradient). The nano-nodes (bio-nanomachines) form clusters and sense the data locally. A bio-nanomachine collects the health parameters, modulates the data, and transmits the information to the other bio-nanomachines (that act as relays). Now, for delivering gathered information to a receiver outside the human body, a graphene-based nano-device is implanted into the body. This implantable nano-device is made up of a chemical nanosensor, a transceiver, and the battery. Based on the concentration of information molecules transmitted by the bio-nanomachines to the implantable

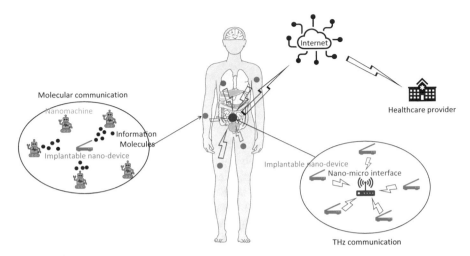

Fig. 6.4 A hybrid nano communication network showing an eco-system consisting of the biological and artificial components, where various communication technologies including molecular and THz communication may co-exist together [16]

nano-device, the concentration is converted to a corresponding electrical signal. Now the implantable nano-devices communicate to the nano-micro interface via THz waves. This interface can be a dermal display or a micro-gateway to connect to the internet. *This type of hybrid communication network is bio-compatible due to MC technology and well connected to the outside world via THz communication.*

6 Standardization Efforts

We conclude this chapter by discussing the key standardization efforts for both the mmWave and THz communications systems.

6.1 Standardization Efforts for mmWave Communications

Given the increasing interest in mmWave communications over the past decade, it is not surprising to note that several industrial standards have been developed for its use. Some of these standards are discussed below.

IEEE 802.11ad This standard is focused on enabling wireless communications in the 60 GHz band. It specifies amendments to the 802.11 physical and MAC layers to support multi-gigabit wireless applications in the 60 GHz band. The unlicensed spectrum around 60 GHz has approximately 14 GHz bandwidth, which is divided into channels of 2.16, 4.32, 6.48, and 8.64 GHz bandwidth. The IEEE 802.11ad standard supports transmission rates of up to 8 Gbps using single-input-single-output (SISO) wireless transmissions over a single 2.16 GHz channel. It supports backward compatibility with existing Wi-Fi standards in the 2.4 and 5 GHz bands. Therefore, future handsets may have three transceivers operating at 2.4 GHz (for general use), 5 GHz (for higher speed applications), and 60 GHz (for ultra-high-speed data applications) [106, 110, 135].

IEEE 802.11ay This standard is the enhancement of IEEE 802.11ad to support fixed point-to-point (P2P) and point-to-multipoint (P2MP) ultra-high-speed indoor and outdoor mmWave communications. It supports channel bonding and aggregation to enable 100 Gbps data rate. Channel bonding allows a single waveform to cover at least two or more contiguous 2.16 GHz channels whereas channel aggregation allows a separate waveform for each aggregated channel [111, 136].

IEEE 802.15.3c This standard defines the physical and MAC layers for the indoor 60 GHz WPANs. In this standard, the MAC implements a random channel access and time division multiple access approaches to support the directional and the quasi-omnidirectional transmissions [135].

ECMA-387 This standard proposed by the European computer manufacturers association (ECMA) specifies the physical layer, the MAC layer and the high-definition multimedia interface (HDMI) protocol adaptation layer (PAL) for the 60 GHz wireless networks. The ECMA-387 standards can be applied to the handheld devices used for low data rate transfer at short distances and to the devices equipped with adaptive antennas used for high data rate multimedia streaming at longer distances [135].

5G NR mmWave Standard This is a global standard platform for the 5G wireless air interfaces connecting mobile devices to the 5G base stations. Together with the efforts of IMT2020, the 3GPP Release 15 gave the first set of standards detailing 5G NR use cases, broadly categorized as the eMBB, the uRLLC and the mMTC. However, Release 15 is solely dedicated to the non-standalone (NSA) operation of 5G NR in which 4G LTE networks and 5G mobile technology co-exist [137]. The 3GPP Release 16 (completed in July 2020) targeted new enhancements for the better performance of standalone (SA) 5G NR networks (operating in the range of 1–52.6 GHz) in terms of increased capacity, improved reliability, reduced latency, better coverage, easier deployment, power requirements, and mobility. It also includes discussions on multi-beam management, over the air (OTA) synchronization to support multi-hop backhauling, integrated access and backhaul (IAB) enhancements, remote interference management (RIM) reference signals, UE power savings and mobility enhancements [137]. The next release is expected to be completed in September 2021, which will address the further enhancements for the 5G NR. It will include the support for new services like critical medical applications, NR broadcast, multicast and multi SIM devices, mission critical applications, cyber security applications, and dynamic spectrum sharing improvements, to name a few [137].

114 S. Tripathi et al.

6.2 Standardization Efforts for THz Communications

Given that the THz communications is still in its nascent phase, its standardization efforts are just beginning. The IEEE 802.15.3d-2017 was proposed in 2017, which is the first standard for THz fixed point-to-point links operating at carrier frequencies between 252 and 321 GHz using eight different channel bandwidths (2.16 GHz to 69.12 GHz and multiples of 2.16 GHz). For the development of nano-network standards at THz frequencies, *IEEE P1906.1/Draft 1.0* discusses recommended practices for nano-scale and molecular communication frameworks [19, 138, 139].

7 Conclusion

The main claim to fame for the 5G communications systems is to demonstrate that mmWave frequencies can be efficiently used for commercial wireless communications systems, which until only a few years back was considered unrealistic because of the unfavorable propagation characteristics of these frequencies. Even though the 5G systems are still being rolled out, it is argued that the gigabit-per-second rates to be supported by the 5G mmWave systems may fall short in supporting many emerging applications, such as 3D gaming and extended reality. Such applications will require several hundreds of gigabits per second to several terabits per second data rates with low latency and high reliability, which are currently considered to be the design goals for the next generation 6G communications systems. Given the potential of THz communications systems to provide such rates over short distances, they are currently considered to be the next frontier for wireless communications research. Given the importance of both mmWave and THz bands in 6G and beyond systems, this chapter has provided a unified treatment of these bands with particular emphasis on their propagation characteristics, channel models, design and implementation considerations, and potential applications. The chapter was concluded with a brief survey of the current standardization activities for these bands.

References

1. T.S. Rappaport, S. Sun, R. Mayzus, H. Zhao, Y. Azar, K. Wang, G.N. Wong, J.K. Schulz, M. Samimi, F. Gutierrez, Millimeter wave mobile communications for 5G cellular: It will work!. IEEE Access 1, 335–349 (2013)
2. H. Sarieddeen, N. Saeed, T.Y. Al-Naffouri, M.-S. Alouini, Next generation terahertz communications: A rendezvous of sensing, imaging, and localization. IEEE Commun. Mag. 58(5), 69–75 (2020)
3. F. Khan, Z. Pi, S. Rajagopal, Millimeter-wave mobile broadband with large scale spatial processing for 5G mobile communication, in *2012 50th Annual Allerton Conference on Communication, Control, and Computing (Allerton)* (IEEE, 2012), pp. 1517–1523

4. K.L. Lueth, *IoT 2019 in Review: The 10 Most Relevant IoT Developments of the Year* (2020)
5. International Telecommunication Union, IMT traffic estimates for the years 2020 to 2030. Report ITU, pp. 2370–0 (2015)
6. N.M. Karie, N.M. Sahri, P. Haskell-Dowland, IoT threat detection advances, challenges and future directions, in *2020 Workshop on Emerging Technologies for Security in IoT* (IEEE, 2020), pp. 22–29
7. H.S. Dhillon, H. Huang, H. Viswanathan, Wide-area wireless communication challenges for the internet of things. IEEE Commun. Mag. **55**(2), 168–174 (2017)
8. A.K. Gupta, A. Banerjee, Spectrum above radio bands, *Spectrum Sharing: The Next Frontier in Wireless Networks*, pp. 75–96 (2020)
9. J.G. Andrews, S. Buzzi, W. Choi, S.V. Hanly, A. Lozano, A.C. Soong, J.C. Zhang, What will 5G be? IEEE J. Sel. Areas Commun. **32**(6), 1065–1082 (2014)
10. A.K. Gupta, N.V. Sabu, H.S. Dhillon, Fundamentals of network densification, in *5G and Beyond: Fundamentals and Standards* (Springer, 2020)
11. J.G. Andrews, X. Zhang, G.D. Durgin, A.K. Gupta, Are we approaching the fundamental limits of wireless network densification? IEEE Commun. Mag. **54**, 184–190 (2016)
12. C. Han, Y. Wu, Z. Chen, X. Wang, *Terahertz Communications (TeraCom): Challenges and Impact on 6G Wireless Systems*, pp. 1–8 (2019). arXiv:1912.06040
13. W. Saad, M. Bennis, M. Chen, A vision of 6G wireless systems: Applications, trends, technologies, and open research problems. IEEE Network **34**(3), 134–142 (2019)
14. J.G. Andrews, T. Bai, M.N. Kulkarni, A. Alkhateeb, A.K. Gupta, R.W. Heath, Modeling and analyzing millimeter wave cellular systems. IEEE Trans. Commun. **65**(1), 403–430 (2016)
15. I. Akyildiz, J. Jornet, The internet of nano-things. IEEE Wireless Commun. **17**(6), 58–63 (2010)
16. K. Yang, D. Bi, Y. Deng, R. Zhang, M.M.U. Rahman, N.A. Ali, M.A. Imran, J.M. Jornet, Q.H. Abbasi, A. Alomainy, A comprehensive survey on hybrid communication in context of molecular communication and terahertz communication for body-centric nanonetworks. IEEE Trans. Mol. Biol. Multi-Scale Commun., 1–1 (2020)
17. N.V. Sabu, A.K. Gupta, Analysis of diffusion based molecular communication with multiple transmitters having individual random information bits. IEEE Trans. Mol. Biol. Multi-Scale Commun. **5**(3), 176–188 (2019)
18. J. ITU, Provisional final acts, in *World Radiocommunication Conference 2019* (ITU Publications, 2019)
19. T. Kurner, A. Hirata, On the impact of the results of WRC 2019 on THz communications, in *Third International Workshop on Mobile Terahertz System* (IEEE, 2020), pp. 1–3
20. P.F. Acts, in *World Radiocommunication Conference (WRC-15)* (ITU, 2015)
21. E. Dahlman, S. Parkvall, J. Skold, *5G NR: The Next Generation Wireless Access Technology* (Academic Press, 2020)
22. F. Boccardi, R.W. Heath, A. Lozano, T.L. Marzetta, P. Popovski, Five disruptive technology directions for 5G. IEEE Commun. Mag. **52**(2), 74–80 (2014)
23. I.A. Hemadeh, K. Satyanarayana, M. El-Hajjar, L. Hanzo, Millimeter-wave communications: Physical channel models, design considerations, antenna constructions, and link-budget. IEEE Commun. Surv. Tutorials **20**(2), 870–913 (2017)
24. S. Rangan, T.S. Rappaport, E. Erkip, Millimeter-wave cellular wireless networks: Potentials and challenges. Proc. IEEE **102**(3), 366–385 (2014)
25. V. Petrov, M. Komarov, D. Moltchanov, J.M. Jornet, Y. Koucheryavy, Interference and SINR in millimeter wave and terahertz communication systems with blocking and directional antennas. IEEE Trans. Wireless Commun. **16**(3), 1791–1808 (2017)
26. M.L. Attiah, A.A.M. Isa, Z. Zakaria, M. Abdulhameed, M.K. Mohsen, I. Ali, A survey of mmWave user association mechanisms and spectrum sharing approaches: An overview, open issues and challenges, future research trends. J. Wireless Netw. **26**(4), 2487–2514 (2020)
27. M. Marcus, B. Pattan, Millimeter wave propagation: Spectrum management implications. IEEE Microwave Mag. **6**(2), 54–62 (2005)

28. S. Tripathi, A. Gupta, Measurement efforts at mmWave indoor and outdoor environments. [Online]. Available: https://home.iitk.ac.in/~gkrabhi/memmwave
29. A. Tharek, J. McGeehan, Propagation and bit error rate measurements within buildings in the millimeter wave band about 60 GHz, in *8th European Conference on Electrotechnics, Conference Proceedings on Area Communication* (IEEE, 1988), pp. 318–321
30. G. Allen, A. Hammoudeh, 60 GHz propagation measurements within a building, in *1990 20th European Microwave Conference*, vol. 2 (IEEE, 1990), pp. 1431–1436
31. G. Allen, A. Hammoudeh, Frequency diversity propagation measurements for an indoor 60 GHz mobile radio link, in *1991 Seventh International Conference on Antennas and Propagation* (IET, 1991), pp. 298–301
32. R. Davies, M. Bensebti, M. Beach, J. McGeehan, Wireless propagation measurements in indoor multipath environments at 1.7 GHz and 60 GHz for small cell systems, in *[1991 Proceedings] 41st IEEE Vehicular Technology Conference* (IEEE, 1991), pp. 589–593
33. H. Yang, M.H. Herben, P.F. Smulders, Impact of antenna pattern and reflective environment on 60 GHz indoor radio channel characteristics. IEEE Antennas Wireless Propag. Lett. **4**, 300–303 (2005)
34. G. Allen, A. Hammoudeh, Outdoor narrow band characterisation of millimetre wave mobile radio signals, in *IEEE Colloquium on Radiocommunications in the Range 30–60 GHz* (IET, 1991), pp. 4–1
35. N. Daniele, D. Chagnot, C. Fort, Outdoor millimetre-wave propagation measurements with line of sight obstructed by natural elements. Electronics Letters **30**(18), 1533–1534 (1994)
36. F. Wang, K. Sarabandi, An enhanced millimeter-wave foliage propagation model. IEEE Trans. Antennas Propag. **53**(7), 2138–2145 (2005)
37. B. Fong, A. Fong, G. Hong, H. Ryu, Measurement of attenuation and phase on 26-GHz wideband point-to-multipoint signals under the influence of rain. IEEE Antennas Wireless Propag. Lett. **4**, 20–21 (2005)
38. E. Ben-Dor, T.S. Rappaport, Y. Qiao, S.J. Lauffenburger, Millimeter-wave 60 GHz outdoor and vehicle AOA propagation measurements using a broadband channel sounder, in *2011 IEEE Global Telecommunications Conference-GLOBECOM 2011* (IEEE, 2011), pp. 1–6
39. T.S. Rappaport, F. Gutierrez, E. Ben-Dor, J.N. Murdock, Y. Qiao, J.I. Tamir, Broadband millimeter-wave propagation measurements and models using adaptive-beam antennas for outdoor urban cellular communications. IEEE Trans. Antennas Propag. **61**(4), 1850–1859 (2012)
40. A. Hirata, H. Takahashi, J. Takeuchi, N. Kukutsu, D. Kim, J. Hirokawa, 120-GHz-band antenna technologies for over-10-Gbps wireless data transmission, in *2012 6th European Conference on Antennas and Propagation (EUCAP)*, pp. 2564–2568 (2012)
41. W. Keusgen, R.J. Weiler, M. Peter, M. Wisotzki, B. Göktepe, Propagation measurements and simulations for millimeter-wave mobile access in a busy urban environment, in *2014 39th International Conference on Infrared, Millimeter, and Terahertz waves (IRMMW-THz)* (IEEE, 2014), pp. 1–3
42. R.J. Weiler, M. Peter, W. Keusgen, M. Wisotzki, Measuring the busy urban 60 GHz outdoor access radio channel, in *2014 IEEE International Conference on Ultra-WideBand (ICUWB)* (IEEE, 2014), pp. 166–170
43. K. Guan, B. Ai, B. Peng, D. He, G. Li, J. Yang, Z. Zhong, T. Kürner, Towards realistic high-speed train channels at 5G millimeter-wave band - part I: Paradigm, significance analysis, and scenario reconstruction. IEEE Trans. Veh. Technol. **67**(10), 9112–9128 (2018)
44. C.U. Bas, R. Wang, S. Sangodoyin, T. Choi, S. Hur, K. Whang, J. Park, C.J. Zhang, A.F. Molisch, Outdoor to indoor propagation channel measurements at 28 GHz. IEEE Trans. Wireless Commun. **18**(3), 1477–1489 (2019)
45. J. Ma, R. Shrestha, L. Moeller, D.M. Mittleman, Invited article: Channel performance for indoor and outdoor terahertz wireless links. APL Photonics **3**(5), 1–12 (2018)
46. J.F. Federici, J. Ma, L. Moeller, Review of weather impact on outdoor terahertz wireless communication links. Nano Commun. Netw. **10**, 13–26 (2016)

47. S. Priebe, D.M. Britz, M. Jacob, S. Sarkozy, K.M.K.H. Leong, J.E. Logan, B.S. Gorospe, T. Kurner, Interference investigations of active communications and passive earth exploration services in the THz frequency range. IEEE Trans. Terahertz Sci. Technol. **2**(5), 525–537 (2012)
48. B. Heile, ITU-R liaison request RE: active services in the band above 275 GHz, IEEE standard 802.15-14-439-00-0THz (2015)
49. K. Guan, B. Peng, D. He, J.M. Eckhardt, S. Rey, B. Ai, Z. Zhong, T. Kurner, Channel characterization for intra-wagon communication at 60 and 300 GHz bands. IEEE Trans. Veh. Technol. **68**(6), 5193–5207 (2019)
50. J.M. Jornet, I.F. Akyildiz, Channel modeling and capacity analysis for electromagnetic wireless nanonetworks in the terahertz band. IEEE Trans. Wireless Commun. **10**(10), 3211–3221 (2011)
51. T.S. Rappaport, Y. Xing, O. Kanhere, S. Ju, A. Madanayake, S. Mandal, A. Alkhateeb, G.C. Trichopoulos, Wireless communications and applications above 100 GHz: Opportunities and challenges for 6G and beyond. IEEE Access **7**, 78,729–78,757 (2019)
52. J. Kokkoniemi, J. Lehtomäki, M. Juntti, A discussion on molecular absorption noise in the terahertz band. Nano Commun. Netw. **8**, 35–45 (2016)
53. A.H. Lettington, I.M. Blankson, M.F. Attia, D. Dunn, Review of imaging architecture. Infrared Passive Millimeter Wave Imag. Syst. Des. Anal. Modell. Test. **4719**, 327 (2002)
54. K. Haneda, J. Zhang, L. Tan, G. Liu, Y. Zheng, H. Asplund, J. Li, Y. Wang, D. Steer, C. Li, et al., 5G 3GPP-like channel models for outdoor urban microcellular and macrocellular environments, in *2016 IEEE 83rd Vehicular Technology Conference (VTC Spring)* (IEEE, 2016), pp. 1–7
55. M.K. Samimi, T.S. Rappaport, G.R. MacCartney, Probabilistic omnidirectional path loss models for millimeter-wave outdoor communications. IEEE Wireless Commun. Lett. **4**(4), 357–360 (2015)
56. T. Bai, R. Vaze, R.W. Heath, Analysis of blockage effects on urban cellular networks. IEEE Trans. Wireless Commun. **13**(9), 5070–5083 (2014)
57. T. Bai, R.W. Heath, Coverage and rate analysis for millimeter-wave cellular networks. IEEE Trans. Wireless Commun. **14**(2), 1100–1114 (2014)
58. M. Gapeyenko, A. Samuylov, M. Gerasimenko, D. Moltchanov, S. Singh, E. Aryafar, S. Yeh, N. Himayat, S. Andreev, Y. Koucheryavy, Analysis of human-body blockage in urban millimeter-wave cellular communications, in *2016 IEEE International Conference on Communications (ICC)* (IEEE, 2016), pp. 1–7
59. K. Venugopal, R.W. Heath, Millimeter wave networked wearables in dense indoor environments. IEEE Access **4**, 1205–1221 (2016)
60. T. Bai, R.W. Heath, Analysis of self-body blocking effects in millimeter wave cellular networks, in *2014 48th Asilomar Conference on Signals, Systems and Computers* (IEEE, 2014), pp. 1921–1925
61. S. Ju, S.H.A. Shah, M.A. Javed, J. Li, G. Palteru, J. Robin, Y. Xing, O. Kanhere, T.S. Rappaport, Scattering mechanisms and modeling for terahertz wireless communications, in *Proceedings of ICC* (IEEE, 2019), pp. 1–7
62. J. Jarvelainen, K. Haneda, M. Kyro, V.-M. Kolmonen, J.-i. Takada, H. Hagiwara, 60 GHz radio wave propagation prediction in a hospital environment using an accurate room structural model, in *2012 Loughbrgh. Antennas and Propagation Conference* (IEEE, 2012), pp. 1–4
63. V. Degli-Esposti, F. Fuschini, E.M. Vitucci, G. Falciasecca, Measurement and modelling of scattering from buildings. IEEE Trans. Antennas Propag. **55**(1), 143–153 (2007)
64. M.N. Kulkarni, E. Visotsky, J.G. Andrews, Correction factor for analysis of MIMO wireless networks with highly directional beamforming. IEEE Wireless Commun. Lett. **7**(5), 756–759 (2018)
65. J. Kokkoniemi, P. Rintanen, J. Lehtomaki, M. Juntti, Diffraction effects in terahertz band - Measurements and analysis, in *Proceedings in GLOBECOM* (IEEE, 2016), pp. 1–6
66. J. Federici, L. Moeller, Review of terahertz and subterahertz wireless communications. J. Appl. Phys. **107**(11), 1–23 (2010)

67. L. Bao, H. Zhao, G. Zheng, X. Ren, Scintillation of THz transmission by atmospheric turbulence near the ground, in *Fifth International Conference on Advanced Computational Intelligence* (IEEE, 2012), pp. 932–936
68. C.A. Balanis, *Antenna Theory: Analysis and Design* (Wiley, 2016)
69. X. Yu, J. Zhang, M. Haenggi, K.B. Letaief, Coverage analysis for millimeter wave networks: The impact of directional antenna arrays. IEEE J. Sel. Areas Commun. **35**(7), 1498–1512 (2017)
70. W. Lu, M. Di Renzo, Stochastic geometry modeling of cellular networks: Analysis, simulation and experimental validation, in *Proceedings of the 18th ACM International Conference on Modeling, Analysis and Simulation of Wireless and Mobile Systems*, pp. 179–188 (2015)
71. M. Di Renzo, W. Lu, P. Guan, The intensity matching approach: A tractable stochastic geometry approximation to system-level analysis of cellular networks. IEEE Trans. Wireless Commun. **15**(9), 5963–5983 (2016)
72. A. Maltsev, A. Pudeyev, I. Bolotin, G. Morozov, I. Karls, M. Faerber, I. Siaud, A. Ulmer-Moll, J. Conrat, R. Weiler, et al., MiWEBA D5.1: Channel modeling and characterization. Tech. Rep. (2014)
73. A. Thornburg, R.W. Heath, Ergodic capacity in mmWave Ad Hoc network with imperfect beam alignment, in *MILCOM 2015-2015 IEEE Military Communications Conference* (IEEE, 2015), pp. 1479–1484
74. N. Deng, M. Haenggi, A novel approximate antenna pattern for directional antenna arrays. IEEE Wireless Commun. Lett. **7**(5), 832–835 (2018)
75. Z. Chen, X. Ma, B. Zhang, Y. Zhang, Z. Niu, N. Kuang, W. Chen, L. Li, S. Li, A survey on terahertz communications. China Communication **16**(2), 1–35 (2019)
76. M.A. Jamshed, A. Nauman, M.A.B. Abbasi, S.W. Kim, Antenna selection and designing for THz applications: Suitability and performance evaluation: A survey. IEEE Access **8**, 113,246–113,261 (2020)
77. Y. He, Y. Chen, L. Zhang, S.-W. Wong, Z.N. Chen, An overview of terahertz antennas. China Communication **17**(7), 124–165 (2020)
78. S. Sun, G.R. MacCartney, T.S. Rappaport, A novel millimeter-wave channel simulator and applications for 5G wireless communications, in *2017 IEEE International Conference on Communications (ICC)* (IEEE, 2017), pp. 1–7
79. A. Faisal, H. Sarieddeen, H. Dahrouj, T.Y. Al-Naffouri, M.-S. Alouini, Ultramassive MIMO systems at terahertz bands: Prospects and challenges. IEEE Veh. Technol. Mag. **15**(4), 33–42 (2020)
80. K. Tekbiyik, A.R. Ekti, G.K. Kurt, A. Gorcin, Terahertz band communication systems: Challenges, novelties and standardization efforts. J. Phys. Commun. **35**, 53–62 (2019)
81. C. Han, A.O. Bicen, I.F. Akyildiz, Multi-ray channel modeling and wideband characterization for wireless communications in the terahertz band. IEEE Trans. Wireless Commun. **14**(5), 2402–2412 (2015)
82. S. Priebe, M. Kannicht, M. Jacob, T. Kurner, Ultra broadband indoor channel measurements and calibrated ray tracing propagation modeling at THz frequencies. J. Commun. Netw. **15**(6), 547–558 (2013)
83. A. Moldovan, M.A. Ruder, I.F. Akyildiz, W.H. Gerstacker, LOS and NLOS channel modeling for terahertz wireless communication with scattered rays, in *IEEE GLOBECOM Workshop* (IEEE, 2014), pp. 388–392
84. Z. Hossain, C. Mollica, J.M. Jornet, Stochastic multipath channel modeling and power delay profile analysis for terahertz-band communication, in *Proceedings of the 4th ACM International Conference on Nanoscale Computing and Communication*, ser. NanoCom '17 (Association for Computing Machinery, New York, NY, USA, 2017). [Online]. Available: https://doi.org/10.1145/3109453.3109473
85. S. Priebe, T. Kurner, Stochastic modeling of THz indoor radio channels. IEEE Trans. Wireless Commun. **12**(9), 4445–4455 (2013)
86. S. Kim, A. Zajic, Statistical modeling of THz scatter channels, in *Ninth European Conference on Antennas and Propagation* (2015)

87. S. Kim, A. Zajic, Statistical modeling and simulation of short-range device-to-device communication channels at sub-THz frequencies. IEEE Trans. Wireless Commun. **15**(9), 6423–6433 (2016)
88. H. Elayan, R.M. Shubair, J.M. Jornet, P. Johari, Terahertz channel model and link budget analysis for intrabody nanoscale communication. IEEE Trans. Nanobioscience **16**(6), 491–503 (2017)
89. H. Elayan, C. Stefanini, R.M. Shubair, J.M. Jornet, End-to-end noise model for intra-body terahertz nanoscale communication. IEEE Trans. Nanobioscience **17**(4), 464–473 (2018)
90. C.H. Y. Chen, Channel modeling and analysis for wireless networks-on-chip communications in the millimeter wave and terahertz bands, in *Proceedings of INFOCOM*, pp. 651–656 (2018)
91. R.W. Heath, N. Gonzalez-Prelcic, S. Rangan, W. Roh, A.M. Sayeed, An overview of signal processing techniques for millimeter wave MIMO systems. IEEE J. Sel. Top. Sig. Process. **10**(3), 436–453 (2016)
92. M.N. Kulkarni, A. Ghosh, J.G. Andrews, A comparison of MIMO techniques in downlink millimeter wave cellular networks with hybrid beamforming. IEEE Trans. Commun. **64**(5), 1952–1967 (2016)
93. J. Kokkoniemi, J. Lehtomaki, M. Juntti, Stochastic geometry analysis for mean interference power and outage probability in THz networks. IEEE Trans. Wireless Commun. **16**(5), 3017–3028 (2017)
94. J. Kokkoniemi, J. Lehtomaeki, M. Juntti, Stochastic geometry analysis for band-limited terahertz band communications. IEEE Veh. Technol. Conf., 1–5 (2018)
95. Z. Pi, F. Khan, An introduction to millimeter-wave mobile broadband systems. IEEE Commun. Mag. **49**(6), 101–107 (2011)
96. R. Sun, P.B. Papazian, J. Senic, C. Gentile, K.A. Remley, in *Angle- and Delay-Dispersion Characteristics in a Hallway and Lobby at 60 GHz* (2018)
97. R. Sun, C.A. Gentile, J. Senic, P. Vouras, P.B. Papazian, N.T. Golmie, K.A. Remley, Millimeter-wave radio channels vs. synthetic beamwidth. IEEE Commun. Mag. **56**(12), 53–59 (2018)
98. N. Yang, L. Wang, G. Geraci, M. Elkashlan, J. Yuan, M. Di Renzo, Safeguarding 5G wireless communication networks using physical layer security. IEEE Commun. Mag. **53**(4), 20–27 (2015)
99. C. Wang, H.-M. Wang, Physical layer security in millimeter wave cellular networks. IEEE Trans. Wireless Commun. **15**(8), 5569–5585 (2016)
100. Y. Zhu, L. Wang, K.-K. Wong, R.W. Heath, Secure communications in millimeter wave Ad Hoc networks. IEEE Trans. Wireless Commun. **16**(5), 3205–3217 (2017)
101. Q. Xue, P. Zhou, X. Fang, M. Xiao, Performance analysis of interference and eavesdropping immunity in narrow beam mmWave networks. IEEE Access **6**, 67,611–67,624 (2018)
102. A.K. Gupta, J.G. Andrews, R.W. Heath, Macrodiversity in cellular networks with random blockages. IEEE Trans. Wireless Commun. **17**(2), 996–1010 (2017)
103. I.K. Jain, R. Kumar, S.S. Panwar, The impact of mobile blockers on millimeter wave cellular systems. IEEE J. Sel. Areas Commun. **37**(4), 854–868 (2019)
104. Y. Zhu, Q. Zhang, Z. Niu, J. Zhu, Leveraging multi-AP diversity for transmission resilience in wireless networks: Architecture and performance analysis. IEEE Trans. Wireless Commun. **8**(10), 5030–5040 (2009)
105. M. Giordani, M. Polese, A. Roy, D. Castor, M. Zorzi, Standalone and non-standalone beam management for 3GPP NR at mmWaves. IEEE Commun. Mag. **57**(4), 123–129 (2019)
106. Y. Niu, Y. Li, D. Jin, L. Su, A.V. Vasilakos, A survey of millimeter wave (mmWave) communications for 5G: Opportunities and challenges. J. Wireless Netw. **21**(8), 2657–2676 (2015)
107. A.K. Gupta, J.G. Andrews, R.W. Heath, On the feasibility of sharing spectrum licenses in mmWave cellular systems. IEEE Trans. Commun. **64**, 3981–3995 (2016)
108. A.K. Gupta, A. Alkhateeb, J.G. Andrews, R.W. Heath, Gains of restricted secondary licensing in millimeter wave cellular systems. IEEE J. Sel. Areas Commun. **34**(11), 2935–2950 (2016)

109. A. Alkhateeb, S. Alex, P. Varkey, Y. Li, Q. Qu, D. Tujkovic, Deep learning coordinated beamforming for highly-mobile millimeter wave systems. IEEE Access **6**, 37,328–37,348 (2018)
110. T. Nitsche, C. Cordeiro, A.B. Flores, E.W. Knightly, E. Perahia, J.C. Widmer, IEEE 802.11 ad: Directional 60 GHz communication for multi-Gigabit-per-second Wi-Fi. IEEE Commun. Mag. **52**(12), 132–141 (2014)
111. Y. Ghasempour, C.R. da Silva, C. Cordeiro, E.W. Knightly, IEEE 802.11 ay: Next-generation 60 GHz communication for 100 Gb/s Wi-Fi. IEEE Commun. Mag. **55**(12), 186–192 (2017)
112. Y. Liu, Y. Jian, R. Sivakumar, D.M. Blough, On the potential benefits of mobile access points in mmWave wireless LANs, in *2020 IEEE International Symposium on Local and Metropolitan Area Networks (LANMAN)* (IEEE, 2020), pp. 1–6
113. K. Aldubaikhy, W. Wu, N. Zhang, N. Cheng, X.S. Shen, Mmwave IEEE 802.11 ay for 5G fixed wireless access. IEEE Wireless Commun. **27**(2), 88–95 (2020)
114. H.S. Dhillon, G. Caire, Wireless backhaul networks: Capacity bound, scalability analysis and design guidelines. IEEE Trans. Wireless Commun. **14**(11), 6043–6056 (2015)
115. C. Saha, M. Afshang, H.S. Dhillon, Bandwidth partitioning and downlink analysis in millimeter wave integrated access and backhaul for 5G. IEEE Trans. Wireless Commun. **17**(12), 8195–8210 (2018)
116. C. Saha, H.S. Dhillon, Millimeter wave integrated access and backhaul in 5G: Performance analysis and design insights. IEEE J. Sel. Areas Commun. **37**(12), 2669–2684 (2019)
117. R. Lombardi, Wireless backhaul for IMT 2020 / 5G: Overview and introduction, in *In Proceedings of the Workshop on Evolution of Fixed Service in Backhaul Support of IMT 2020/5G, Geneva, Switzerland*, 29 (2019)
118. M. Jaber, M.A. Imran, R. Tafazolli, A. Tukmanov, 5G backhaul challenges and emerging research directions: A survey. IEEE Access **4**, 1743–1766 (2016)
119. T.S. Rappaport, R.W. Heath Jr., R.C. Daniels, J.N. Murdock, *Millimeter Wave Wireless Communications* (Pearson Education, 2015)
120. S. Barberis, D. Disco, R. Vallauri, T. Tomura, J. Hirokawa, Millimeter wave antenna for information shower: Design choices and performance, in *2019 European Conference on Networks and Communications (EuCNC)* (IEEE, 2019), pp. 128–132
121. S. Jaswal, D. Yadav, D.P. Bhatt, M. Tiwari, in *MmWave Technology: An Impetus for Smart City Initiatives* (IET, 2019)
122. H.S. Dhillon, V.V. Chetlur, *Poisson Line Cox Process: Foundations and Applications to Vehicular Networks* (Morgan & Claypool, 2020)
123. P. Kumari, N. Gonzalez-Prelcic, R.W. Heath, Investigating the IEEE 802.11 ad standard for millimeter wave automotive radar, in *2015 IEEE 82nd Vehicular Technology Conference (VTC2015-Fall)* (IEEE, 2015), pp. 1–5
124. J. Hasch, E. Topak, R. Schnabel, T. Zwick, R. Weigel, C. Waldschmidt, Millimeter-wave technology for automotive radar sensors in the 77 GHz frequency band. IEEE Trans. Microwave Theory Tech. **60**(3), 845–860 (2012)
125. Y. Han, E. Ekici, H. Kremo, O. Altintas, Automotive radar and communications sharing of the 79-GHz band, in *Proceedings of the First ACM International Workshop on Smart, Autonomous, and Connected Vehicular Systems and Services*, pp. 6–13 (2016)
126. V. Petrov, J. Kokkoniemi, D. Moltchanov, J. Lehtomäki, M. Juntti, Y. Koucheryavy, The impact of interference from the side lanes on mmWave/THz band V2V communication systems with directional antennas. IEEE Trans. Veh. Technol. **67**(6), 5028–5041 (2018)
127. V. Petrov, G. Fodor, J. Kokkoniemi, D. Moltchanov, J. Lehtomaki, S. Andreev, Y. Koucheryavy, M. Juntti, M. Valkama, On unified vehicular communications and radar sensing in millimeter-wave and low terahertz bands. IEEE Wireless Commun. **26**(3), 146–153 (2019)
128. I.F. Akyildiz, J.M. Jornet, C. Han, Terahertz band: Next frontier for wireless communications. J. Phys. Commun. **12**, 16–32 (2014)
129. R. Singh, D. Sicker, Parameter modeling for small-scale mobility in indoor THz communication, in *Proceedings of GLOBECOM* (IEEE, 2019), pp. 1–6

130. T. Kürner, S. Priebe, Towards THz communications - Status in research, standardization and regulation. J. Infrared Millimeter Terahertz Waves **35**(1), 53–62 (2014)
131. A. Saeed, O. Gurbuz, M.A. Akkas, Terahertz communications at various atmospheric altitudes. J. Phys. Commun. **41**, 101–113 (2020)
132. I. Rasheed, F. Hu, Intelligent super-fast vehicle-to-everything 5G communications with predictive switching between mmWave and THz links. Vehicular Communication, 100303 (2020)
133. I. Akyildiz, M. Pierobon, S. Balasubramaniam, Y. Koucheryavy, The internet of bio-nano things. IEEE Commun. Mag. **53**(3), 32–40 (2015)
134. S. Abadal, E. Alarcón, A. Cabellos-Aparicio, M. Lemme, M. Nemirovsky, Graphene-enabled wireless communication for massive multicore architectures. IEEE Commun. Mag. **51**(11), 137–143 (2013)
135. N. Al-Falahy, O.Y. Alani, Millimetre wave frequency band as a candidate spectrum for 5G network architecture: A survey. J. Phys. Commun. **32**, 120–144 (2019)
136. P. Zhou, K. Cheng, X. Han, X. Fang, Y. Fang, R. He, Y. Long, Y. Liu, IEEE 802.11 ay-based mmWave WLANs: Design challenges and solutions. IEEE Commun. Surv. Tutorials **20**(3), 1654–1681 (2018)
137. J. Peisa, P. Persson, S. Parkvall, E. Dahlman, A. Grovlen, C. Hoymann, D. Gerstenberger, 5G evolution: 3GPP releases 16 & 17 overview. Ericsson Technology Revisions **9**, 1–5 (2020)
138. IEEE P1906.1/Draft 1.0, *Recommended Practice for Nanoscale and Molecular Communication Framework* (2014)
139. H. Elayan, O. Amin, B. Shihada, R.M. Shubair, M.-s. Alouini, Terahertz band: The last piece of RF spectrum puzzle for communication systems. IEEE Open J. Commun. Soc. **1**, 1–32 (2020)

Chapter 7
Challenges in Transport Layer Design for Terahertz Communication-Based 6G Networks

Madhan Raj Kanagarathinam and Krishna M. Sivalingam

Abstract With the launch of 3GPP fifth-generation (5G) commercial cellular networks around the world, the research community has started focusing on the design of the sixth-generation (6G) system. One of the considerations is the use of Terahertz communications that aims to provide 1 Tbps (terabits per second) and air latency less than 100 μs. Further, 6G networks are expected to provide for more stringent Quality of Service (QoS) and mobility requirements. While addition to innovations at the physical layer and radio technologies can achieve these goals to a great extent, the end-to-end applications would still face challenges to fully utilize the network capacity due to limitations of the current transport layer protocols. In this chapter, we explore the challenges in the design of next-generation transport layer protocols (NGTP) in 6G Terahertz communication-based networks. Some of the challenges are due to user mobility, high-speed and high-bitrate communications, and other issues. The impact of these issues and potential approaches to mitigate these challenges are also discussed.

1 Introduction

Cellular communication technologies have evolved aggressively over the past decade. The Fifth Generation (5G) and beyond (B5G) networks are required to support massive connectivity, high reach-ability, ultra-low latency, and high data

M. R. Kanagarathinam
Mobile Communication R&D Department, Samsung R&D Institute India-Bangalore, Bangalore, India

Department of Computer Science and Engineering, Indian Institute of Technology Madras, Chennai, India
e-mail: madhan.raj@samsung.com; cs20s021@smail.iitm.ac.in

K. M. Sivalingam (✉)
Department of Computer Science and Engineering, Indian Institute of Technology Madras, Chennai, India
e-mail: skrishnam@cse.iitm.ac.in

© The Author(s), under exclusive license to Springer Nature Switzerland AG 2021
Y. Wu et al. (eds.), *6G Mobile Wireless Networks*, Computer Communications and Networks, https://doi.org/10.1007/978-3-030-72777-2_7

rates [1]. The usage of millimeter-wave (mmWave) communications has enabled higher data rates that 5G boasts today. However, it also results in challenges due to the high variability of channel quality. Substantial research and development have been directed towards the design of efficient physical (PHY) and Medium Access Control (MAC) protocol layers. However, the effect of these challenges on the transport and application layer protocols has been comparatively not explored. Transport Control Protocol (TCP) is a predominantly used protocol in the Internet protocol suite for several decades [2]. In TCP, the congestion control algorithm is sensitive to blockages and drastic data rate changes due to transitions from line-of-sight (LOS) to non-line-of-sight (NLOS) communications and vice-versa. It impacts the good-put during channel variations [3].

The European Telecommunications Standards Institute (ETSI) formed a working group Next Generation Protocol (NGP) to standardize the requirements of future Internet architecture [4]. The Industry Specification Group (ISG) on NGP sets its vision to create an efficient internet architecture with mobility, multi-homing, and multi-path solutions to efficiently cater to the use cases that will be enabled by 5G and beyond networks. TCP was developed more than 45 years ago [2] when the end hosts had static addresses and the links were wired. Today, wireless networking predominates end-user access and the pace with which varied types of devices are being connected to the Internet has increased enormously, leading to billions of connected devices, including machines and sensor nodes. At the same time, TCP has evolved all these years and remains to be the most used reliable transport layer protocol.

There are several major improvements in the transport layer protocols. The enhancements can be broadly categorized into three types: (a) Plug-in solutions, which are upgrades to the current TCP/IP stack, including Multipath TCP [5], Congestion control variants such as TCP BBR [6], TCP Vegas [7]; (b) Hybrid solutions that introduce a Non-IP layer network mapping and management, such as QUIC [8]; and (c) Clean-slate redesigned solutions, that are completely non-IP based. Information-Centric Networks (ICN) [9] and Delay tolerant networks (DTN) [10] can fall under this category. Though it provides more flexibility and advantages to the NextGen Transport Protocols, it needs substantial revisions in the Network entities to visualize the end-to-end advantage.

Most modern devices comprise multiple networking communication interfaces, which can boost throughput and reliability. It requires smooth operations during vertical and horizontal handover conditions. The signaling should not impact the end-to-end user performance. Various solutions have been proposed at different layers to ensure resilience to network handover. Multipath TCP (MPTCP) emerged as one of the promising paradigms to address the mobility issue. MPTCP increases the network capacity and reliability. It provides seamless fail-over by facilitating Multipath operation at the transport layer. MPTCP connection starts with an initial subflow, which is similar to a regular TCP connection. Multiple connections, via subflows, can be created using MPTCP on the available network interfaces. Data can flow over any of the active and capable subflows.

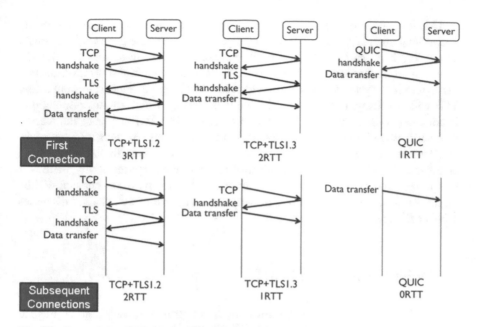

Fig. 7.1 Comparison of TCP (TLS 1.2 and TLS 1.3) with QUIC

TCP was primarily designed for providing reliable end-to-end sessions, while UDP was designed for real-time sessions. One of the major issues with TCP is the initial overhead in the three-way handshake protocol during connection establishment. An alternate to TCP, called Quick UDP Internet Connections (QUIC) [8], was developed and deployed by Google in 2012. QUIC is built on the top of User Datagram Protocol (UDP), making the revisions substantial to the NextGen Networks (NGN). It aims to reduce the connection establishment duration using a 0-RTT/1-RTT handshake. As shown in Fig. 7.1, QUIC uses one RTT for the initial connection and zero RTT for the subsequent connection. QUIC can support multipath and enhance the experience similar to Multipath TCP. Multipath QUIC is under evaluation for an extension over the IETF QUIC [11]. As of April 2020, QUIC accounts for 4.7% of the total website traffic volume, of which Google contributes 98% [12]. Though it improves the page loading time and improves security by coupling the TLS 1.3, the congestion and flow control algorithm are adapted from the traditional TCP protocol.

On the other hand, a Non-IP solution such as Information-Centric Networking (ICN) [13] can provide an additional advantage over the current IP suite. The legacy IP suite was designed around host-to-host applications such as FTP. The B5G/ 6G network is service access-based. ICN's name-based forwarding, multi-path solutions, and on-node dynamic caching can help the B5G/6G reduce the end-to-end latency. The layered network architecture of the current IP suite needs to be enhanced to reduce the processing delay, and to provide better Quality of Experience (QoE).

Despite these advances, TCP still has limitations that will critically impact 6G networking and the services that it aims to provide. One of the issues is the planned use of Terahertz communications in 6G networks, which introduces significantly new challenges that have to be effectively dealt with-in the Transport layer design. First, Terahertz systems enable link bandwidths of 1 Tbps and higher; to what extent TCP will effectively operate at this bitrate to deliver very high end-user throughput is not evident. Such high bitrates require more efficient TCP sequence numbering mechanisms and improved congestion control schemes. Second, 6G networks are expected to support mobility of up to 1000 Km/h. Due to the propagation characteristics of Terahertz communications such as blockage sensitivity, there will be sporadic and spurious losses on the wireless channel. Also, the channel will be highly variable resulting in remiss flow control. The rest of this chapter discusses these challenges in detail.

2 Mobility Challenges in TCP

The 6G KPIs (Key Performance Indicators) aim to achieve a very high user speed with mobility requirement of up to 1000 Km/h. This leads to specific challenges as described below.

2.1 Impact on Spurious Wireless Loss

The current TCP protocol interprets packet loss as a congestion event, by default. Due to Terahertz communication's propagation characteristics, such as blockage sensitivity leading to higher outages, there would be sporadic losses and multiple re-transmissions. Over the past few decades, various algorithms have been proposed for detecting and estimating congestion. These algorithms estimate the congestion based on checking the loss event (timeout, explicit notification, or duplicate acknowledgment) or calculating Round Trip Time (RTT) and detect the congestion if there is a dramatic increase in RTT. These algorithms are categorized as: (1) Loss-based (2) Delay-based and (3) Hybrid algorithm that considers both packet loss and queuing delay.

As shown in Fig. 7.2, momentary blockage loss can drastically degrade the overall throughput of the network. The conventional TCP congestion control algorithm, which uses Addictive increase and Multiplicative Decrease (AIMD) scheme, can reduce the bitrate by up to 72% relative to non-blocking idle upload.

The TCP congestion control window mechanism should be substantially revised to handle the traffic dynamics in the 6G Terahertz communication networks. The congestion control scheme should understand the Terahertz characteristics, thereby differentiating actual congestion and spurious wireless loss due to blockage or momentary outage.

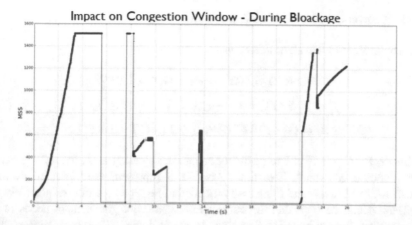

Fig. 7.2 Impact on Blockage: cubic congestion window variation with 1 s blockage

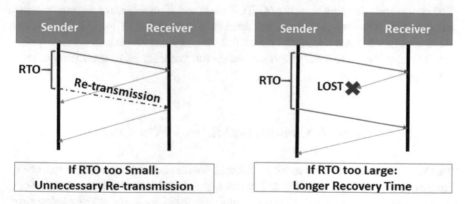

Fig. 7.3 Re-transmission timeout (RTO) selection

2.2 Inefficient Re-transmission Timers

Re-transmission Timeout (RTO) plays a vital role in the performance of the TCP. As shown in Fig. 7.3, if the RTO is too small, then it causes unnecessary re-transmissions, and if too large, it causes slow reactions to losses.

High scattering sensitivity of the Terahertz link causes fluctuations in the throughput and preventing it from achieving throughput exceeding 100 Gbps. The Conventional TCP protocol uses Karn and Jacobson's algorithm [14], which is inaccurate in fluctuating networks. The QUIC protocol also follows the traditional RTO algorithm [8].

2.2.1 Conventional TCP RTO Algorithm

The conventional RTO algorithm is written as:

$$RTO = SRTT + max(G, 4 \times RTTVAR)$$

$$SRTT_t = (1 - \alpha)SRTT_{t-1} + \alpha ERTT_{t-1} \tag{7.1}$$

$$RTTVAR_t = (1 - \beta)RTTVAR_{t-1} + \beta|SRTT - ERTT|$$

where G, RTTVAR, SRTT respectively denote the clock granularity, RTT variance and Smoothened Round Trip time. Also, α, β represent the weighted average constants. In Linux, α and β are recommended to be set as $\frac{1}{8}$ and $\frac{1}{4}$ respectively.

Thus, it can be seen that the current re-transmission mechanism needs to be reconsidered for the NextGen Transport Protocol design. There are some studies that have explored an adaptive timer based on the wireless conditions [15, 16]. The re-transmission timers in the 6G TCP must avoid unnecessary re-transmission using an efficient way to estimate the Smoothen Round Trip Time (SRTT) and the Re-transmission Time Out (RTO). For instance, a machine learning based timer prediction, could have a very high impact on the accuracy of re-transmission time estimation.

2.3 Highly Variable Channel and Remiss Flow Control

Terahertz communications can cause channel fluctuations due to frequent transition from Non-line of Sight (NLOS) to Line of sight (LOS) communications and vice-versa. If the user equipment (UE) sends multiple triple duplicate acknowledgments (TDA), the server would reduce the throughput drastically, in response. It would take multiple rounds to recover the previous maximum, due to TCP's fairness policy. A UE-driven intelligent flow control mechanism, which can control the congestion window of the server, without any changes in the server, can have a huge impact on the client side architecture. The sender might send the packet to both wireless and non-wireless clients. If the client are connected to 6G network, then the clients using the intelligent flow control mechanism can indirectly control the download capacity.

3 Achieving Tbps Bitrate in TCP

Considering the extremely high data rates (100 Gbps) in Terahertz communications, and super-low end-to-end latency (0.1 ms) requirements in 6G networks, simplifying interactions across network layers will be an important design principle. The following are important challenges in TCP to achieve very high data rates.

3.1 Wireless TCP Adaptive Congestion Control

The congestion control algorithm of TCP plays a vital role in the upload bitrate of the application in the client. The congestion control algorithm in the end-server impacts the download bitrate of all the applications in the client. Hence, in addition to bitrates, other parameters such as inter-fairness, intra-fairness are key parameter indices to measure the effectiveness of a congestion control algorithm.

A congestion control algorithm is said to be aggressive, if it increases the overall good-put by impacting the good-put of other connected clients. Hence, the congestion control algorithm has to be fair and efficient. To operate fairly, Addictive Increase Multiplicative Decrease (AIMD) algorithm was proposed and implemented in the TCP Reno congestion control algorithm [17, 18]. There have been numerous variants till now, each with its own advantages and shortcomes. A summary of popular TCP congestion Control algorithm is presented is Table 7.1.

Though there are numerous algorithms, TCP congestion control algorithm for 6G must be able to differentiate wireless congestion from packet loss. In the wireless environment, re-transmissions are not always due to congestion. There may be momentary wireless channel conditions leading to re-transmission. However, the TCP congestion control algorithm is not aware of the wireless channel medium. The 6G channel conditions, including the key choice of parameters such as signal to noise ratio, should be considered in TCP for achieving the latency goal of 0.1 ms.

The congestion control algorithm should estimate the currently available bandwidth and adaptively increase or decrease the Congestion Window ($CWND$), based on the wireless estimation. If the current bandwidth, BW_C, is less than the estimated bandwidth, BW_C, the CWND is increased aggressively to avoid under-utilization. If the BW_C is closer or above BW_C, the CWND is increased in additive steps to overcome packet loss. As shown in Fig. 7.4, the CWND should adapt as follows:

$$\text{For each RTT: CWND} \leftarrow \text{CWND} + \text{BWF} \qquad (7.2)$$

$$\text{For each Loss: CWND} \leftarrow \text{CWND} - \frac{\text{BWF}}{\text{CWND}} \qquad (7.3)$$

Here, BWF denotes a bandwidth factor that is determined by wireless conditions, based on which the congestion window is adaptively reduced or increased. There are many cross-layer congestion control algorithms proposed in literature [24–29]. However, these need to be suitably adapted and modified according to the 6G KPIs.

3.2 Multi-Core Awareness

As per the current core utilization, the maximum throughput supported by IP per core can not exceed more than 100 Gbps [30, 31]. Multiple streams have to be used from TCP or UDP to increase the overall speed. These parallel streams should

Table 7.1 Summary of Current TCP congestion control

Algo.	Ref.	Work summary	Objective(s)
TCP Reno	[17, 18]	• Congestion window increase by 1 in case of ACK	• AIMD congestion control
		• Reduce to half in case of 3 duplicate ACK's	• Fast Re-transmit and fast recovery in new reno
TCP Veno	[19]	• Monitors the congestion level and decide packet loss are owing to congestion or random bit error	
		• Refines the multiplicative decrease and linear increase of TCP Reno	• Efficient in handling random packet loss
TCP cubic	[20]	• Modifies linear window growth to a cubic function	• Default in Linux kernels between versions 2.6.19 and 3.2
			• Equal bandwidth allocations between flows with different RTTs
TCP vegas	[7]	• New time out mechanism operation	
		• Use RTT estimate to decide for re-transmission	• Proactive congestion detection
		• Overcomes the problem of duplicate ACKs	• Better throughput
FAST TCP	[21]	• Each source tries to keep a constant number of packets in the queue	• Congestion algorithm for high-speed long-latency network
		• Packets in queues are calculated using observed RTT and base RTT	
TCP westwood	[22, 23]	• Filter based bandwidth estimation based on ACK	• Handles large BDP connections
TCP-BBR	[6]	• BBR calculates a continuous estimate of the flow's RTT and bottleneck capacity	• Linux default congestion control algorithm
		• Estimates the bandwidth delay product and adapts accordingly	• BBR is efficient and fast, but its fairness to non-BBR streams is disputed

be mapped to different radio bearers (RB) to scale to 100 Gbps. For a single RB to support over 10 Gbps is a major design challenge. It would require protocol specification changes to support parallel flows per RB. Hence, NextGen Transport Protocols should be re-designed to be core-aware for achieving bitrates higher than 100 Gbps. The role of Multipath Protocols [5, 11, 32–34] could also be vital here. If the Multipath TCP Path Manager and Scheduler is built core-aware, then it can schedule the packet amongst the cores, in a fair and efficient paradigm.

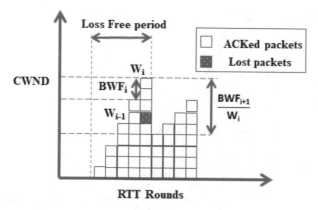

Fig. 7.4 Congestion window growth during every RTT round

4 Other Challenges in TCP

There are a few other important considerations that the NextGen Transport Protocol has to consider for 6G networks.

4.1 TCP Sequence Number Limits

The current TCP header uses a default 32-bit Sequence and Acknowledgement number field. The Sequence number changes from 0 to 2^{31} (assuming Selective Repeat option), and then the counter resets to 0 (the starting sequence number is usually a random number in this range). Table 7.2 outlines the time taken for sequence number wraparound with 32 bits.

In RFC 1323 [35], for a high-performance network, Protection Against Wrapped Sequence number (PAWS) algorithm is proposed. PAWS uses the TCP timestamp option to prevent old duplicates from the same connection. In a Terabit network, the 32-bit sequence number can wraparound within 16–160 ms. Hence, even PAWS would not be much efficient, since identical sequence numbers can contain the same timestamp. Hence, NextGen Transport Protocols should extend the TCP Sequence as part of TCP OPTION extension, which would improve the roll time up-to 795 days.

For the 6G networks and Beyond, at least a 64-bit TCP sequence number is required. The time for 64-bit sequence number to wraparound (assuming Selective Repeat option) is given as $\frac{(2^{63})}{B} > MSL(sec)$, where MSL denotes Maximum Segment Lifetime (usually taken to be 120 s). Table 7.3 presents the sequence number wraparound time for 6G and future networks, with 64-bit sequence numbers.

The sequence number extension can be done using either a Header Change or via TCP Options. Figure 7.5 shows a 64-bit SEQ and 64-bit ACK number in the

Table 7.2 Time to wraparound with a 32-Bit TCP sequence number

Type	Speed (bps)	Speed (Bps)	Time to wrap
ARPANET	56 Kbps	7 kBps	3.1 days
Giganet or 5G network	1 Gbps	125 MBps	16.384 s
TeraBit or 6G network	1 Tbps	125 GBps	0.016 s (16 ms)
TeraBit real-time	100 Gbps	12.5 GBps	0.16 s (160 ms)

Table 7.3 Time to rollover with a 64-Bit TCP sequence number

Type	Speed (bps)	Speed (Bps)	Time to wrap
Terabit network end user	100 Gbps	12.5 GBps	7950 days
Terabit or 6G network	1 Tbps	125 GBps	795 days
Petabit network	1024 Tbps	128 TBps	18.66 h

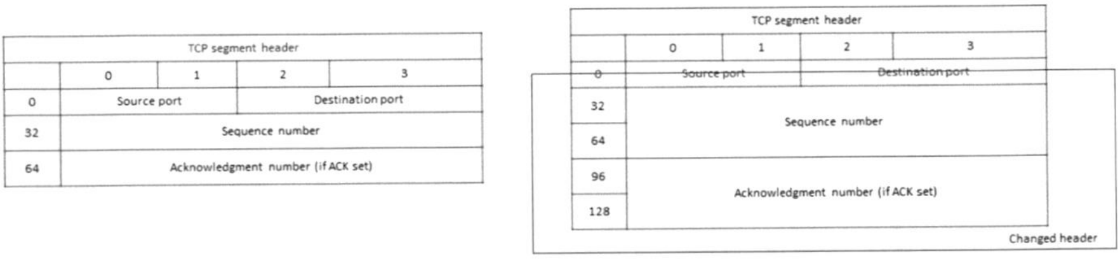

Fig. 7.5 TCP header change to support 64-bit SEQ/ACK number field

TCP header. The traditional TCP header might require a change, as the demand for the speed increases. However, changing the TCP Header is not as simple as it sounds. This method would lead to backward compatibility issues. The old 32-bit SEQ number might intercept the Least Significant 32 bits of SEQ number field as ACK number. Hence, an overall change in the network is required. Not just the client and the sender, but also the nodes in middles (middlebox), look into the TCP SEQ/ACK number for certain operations.

TCP Options is a more extensible method and is backward compatible. Figure 7.6 presents the method for extension of SEQ/ ACK number using the TCP OPTIONS field.

4.2 TCP Buffer Tuning

TCP behavior can be tuned from user-space without any modifications in TCP/IP kernel stack. One of the important parameters is the TCP buffer size. The send-receive buffer size of TCP comprises of three parameters. The first parameter is used to set the minimum size of the send-receive buffer of the TCP socket. The second parameter defines the default size of the send-receive buffer of the TCP socket. The third parameter controls the maximum buffer size per each socket.

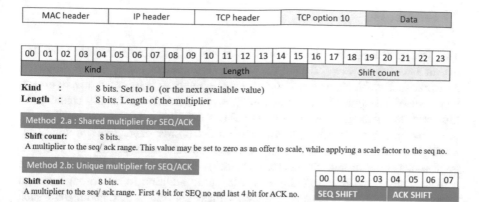

Fig. 7.6 Extension via TCP option to support 64b SEQ/ACK numbers

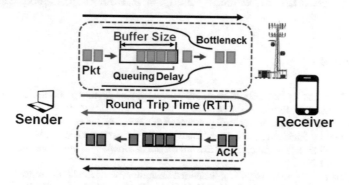

Fig. 7.7 TCP buffer queuing mechanism

The last parameter, the maximum buffer size, plays a critical role in the maximum achievable throughput. As shown in Fig. 7.7, the buffer size depends on the Bandwidth and the Round Trip Time. Hence, the Maximum value is set based on the Bandwidth Delay Product (BDP). The relationship between the maximum receive window (RWIN) and the throughput is given as the following:

$$\text{Throughput} \leq \frac{\text{RWIN}}{\text{RTT}} \tag{7.4}$$

The Linux kernel uses the Dynamic Right-Sizing (DRS) paradigm to control the buffer size by dynamically tuning the advertised window. However, the value of the maximum parameter has to be set to a large value [36], to ensure a bitrate up-to 100 Gbps in real-time scenarios. A bigger buffer size might create a buffer-bloating issue [37]. On the other hand, a smaller buffer might throttle the maximum achievable throughput in different RTTs. Hence, the NextGen Transport Protocol must consider an optimal value, such that it operates efficiently. The application also can control its maximum buffer size using TCP socket options. Thin-stream applications should set smaller buffer values for better performance.

5   Summary

The Terahertz frequency band can support up to 1 terabit data rates and 100-ms air latency. However, due to its propagation characteristics of the Terahertz network, such as blockage sensitivity causing a higher outage, beam misalignment, there would be sporadic loss and multiple re-transmission. In this chapter, we discussed the challenges in the transport layer on Terahertz's characteristics. We presented how a transport layer with a centralized cross-layer (add-on approach) intelligence can help the end-to-end network operate more efficiently. The new approaches (built-in and clean-slate approach) should study the 6G properties to achieve a high-data-rate and ultra-low latency. In addition to the throughput and latency, the NextGen Transport Protocols must consider other parameters such as mobility, high-density network, handover-resilient design, parallel processing, adaptive framework, core awareness, multipath TCP and multi-homing capability.

References

1. 3GPP Specification Set: 5G (2020). https://www.3gpp.org/dynareport/SpecList.htm?release= Rel-15%26tech=4. Retrieved 6 Dec 2020
2. J. Postel (Ed.),Transmission control protocol. RFC 793 (1981). Retrieved 6 Dec 2020
3. P.J. Mateo, C. Fiandrino, J. Widmer, Analysis of TCP performance in 5G mmWave mobile networks, in *Proceedings of the IEEE International Conference on Communications (ICC)* (2019), pp. 1–7
4. Next Generation Protocols (NGP); Scenarios Definition, V 1.1.1 (2016). https://www.etsi.org/deliver/etsi_gs/NGP/001_099/001/01.01.01_60/gs_NGP001v010101p.pdf. Retrieved 6 Dec 2020
5. A. Ford, C. Raiciu, M.J. Handley, O. Bonaventure, C. Paasch, TCP Extensions for Multipath Operation with Multiple Addresses. RFC 8684 (2020). Retrieved 6 Dec 2020
6. N. Cardwell, Y. Cheng, C.S. Gunn, S.H. Yeganeh, V. Jacobson, BBR: congestion-based congestion control. ACM Queue **14**, 20–53 (2016)
7. L.S. Brakmo, L.L. Peterson, TCP Vegas: end to end congestion avoidance on a global internet. IEEE J. Selec. Areas Commun. **13**, 1465–1480 (2006)
8. A. Langley, A. Riddoch, A. Wilk, A. Vicente, C. Krasic, D. Zhang, F. Yang, F. Kouranov, I. Swett, J. Iyengar, et al., The QUIC transport protocol: Design and internet-scale deployment, in *Proceedings of ACM SIGCOMM* (2017), pp. 183–196
9. V. Jacobson, D.K. Smetters, J.D. Thornton, M.F. Plass, N.H. Briggs, R.L. Braynard, Networking named content, in *Proceedings of the 5th International Conference on Emerging Networking Experiments and Technologies* (2009), pp. 1–12
10. E.P. Jones, L. Li, J.K. Schmidtke, P.A. Ward, Practical routing in delay-tolerant networks. IEEE Trans. Mobile Comput. **6**(8), 943–959 (2007)
11. Q.D. Coninck, O. Bonaventure, Multipath Extensions for QUIC (MP-QUIC). Internet-Draft draft-deconinck-quic-multipath-05, Internet Engineering Task Force (2020). Retrieved 6 Dec 2020
12. Usage statistics of QUIC for websites. https://w3techs.com/technologies/details/ce-quic. Retrieved 6 Dec 2020
13. IRTF Information-Centric Networking Research Group (ICNRG). https://irtf.org/icnrg. Retrieved 6 Dec 2020

14. P. Karn, C. Partridge, Improving round-trip time estimates in reliable transport protocols, in *Proceedings of the ACM Workshop on Frontiers in Computer Communications Technology*, SIGCOMM '87 (Association for Computing Machinery, New York, 1987), p. 2–7
15. Z. Bazzal, A.M. Ahmad, I. El Bitar, M. Rizk, M. Raad, Proposition of an adaptive retransmission timeout for TCP in 802.11 wireless environments. Int. J. Eng. Res. Appl. **7**, 64–71 (2017)
16. M. Larsson, A. Silfver, Signal-aware adaptive timeout in cellular networks: Analysing predictability of link failure in cellular networks based on network conditions (2017). http://urn.kb.se/resolve?urn=urn:nbn:se:liu:diva-138128. Retrieved 6 Dec 2020
17. V. Jacobson, Congestion avoidance and control. SIGCOMM Comput. Commun. Rev. **18**, 314–329 (1988)
18. S. Floyd, T. Henderson, RFC 2582: The NewReno Modification to TCP's Fast Recovery Algorithm (1999). Retrieved 6 Dec 2020
19. C.P. Fu, S.C. Liew, TCP Veno: TCP enhancement for transmission over wireless access networks. IEEE J. Sel. Areas. Commun. **21**, 216–228 (2006)
20. S. Ha, I. Rhee, L. Xu, CUBIC: a new TCP-friendly high-speed TCP variant. ACM SIGOPS Operat. Syst. Rev. **42**, 64–74 (2008)
21. D.X. Wei, C. Jin, S.H. Low, S. Hegde, FAST TCP: motivation, architecture, algorithms, performance. IEEE/ACM Trans. Netw. **14**, 1246–1259 (2006)
22. C. Claudio, G. Mario, M. Saverio, M. Sanadidi, W. Ren, TCP westwood: end-to-end congestion control for wired/wireless networks. Wireless Netw. **8**, 1572–8196 (2002)
23. L.A. Grieco, S. Mascolo, Performance evaluation and comparison of westwood+, New Reno, and Vegas TCP congestion control. ACM SIGCOMM Comput. Commun. Rev. **34**, 25–38 (2004)
24. D. Kliazovich, F. Granelli, Cross-layer congestion control in ad hoc wireless networks. Ad Hoc Netw. **4**(6), 687–708 (2006)
25. M.R. Kanagarathinam, S. Singh, I. Sandeep, A. Roy, N. Saxena, D-TCP: dynamic TCP congestion control algorithm for next generation mobile networks, in *Proceedings of the IEEE Annual Consumer Communications Networking Conference (CCNC)* (2018), pp. 1–6
26. F. Lu, H. Du, A. Jain, G.M. Voelker, A.C. Snoeren, A. Terzis, CQIC: revisiting cross-layer congestion control for cellular networks, in *Proceedings of the 16th International Workshop on Mobile Computing Systems and Applications* (2015), pp. 45–50
27. M.R. Kanagarathinam, S. Singh, I. Sandeep, H. Kim, M.K. Maheshwari, J. Hwang, A. Roy, N. Saxena. NexGen D-TCP: next generation dynamic TCP congestion control algorithm. IEEE Access **8**, 164482–164496 (2020)
28. T. Azzino, M. Drago, M. Polese, A. Zanella, M. Zorzi, X-TCP: a cross layer approach for TCP uplink flows in mmWave networks, in *Proceedings of the Annual Mediterranean Ad Hoc Networking Workshop (Med-Hoc-Net)* (IEEE, Piscataway, 2017), pp. 1–6
29. T. Zhang, S. Mao, Machine learning for end-to-end congestion control. IEEE Commun. Mag. **58**(6), 52–57 (2020)
30. Achieving >10 Gbps Network Throughput on Dedicated Host Instances. https://tinyurl.com/y48yrpym. Retrieved 6 Dec 2020
31. S. Yu, J. Chen, J. Mambretti, F. Yeh, Analysis of CPU pinning and storage configuration in 100 gbps network data transfer, in *2018 IEEE/ACM Innovating the Network for Data-Intensive Science (INDIS)* (2018), pp. 64–74
32. G.K. Choudhary, M.R. Kanagarathinam, H. Natarajan, K. Arunachalam, G. Monty, R.S. Lingappa, J.M. Ppallan, S.R. Jayaseelan, C. Bharti, Method and system for handling data path creation in wireless network system. US Patent App. 16/384,040 (2019)
33. E. Altman, D. Barman, B. Tuffin, M. Vojnovic, Parallel TCP sockets: Simple model, throughput and validation, in *INFOCOM*, vol. 2006 (2006), pp. 1–12
34. G.K. Choudhary, M.R. Kanagarathinam, H. Natarajan, K. Arunachalam, S.R. Jayaseelan, G. Sinha, D. Das, Novel multipipe quic protocols to enhance the wireless network performance, in *2020 IEEE Wireless Communications and Networking Conference (WCNC)* (2020), pp. 1–7

35. D. Borman, R. Braden, V. Jacobson, R. Scheffenegger, TCP extensions for high performance, *Request for Comments (Proposed Standard) RFC*, vol. 1323 (1992)
36. G. Appenzeller, I. Keslassy, N. McKeown, Sizing router buffers, in *Proceedings of the ACM SIGCOMM*, pp. 281–292 (2004)
37. J. Gettys, K. Nichols, Bufferbloat: Dark buffers in the internet. Commun. ACM **55**, 57–65 (2012)

Chapter 8
Mode Hopping for Anti-Jamming in 6G Wireless Communications

Wenchi Cheng, Liping Liang, Wei Zhang, and Hailin Zhang

Abstract Frequency hopping (FH) has been widely used as a powerful technique for anti-jamming in wireless communications. However, the existing radio wireless communication bearing form can no longer satisfied the requirements of business exponential growth in the frequency domain, which increases the difficulty to achieve efficient anti-jamming results with FH based schemes in sixth generation (6G) wireless communications. Orbital angular momentum (OAM), which provides the new angular/mode dimension for wireless communications, offers an intriguing way for anti-jamming. In this paper, we propose to use the orthogonality of OAM-modes for anti-jamming in wireless communications. In particular, we propose the mode hopping (MH) scheme for anti-jamming within the narrow frequency band. We derive the closed-form expression of bit error rate (BER) for multiple users scenario with our developed MH scheme. Our developed MH scheme can achieve the same anti-jamming results within the narrow frequency band as compared with the conventional wideband FH scheme. Furthermore, we propose mode-frequency hopping (MFH) scheme, which jointly uses our developed MH scheme and the conventional FH scheme to further decrease the BER for wireless communication. Taking 6G broadband wireless communication and anti-jamming transmission as backgrounds, OAM will demonstrate the efficient application, thus providing a basic theoretical support for greatly improving the anti-jamming capability of the wireless transmission systems.

Digital Object Identifier 10.1109/TVT.2018.2825539. ©2018 IEEE.

W. Cheng (✉) · L. Liang · H. Zhang
The State Key Laboratory of Integrated Services Networks, Xidian University, Xian, China
e-mail: wccheng@xidian.edu.cn

W. Zhang
The University of New South Wales, Sydney, NSW, Australia
e-mail: wzhang@ee.unsw.edu.au

137

1 Introduction

Frequency hopping (FH), which is a solid anti-jamming technique, has been extensively used in wireless communications. There exist some typical FH schemes such as adaptive FH [1], differential FH [2], uncoordinated FH [3], adaptive unco-ordinated FH [4], and message-driven FH [5], etc. These FH schemes can achieve efficient anti-jamming results for various wireless communications scenarios.

However, as the wireless spectrum becoming more and more crowded, it is very difficult for FH schemes to satisfy the reliability requirements of wireless communications. The authors of [6] pointed out that if the interference covers the whole frequency band, it is very hard to guarantee the reliability of wireless communications with FH. Also, when the number of channels with partial-band-noise jamming increases to a certain proportion of the total number of channels, the bit error rate (BER) of FH based wireless communications is relatively large [7, 8]. When the available number of hopping channel is small, the probability jammed by interfering users becomes very high, thus severely downgrading the spectrum efficiency of wireless communications [9]. With these limitations in mind, it is highly demanded to achieve highly efficient anti-jamming results for future wireless communications. How to guarantee the reliability of wireless communications still is an open challenge [10, 11].

Recently, more and more academic researchers show their interests in orbital angular momentum (OAM), which is another important property of electromagnetic waves and a result of signal possessing helical phase fronts. There are some studies on OAM in radio wireless communications [12–19]. The authors of [12] performed the first OAM experiment in the low frequency domain and showed that OAM is not restricted to the very high frequency range. Then, the authors of [13, 14, 16–18] started to apply OAM based transmission in radio wireless communications. Also, the OAM-based vortex beams can be generated by designing coding metasurfaces based on the Pancharatnam–Berry phase [20] and combined with orthogonal polar-izations to encode information of the matesurfaces at the transmitter, thus reducing the information loss [21]. In addition, the authors proposed the radio system based OAM [22], vortex multiple-input-multiple-output (MIMO) communication system [23], and OAM-embedded-MIMO system [24] for achieving higher capacity without increasing the bandwidth. The OAM-based vortex waves with different topological charges are mutually orthogonal when they propagate along the same spatial axis [25, 26], and thus can carry a number of independent data streams within a narrow frequency band. Also, we have studied the OAM-mode modulation and OAM waves converging [27, 28].

From the generic viewpoint, OAM can be considered as line of sight (LOS) MIMO because there are multiple antennas/array-elements at the transmitter and receiver. However, there are some differences between OAM and MIMO. As compared with MIMO in the conventional space domain, OAM provides a new mode domain. OAM is the mode multiplexing technique in the field of vortex-electromagnetic beams while MIMO is the spatial multiplexing technique in the field of plane-electromagnetic beams [29]. Mode multiplexing utilizes the orthogonality of OAM beams to minimize interchannel crosstalk and recover different data streams, thereby avoiding the use of MIMO processing. However, each data stream is received by multiple spatially separated receivers using MIMO spatial multiplexing technique. As compared with the conventional FH schemes in frequency domain, OAM, which provides the angular/mode dimension for wireless communication, offers a new way to achieve efficient anti-jamming results for wireless communications.

The authors of [30–33] showed that OAM has the potential anti-jamming for wireless communications. In the experiment [30], any attempt to sample the OAM based vortex wave will be subject to an angular restriction and a lateral offset, both of which result in inherent uncertainty in the measurement. Thus, the information encoded with OAM-modes is resistant to eavesdropping. Using the OAM to encode the data offers an inherent security enhancement for OAM-based millimeter wave wireless communications [31]. Moreover, the OAM-mode division multiplexing technique can provide high security for wireless communications [33]. However, how to hop among different OAM-modes for anti-jamming is still an open and challenging problem.

To achieve efficient hopping performance for wireless communications, in this paper we utilize the orthogonality of OAM-modes for anti-jamming. First, we propose the mode hopping (MH) scheme for anti-jamming within the narrow frequency band. We derive the closed-form expression of BER for multiple users scenario with our developed MH scheme. Our developed MH schemes can achieve the same anti-jamming results within the narrow frequency band as compared with the conventional wideband FH scheme. Furthermore, we propose mode-frequency hopping (MFH) scheme, which jointly uses our developed MH and the conventional FH scheme, for better anti-jamming results. We also derive the closed-form expression of BER for multiple users scenario with our developed MFH scheme, which can further significantly decrease the BER of wireless communications. We conduct extensive numerical results to evaluate our developed schemes, showing that our proposed MH and MFH schemes are superior than the conventional FH schemes.

The rest of this paper is organized as follows. Section 2 gives the MH and MFH system models. Section 3 de-hops OAM-mode, decomposes OAM-mode, and derives the closed-form expression of BER for multiple users scenario with our developed MH scheme. Based on the MH scheme, Sect. 4 derives the closed-form expression of BER for multiple users scenario with our developed MFH scheme. Section 5 evaluates our developed MH and MFH schemes, and compares the BERs of our developed schemes with that of the conventional FH scheme. The paper concludes with Sect. 6.

2   MFH System Model

In this section, we build up the MFH system model (including the MH system), an example of which is shown in Fig. 8.1. The MH system consists of OAM-transmitter, mode synthesizer, pseudorandom noise sequence generator (PNG), band pass filter (BPF), OAM-receiver, integrator, and low pass filter (LPF). The MFH system adds two frequency synthesizers based on the MH system. The OAM-transmitter and OAM-receiver can be uniform circular array (UCA) antenna, which consists of N array-elements distributed equidistantly around the perimeter of circle [34]. For the OAM-transmitter, the N array-elements are fed with the same input signal, but with a successive delay from array-element to array-element such that after a full turn the phase has been incremented by an integer multiple l of 2π, where l is the OAM-mode and satisfies with $-N/2 < l \leq N/2$. We have K interfering users which may use the same OAM-modes with the desired user, thus causing interference on the desired user.

In our proposed system, one data symbol experiences U OAM-mode hops or frequency hops. At the transmitter, the mode/frequency synthesizer, which is controlled by PNG, selects an OAM-mode or a range of frequency band. To de-hop OAM-mode at the receiver, PNGs are identical to those used in the transmitter. The integrator and low pass filter are used at the receiver to recover the transmit signal.

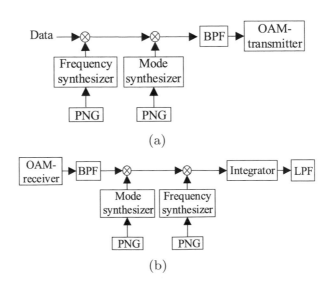

Fig. 8.1 The MFH system model. (**a**) Transmitter. (**b**) Receiver

2.1   MH Pattern

An example of MH pattern is illustrated in Fig. 8.2a, where the OAM-mode resource is divided into N OAM-modes and the time resource is divided into U time-slots. For the MH system, we denote by t_h the duration of one time slot, which is also called hop duration. We integrate OAM-mode and time into a two-dimension time-mode resource block. Each hop corresponds to a time-mode resource block. For the u-th hop, we denote by l_u ($1 \leq u \leq U, -N/2 < l_u \leq N/2$) the corresponding OAM-mode.

For comparison purpose, we also plot an example of FH pattern in Fig. 8.2b, where the frequency resource is divided into Q frequency bands and the time resource is divided into U time-slots. As shown in Fig. 8.2b, the frequency and time are integrated to form a two-dimension time-frequency resource blocks. Each hop corresponds to a time-frequency resource block. We denote by q ($1 \leq q \leq Q$) the index of frequency band. For the q-th frequency band, we denote by F_q the corresponding carrier frequency. For the u-th hop, we denote by f_u ($f_u \in \{F_1, \ldots, F_q, \ldots, F_Q\}$) the corresponding carrier frequency.

2.2   MFH Pattern

An example of MFH pattern is illustrated in Fig. 8.2c, where the OAM-mode resource is divided into N OAM-modes, the frequency resource is divided into Q frequency bands, and the time resource is divided into U time-slots. A cube denotes a hop with respect to the carrier frequency, OAM-mode, and time-slot. Each hop is identified by the specified color. For the u-th hop, the corresponding frequency band and OAM-mode are f_u ($f_u \in \{F_1, \ldots, F_q, \ldots, F_Q\}$) and l_u ($-N/2 < l_u \leq N/2$), respectively. We denote by u the index of hop.

As shown in Figs. 8.2a,c, one data symbol carrying different OAM-modes can be transmitted within U hops. For each hop, the corresponding OAM-mode is one of N OAM-modes controlled by PNG. Also, any interference to impact the desired OAM signal should be with the same azimuthal angle. However, being with the same azimuthal angle is a small probability event. Thus, the desired OAM signal is resistant to jamming caused by interfering users.

When the available frequency band is relatively narrow, the conventional FH scheme cannot be efficiently used. However, our developed MH scheme can solve the problem mentioned above for anti-jamming without increasing the frequency bands. Using our developed MH scheme, signal can be transmitted using the new mode dimension within the narrow frequency band, thus achieving efficient anti-jamming results in wireless communications. On the other hand, when frequency band is relatively wide, signal can be transmitted within the angular domain and the frequency domain simultaneously by using the MFH scheme, which jointly uses the MH scheme and the FH scheme, to further achieve better anti-jamming results than

Fig. 8.2 The MH, FH, and MFH patterns. (**a**) MH pattern. (**b**) FH pattern. (**c**) MFH pattern

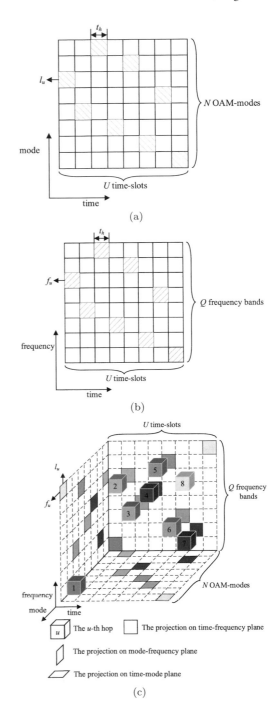

the conventional FH scheme. In addition, signal can be encoded with orthogonal polarized bit and OAM mode bit [21]. Our developed MH and MFH scheme can add orthogonal polarization parameter to achieve better anti-jamming results.

In the following, we first study the MH system within the narrow frequency band. Then, we investigate the MFH system by jointly using our developed MH scheme and the conventional FH scheme.

3 MH Scheme

In this section, we propose the MH scheme and derive the corresponding BER for our developed MH scheme. First, we give the transmit signal of MH scheme and derive the channel amplitude gain for UCA antenna based tranceiver. Then, at the receiver we de-hop and decompose OAM mode, to recover the transmit signal. Finally, we derive the BER of our developed MH scheme to analyze the anti-jamming performance for MH communications.

3.1 Transmit Signal

For MH communications, we denote by t the time variable. The transmit signal, denoted by $x_1(u, t)$, for the desired user corresponding to the u-th hop can be expressed as follows:

$$x_1(u, t) = s(t)\varepsilon_u(t - ut_h)e^{j\varphi l_u}, \tag{8.1}$$

where $s(t)$ represents the transmit signal within a symbol duration, φ is the azimuthal angle for all users, and $\varepsilon_u(t)$ is the rectangular pulse function given by

$$\varepsilon_u(t) = \begin{cases} 1 & \text{if } (u-1)t_h \leq t < ut_h; \\ 0 & \text{otherwise.} \end{cases} \tag{8.2}$$

We denote by h_{l_u} the channel gain corresponding to the channel from the desired OAM-transmitter to the OAM-receiver for the u-th hop, $h_{l_u,k}$ the channel gain corresponding to the channel from the k-th ($1 \leq k \leq K$) interfering user's OAM-transmitter to the OAM-receiver for the u-th hop, $s_k(t)$ the transmit signal for the k-th interfering user within a symbol duration, $l_{u,k}$ ($-N/2 < l_{u,k} \leq N/2$) the hopping OAM-mode for the k-th interfering user corresponding to the u-th hop, and $n(u, t)$ the received noise for the u-th hop. Then, the received signal, denoted by $r_1(u, t)$, for the desired user corresponding to the u-th hop can be obtained as follows:

$$r_1(u, t) = h_{l_u} x_1(u, t) + n(u, t)$$

$$+ \sum_{k=1}^{K} h_{l_{u,k}} s_k(t) \varepsilon(t - u t_h) e^{j\varphi l_{u,k}}. \tag{8.3}$$

For UCA antenna based transceiver, the pathloss, denoted by h_d, can be given as follows [35]:

$$h_d = \beta \frac{\lambda}{4\pi \mid \vec{d} - \vec{r}_n \mid} e^{-j\frac{2\pi \mid \vec{d} - \vec{r}_n \mid}{\lambda}}, \tag{8.4}$$

where $\vec{d}$ denotes the position vector from the OAM-transmitter to the OAM-receiver in free space, $\vec{r}_n$ denotes the position vector from the n-th ($1 \leq n \leq N$) array-element of the OAM-transmitter to the center of OAM-transmitter, β contains all relevant constants such as attenuation and phase rotation caused by antennas and their patterns on both sides, and λ is the wave length of carrier. Thus, the channel amplitude gain, denoted by h_l, from the OAM-transmitter to the OAM-receiver for the l-th OAM-mode can be derived as follows [35]:

$$h_l = \sum_{n=1}^{N} \beta \frac{\lambda}{4\pi \mid \vec{d} - \vec{r}_n \mid} e^{-j\frac{2\pi \mid \vec{d} - \vec{r}_n \mid}{\lambda}} e^{j\frac{2\pi(n-1)}{N}l}$$

$$= \sum_{n=1}^{N} \beta \frac{\lambda}{4\pi d} e^{-j\frac{2\pi d}{\lambda}} e^{j\frac{2\pi \mid \vec{d} \cdot \vec{r}_n \mid}{\lambda}} e^{j\frac{2\pi(n-1)}{N}l}$$

$$= \beta \frac{\lambda}{4\pi d} e^{-j\frac{2\pi d}{\lambda}} \sum_{n=1}^{N} e^{j\frac{2\pi \mid \vec{d} \cdot \vec{r}_n \mid}{\lambda}} e^{j\frac{2\pi(n-1)}{N}l}, \tag{8.5}$$

where we have used $\mid \vec{d} - \vec{r}_n \mid \approx d$ for amplitudes and $\mid \vec{d} - \vec{r}_n \mid \approx d - \mid \vec{d} \cdot \vec{r}_n \mid$ for phases [36].

When $N \rightarrow \infty$, we have

$$\sum_{n=1}^{N} e^{j\frac{2\pi \mid \vec{d} \cdot \vec{r}_n \mid}{\lambda}} e^{j\frac{2\pi(n-1)}{N}l}$$

$$= \sum_{n=1}^{N} e^{j\frac{2\pi}{\lambda} R \sin\theta \cos\varphi} e^{j\frac{2\pi(n-1)}{N}l}$$

$$\approx \frac{N e^{j\theta l}}{2\pi} \int_0^{2\pi} e^{j\frac{2\pi}{\lambda} R \sin\theta \cos\varphi'} e^{-j\varphi' l} d\varphi'$$

$$= N j^{-l} e^{j\varphi l} J_l \left(\frac{2\pi}{\lambda} R \sin\theta \right), \tag{8.6}$$

where R denotes the radius of the UCA antenna, θ denotes the included angle between the normal line of the transmit UCA and the line from the center of the OAM-receiver to the center of OAM-transmitter, and

$$J_l(x) = \frac{1}{2\pi j^{-l}} \int_0^{2\pi} e^{j(x\cos(\varphi')-l\varphi')} d\varphi', \qquad (8.7)$$

is the first kind Bessel function with order l. Thus, the expression of channel amplitude gain h_l can be re-written as follows:

$$h_l = \frac{\beta \lambda N j^{-l}}{4\pi d} e^{-j\frac{2\pi}{\lambda}d} e^{j\varphi l} J_l \left(\frac{2\pi}{\lambda} R \sin\theta \right). \qquad (8.8)$$

Based on Eq. (8.8), we can find that h_l increases as the number of array-elements increases.

3.2 Received Signal

Replacing l by l_u and $l_{u,k}$, respectively, in Eq. (8.8), and substituting h_{l_u} and $h_{l_{u,k}}$ into Eq. (8.3) as well as multiplying $r_1(u,t)$ with $e^{j\varphi l_u}$, we can obtain the de-hopping signal, denoted by $\tilde{r}_1(u,t)$, for the u-th hop as follows:

$$\tilde{r}_1(u,t) = r_1(u,t)e^{j\varphi l_u}. \qquad (8.9)$$

Our goal aims to recover the transmit signal for the desired user. However, $\tilde{r}_1(u,t)$ carries OAM-mode and thus needs to be decomposed. Using the integrator, we can obtain the decomposed signal, denoted by $r_1'(u,t)$, as follows:

$$r_1'(u,t) = \frac{1}{2\pi} \int_0^{2\pi} \tilde{r}_1(u,t) \left(e^{j2\varphi l_u} \right)^* d\varphi$$

$$= \begin{cases} h_{l_u}s(t)\varepsilon(t-ut_h) + \tilde{n}(u,t), & l_{u,k} \neq l_u; \\ h_{l_u}s(t)\varepsilon(t-ut_h) + \sum_{k=1}^{D_u} h_{l_u,k}s_k(t)\varepsilon(t-ut_h) + \tilde{n}(u,t), & l_{u,k} = l_u, \end{cases} \qquad (8.10)$$

where $(\cdot)^*$ represents the complex conjugate operation, $\tilde{n}(u,t)$ represents the received noise for the u-th hop after OAM-mode decomposition, and $D_u \subseteq \{1, 2, \cdots, K\}$.

Based on Eq. (8.10), we can calculate the signal-to-noise ratios (SNRs) for the scenarios with $l_{u,k} \neq l_u$ and signal-to-interference-plus-noise ratios (SINRs) for the scenarios with $l_{u,k} = l_u$. We denote by E_h the transmit power for each hop. For the scenario with $l_{u,k} \neq l_u$, the received instantaneous SNR, denoted by γ_u, after OAM-mode decomposition for the u-th hop can be expressed as follows:

$$\gamma_u = \frac{E_h h_{l_u}^2}{\sigma_{l_u}^2}, \tag{8.11}$$

where $\sigma_{l_u}^2$ is the variance of the received noise after OAM-mode decomposition for the u-th hop corresponding to the l_u-th OAM-mode. Under the scenario with $l_u \neq l_{u,k}$, we assume the average SNRs for each hop are the same. Thus, the average SNR, denoted by ζ, can be expressed as follows:

$$\zeta = E_h \mathbb{E} \left(\frac{h_{l_u}^2}{\sigma_{l_u}^2} \right), \tag{8.12}$$

where $\mathbb{E}(\cdot)$ represents the expectation operation.

For the scenario with $l_{u,k} = l_u$, we assume that there are V ($1 \leq V \leq U$) hops for the signal of desired user jammed by interfering users. We also assume that there are corresponding D_v ($1 \leq D_v \leq K, 1 \leq v \leq V$) interfering users for the v-th hop. The received instantaneous SINR, denoted by δ_v, with D_v interfering users after OAM-mode decomposition for the v-th hop in MH communications can be derived as follows:

$$\delta_v = \frac{h_{\tilde{l}_v}^2 E_h}{\sigma_{\tilde{l}_v}^2 + \sum_{k=1}^{D_v} E_h h_{\tilde{l}_{v,k}}^2}, \tag{8.13}$$

where $h_{\tilde{l}_v}$ is the channel gain corresponding to the channel from the desired OAM-transmitter to the OAM-receiver for the v-th hop and can be obtained by replacing l by $\tilde{l}_v$ in Eq. (8.8), $h_{\tilde{l}_{v,k}}$ is the channel gain corresponding to the channel from the interfering users' OAM-transmitter to the OAM-receiver for the v-th hop and can be obtained by replacing l by $\tilde{l}_{v,k}$ in Eq. (8.8), and $\sigma_{\tilde{l}_v}^2$ is the variance of the received noise after OAM-mode decomposition for the v-th hop corresponding to the $\tilde{l}_v$-th OAM-mode.

For the scenario with $l_{u,k} = l_u$, the average SINR, denoted by $\bar{\delta}_v$, with D_v interfering users corresponding to the v-th hop can be expressed as follows:

$$\bar{\delta}_v = \mathbb{E} \left(\frac{E_h h_{\tilde{l}_v}^2}{\sum_{k=1}^{D_v} E_h h_{\tilde{l}_{v,k}}^2 + \sigma_{\tilde{l}_v}^2} \right). \tag{8.14}$$

We employ the equal gain combining (EGC) diversity reception. Thus, the received instantaneous SINR, denoted by γ_s, for the desired user for U hops at the output of EGC diversity reception can be obtained as follows:

$$\gamma_s = \frac{E_h \left(\sum_{u=1}^{U-V} h_{\tilde{l}_u}^2 + \sum_{v=1}^{V} h_{\tilde{l}_v}^2 \right)}{\sum_{u=1}^{U-V} \sigma_{\tilde{l}_u}^2 + \sum_{v=1}^{V} \left(\sum_{k=1}^{D_v} E_h h_{\tilde{l}_{v,k}}^2 + \sigma_{\tilde{l}_{v,k}}^2 \right)}. \tag{8.15}$$

In the MH system, the complexity of MH receiver mainly depends on OAM-receiver, mode synthesizer, integrator, and EGC. The complexity of FH receiver mainly depends on receive antenna, frequency synthesizer, and EGC. The complexity of mode synthesizer in the MH system is similar to the complexity of frequency synthesizer in the FH system. Also, the complexity of EGC used in the MH system is similar to the complexity of EGC used in the FH system. Although the FH receiver uses single receive antenna, the MH system uses OAM-receiver based UCA which can be considered as single radio frequency chain antenna. In addition, the MH system adds a simple integrator.

3.3   Performance Analysis

To analyze the performance of our developed MH scheme, we employ binary differential phase shift keying (DPSK) and binary non-coherent frequency shift keying (FSK) modulation. We introduce a constant denoted by μ. If $\mu = 1$, it means that we employ binary DPSK modulation. If $\mu = 1/2$, it implies that we employ binary non-coherent FSK modulation. In addition, Nakagami-m fading can be used to analyze the system performance in radio vortex wireless communications [37].

Then, we assume that the number of interfering users is L $(1 \leq L \leq K)$ for a $(0 \leq a \leq V)$ hops while the numbers of interfering users for the other $(V - a)$ hops are different. We denote by $\bar{\delta}_L$ the average SINR with L interfering users. Given V hops jammed by corresponding D_v interfering users, the BER, denoted by $P_b(\gamma_s, V, D_v|U)$, for U hops with EGC reception at the receiver is given as follows [38]:

$$P_b(\gamma_s, V, D_v|U) = 2^{1-2U} e^{-\mu\gamma_s} \sum_{v_1=0}^{U-1} c_{v_1} \gamma_s^{v_1},$$

$$\tag{8.16}$$

where c_{v_1} is given by

$$c_{v_1} = \frac{1}{v_1!} \sum_{v_2=0}^{U-v_1-1} \binom{2U-1}{v_2}. \tag{8.17}$$

For Nakagami-m fading, the probability density function (PDF), denoted by $p_\gamma(\gamma)$, can be expressed as follows:

$$p_\gamma(\gamma) = \frac{\gamma^{m-1}}{\Gamma(m)} \left(\frac{m}{\overline{\gamma}}\right)^m e^{-m\frac{\gamma}{\overline{\gamma}}}, \tag{8.18}$$

where m is the fading parameter, $\Gamma(\cdot)$ is Gamma function, γ represents the SINR of channel, and $\overline{\gamma}$ represents the average SINR of channel. Then, $P_e(V, D_v|U)$ can be expressed as follows:

$$P_e(V, D_v|U) = \underbrace{\int_0^\infty \int_0^\infty \cdots \int_0^\infty}_{U-fold} P_b(\gamma_s, V, D_v|U) \left[\prod_{u=1}^{U-V} p_{\gamma_u}(\gamma_u) \prod_{v=1}^{V} p_{\delta_v}(\delta_v)\right] \left[\prod_{u=1}^{U-V} d\gamma_u \prod_{v=1}^{V} d\delta_v\right], \tag{8.19}$$

which is an U-fold integral. For convenience to calculate $P_e(V, D_v|U)$, we can express the right hand of Eq. (8.19) into an integral. Thus, $P_e(V, D_v|U)$ can be expressed as follows:

$$P_e(V, D_v|U) = \int_0^\infty P_b(\gamma_s, V, D_v|U) p_{\gamma_s}(\gamma_s) d\gamma_s, \tag{8.20}$$

where $p_{\gamma_s}(\gamma_s)$ is the PDF of γ_s. Based on the joint PDF, namely $\left[\prod_{u=1}^{U-V} p_{\gamma_u}(\gamma_u)\right.$ $\left.\prod_{v=1}^{V} p_{\delta_v}(\delta_v)\right]$ in Eq. (8.19), we need to derive the expression of $p_{\gamma_s}(\gamma_s)$.

Based on the PDFs of γ_u and δ_v, we can calculate the characteristic functions, denoted by $\Phi_{\gamma_u}(w)$ and $\Phi_{\gamma_u}(w)$, for γ_u and δ_v, respectively, with the Fourier transform as follows:

$$\begin{cases} \Phi_{\gamma_u}(w) = \left(\dfrac{m}{m - jw\zeta}\right)^m ; & (8.21a) \\[3mm] \Phi_{\delta_v}(w) = \left(\dfrac{m}{m - jw\overline{\delta_v}}\right)^m . & (8.21b) \end{cases}$$

Since instantaneous SINR experiences Nakagami-m fading and fadings on the U channels are mutually statistically independent, instantaneous SINRs are statistically independent. Hence, the characteristic function, denoted by $\Phi_{\gamma_s}(w)$, for γ_s in MH communications can be obtained as follows:

$$\begin{aligned} \Phi_{\gamma_s}(w) &= \left[\Phi_{\gamma_u}(w)\right]^{U-V} \prod_{v=1}^{V} \Phi_{\delta_v}(w) \\ &= \prod_{u=1}^{U-V} \left(\frac{m}{m - jw\zeta}\right)^m \prod_{v=1}^{V} \left(\frac{m}{m - jw\overline{\delta_v}}\right)^m . \end{aligned} \tag{8.22}$$

Then, by using partial fraction decomposition algorithm, the Eq. (8.22) can be re-written as follows:

$$\Phi_{\gamma_s}(w) = m^{mU} \prod_{u=1}^{U-V} \left(\frac{1}{m-jw\zeta} \right)^m \prod_{v=1}^{V} \left(\frac{1}{m-jw\bar{\delta}_v} \right)^m$$

$$= \begin{cases} m^{mU} \left[\sum\limits_{u_1=1}^{m(U-V)} \frac{P_{u_1}}{(m-jw\zeta)^{m(U-V)-u_1+1}} + \sum\limits_{u_2=1}^{ma} \frac{Q_{u_2}}{(m-jw\bar{\delta}_L)^{ma-u_2+1}} + \sum\limits_{v=1}^{V-a} \sum\limits_{u_3=1}^{m} \frac{W_{vu_3}}{(m-jw\bar{\delta}_v)^{m-u_3+1}} \right], \\ \qquad\qquad\qquad\qquad\qquad\qquad\qquad\qquad\qquad\qquad\qquad\qquad\qquad\qquad a \geq 1; \\ m^{mU} \left[\sum\limits_{u_1=1}^{m(U-V)} \frac{P_{u_1}}{(m-jw\zeta)^{m(U-V)-u_1+1}} + \sum\limits_{v=1}^{V} \sum\limits_{u_3=1}^{m} \frac{W_{vu_3}}{(m-jw\bar{\delta}_v)^{m-u_3+1}} \right], a = 0, \end{cases}$$

(8.23)

where P_{u_1}, Q_{u_2}, and W_{vu_3} are given as follows:

$$\begin{cases} P_{u_1} = \frac{1}{(u_1-1)!} \frac{d^{u_1-1}}{d(jw)^{u_1-1}} \Big[(m-jw\zeta)^{m(U-V)} \Phi_{\gamma_s}(w) \Big] \Big|_{jw=\frac{m}{\zeta}}; & (8.24a) \\[4mm] Q_{u_2} = \frac{1}{(u_2-1)!} \frac{d^{u_2-1}}{d(jw)^{u_2-1}} \Big[(m-jw\bar{\delta}_L)^{ma} \Phi_{\gamma_s}(w) \Big] \Big|_{jw=\frac{m}{\bar{\delta}_L}}; & (8.24b) \\[4mm] W_{vu_3} = \frac{1}{(u_3-1)!} \frac{d^{u_3-1}}{d(jw)^{u_3-1}} \Big[(m-jw\bar{\delta}_v)^{m} \Phi_{\gamma_s}(w) \Big] \Big|_{jw=\frac{m}{\bar{\delta}_v}}. & (8.24c) \end{cases}$$

Thus, we can obtain the generic PDF of γ_s as follows:

$$p_{\gamma_s}(\gamma_s) = \frac{1}{2\pi} \int_{\infty}^{\infty} \Phi_{\gamma_s}(w) e^{-jw\gamma_s} dw$$

$$= \begin{cases} \frac{m^{mU}}{2\pi j} \int_{-j\infty}^{+j\infty} \sum\limits_{u_1=1}^{m(U-V)} \frac{P_{u_1} e^{-jw\gamma_s} d(jw)}{(m-jw\zeta)^{m(U-V)-u_1+1}} + \frac{m^{mU}}{2\pi j} \int_{-j\infty}^{+j\infty} \sum\limits_{u_2=1}^{ma} \frac{Q_{u_2} e^{-jw\gamma_s} d(jw)}{(m-jw\bar{\delta}_L)^{ma-u_2+1}} \\ + \frac{m^{mU}}{2\pi j} \int_{-j\infty}^{+j\infty} \sum\limits_{v=1}^{V-a} \sum\limits_{u_3=1}^{m} \frac{W_{vu_3} e^{-jw\gamma_s} d(jw)}{(m-jw\bar{\delta}_v)^{m-u_3+1}}, \qquad a \geq 1; \\ \frac{m^{mU}}{2\pi j} \int_{-j\infty}^{+j\infty} \sum\limits_{u_1=1}^{m(U-V)} \frac{P_{u_1} e^{-jw\gamma_s} d(jw)}{(m-jw\zeta)^{m(U-V)-u_1+1}} + \frac{m^{mU}}{2\pi j} \int_{-j\infty}^{+j\infty} \sum\limits_{v=1}^{V} \sum\limits_{u_3=1}^{m} \frac{W_{vu_3} e^{-jw\gamma_s} d(jw)}{(m-jw\bar{\delta}_v)^{m-u_3+1}}, \\ \qquad\qquad\qquad\qquad\qquad\qquad\qquad\qquad\qquad\qquad\qquad\qquad\qquad\qquad a = 0, \end{cases}$$

(8.25)

When $a \geq 1$, the first term on the right hand of Eq. (8.25) can be derived as follows:

$$\frac{m^{mU}}{2\pi j} \int_{-j\infty}^{+j\infty} \sum_{u_1=1}^{m(U-V)} \frac{P_{u_1}}{(m-jw\zeta)^{m(U-V)-u_1+1}} e^{-jw\gamma_s} d(jw)$$

$$= \sum_{u_1=1}^{m(U-V)} \frac{m^{mU} P_{u_1} e^{-m\gamma_s/\zeta}}{\zeta^{m(U-V)-u_1+1} \times 2\pi j} \int_{-j\infty+\frac{m}{\zeta}}^{+j\infty\frac{m}{\zeta}} \frac{e^{\gamma_s z}}{z^{m(U-V)-u_1+1}} dz.$$

(8.26)

In the integral part, $1/z$ approaches to 0 as $Re(z)$ approaches to ∞. Using Cauchy's theorem and Residue theorem [39], we can obtain

$$\frac{1}{2\pi j}\int_{-j\infty+\frac{m}{\zeta}}^{+j\infty+\frac{m}{\zeta}}\frac{e^{\gamma_s z}}{z^{m(U-V)-u_1+1}}dz=\frac{1}{2\pi j}\int_{C}\frac{e^{\gamma_s z}}{z^{m(U-V)-u_1+1}}dz, \tag{8.27}$$

where C is an open contour from initial point to negative infinity on the real axis. According to the characteristics of Gamma function [40], we have

$$\frac{1}{2\pi j}\int_{C}\frac{e^{\gamma_s z}}{z^{m(U-V)-u_1+1}}dz=\frac{\gamma_s^{m(U-V)-u_1}}{\Gamma[m(U-V)-u_1+1]}, \tag{8.28}$$

where $\Gamma(\cdot)$ is the Gamma function and given as follows:

$$\Gamma[m(U-V)-u_1+1]=[m(U-V)-u_1+1]!. \tag{8.29}$$

Thus, Eq. (8.26) can be re-written as follows:

$$\frac{m^{mU}}{2\pi j}\int_{-j\infty}^{+j\infty}\sum_{u_1=1}^{m(U-V)}\frac{P_{u_1}}{(m-jw\zeta)^{m(U-V)-u_1+1}}e^{-jw\gamma_s}d(jw)$$

$$=\sum_{u_1=1}^{m(U-V)}\frac{m^{mU}P_{u_1}e^{-m\gamma_s/\zeta}}{\zeta^{m(U-V)-u_1+1}}\frac{\gamma_s^{m(U-V)-u_1}}{\Gamma[m(U-V)-u_1+1]}. \tag{8.30}$$

Similar to the analysis above, the other terms on the right hand of Eq. (8.25) also can be derived. Thus, the generic PDF of γ_s corresponding to Eq. (8.25) can be re-written as follows:

$$P_{\gamma_s}(\gamma_s)=\begin{cases}m^{mU}\left\{\displaystyle\sum_{u_1=1}^{m(U-V)}\frac{P_{u_1}e^{-\frac{m}{\zeta}\gamma_s}\gamma_s^{m(U-V)-u_1}}{\zeta^{m(U-V)-u_1+1}\Gamma[m(U-V)-u_1+1]}+\sum_{u_2=1}^{ma}\frac{Q_{u_2}e^{-\frac{m}{\delta_L}\gamma_s}\gamma_s^{ma-u_2}}{\delta_L^{ma-u_2+1}\Gamma[ma-u_2+1]}\right.\\ \left.+\displaystyle\sum_{v=1}^{V-a}\sum_{u_3=1}^{m}\frac{W_{vu_3}e^{-\frac{m}{\delta_v}\gamma_s}\gamma_s^{m-V}}{\delta_v^{m-V+1}\Gamma[m-V+1]}\right\}, & a\ge 1; \\[2em] m^{mU}\left\{\displaystyle\sum_{u_1=1}^{m(U-V)}\frac{P_{u_1}e^{-\frac{m}{\zeta}\gamma_s}\gamma_s^{m(U-V)-u_1}}{\zeta^{m(U-V)-u_1+1}\Gamma[m(U-V)-u_1+1]}+\sum_{v=1}^{V}\sum_{u_3=1}^{m}\frac{W_{vu_3}e^{-\frac{m}{\delta_v}\gamma_s}\gamma_s^{m-V}}{\delta_v^{m-V+1}\Gamma[m-V+1]}\right\}, & a=0.\end{cases} \tag{8.31}$$

Then, substituting Eqs. (8.16) and (8.31) into Eq. (8.20), we can obtain the following Theorem 1.

Theorem 1 *The average BER, denoted by $P_e(V, D_v|U)$, of MH scheme given V hops jammed by corresponding D_v interfering users with Nakagami-m fading channel is given by*

$$
P_e(V, D_v|U) = \begin{cases} m^{mU} \left\{ \displaystyle\sum_{v_1=0}^{U-V-1} \sum_{u_1=1}^{m(U-V)} \frac{2^{1-2(U-V)} P_{u_1} c_{v_1} \Gamma[m(U-V)-u_1+v_1+2]}{\zeta^{m(U-V)-u_1+1} \Gamma[m(U-V)-u_1+1]\left(\mu+\frac{m}{\zeta}\right)^{m(U-V)-u_1+v_1+2}} \right. \\[2ex] \quad + \displaystyle\sum_{v_1=0}^{a-1} \sum_{u_2=1}^{ma} \frac{2^{1-2a} Q_{u_2} c_{v_1} \Gamma(ma-u_2+v_1+2)}{\bar{\delta}_L^{ma-u_2+1} \Gamma(ma-u_2+1)\left(\mu+\frac{m}{\bar{\delta}_L}\right)^{ma-u_2+v_1+2}} \\[2ex] \left. \quad + \displaystyle\sum_{v_1=0}^{V-a-1} \sum_{v=1}^{V-a} \sum_{u_3=1}^{m} \frac{2^{1-2(V-a)} W_{vu_3} c_{v_1} \Gamma(m-u_3+v_1+2)}{\bar{\delta}_v^{m-u_3+1} \Gamma(m-u_3+1)\left(\mu+\frac{m}{\bar{\delta}_v}\right)^{m-u_3+v_1+2}} \right\}, & a \geq 1; \\[3ex] m^{mU} \left\{ \displaystyle\sum_{v_1=0}^{U-V-1} \sum_{u_1=1}^{m(U-V)} \frac{2^{1-2(U-V)} P_{u_1} c_{v_1} \Gamma[m(U-V)-u_1+v_1+2]}{\zeta^{m(U-V)-u_1+1} \Gamma[m(U-V)-u_1+1]\left(\mu+\frac{m}{\zeta}\right)^{m(U-V)-u_1+v_1+2}} \right. \\[2ex] \left. \quad + \displaystyle\sum_{v_1=0}^{V-1} \sum_{v=1}^{V} \sum_{u_3=1}^{m} \frac{2^{1-2V} W_{vu_3} c_{v_1} \Gamma(m-u_3+v_1+2)}{\bar{\delta}_v^{m-u_3+1} \Gamma(m-u_3+1)\left(\mu+\frac{m}{\bar{\delta}_v}\right)^{m-u_3+v_1+2}} \right\}, & a = 0. \end{cases} \tag{8.32}
$$

When $V = U$, $a = 0$, and $m = 1$, Eq. (8.22) can be reduced to

$$
\Phi_{\gamma_s}(w) = \sum_{v=1}^{U} M_u y_v, \tag{8.33}
$$

where

$$
\begin{cases} M_v = \displaystyle\prod_{\substack{i=1 \\ i \neq v}}^{U} \frac{\bar{\delta}_v}{\bar{\delta}_v - \bar{\delta}_i}; & (8.34a) \\[3ex] y_v = \dfrac{1}{1 - jw\bar{\delta}_v}. & (8.34b) \end{cases}
$$

Thus, when $V = U$, $a = 0$, *and* $m = 1$, *the PDF of* γ_s *can be reduced to*

$$
p_{\gamma_s}(\gamma_s) = \sum_{v=1}^{U} M_v \frac{e^{-\gamma_s/\bar{\delta}_v}}{\bar{\delta}_v}. \tag{8.35}
$$

Then, when $V = U$, $a = 0$, *and* $m = 1$, *the BER* $P_e(V, D_v|U)$ *can be reduced to*

$$
P_e(V, D_v|U) = \left(\frac{1}{2}\right)^{2U-1} \sum_{v=1}^{U} \sum_{v_1=0}^{U-1} \frac{M_v \Gamma(v_1-1) c_{v_1}}{\bar{\delta}_v} \left(\frac{\bar{\delta}_v}{1+\mu\bar{\delta}_v}\right)^{v_1+1}. \tag{8.36}
$$

For the scenario with $l_u \neq l_{u,k}$ corresponding to U hops, we assume that interfering users use different OAM-modes from the desired user. The characteristic function of γ_s corresponding to U hops can be re-written as follows:

$$
\Phi_{\gamma_s}(w) = \left(\frac{m}{m - jw\zeta}\right)^{mU}. \tag{8.37}
$$

Thus, the corresponding PDF with Nakagami-m fading for γ_s can be re-written as follows:

$$p_{\gamma_s}(\gamma_s) = \left(\frac{m}{\zeta}\right)^{mU} \frac{\gamma_s^{mU-1}}{\Gamma(mU)} e^{-\frac{m\gamma_s}{\zeta}}. \tag{8.38}$$

Then, we have the following Theorem 2.

Theorem 2 *The average BER, denoted by $P_e(U)$, of MH scheme for the scenario with $l_u \neq l_{u,k}$ corresponding to U hops is given by*

$$P_e(U) = \frac{2^{1-2U}}{\Gamma(mU)} \left(\frac{m}{m+\mu\zeta}\right)^{mU} \sum_{v_1=0}^{U-1} c_{v_1} \Gamma(mU+v_1) \left(\frac{\zeta}{m+\mu\zeta}\right)^{v_1}. \tag{8.39}$$

When the OAM-mode of interfering users is different from that of the desired user in MH communications, the system is equivalent to the single user system. Thus, the average BER of single user system can be expressed as Eq. (8.39).

Then, we denote by P_0 the probability that the signal of desirable user is jammed by an interfering user. Assuming that the probability of an OAM-mode carried by transmit signal for each hop is equal to $1/N$ in MH communications, the probability P_0 is equal to $1/N$ for each hop. Thereby, the probability, denoted by $p(U)$, for the scenario with $l_u \neq l_{u,k}$ corresponding to U hops can be obtained as follows:

$$p(U) = (1 - P_0)^{KU}. \tag{8.40}$$

Also, we can calculate the probability, denoted by $p(V|U)$, given V hops jammed by interfering users for U hops as follows:

$$p(V|U) = \underbrace{\sum_{D_1=1}^{K} \sum_{D_2=1}^{K} \cdots \sum_{D_V=1}^{K}}_{V-fold} \binom{U}{V}(1-P_0)^{K(U-V)} \prod_{v=1}^{V} \binom{K}{D_v} P_0^{D_v}(1-P_0)^{K-D_v}. \tag{8.41}$$

Then, we can calculate the average BER, denoted by P_s, considering all possible cases in MH communications as follows:

$$P_s = p(U) P_e(U) + \sum_{V=1}^{U} p(V|U) P_e(V, D_v|U). \tag{8.42}$$

Substituting Eqs. (8.32), (8.39), (8.40), and (8.41) into Eq. (8.42), we can obtain the average BER for all possible cases in MH communications.

Observing the BER of MH scheme, we can find that the average SINR, the number of interfering users, the number of OAM-modes, and the number of hops make impact on the value of BER. Although the BER has complex form, we can directly obtain some results. First, the BER monotonically increases as the number of interfering users increases. Second, the BER monotonically decreases as the average SINR increases. Thus, increasing the transmit power and mitigating interference can decrease the BER. In addition, the BER monotonically decreases as the number of OAM-modes increases. Generally, the number of OAM-modes mainly depends on the number of array-elements used by the OAM-transmitter. Hence, increasing the number of array-elements corresponding to the OAM-transmitter can decrease the BER in the MH system. Also, the BER monotonically decreases as the number of hops increases. Thus, increasing the number of hops can decrease the BER in the MH system.

4   MFH Scheme

For our developed MFH communications, the transmit signal, denoted by $x_2(u,t)$, for the desired user corresponding to the u-th hop can be expressed as follows:

$$x_2(u,t) = s(t)\varepsilon(t - ut_h)e^{j\varphi l_u}\cos(2\pi f_u t + \alpha_u), \tag{8.43}$$

where α_u $(0 \le \alpha_u \le 2\pi)$ denotes the initial phase corresponding to the u-th hop. We denote by $f_{u,k}$ the carrier frequency and $\alpha_{u,k}$ $(0 \le \alpha_{u,k} \le 2\pi)$ the initial phase distributed of the k-th interfering user for the u-th hop, respectively. Thus, the received signal, denoted by $r_2(u,t)$, for the desired user corresponding to the u-th hop can be derived as follows:

$$r_2(u,t) = h_{l_u}x_2(u,t) + n(u,t) + \sum_{k=1}^{K} h_{l_u,k}s_k(t)\varepsilon(t-ut_h)e^{j\varphi l_{u,k}}\cos(2\pi f_{u,k}t+\alpha_{u,k}). \tag{8.44}$$

Then, multiplying $r_2(u,t)$ with $e^{j\varphi l_u}$ and $\cos(2\pi f_u t + \alpha_u)$, we can obtain the de-hopping signal, denoted by $\tilde{r}_2(u,t)$, for the u-th hop as follows:

$$\tilde{r}_2(u,t) = r_2(u,t)e^{j\varphi l_u}\cos(2\pi f_u t + \alpha_u). \tag{8.45}$$

Using the integrator, we can obtain the decomposed signal, denoted by $r_2'(u,t)$, as follows:

$$r_2'(u,t) = \frac{1}{2\pi}\int_0^{2\pi} \tilde{r}_2(u,t)\left(e^{j2\varphi l_u}\right)^* d\varphi$$

$$= \begin{cases} h_{l_u}s(t)\varepsilon(t - ut_h)\cos^2(2\pi f_u t + \alpha_u) + \tilde{n}(u,t), & l_{u,k} \ne l_u; \\ \sum_{k=1}^{D_u} h_{l_u,k}s_k(t)\varepsilon(t-ut_h)\cos(2\pi f_{u,k}t+\alpha_{u,k})\cos(2\pi f_u t+\alpha_u) \\ +h_{l_u}s(t)\varepsilon(t - ut_h)\cos^2(2\pi f_u t + \alpha_u) + \tilde{n}(u,t), & l_{u,k} = l_u. \end{cases} \tag{8.46}$$

Then, using the low pass filter after OAM-mode decomposition, we can obtain the received signal, denoted by $y(u, t)$, for the desired user corresponding to the u-th hop as follows:

$$
y(u, t) =
\begin{cases}
h_{l_u} s(t) \varepsilon(t - ut_h) + \tilde{n}(u, t), & f_{u,k} \neq f_u, l_{u,k} \neq l_u; & (8.47a) \\[2mm]
h_{l_u} s(t) \varepsilon(t - ut_h) + \tilde{n}(u, t), & f_{u,k} = f_u, l_{u,k} \neq l_u; & (8.47b) \\[2mm]
h_{l_u} s(t) \varepsilon(t - ut_h) + \tilde{n}(u, t), & f_{u,k} \neq f_u, l_{u,k} = l_u; & (8.47c) \\[2mm]
h_{l_u} s(t) \varepsilon(t - ut_h) + \tilde{n}(u, t) + \sum_{k=1}^{D_u} h_{l_u,k} s_k(t) \varepsilon(t - ut_h), & f_{u,k} = f_u, l_{u,k} = l_u. & (8.47d)
\end{cases}
$$

Clearly, there are four cases, which are described in the following.

Case 1: If both OAM-modes and carrier frequencies of K interfering users are different from that of desired user for each mode/frequency hop, the received signal can be obtained as the right hand of the Eq. (8.47a). Thus, we can obtain that the interfering signals can be entirely removed.

Case 2: If carrier frequencies of the K interfering users are the same with that of desired user while OAM-modes of the K interfering users are different from that of the desired user for each hop, we can obtain the signal as the right hand of the Eq. (8.47b). Clearly, the interfering signals can also be entirely removed.

Case 3: If the OAM-modes of K interfering users are the same modes with that of desired user while carrier frequencies of interfering users are different from that of the desired user for each hop, the received signal can be expressed as the right hand of Eq. (8.47c). In this case, the integrator mentioned above doesn't work for interfering signals while the low pass filter can filter out interfering signals.

Case 4: If both the carrier frequencies and OAM-modes of K interfering users are the same with that of the desired user, the received signal can be given as the right hand of the Eq. (8.47d). The integrator and low pass filter cannot cancel interfering signals.

Observing the above four cases, we can know that the interfering signals make impact on the performance of the MFH system only when both the OAM-modes and carrier frequencies of interfering users are the same with that of the desired user. The interfering signals can be canceled by the integrator or low pass filter for the first three cases. Only Case 4 degrades the performance of system.

For Case 1, 2 and 3, the instantaneous SNR, denoted by ρ_u, after OAM decomposition for the u-th hop in MFH communications, can be expressed as follows:

$$
\rho_u = \frac{h_{l_u}^2 E_h}{\Omega_{l_u}^2}, \tag{8.48}
$$

where $\Omega^2_{\tilde{l}_u}$ denotes the variance of received noise after OAM decomposition for the u-th hop in MFH communications.

For Case 4, the instantaneous SINR, denoted by ϱ_v, with D_v interfering users after OAM decomposition for the v-th hop in MFH communications, can be expressed as follows:

$$\varrho_v = \frac{h^2_{\tilde{l}_v} E_h}{\Omega^2_{\tilde{l}_v} + \sum\limits_{u=1}^{D_v} E_h h^2_{\tilde{l}_{v,k}}}, \tag{8.49}$$

where $\Omega^2_{\tilde{l}_v}$ is the variance of received noise after OAM decomposition corresponding to the v-th hop for Case 4 in MFH communications.

In MFH communications, signal can be transmitted in both angle/mode domain and frequency domain simultaneously by jointly using MH and FH schemes. Thus, MH and FH are mutually independent to each other. Hence, the processing gain [41] of MFH scheme is the product of the processing gains of FH scheme and MH scheme. We denote by G (which approximates to Q) the processing gain of FH scheme. Given a fix transmit SNR, the receive SNR of MFH scheme is G times that of MH scheme. Thus, we can re-write Eq. (8.48) as follows:

$$\rho_u = G\gamma_u. \tag{8.50}$$

Also, we can re-write Eq. (8.49) as follows:

$$\varrho_v = \delta_v + \frac{(G-1)\Omega^2_{\tilde{l}_v}}{\Omega^2_{\tilde{l}_v} + \sum\limits_{u=1}^{D_v} E_h h^2_{\tilde{l}_{v,k}}} \delta_v. \tag{8.51}$$

Based on Eq. (8.51), we can find that the received average SNR or SINR of MFH scheme is always larger than that of MH scheme. For Case 1, 2, and 3, the average SNR, denoted by ξ, can be calculated as follows:

$$\xi = G\zeta. \tag{8.52}$$

For Case 4, the average SINR, denoted by $\overline{\varrho}_v$, with D_v interfering users for the v-th hop can be derived as follows:

$$\overline{\varrho}_v = \overline{\delta}_v + \mathbb{E}\left[\frac{(G-1)\Omega^2_{\tilde{l}_v}}{\Omega^2_{\tilde{l}_v} + \sum\limits_{u=1}^{D_v} E_h h^2_{\tilde{l}_{v,k}}} \right] \overline{\delta}_v. \tag{8.53}$$

Replacing D_v by L in Eq. (8.53), we can obtain the average SINR, denoted by $\overline{\varrho}_L$, with L interfering users.

Similar to the analyses of the BER in MH communications, the BER, denoted by $P_e'(U)$, for U hops without interfering users in MFH communications can be derived as follows:

$$P_e'(U) = \frac{2^{1-2U}}{\Gamma(mU)} \left(\frac{m}{m+G\mu\zeta}\right)^{mU} \sum_{v_1=0}^{U-1} c_{v_1} \Gamma(mU+v_1) \left(\frac{G\zeta}{m+G\mu\zeta}\right)^{v_1} \tag{8.54}$$

We denote by ρ_s the received instantaneous SINR for U hops at the output of EGC diversity reception in MFH communications. Then, we have

$$\rho_s = \frac{E_h \left(\sum\limits_{u=1}^{U-V} h_{l_u}^2 + \sum\limits_{v=1}^{V} h_{\bar{l}_v}^2\right)}{\sum\limits_{u=1}^{U-V} \Omega_{l_u}^2 + \sum\limits_{v=1}^{V} \left(\sum\limits_{k=1}^{D_v} E_h h_{\bar{l}_{v,k}} + \Omega_{\bar{l}_{v,k}}^2\right)}. \tag{8.55}$$

We denote by $P_e'(V, D_v|U)$ the average BER of MFH system given V hops jammed by corresponding D_v interfering users with Nakagami-m fading channel. Then, we can obtain the average BER $P_e'(V, D_v|U)$ in MFH communications as follows:

$$P_e'(V, D_v|U) = \begin{cases} m^{mU}\left\{ \sum\limits_{v_1=0}^{U-V-1} \sum\limits_{u_1=1}^{m(U-V)} \frac{2^{1-2(U-V)} P_{u_1} c_{v_1} \Gamma[m(U-V)-u_1+v_1+2]}{\xi^{m(U-V)-u_1+1}\Gamma[m(U-V)-u_1+1]\left(\mu+\frac{m}{\xi}\right)^{m(U-V)-u_1+v_1+2}} \right. \\ \quad + \sum\limits_{v_1=0}^{a-1} \sum\limits_{u_2=1}^{ma} \frac{2^{1-2a} Q_{u_2} c_{v_1} \Gamma(ma-u_2+v_1+2)}{\overline{\varrho}_L^{ma-u_2+1}\Gamma(ma-u_2+1)\left(\mu+\frac{m}{\overline{\varrho}_L}\right)^{ma-u_2+v_1+2}} \\ \quad \left. + \sum\limits_{v_1=0}^{V-a-1} \sum\limits_{v=1}^{V-a} \sum\limits_{u_3=1}^{m} \frac{2^{1-2(V-a)} W_{vu_3} c_{v_1} \Gamma(m-u_3+v_1+2)}{\overline{\varrho}_v^{m-u_3+1}\Gamma(m-u_3+1)\left(\mu+\frac{m}{\overline{\varrho}_v}\right)^{m-u_3+v_1+2}} \right\}, \quad a \geq 1; \\ m^{mU}\left\{ \sum\limits_{v_1=0}^{U-V-1} \sum\limits_{u_1=1}^{m(U-V)} \frac{2^{1-2(U-V)} P_{u_1} c_{v_1} \Gamma[m(U-V)-u_1+v_1+2]}{\xi^{m(U-V)-u_1+1}\Gamma[m(U-V)-u_1+1]\left(\mu+\frac{m}{\xi}\right)^{m(U-V)-u_1+v_1+2}} \right. \\ \quad \left. + \sum\limits_{v_1=0}^{V-1} \sum\limits_{v=1}^{V} \sum\limits_{u_3=1}^{m} \frac{2^{1-2V} W_{vu_3} c_{v_1} \Gamma(m-u_3+v_1+2)}{\overline{\varrho}_v^{m-u_3+1}\Gamma(m-u_3+1)\left(\mu+\frac{m}{\overline{\varrho}_v}\right)^{m-u_3+v_1+2}} \right\}, \quad a = 0. \end{cases} \tag{8.56}$$

$$P_s' = \begin{cases} (1-P_1)^{KU} \frac{2^{1-2U}}{\Gamma(mU)} \left(\frac{m}{m+G\mu\zeta}\right)^{mU} \sum_{v_1=0}^{U-1} c_{v_1} \Gamma(mU+v_1) \left(\frac{G\zeta}{m+G\mu\zeta}\right)^{v_1} \\[2mm] + \left[\sum_{D_1=1}^{K} \sum_{D_2=1}^{K} \cdots \sum_{D_V=1}^{K} \binom{U}{V}(1-P_1)^{K(U-V)} \prod_{v=1}^{V} \binom{K}{D_v} P_1^{D_i}(1-P_1)^{K-D_v} \right] \\[2mm] m^{mU} \left\{ \sum_{v_1=0}^{U-V-1} \sum_{u_1=1}^{m(U-V)} \frac{2^{1-2(U-V)} P_{u_1} c_{v_1} \Gamma[m(U-V)-u_1+v_1+2]}{(G\zeta)^{m(U-V)-u_1+1} \Gamma[m(U-V)-u_1+1]\left(\mu+\frac{m}{G\zeta}\right)^{m(U-V)-u_1+v_1+2}} \right. \\[2mm] + \sum_{v_1=0}^{a-1} \sum_{u_2=1}^{ma} \frac{2^{1-2a} Q_{u_2} c_{v_1} \Gamma(ma-u_2+v_1+2)}{\varrho_L^{ma-u_2+1} \Gamma(ma-u_2+1)\left(\mu+\frac{m}{\varrho_L}\right)^{ma-u_2+v_1+2}} \\[2mm] \left. + \sum_{v_1=0}^{V-a-1} \sum_{v=1}^{V-a} \sum_{u_3=1}^{m} \frac{2^{1-2(V-a)} W_{vu_3} c_{v_1} \Gamma(m-u_3+v_1+2)}{\varrho_v^{m-u_3+1} \Gamma(m-u_3+1)\left(\mu+\frac{m}{\varrho_v}\right)^{m-u_3+v_1+2}} \right\}, \\ \hspace{8cm} a \geq 1; \\[4mm] (1-P_1)^{KU} \frac{2^{1-2U}}{\Gamma(mU)} \left(\frac{m}{m+G\mu\zeta}\right)^{mU} \sum_{v_1=0}^{U-1} c_{v_1} \Gamma(mU+v_1) \left(\frac{G\zeta}{m+G\mu\zeta}\right)^{v_1} \\[2mm] + \left[\sum_{D_1=1}^{K} \sum_{D_2=1}^{K} \cdots \sum_{D_V=1}^{K} \binom{U}{V}(1-P_1)^{K(U-V)} \prod_{v=1}^{V} \binom{K}{D_v} P_1^{D_i}(1-P_1)^{K-D_v} \right] \\[2mm] m^{mU} \left\{ \sum_{v_1=0}^{U-V-1} \sum_{u_1=1}^{m(U-V)} \frac{2^{1-2(U-V)} P_{u_1} c_{v_1} \Gamma[m(U-V)-u_1+v_1+2]}{(G\zeta)^{m(U-V)-u_1+1} \Gamma[m(U-V)-u_1+1]\left(\mu+\frac{m}{G\zeta}\right)^{m(U-V)-u_1+v_1+2}} \right. \\[2mm] \left. + \sum_{v_1=0}^{V-1} \sum_{v=1}^{V} \sum_{u_3=1}^{m} \frac{2^{1-2V} W_{vu_3} c_{v_1} \Gamma(m-u_3+v_1+2)}{\varrho_v^{m-u_3+1} \Gamma(m-u_3+1)\left(\mu+\frac{m}{\varrho_v}\right)^{m-u_3+v_1+2}} \right\}, \hspace{1cm} a=0. \end{cases}$$

$$\hspace{13cm} (8.57)$$

Then, we need to calculate the average BER, denoted by P_s', for all possible cases. We denote by P_1 the probability that the signal of desirable user is jammed by an interfering user in MFH communications. Since mode and frequency are two disjoint events, the probability P_1 is equal to $1/(NQ)$. Thus, we can obtain P_s' for all possible cases as Eq. (8.57).

Observing the BER of MFH scheme, we can find that the average SINR, the number of interfering users, the number of OAM-modes, the number of available frequency bands, and the number of hops impact on the BER. Although the BER has complex form in the MFH system, we can also directly obtain some results. First, the BER monotonically increases as the average SINR increases. Hence, increasing the transmit power and mitigating interference can decrease the BER of MFH scheme. Second, the BER monotonically increases as the number of interfering users increases. Also, the BER monotonically decreases as the number of OAM-modes increases. In addition, the BER monotonically decreases as the number of available frequency bands increases. Thus, increasing the number of frequency bands can decrease the BER in the MFH system. Moreover, the BER monotonically decreases as the number of hops increases in the MFH system.

In the MFH system, the complexity of MFH system mainly depends on OAM-transmitter, mode synthesizer, frequency synthesizer, OAM-receiver, integrator, and EGC. The OAM-transmitter and OAM-receiver can be UCA antenna, which consists of several array-elements distributed equidistantly around the perimeter of receive circle. Since OAM signal can be transmitted within one antenna which equips several array-elements while using a single radio frequency chain, OAM-transmitter and OAM-receiver based UCA can be considered as single radio frequency chain antenna. Mode synthesizer and frequency synthesizer select an OAM-mode and a range of frequency, respectively, to de-hop the received OAM signal. Similar to the integrator in the MH system, integrator in the MFH system can also decompose the received OAM signal. EGC reception which has low complexity among existing receiver models, such as maximal ratio combining, EGC, and selection combining, is used for summing up the square of the signal with equal probability, thus obtaining the received instantaneous SINR. The complexity of FH mainly depends on transmit antenna, receive antenna, frequency synthesizer, and EGC. The complexity of EGC used in MFH system is similar to the complexity of EGC used in the FH system. Although the FH system uses single transmit antenna and receive antenna, the MH system uses OAM-transmitter based UCA and OAM-receiver based UCA which can be considered as single radio frequency chain antenna. Therefore, our developed MFH system and the conventional FH system belong to the category of single radio frequency chain. In addition, the MFH system adds a simple integrator and two mode synthesizers. The complexity of mode synthesizer is similar to the complexity of frequency synthesizer.

Since signal can be transmitted within the new mode dimension and the frequency dimension, our developed MFH scheme can be used for achieving better anti-jamming results for various interfering waveforms such as the wideband noise interference, partial-band noise interference, single-tone interference, and multitone interference. Thus, our developed MFH scheme can potentially be applied into various scenarios, such as wireless local area networks, indoor wireless communication, satellite communication, underwater communication, radar, microwave and so on.

5   Performance Evaluations

In this section, we evaluate the performances for our developed MH and MFH schemes and also compare the BERs of our developed schemes with the conventional wideband FH scheme. Numerical results for anti-jamming evaluated with different mode/frqeucny hops, SINR, and different interfering users are presented. In Figs. 8.3, 8.4, 8.5, 8.6, and 8.7, we employ binary DPSK modulation to evaluate the BER of the systems. Figures 8.8 and 8.9 depict the effect of binary DPSK modulation and non-coherent binary FSK modulation on the BER of our developed schemes. Throughout the simulations, we set m as 1. The numerical results prove that our developed MH scheme within the narrow frequency band can achieve the same BER as the conventional FH scheme and our developed MFH scheme can achieve the smallest BER among these three schemes.

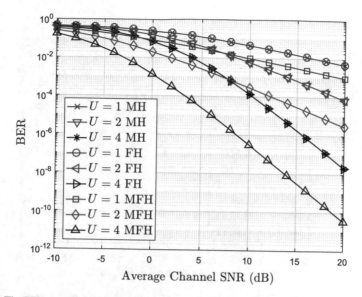

Fig. 8.3 The BERs of FH, MH, and MFH schemes using binary DPSK modulation versus average channel SNR, respectively

5.1 BERs for Single User Scenario

Figure 8.3 shows the BERs of FH, MH, and MFH schemes using binary DPSK modulation for single user scenario versus average channel SNR with different hops, where we set U as 1, 2, and 4, respectively. Clearly, the BERs of FH and MH schemes are the same with each other given a fixed hop number. The BER of our developed MFH scheme is the smallest one among the three schemes. The BERs of the three schemes decrease as the average channel SNR increases. Given the hop number, the BERs of our developed MH and MFH schemes are very close to each other in the low SNR region while the differences are widening in the high SNR region. In addition, the BERs of the three schemes decrease as U increases. What's more, the curve of BER falls much faster with higher number of U. For example, comparing the BERs with the mode/frequency hops $U = 2$ and $U = 4$, we can find that the BER of our developed MH scheme for single user scenario decreases from 5.4×10^{-3} to 1.3×10^{-4} with $U = 4$ hops when the average channel SNR increases from 5 to 10 dB. Also, the BER decreases from 4.0×10^{-2} to 6.0×10^{-3} with $U = 2$ hops. As SNR increases from 10 dB to 15 dB, the BER reduces by about 10^{-2} with $U = 4$ hops and 10^{-1} with $U = 2$ hops. As shown in Fig. 8.3, the BER of MH scheme is smaller than that of the scenario with $U = 1$. Observing Eqs. (21) and (37), we have $\zeta/(m + \mu\zeta) = 1/(\frac{m}{\zeta} + \mu)$ and $G\zeta/(m + G\mu\zeta) = G/(\frac{m}{\zeta} + G\mu)$. Thus, for both MH and MFH schemes the BER decreases as the average SNR increases. Since the received SNR of MFH scheme is G times the received SNR of MH scheme, the BER of MFH scheme is smaller than

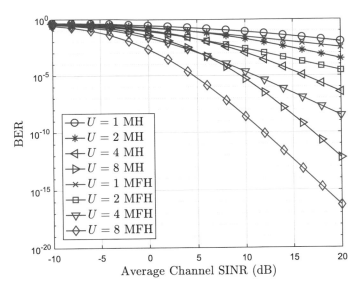

Fig. 8.4 The BERs of our developed MH and MFH schemes using binary DPSK modulation versus average channel SINR for multiple users scenario, respectively

that of MH scheme, which is consistent with the numerical results. Obtained results verify that the anti-jamming performance of our developed schemes can be better with the higher number of hops and higher SNR. Also, MFH scheme can achieve the best anti-jamming performance among the three schemes.

5.2 BERs for Multiple Users Scenario

Figure 8.4 compares the BERs of our developed MH and MFH schemes using DPSK modulation versus average channel SINR, where we set the number of interfering users as 10, the available number of FH as 5, available number of MH 10, and the number of mode/frequency hops as 1, 2, 4, and 8, respectively. Given fixed number of hops, the BER of our developed MFH scheme is smaller than that of MH scheme in the average channel SINR region for multiple users scenario. Only in low SINR region, the BERs between MFH system and MH system are close to each other. This is because the probability jammed by interfering users in the MFH scheme is smaller than that in the MH scheme. Results prove that our developed MH scheme can be jointly used with the conventional FH scheme to achieve lower BER.

Figure 8.5 shows the BERs of FH, MH, and MFH schemes using binary DPSK modulation versus the number of interfering users with different hops for interfering multiple users scenario, where we set average SINR as 10 dB, and U as 1, 2, and 4, respectively. As shown in Fig. 8.5, the BERs of FH, MH, and MFH schemes for interfering multiple users scenario are in proportion to the number of interfering

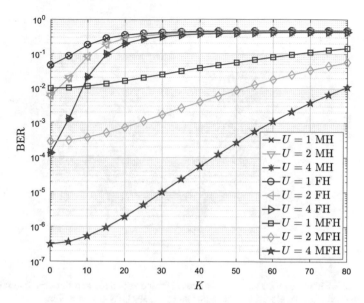

Fig. 8.5 The BERs versus the number of interfering users with FH, MH, and MFH schemes using binary DPSK modulation, respectively

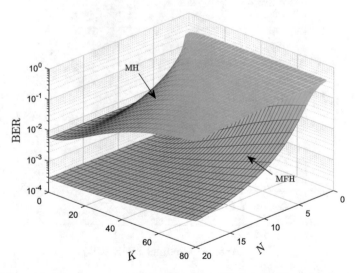

Fig. 8.6 The BERs versus different number of interfering users and the available number of OAM-modes using our developed MH and MFH schemes using binary DPSK modulation, respectively

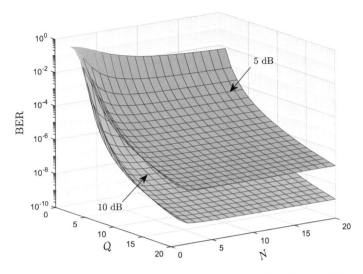

Fig. 8.7 The BER versus the available number of frequency bands and the available number of OAM-modes with our developed MFH scheme using binary DPSK modulation

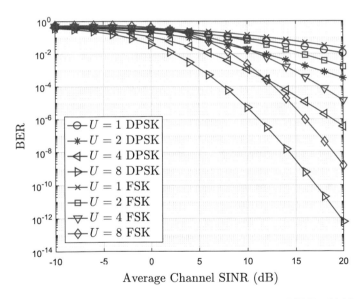

Fig. 8.8 The BER of our developed MH scheme versus average channel SINR with binary DPSK and FSK modulation, respectively

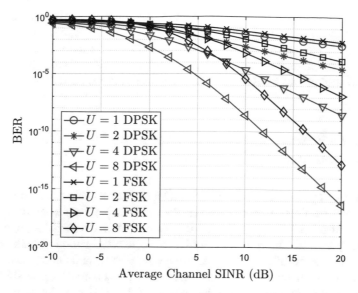

Fig. 8.9 The BER of our developed MFH scheme versus average channel SINR with binary DPSK and FSK modulation, respectively

users. As the number of interfering users increases, the BERs of the three schemes increase until BERs are very close to a fixed value. In addition, comparing BERs of our developed MH scheme and MFH scheme, we can obtain that the BER of MFH scheme is much smaller than that of the MH scheme. Results coincide with the fact that the anti-jamming performance of our developed MH and MFH schemes are better with smaller number of interfering users. Moreover, our developed MFH scheme can significantly improve the anti-jamming performance.

Figure 8.6 shows the BERs of our developed MH and MFH schemes using binary DPSK modulation versus the number of interfering users and the number of OAM-modes, where we set SINR as 10 dB, the number of OAM-mode hops as $U = 2$, and the available number of FH as 5. We can obtain that the BERs decrease as the number of OAM-modes increases and the number of interfering users decreases. This is because the number of OAM-modes increases as the probability jammed by interfering users decreases. The number of interfering users decreases as the probability jammed by interfering users decreases, thus increasing the received SNR. Moreover, the BER of our developed MFH scheme is much smaller than the MH scheme. Results prove that the probability of desired signal jammed by interfering user decreases as the number of OAM-modes increases and the number of interfering users decreases, thus guaranteeing the transmission reliability.

Figure 8.7 shows the BERs of our developed MFH scheme using binary DPSK modulation versus the available number of frequency bands and the available number of OAM-modes, where we set the number of interfering users as 10, the number of hops as 4, and the average channel SINR as 5 dB and 10 dB.

The BER decreases as the available number of frequency bands and the available number of OAM-modes increase. This is because the probability that the signal of desired user is jammed by interfering users decreases as the number of OAM-modes and frequency bands increase, leading to the downgrading of corresponding BER. Figure 8.7 also proves that the BER decreases as the average SINR increases.

5.3 Comparison Between Binary DPSK and FSK Modulations for MH and MFH Schemes

Figure 8.8 compares the BERs of our developed MH scheme using binary DPSK modulation and binary FSK modulation, for the multiple users scenario versus the average channel SINR, where we set the number of OAM-mode hops as 1, 2, 4, and 8, respectively. The results in Fig. 8.8 show that the BER is smaller of using binary DPSK modulation than that of using binary FSK modulation. The BERs of using binary DPSK and FSK modulation are close to each other in low SINR region while the BER difference between binary DPSK and FSK modulation increases in the high SINR region. This is because $\zeta/(m + \mu\zeta)$ for MH scheme increases as μ decreases. Therefore, the BER corresponding to binary DPSK modulation ($\mu = 1$) is smaller than that corresponding to binary FSK modulation ($\mu = 1/2$). Clearly, the BER of the MH scheme using binary DPSK modulation decreases as the number of hops increases.

Figure 8.9 shows the BERs of our developed MFH scheme using binary DPSK and BFSK versus the average channel SINR for multiple users scenario, where we set the number of interfering users as 10, the available number of FH as 2, the available number of MH as 10, the number of OAM-mode hops as 1, 2, 4, and 8, respectively. As shown in Fig. 8.9, we also can obtain that the BER of MFH scheme using binary DPSK modulation is smaller than that using binary FSK modulation. This is because $G\zeta/(m + G\mu\zeta)$ for MFH scheme increases as μ decreases. Observing Figs. 8.8 and 8.9, the BERs of MFH scheme using binary FSK modulation are larger than that of MH scheme.

6   Conclusions

In this paper, we proposed the MH scheme which is expected to be a new technique for anti-jamming in wireless communications. To evaluate the anti-jamming performance, we derived the generic closed-form expression of BER with our developed MH scheme. Since our developed MH scheme provides a new angular/mode dimension within narrow frequency band for wireless communications, the anti-jamming results of our developed MH scheme can be the same with that of the conventional wideband FH schemes. Furthermore, we proposed MFH scheme to further enhance the anti-jamming performance for wireless communications. We also derived the closed-form expression of BER with our developed MFH scheme to analyze the anti-jamming results. Numerical results show that our developed MH scheme within the narrow frequency has the same anti-jamming results as compared with the conventional wideband FH schemes and that the BER of our developed MH scheme decreases as the number of hops increases, the number of interfering users decreases, and the average channel SINR increases. Moreover, our developed MFH scheme outperforms the conventional FH schemes. In addition, our developed MH and MFH schemes using binary DPSK modulation can achieve better anti-jamming results than those using binary FSK modulation.

References

1. P. Popovski, H. Yomo, R. Prasad, Dynamic adaptive frequency hopping for mutually interfering wireless personal area networks. IEEE Trans Mobile Comput. 5(8), 991–1003 (2006)
2. C. Zhi, S. Li, B. Dong, Performance analysis of differential frequency hopping system in AWGN. Signal Process. 22(6), 891–894 (2006)
3. C. Popper, M. Strasser, S. Capkun, Anti-jamming broadcast communication using uncoordinated spread spectrum techniques. IEEE J. Sel. Areas Commun. 28(5), 703–715 (2010)
4. Q. Wang, P. Xu, K. Ren, X.Y. Li, Towards optimal adaptive UFH-based anti-jamming wireless communication. IEEE J. Sel. Areas Commun. 30(1), 16–30 (2012)
5. L. Zhang, H. Wang, T. Li, Anti-jamming message-driven frequency hopping-part I: system design. IEEE Trans. Wireless Commun. 12(1), 70–79 (2013)
6. J. Zander, G. Malmgren, Adaptive frequency hopping in HF communications. IEE Proc. Commun. 142(2), 99–105 (1995)
7. J. Zhang, K.C. Teh, K.H. Li, Error probability analysis of FFH/MFSK receivers over frequency-selective Rician-fading channels with partial-band-noise jamming. IEEE Trans. Commun. 57(10), 2880–2885 (2009)
8. J. Zhang, K.C. Teh, K.H. Li, Maximum-likelihood FFH/MFSK receiver over Rayleigh-fading channels with composite effects of MTJ and PBNJ. IEEE Trans. Commun. 59(3), 675–679 (2011)
9. K. Pelechrinis, C. Koufogiannakis, Krishnamurthy. S.V., On the efficacy of frequency hopping in coping with jamming attacks in 802.11 networks. IEEE Trans. Wireless Commun. 9(10), 3258–3271 (2010)
10. H. He, P. Ren, Q. Du, L. Sun, Y. Wang, Enhancing physical-layer security via big-data-aided hybrid relay selection. J. Commun. Inf. Netw. 2(1), 97–110 (2017)
11. J. Chen, G. Mao, On the security of warning message dissemination in vehicular Ad hoc networks. J. Commun. Inf. Netw. 2(2), 46–58 (2017)
12. B. Thidé, H. Then, J. Sjöholm, K. Palmer, J. Bergman, T.D. Carozzi, Y.N. Istomin, N.H. Ibragimov, R. Khamitova, Utilization of photon orbital angular momentum in the low-frequency radio domain. Phys. Rev. Lett. 99(8), 087701 (2007)
13. F.E. Mahmouli, S.D. Walker, 4-Gbps uncompressed video transmission over a 60-Ghz orbital angular momentum wireless channel. IEEE Wireless Commun. Lett. 2(2), 223–226 (2013)
14. F. Tamburini, E. Mari, A. Sponselli, et al., Encoding many channels in the same frequency through radio vorticity: first experimental test. IEEE Antennas Wireless Propag. Lett. 14(3), 033001 (2013)
15. W. Cheng, W. Zhang, H. Jing, S. Gao, H. Zhang, Orbital angular momentum for wireless communications. IEEE Wireless Commun. Mag. 26, 100–107 (2019)

16. K. Liu, Y. Cheng, Z. Yang, H. Wang, Y. Qin, X. Li, Orbital-angular-momentum-based electromagnetic vortex imaging. IEEE Antennas Wireless Propag. Lett. **14**, 711–714 (2015)
17. K.A. Opare, Y. Kuang, J.J. Kponyo, Mode combination in an ideal wireless OAM-MIMO multiplexing system. IEEE Wireless Commun. Lett. **4**(4), 449–452 (2015)
18. T. Yuan, H. Wang, Y. Qin, Y. Cheng, Electromagnetic vortex imaging using uniform concentric circular arrays. IEEE Antennas Wireless Propag. Lett. **15**, 1024–1027 (2016)
19. H. Jing, W. Cheng, X.G. Xia, Concentric UCAs based low-order OAM for high capacity in radio vortex wireless communications. IEEE Trans. Commun. **3**, 85–100 (2018)
20. L. Zhang, S. Liu, L. Li, T.J. Cui, Spin-controlled multiple pencil beams and vortex beams with different polarizations generated by pancharatnam-berry coding metasurfaces. ACS Appl. Mater. Interfaces **9**(41), 36447–36455 (2017)
21. Q. Ma, C.B. Shi, G.D. Bai, T. Chen, A. Noor, T.J. Cui, Beam-editing coding metasurfaces based on polarization bit and orbital-angular-momentum-mode bit. Adv. Opt. Mater. **5**(23), 1–7 (2017)
22. Q. Zhu, T. Jiang, Y. Cao, K. Luo, N. Zhou, Radio vortex for future wireless broadband communications with high capacity. IEEE Wireless Commun. **22**(6), 98–104 (2015)
23. Q. Zhu, T. Jiang, D. Qu, D. Chen, N. Zhou, Radio vortex-multiple-input multiple-output communication systems with high capacity. IEEE Access **3**, 2456–2464 (2015)
24. W. Cheng, H. Zhang, L. Liang, H. Jing, Z. Li, Orbital-angular-momentum embedded massive MIMO: achieving multiplicative spectrum-efficiency for mmwave communications. IEEE Access **6**, 2732–2745 (2018)
25. J. Wang, J.Y. Yang, I.M. Fazal, et al., Terabit free-space data transmission employing orbital angular momentum multiplexing. Nat. Photon. **6**, 488–496 (2012)
26. N. Bozinovic, S. Ramachandran, Terabit-scale orbital angular momentum mode division multiplexing in fibers. Science **340**(6140), 1545 (2013)
27. Y. Yang, W. Cheng, W. Zhang, H. Zhang, Mode modulation for orbital-angular-momentum based wireless vorticose communications, in *2017 IEEE Global Communications Conference (GLOBECOM), Singapore* (2017), pp. 1–7
28. S. Gao, W. Cheng, H. Zhang, Z. Li, High-efficient beam-converging for UCA based radio vortex wireless communications, in *2017 IEEE/CIC International Conference on Communications in China (ICCC), Qingdao* (2017), pp. 1–7
29. Y. Ren, L. Li, G. Xie, Y. Yan, Y. Cao, H. Huang, N. Ahmed, Z. Zhao, P. Liao, C. Zhang, G. Caire, A.F. Molisch, M. Tur, A.E. Willner, Line-of-sight millimeter-wave communications using orbital angular momentum multiplexing combined with conventional spatial multiplexing. IEEE Trans. Wireless Commun. **16**(5), 3151–3161 (2017)
30. G. Gibson, J. Courtial, M. Padgett, M. Vasnetsov, V. Pas' Ko, S. Barnett, S. Franke-Arnold, Free-space information transfer using light beams carrying orbital angular momentum. Opt. Exp. **12**(22), 5448 (2004)
31. Y. Jiang, Y. He, F. Li, Wireless communications using millimeter-wave beams carrying orbital angular momentum, in *2009 WRI International Conference on Communications and Mobile Computing*, vol. 1 (2009), pp. 495–500
32. E. Cano, B. Allen, Q. Bai, A. Tennant, Generation and detection of OAM signals for radio communications, in *2014 Loughborough Antennas and Propagation Conference (LAPC)* (2014), pp. 637–640
33. W. Zhang, S. Zheng, X. Hui, R. Dong, X. Jin, H. Chi, X. Zhang, Mode division multiplexing communication using microwave orbital angular momentum: an experimental study. IEEE Trans. Wireless Commun. **16**(2), 1308–1318 (2017)
34. B. Thidé, *Eletromagnetic Field Theory*, 2nd edn. (Dov, Mineola, 2011)
35. O. Edfors, A.J. Johansson, Is orbital angular momentum (OAM) based radio communication an unexploited area? IEEE Trans. Antennas Propag. **60**(2), 1126–1131 (2012)

36. S.M. Mohammadi, L.K.S. Daldorff, J.E.S. Bergman, R.L. Karlsson, B. Thidé, K. Forozesh, T.D. Carozzi, B. Isham, Orbital angular momentum in radio: a system study. IEEE Trans. Antennas Propag. **58**(2), 565–572 (2010)
37. M. Alfowzan, J.A. Anguita, B. Vasic, Joint detection of multiple orbital angular momentum optical modes, in *2013 IEEE Global Communications Conference (GLOBECOM)* (2013), pp. 2388–2393
38. J.G. Proakis, *Digital Communications* (McGraw-Hill, New York, 1989)
39. Folland, G.B., *Fourier Analysis and Its Applications* (Springer, New York, 2010)
40. G.E. Andrews, R. Askey, R. Roy, *Special Functions* (Cambridge University Press, London, 1999)
41. R. Pickholtz, D. Schilling, L. Milstein, Theory of spread-spectrum communications - a tutorial. IEEE Trans. Commun. **30**(5), 855–884 (1982)

Chapter 9
Hybrid Lightwave/RF Connectivity for 6G Wireless Networks

Vasilis K. Papanikolaou, Panagiotis D. Diamantoulakis, and George K. Karagiannidis

Abstract The integration of lightwave technology in wireless access has been groomed, due to its clear superiority in various aspects of wireless networking. Despite its many advantages, though, its very nature renders it a dependent technology, to be used complimentary with conventional RF networks. Therefore, the notion of hybrid lightwave/RF networking was established as a promising solution to bring the best of both worlds. The synergy between the two networks can offer significant gains in spectral efficiency and network densification, since optical networks can deliver much higher data rates per area. However, because of the heterogeneous networking, several challenges arise, as such hybrid lightwave/RF networks have been studied for resource allocation, user scheduling and integration with other promising technologies of the next generations of wireless access, such as non-ortogonal multiple access (NOMA). Apart from that, the simultaneous lightwave information and power transfer (SLIPT) has also attracted the interest of academia and industry alike as a means for more energy sustainable networks. In this chapter, the challenges of hybrid lightwave/RF networking are detailed and the interplay between hybrid lightwave/RF networks and enabling technologies of the next generations of wireless access is discussed. Finally, the concept of cross-band network design is presented as an enabling technology for the sixth generation of wireless networks (6G).

1 Introduction

The current advances in wireless communications have led to the fifth generation (5G) networking to satisfy the growing demand of connectivity, low latency, and high data rate that emerging services and applications require. Following this, the next generation of wireless networking will be a key enabler for intelligent

V. K. Papanikolaou (✉) · P. D. Diamantoulakis · G. K. Karagiannidis
Department of Electrical and Computer Engineering, Aristotle University of Thessaloniki, Thessaloniki, Greece
e-mail: vpapanikk@auth.gr; padiaman@ieee.org; geokarag@auth.gr

information sharing across a myriad of devices with different capabilities and requirements [1]. In recent years, the conventional radio-frequency band has been proven insufficient to support the plethora of demands of future networks. The vast number of devices and their required data rate will inevitably reach a condition called "spectrum crunch", where no new devices would be able to connect to the network due to a lack of bandwidth. Therefore, different areas of the electromagnetic spectrum have attracted attention from the research community and the industry as an alternative to conventional RF systems. Regions in higher frequencies offer ample bandwidth for the emerging applications with the most prominent being the mmWave region that has already been included in the 5G standardization [2, 3]. In this setting, use of the optical region of the spectrum becomes very enticing, since lightwave technology is already mature from the use of optical fiber communications in the last 50 years. Optical wireless communications (OWC) and more specifically, visible light communications (VLC), that make use of the existing room illumination to provide data access, have been examined extensively the past 5 years mostly for indoor wireless access [4–7].

VLC's main advantages are their ability to achieve very high data rates, their low cost and energy efficiency, since they make use of LED bulbs, their high frequency reuse factor and their inherent high physical layer security since light cannot penetrate walls or other opaque objects. On the other hand, this characteristic can be problematic since this kind of line-of-sight (LoS) based communications can be blocked by movement or rotation of the receiver or from a passing object. Moreover, VLC is dependent on the room illumination levels. While this latter problem can be efficiently countered by the use of infrared wavelengths, the uplink in this type of network remains challenging as a moving terminal has to be pointed directly at the access point to communicate. The above reasons have led to the examination of a hybrid heterogeneous solution with both lightwave and RF technologies. The hybrid lightwave/RF network can provide ubiquitous coverage thanks to the RF counterpart, while offering high data rate to its users with its lightwave subsystem.

In such a heterogeneous network, problems arise when resource allocation needs optimization since the two different networks have different properties and it is crucial that wireless access and fairness among users is preserved [7–10]. Furthermore, it is imperative to examine how this promising networking solution interacts with other attractive technologies to be used for 5G and beyond networks, such as non-orthogonal multiple access (NOMA) [11, 12]. A fundamental problem of NOMA networks is the user grouping, since the partitioning of the user set affects the achieved data rates [13, 14].

A different aspect of lightwave networks has been their implementation in wireless powered networks (WPN) [15]. Wireless power transfer is a key technology to extend the battery lifetime of mobile devices by charging them through artificially created radio-signals. Energy harvesting via solar radiation has attracted attention as well, since it is essential free energy for the mobile terminal. However, its stochastic nature makes it inconsistent, paving the way for lightwave energy harvesting methods with artificial sources, such as LED bulbs [16]. The simultaneous lightwave information and power transfer (SLIPT) concept has been investigated as a viable alternative to RF energy harvesting [17]. A lightwave power transfer solution is beneficial over the RF option since artificial light sources are abundant in every indoor space at a conveniently lower cost, while RF energy harvesting signals can lead to RF pollution and increased interference in the network.

In this chapter, we take a look at common use scenarios of hybrid lightwave/RF networks, examining fundamental problems such as resource allocation. Moreover, the integration of hybrid lightwave/RF networks with other promising state-of-the-art technologies such as NOMA is presented and a brief look into wireless powered networks with lightwave technology. Finally, future challenges are identified for the area of hybrid lightwave/RF networking.

2 Channel Model

2.1 VLC Channel Model

For indoor VLC systems, the channel gain is modeled with the Lambertian emission. Without loss of generality, a LoS is assumed to be always available for VLC users. The NLoS is modeled with the use of Lambertian reflections, however, for VLC users it can safely be assumed that a LoS is always available. This assumption is fueled by the fact that users are serviced by the VLC in order to increase their achieved data rates. A NLoS link cannot, in general, provide enough power for this use [8, 18, 19]. As such, the channel gain is given by [20, 21]

$$h = A_r \frac{m+1}{2\pi d^2} T_s(\psi) g(\psi) \cos^m(\phi) \cos(\psi), \tag{9.1}$$

where A_r denotes the optical detector size, d the distance between transmitter and receiver, and ϕ and ψ are the irradiance and the incident angles, respectively. Additionally, $T_s(\psi)$ is the optical filter gain, while $g(\psi)$ represents the gain of the concentrator given by [20, 22]

$$g(\psi) = \begin{cases} \frac{n^2}{\sin^2(\Psi_{FoV})}, & 0 \le \psi \le \Psi_{FoV} \\ 0, & \psi > \Psi_{FoV}, \end{cases} \tag{9.2}$$

where n is the refractive index and Ψ_{FoV} is the field of view. Finally m is the order of the Lambertian model given by

$$m = -\frac{1}{\log_2 \cos(\Phi_{1/2})}, \tag{9.3}$$

where $\Phi_{1/2}$ is the semi-angle at half luminance.

Assuming the access point is placed on the ceiling in a distance L from the floor and the terminals are facing upwards, the channel gain can be expressed in terms of distance as follows [23]

$$h = A_r \frac{m+1}{2\pi} T(\psi) g(\psi) \frac{L^{m+1}}{(r^2 + L^2)^{\frac{m+3}{2}}} = \frac{C(m+1)L^{m+1}}{(r^2 + L^2)^{\frac{m+3}{2}}}, \qquad (9.4)$$

where $C = \frac{A_r T(\psi) g(\psi)}{2\pi}$ and r is the radial distance of the terminal on the ground plane.

2.2 RF Channel Model

RF systems are a well-established technology and depending on the exact use a variety of models can be used in each case that offers improved accuracy. Conventional RF operates at under 6 GHz, e.g. WiFi. For the indoor case a simple logarithmic model can be used for path loss, i.e.,

$$PL(d)[dB] = PL(d_0)[dB] + 10\alpha \log_{10}(d/d_0), \qquad (9.5)$$

where d_0 is the reference distance and α is the path loss exponent. Small-scale fading is modeled via a random variable g that follows most commonly the Rayleigh distribution, but in practice there is a plethora of distributions to model a variety of cases.

3 Resource Allocation

A hybrid VLC/RF network is considered with a single VLC access point (AP) and a single RF AP. Two sets of users are considered, $\mathcal{N} = \{1, \ldots, n, \ldots, N\}$ and $\mathcal{M} = \{1, \ldots, m, \ldots, M\}$ that are served by the VLC AP and the RF AP, respectively. In a practical scenario, the hybrid network will operate with a single backhaul link, that both subsystems share, such as a subscription line to a provider. The backhaul link has a capacity equal to C_0. Moreover, the users are all assumed to have a single antenna and photodetector. An orthogonal multiple access scheme is considered for every subsystem as the most conventional way of accessing the common channel. Without loss of generality, TDMA is considered for VLC, as in [7, 24] and OFDMA is considered for the RF.

Therefore, the received electrical signal-to-noise ratio (SNR) of VLC user n is given by

$$\mathrm{SNR}_n^{\mathrm{VLC}} = \frac{(\eta h_n P_n)^2}{\sigma^2},\tag{9.6}$$

where η denotes the photodetector's responsivity, P_n is the allocated optical power to user n, and σ^2 is the noise variance at the receiver, which is assumed to be the same for every user. Intensity modulation with direct detection (IM/DD) is assumed for the VLC subsystem, therefore Shannon's classic formula for calculating capacity is not applicable in this situation. A lower band of the capacity of user n is used instead, given by Wang et al. [25]

$$R_n^{\mathrm{VLC}} = t_n \log_2\left(1 + \frac{e}{2\pi}\mathrm{SNR}_n^{\mathrm{VLC}}\right),\tag{9.7}$$

where t_n is the timeslot duration assigned to user n in the TDMA scheme. The capacity is given here normalized with bandwidth.

Finally, the power of user n is constrained by the maximum allowed power $P_n \leq P_{\max}$ and by the average optical power, so that the illumination remains constant at all times given by

$$\sum_{n=1}^{N} t_n P_n \leq P_{\mathrm{av}},\tag{9.8}$$

where the total timeslot duration is normalized, i.e., $\sum_{n=1}^{N} t_n \leq 1$, and P_{av} denotes the average optical power.

On the other hand, a user m served by the RF AP has an achievable rate R_m^{RF} given by

$$R_m^{\mathrm{RF}} = B_m \log_2\left(1 + \frac{|g_m|^2 \mathrm{PL}_m p_m}{N_0 B_m}\right),\tag{9.9}$$

where B_m denotes the allocated bandwidth at user m, p_m is the allocated power of user m, and N_0 is the power spectral density of the AWGN. Moreover, the total bandwidth is normalized as well, as is the timeslot duration, while the following constraints need to hold

$$\sum_{m=1}^{M} p_m \leq p_{\max},\tag{9.10}$$

which is the total power constraint and

$$\sum_{m=1}^{M} B_m \leq 1.\tag{9.11}$$

In the hybrid VLC/RF network, the common backhaul link dictates that the sum of all rates is constrained by the backhaul capacity. Therefore, the following constraint needs to hold as well

$$\sum_{n=1}^{N} R_n^{\text{VLC}} + \sum_{m=1}^{M} R_m^{\text{RF}} \leq C_0. \tag{9.12}$$

Finally, resource allocation can be optimized based on an objective function f as follows

$$
\begin{aligned}
\max_{\mathbf{P},\mathbf{p},\mathbf{t},\mathbf{B}} \quad & f \\
\text{s.t.} \quad & C_1 : \sum_{n=1}^{N} R_n^{[\text{VLC}]} + \sum_{m=1}^{M} R_m^{[\text{RF}]} \leq C_0, \\
& C_2 : \sum_{n=1}^{N} t_n P_n \leq P_{\text{av}}, \\
& C_3 : P_n \leq P_{\text{max}}, \quad \forall n \in \mathcal{N} \\
& C_4 : \sum_{m=1}^{M} p_m \leq p_{\text{max}}, \\
& C_5 : \sum_{n=1}^{N} t_n \leq 1, \\
& C_6 : \sum_{m=1}^{M} B_m \leq 1.
\end{aligned}
\tag{9.13}
$$

3.1 The Value of User Fairness

In (9.13), objective function f can be a plethora of objective functions that are common in wireless networks, such as the sum rate or the energy efficiency. However, in this type of network it is imperative to guarantee fairness among users. In the popular case of sum rate maximization, the VLC users, since they can achieve much higher data rates, will accumulate all of the available backhaul capacity leading the RF users to experience outage. Maximizing the minimum rate achieved by users is a way to guarantee fairness, however, in this situation where VLC is implemented to significantly increase the data rates a much more suitable option is optimizing the proportional fairness.

The proportional fairness metric is defined as the sum of the logarithm of a utility function, in this case the data rate [7, 26, 27]. Therefore, the sum-log-rate is a fitting objective function for this type of heterogeneous networks. Proportional fairness has two desirable qualities

- It is an increasing function of data rate
- It tends to negative infinity as the data rate of a user tends to zero.

The last property is the one that guarantees fairness, as the optimization will rarely give a solution where any user has a really low achievable data rate. This objective function is maximized when every user achieves the same data rate. However, due to the hardware constraints of the access points and the backhaul capacity constraint, it is not always feasible to reach such a point. Finally, it is possible to prioritize one subsystem over the other with the use of a weighting factor, $0 \leq a \leq 1$. The objective function f is expressed then as

$$f = a \sum_{n=1}^{N} \log \left(R_n^{\text{VLC}} \right) + (1 - a) \sum_{m=1}^{M} \log \left(R_m^{\text{RF}} \right). \tag{9.14}$$

The solution of this optimization problem provides valuable insights about how hybrid VLC/RF networks operate. When the backhaul capacity is low enough, all users share it equally. However, as the backhaul capacity increases RF users cannot increase their data rates, due to hardware constraints, reaching a ceiling. The VLC users in this case can capitalize on this and reach even higher data rates before reaching their own ceiling as well. When $a \neq 0.5$ is used, therefore one subsystem has a priority over the other in serving its users, the aforementioned phenomenon is shifted to accommodate the different weights, but no essential difference is observed in the behavior of the network.

4 Integration with Non-orthogonal Multiple Access

In the previous section, an orthogonal multiple access scheme was used in the network. The exponentially growing connectivity and spectral efficiency requirements of the next generations have led the research community to the concept of non-orthogonal multiple access (NOMA). According to NOMA, the users' messages are superimposed on the power domain, while successive interference cancellation (SIC) is used at the receiver to distinguish the signals [12]. NOMA has shown superior performance in a variety of scenarios compared to conventional orthogonal schemes. Since hybrid VLC/RF networks are a prime candidate to support higher spectral efficiency, their integration with NOMA has attracted significant attention.

4.1 User Grouping

Several challenges are faced when dealing with a hybrid VLC/RF network due to its heterogeneous nature that lead to an asymmetry in data rates. User grouping in such a network plays a vital role in maximizing the benefits of the hybrid system. User grouping in NOMA is a fundamental problem of combinatorial nature that affects the capacity regions of the users [13]. Empirical methods that divide the users into strong and weak are of little use in the hybrid VLC/RF network, especially with multiple APs, since each user experiences different channel gain from each AP. As it was made clear in the previous section, fairness issues ensue in such networks, therefore a different utility function is needed to offer a trade-off between maximization of individual rates and fairness.

The complicated interactions between multiple users served by multiple APs can be examined with the use of coalitional game theory [28]. Each coalition is assigned to a specific access point, VLC or RF, with users joining the coalition that maximizes their payoff. Each coalition can also be thought as a single resource block. In NOMA, users are superimposed into a single resource block increasing the spectral efficiency, at the cost of increased interference at the weaker users. As the number of users in a single coalition increases, the aggregated interference hinders the data rate of weaker users, significantly lowering the user fairness in the system. Thus, there exists a trade-off between the increased spectral efficiency in the network and fairness. Moreover, with a larger number of users, stronger users' ability to perform SIC becomes much more complex.

We consider a hybrid VLC/RF network with a total of $|\mathcal{M}| = M + 1$ APs, consisting of a single RF AP to provide ubiquitous coverage and M VLC APs to increase the capacity of the network, and a total of $|\mathcal{N}| = N$ users that are equipped with a single antenna and a single photodetector. The operator $|\mathcal{A}|$ denotes the cardinality of set $\mathcal{A}$ and the RF AP is denoted by $m = 0$.

During the transmission phase, each AP transmits a superposition of the users' signals that are assigned to it. It is assumed that N_m users are served by AP m with $\sum_{m=0}^{M} N_m = N$. The baseband equivalent of the received signal at user n served by AP m is expressed as

$$y_{nm} = h_{nm} \sum_{i=1}^{N_m} P_{im}^q s_{im} + n_n, \qquad (9.15)$$

where h_{nm} denotes the channel gain between user n and AP m. Without loss of generality it is assumed that channel gains are orders such as $|h_1| < |h_2| < \ldots |h_{N_m}|$. Also, P_{im} is the allocated power at user i served by AP m, s_{im} is the message from AP m to user i, n_n is the additive white Gaussian noise (AWGN) at the receiver, and q is given as

$$q = \begin{cases} 1/2, \, m = 0 \\ 1, \, m \neq 0, \end{cases} \qquad (9.16)$$

since optical power is considered for the VLC APs.

SIC is performed at the receivers and only the interference of stronger users remain during the decoding phase. The signal-to-interference plus noise ratio (SINR) for VLC user n served by AP m to decode the message for user j also served by the same AP with $j < n$ is given by

$$\mathrm{SINR}_{j \to nm}^{\mathrm{VLC}} = \frac{(\eta h_{nm} P_{jm})^2}{\eta^2 h_{nm}^2 \sum_{i=j+1}^{N_m} P_{im}^2 + \sigma^2}, \qquad (9.17)$$

while, if the SIC is successful, the SINR of user n to decode its own message is given by

$$\text{SINR}_{nm}^{\text{VLC}} = \frac{(\eta h_{nm} P_{nm})^2}{\eta^2 h_{nm}^2 \sum_{i=n+1}^{N_m} P_{im}^2 + \sigma^2},$$
(9.18)

where η is the photodetector's responsivity and σ^2 is the variance of the AWGN. User N_m can perform SIC for every other user that is served by the same AP, therefore their SINR can be expressed as

$$\text{SINR}_{Nm}^{\text{VLC}} = \frac{(\eta h_{nm} P_{nm})^2}{\sigma^2}.$$
(9.19)

The total power constraint needs to be met (lightning constraint), i.e.,

$$\sum_{n=1}^{N_m} P_{nm} \leq P_{\max}.$$
(9.20)

Finally, due to the IM/DD implementation of the VLC system, a lower bound is used for the capacity in this scenario, as well. Thus, the achievable rate is given by

$$R_{nm}^{\text{VLC}} = \log_2 \left(1 + \frac{e}{2\pi} \text{SINR}_{nm}^{\text{VLC}}\right),$$
(9.21)

where the rate is presented normalized with regards to the available bandwidth.

In a similar fashion, the achievable rate of user n served by the RF AP, given that SIC is successful, is given by

$$R_{n0}^{\text{RF}} = \log_2 \left(1 + \frac{|h_{n0}|^2 P_{n0}}{|h_{n0}|^2 \sum_{i=n+1}^{N_0} P_{i0} + \sigma^2}\right).$$
(9.22)

In order to reach a desirable user grouping, a coalitional game is played [14]. The users can switch between coalitions if

- they can increase their utility function
- they have the consent of the rest of the users in that coalition, i.e., they do not lower the utility of the others.

Therefore, a utility function needs to be designed that takes into account that

- higher data rate is the prime concern of the users, so the function is increasing with regards to data rate,
- each user pays a price to join a coalition, based on their ordering, since weaker users increase the complexity of stronger users through SIC, and stronger users increase the interference to weaker users.

The utility function of user n that belongs to coalition S_m can be expressed as

$$u_n(S_m) = R_{nm} - \kappa_{S_m}(n), \tag{9.23}$$

where κ_{S_m} is the cost function of coalition S_m for user n. The total cost of a coalition S_m is the total cost that all users pay to be part of the coalition, expressed as

$$\Xi(S_m) = \sum_{n=1}^{N_m} \kappa_{S_m}(n), \tag{9.24}$$

where $\kappa_{S_m}(n)$ is defined as

$$\kappa_{S_m}(n) = \lambda^{i-1}\kappa_0, \tag{9.25}$$

where i is the order of user n in the coalition and κ_0 is the standard cost that is calculated in a recurring manner via (9.24). Depending on the values of λ, the cost is distributed more to weaker users for $\lambda < 1$, stronger users for $\lambda > 1$, or equally for $\lambda = 1$.

Power allocation is critical in a NOMA system, since it can used to maximize a utility function, as well. In this setting, users in a coalition have to pay a cost for being served by a specific AP. When a new user wants to join the coalition, they need to get consent from the rest of the users, i.e., not to drop their utility function. Since, a new user can take up some of the cost of the coalition, a portion of the total power can now be allocated to the new user without decreasing the utility of the rest of the users.

The most interesting insight from such scenario is that users do not always tend to connect to the VLC APs despite their superior achievable rates. While crowding can occur in a VLC AP leading to low payoffs, the RF AP can provide a profitable alternative for users with low payoff in the VLC APs.

4.2 Cross-Band Selection Combining

Up until now, the RF network plays a complementary role as a stand-alone access point in the hybrid lightwave/RF setup. It is possible to take advantage of RF in a different manner when a VLC AP is in place. We consider an indoor downlink transmission system where a VLC AP is placed on the ceiling at distance L and two users are located on the floor, U_1 and U_2. We assume that U_1 is the strong users, i.e., near user and the U_2 is the weak user, or far user. NOMA is used to transmit information from the AP to the users simultaneously. In order to assist the far user with the detection, a complimentary RF relay link is used from U_1 to U_2.

It is evident that the channel gain depends on the distance between AP and users. U_1 and U_2 are assumed to be uniformly distributed in an disk of radius R_1 and in an annular area bounded in $[R_1, R_2]$, respectively. Following that, the CDF of the channel gain $|h_i|^2$ is given as

$$F_{|h_i|^2}(y) = \Pr\left[\frac{(C(m+1)L^{m+1})^2}{(r_i^2 + L^2)^{m+3}} < y\right] = 1 - \Pr[r_i < T(y)], \qquad (9.26)$$

where $\Pr[\cdot]$ denotes the probability and $T(y)$ is defined as

$$T(y) = \sqrt{\left(\frac{(C(m+1)L^{m+1})^2}{y}\right)^{\frac{1}{m+3}} - L^2}. \qquad (9.27)$$

Since the users are uniformly distributed in their respective areas the CDF of channel gain for U_1 and U_2 can be expressed as

$$F_{|h_1|^2}(y) = \begin{cases} 1 - \frac{T(y_1)}{R_1^2}, & Y_{1,\min} \leq y_1 \leq Y_{1,\max} \\ 1, & y_1 > Y_{1,\max} \\ 0, & y_1 < Y_{1,\min}, \end{cases} \qquad (9.28)$$

and

$$F_{|h_2|^2}(y) = \begin{cases} 1 - \frac{T(y_2)^2 - 2R_1 T(y_2) + R_1^2}{(R_2 - R_1)^2}, & Y_{2,\min} \leq y_2 \leq Y_{2,\max} \\ 1, & y_2 > Y_{2,\max} \\ 0, & y_2 < Y_{2,\min}, \end{cases} \qquad (9.29)$$

with $Y_{1,\min} = h(r_i)|_{r_i=R_1}$, $Y_{1,\max} = h(r_i)|_{r_i=0}$, $Y_{2,\min} = h(r_i)|_{r_i=R_2}$, and $Y_{2,\max} = h(r_i)|_{r_i=R_1}$.

Following that, the achieved rates of the VLC users can also be described by lower bounds proposed in [29] as follows

$$R_{2 \to 1} = \left[B_v \log_2\left(1 + \frac{\eta^2 P_2^2 |h_1|^2}{(\eta^2 P_1^2 |h_1|^2 + 9\sigma^2)(1 + \epsilon_\mu)^2}\right) - \epsilon_\phi\right]^+, \qquad (9.30)$$

$$R_1 = \left[B_v \log_2\left(1 + \frac{\eta^2 P_1^2 |h_1|^2}{9\sigma^2(1 + \epsilon_\mu)^2}\right) - \epsilon_\phi\right]^+, \qquad (9.31)$$

$$R_2 = \left[B_v \log_2\left(1 + \frac{\eta^2 P_2^2 |h_2|^2}{(\eta^2 P_1^2 |h_2|^2 + 9\sigma^2)(1 + \epsilon_\mu)^2}\right) - \epsilon_\phi\right]^+, \qquad (9.32)$$

where B_v is the bandwidth of the VLC system, $\epsilon_\mu = 0.0015$, $\epsilon_\phi = 0.016$, and $[\cdot]^+ = \max(\cdot, 0)$.

On the other hand, the RF link between the two users leads to a rate of

$$R_{2,\mathrm{RF}} = B_r \log_2 \left(1 + \frac{P_{\mathrm{RF}} |h_R|^2}{\sigma_R^2} \right), \tag{9.33}$$

where P_{RF} is the transmitted power, h_R is the fading coefficient, usually formulated as a Nakagami random variable, and σ_R^2 is the variance of the AWGN at the RF receiver.

As far as fixed policies are concerned, U_1 suffers an outage when they cannot decode U_2's message, which needs a rate Γ_2 or when after a successful SIC they cannot decode their own message, which needs a data rate of Γ_1. This is expressed as [23]

$$\begin{aligned} P_{\mathrm{O,VLC}}^1 &= \Pr[R_{2\to1} < \Gamma_2] + \Pr[R_{2\to1} > \Gamma_2, R_1 < \Gamma_1] \\ &= 1 - \Pr[R_{2\to1} > \Gamma_2 \cap R_1 > \Gamma_1]. \end{aligned} \tag{9.34}$$

By combining (9.28), (9.30), (9.31) the outage probability can be computes as

$$P_{\mathrm{O,VLC}}^1 = \begin{cases} 1 - \dfrac{T(\zeta^*)^2}{R_1^2}, & \gamma_2 \leq \dfrac{P_2^2}{P_1^2}, \\ 1, & \text{otherwise}, \end{cases} \tag{9.35}$$

where

$$\zeta^* = \min\{\max\{\zeta, Y_{1,\min}\}, Y_{1,\max}\}, \tag{9.36}$$

$$\zeta = \max\{\zeta_1, \zeta_2\}, \tag{9.37}$$

$$\zeta_1 = \frac{9\gamma_1\sigma^2}{\eta^2 P_1^2}, \tag{9.38}$$

$$\zeta_2 = \frac{9\gamma_2\sigma^2}{\eta^2(P_2^2 - \gamma_2 P_1^2)}, \tag{9.39}$$

$$\gamma_i = (2^{(\Gamma_i + \epsilon_\phi)/B_v} - 1)(1 + \epsilon_\mu)^2. \tag{9.40}$$

Similarly, if the RF link is inactive, U_2 experiences an outage when

$$P_{\mathrm{O,VLC}}^2 = \Pr[R_2 < \Gamma_2]. \tag{9.41}$$

After substituting (9.29) and (9.32), the outage probability can be expressed as

$$P_{\mathrm{O,VLC}}^2 = \begin{cases} 1 - \dfrac{T(\zeta_{2,2}^*)^2 - 2R_1 T(\zeta_{2,2}^*) + R_1^2}{(R_2 - R_1)^2}, & \gamma_2 \leq \dfrac{P_2^2}{P_1^2}, \\ 1, & \text{otherwise}, \end{cases} \tag{9.42}$$

where $\zeta_{2,2}^* = \min\{\max\{\zeta_2, Y_{2,\min}\}, Y_{2,\max}\}$.

If the RF relay link is used, U_2 experiences an outage if the rate at which U_1 decodes U_2's message from the VLC link is lower than Γ_2 or if the RF rate is lower than Γ_2. This can be expressed as follows

$$P^2_{O,VLC/RF} = \Pr\left[\min\{R_{2\rightarrow1}, R_{2,RF}\} < \Gamma_2\right]. \tag{9.43}$$

The mathematical tractability of (9.43) is hard to follow, but a closed-form expression for the high SNR regime is presented in [23]. The throughput of the network then is given by

$$R_{VLC} = (1 - P^1_{O,VLC})\Gamma_1 + (1 - P^2_{O,VLC})\Gamma_2, \tag{9.44}$$

when only the VLC link is utilized and

$$R_{VLC/RF} = (1 - P^1_{O,VLC})\Gamma_1 + (1 - P^2_{O,VLC/RF})\Gamma_2, \tag{9.45}$$

when the hybrid VLC/RF link is utilized for U_2 instead of directly decoding its message from the VLC AP.

Apart from the fixed policies, a cross-band selection combining (CBSC) scheme can be utilized to increase the performance of the network [23]. CBSC is motivated by the two copies of their message U_2 receives and by the fact that due to the heterogeneous nature of the setup better performing combining methods such as MRC are not applicable in this situation. Therefore, the outage probability in this case is expressed as

$$P^2_{O,CBSC} = \Pr[O^{VLC}_2 \cap O^{VLC/RF}] = \Pr[O^{VLC}_2]\Pr[O^{VLC/RF}|O^{VLC}_2], \tag{9.46}$$

where for simplicity O^{VLC}_2 and $O^{VLC/RF}_2$ denote the outage events in the direct VLC and the hybrid VLC/RF link, respectively. A closed-form expression of the outage is provided in [23].

Therefore, the system throughput when CBSC is employed is expressed as

$$R_{CBSC} = (1 - P^1_{O,VLC})\Gamma_1 + (1 - P^2_{O,CBSC})\Gamma_2. \tag{9.47}$$

5   Ultra Small Cells with Lightwave Power Transfer

A different aspect of lightwave technology utilization in modern networks is the ability to use a solar cell in order to harvest lightwave energy as well as decode optical information carrying signals [17]. In general, wireless power transfer (WPT) is well explored subject needed to combat the energy efficiency requirements of the next generations of wireless communications. On top of that, optical WPT and SLIPT are emerging solutions that cause a paradigm shift in energy harvesting methods. Different from conventional solar energy harvest, optical WPT utilizes artificially created optical signals to charge mobile terminals with great success [30].

Motivated by this, we consider a VLC indoor cellular network with ultra-small cells and a single RF AP with multiple antennas. VLC APs can be composed of RGB LEDs, meaning they can opt to transmit information through a single color, leading to no interference to neighboring cells. As such, a LED element i transmits information decoded through color $s \in S$ to a mobile device j. Meanwhile, the rest of the colors $c \in \bar{S}$, $\bar{S} = S - \{s\}$ are also transmitted by i to preserve the white color of the LED and are received by receiver j. The received signal at a mobile device j is then described by

$$y_{i,j} = N_s V_s [A^s_{i,j} x_{i,j} + B^s_{i,j}] h_{i,j} + \sum_{c \in \bar{S}} N_c V_c B^c_{i,j} h_{i,j}, \qquad (9.48)$$

where N_s, N_c are the number of LEDs of the selected color and the other colors, respectively in element i, V_s and V_c denote the respective voltage of the LED, while $x_{i,j}$ is the modulated electrical signal with zero mean and unity variance [31]. Finally, $A^s_{i,j}$ is the AC component of the selected color associated with the transmitted electrical signal, while $B^s_{i,j}$ and $B^c_{i,j}$ denote the DC bias of the selected color and the other colors, respectively. In this context, the input current bias has a minimum of I_L and a maximum of I_H and this constraint is passed to the AC component as [17, 32]

$$A^s_{i,j} \leq \min\{B^s_{i,j} - I_L, I_H - B^s_{i,j}\}. \qquad (9.49)$$

Moreover, in order to guarantee white illumination the average light intensity of all colors need to be equal. If we consider that each color has the same number of LEDs N_{LED} at the same voltage V_{LED} it leads to

$$B^s_{i,j} = B^c_{i,j} = B. \qquad (9.50)$$

The SNR at the receiver is then expressed as

$$\text{SNR}^s_{i,j} = \frac{(\eta N_{\text{LED}} V_{\text{LED}} h_{i,j} A^s_{i,j})^2}{\sigma^2}. \qquad (9.51)$$

On the other hand, the DC component of the VLC signal is used for energy harvesting. More specifically, according to [17], the harvested energy is computed as

$$\text{EH}^{\text{VLC}}_j = f I_{j,G} V_{j,\text{oc}}, \qquad (9.52)$$

where f is the fill factor and $I_{j,G}$ is the light generated current. It is important to note, that all optical radiation that reaches receiver j contributes to the energy harvesting, therefore $I_{j,G}$ is defined as

$$I_{j,G} = 3\eta N_{\text{LED}} V_{\text{LED}} B \sum_i h_{i,j}. \qquad (9.53)$$

Moreover, $V_{j,\text{oc}}$ is the open-circuit voltage at receiver j given by

$$V_{j,\text{oc}} = V_T \ln\left(1 + \frac{I_{j,G}}{I_D}\right), \qquad (9.54)$$

where V_T is the thermal voltage and I_D is the dark saturation current. It is made clear that there exists a trade-off between the achieved SNR and the harvested energy based on the chosen values of AC gain and DC Bias. In order to guarantee that a preset amount of energy can be harvested by the users without lowering the SNR, a backup RF link may be used for wireless power transfer to users.

The RF received energy signal is expressed as

$$y_j^{\text{RF}} = x_j^{\text{RF}} \mathbf{g}_j^H \mathbf{w}_j, \qquad (9.55)$$

where x_j^{RF} is the unit energy signal, $\mathbf{g}_j$ is the channel gain vector and $\mathbf{w}_j$ is the beamforming vector. In the same manner as in optical WPT, the input at the energy harvesting circuit of device j is the sum of all RF energy signals j', therefore

$$\Phi_j^{\text{RF}} = \sum_{j'} |\mathbf{g}_j^H \mathbf{w}_{j'}|^2. \qquad (9.56)$$

In practice, the energy that is harvested depends on the efficiency of the conversion. However, the conversion efficiency is not a constant and depends on the input energy. Therefore, a non-linear function better models the RF energy harvesting as follows [33, 34]

$$\text{EH}_j^{\text{RF}} = \frac{\dfrac{M^{\text{EH}}}{1+\exp\left(-a\left(\Phi_j^{\text{RF}}-b\right)\right)} - \dfrac{M^{\text{EH}}}{1+\exp(ab)}}{1 - \dfrac{1}{1+\exp(ab)}}. \qquad (9.57)$$

It is possible now to formulate an optimization problem to maximize the minimum SNR achieved by all users in the network, while guaranteeing a preset value of energy that can be harvested by all users. The problem can be expressed as follows [32]

$$\max_{B, A^s, EH^{RF}} \quad \min_{i,j} \alpha_{i,j} \frac{(\eta N_{LED} V_{LED} h_{i,j} A_{i,j}^s)^2}{\sigma^2}.$$

$$\text{s.t.} \quad C_1 : EH_j^{VLC} + EH_j^{RF}, \ \forall j,$$
$$C_2 : B_{i,j} = B, \ \forall j, \forall i, \tag{9.58}$$
$$C_3 : A_{i,j}^s \leq I_H - B_{i,j}^s, \ \forall j, \forall i,$$
$$C_4 : \frac{I_H + I_L}{2} \leq B \leq I_H,$$
$$C_5 : EH_j^{RF} \leq \theta^{RF},$$

where $a_{i,j}$ is equal to 1 if element i serves user j and 0 otherwise. Furthermore, θ is the preset value of total harvested energy. Finally, θ^{RF} denotes the energy that the RF AP contributes to the total harvested energy of users. The optimization variable EH_j^{RF}, i.e., the RF energy needed is found through this problem. Afterwards, the RF AP can solve a minimization problem of total power to ensure the optimal value EH_j^{RF*} can be transmitted to all users at a lower energy cost.

To efficiently solve this problem, the maximization of the SNR of the weakest user is considered to find the required energy to be harvested from the RF AP. By considering the optimal value of RF harvested energy, the original problem in the hybrid VLC/RF network (9.58) is reduced to a problem purely in the VLC subsystem.

In this hybrid lightwave/RF setup, the RF AP acts as a backup energy source in order to guarantee a higher SNR to all users while ensuring that the harvested energy of users is over a desired threshold. The increase that is achieved with the help of the RF AP in the achieved SNR of users is proven to be significant.

6 Conclusions and Future Research

In this chapter, we have discussed various hybrid lightwave/RF setups that are crucial to tackle the exponential growing demands of the next generations of wireless access. Fundamental problems of wireless communications such as resource allocation, as well as emerging topics have been examined. The integration of hybrid lightwave/RF networks with NOMA seems to be a very promising candidate for the next generation indoor networks. Future research into 6G networks will build a new arsenal of tools to face the complexity of the ensuing problems of hybrid networking, such as artificial intelligence. Heuristic solutions found in this way can alleviate the high computational cost of optimizing the heterogeneous network with very low performance degradation. Finally, lightwave energy harvesting presents a viable opportunity to develop more energy efficient networks and it can also be paired with its RF counterpart to yield higher performance.

References

1. Z. Zhang, Y. Xiao, Z. Ma, M. Xiao, Z. Ding, X. Lei, G.K. Karagiannidis, P. Fan. 6G wireless networks: vision, requirements, architecture, and key technologies. IEEE Vehic. Technol. Mag. **14**(3), 28–41 (2019)
2. T.S. Rappaport, S. Sun, R. Mayzus, H. Zhao, Y. Azar, K. Wang, G.N. Wong, J.K. Schulz, M. Samimi, F. Gutierrez. Millimeter wave mobile communications for 5G cellular: It will work! IEEE Access **1**, 335–349 (2013)
3. T.S. Rappaport, R.W. Heath Jr., R.C. Daniels, J.N. Murdock. *Millimeter Wave Wireless Communications*. (Pearson Education, London, 2015)
4. M. Kavehrad, Sustainable energy-efficient wireless applications using light. IEEE Commun. Mag. **48**(12), 66–73 (2010)
5. S. Arnon, *Visible Light Communication* (Cambridge University Press, Cambridge, 2015)
6. M. Ayyash, H. Elgala, A. Khreishah, V. Jungnickel, T. Little, S. Shao, M. Rahaim, D. Schulz, J. Hilt, R. Freund, Coexistence of WiFi and LiFi toward 5G: concepts, opportunities, and challenges. IEEE Commun. Mag. **54**(2), 64–71 (2016)
7. V.K. Papanikolaou, P.D. Diamantoulakis, P.C. Sofotasios, S. Muhaidat, G.K. Karagiannidis, On optimal resource allocation for hybrid vlc/RF networks with common backhaul. IEEE Trans. Cog. Commun. Netw. **6**(1), 352–365 (2020)
8. D.A. Basnayaka, H. Haas, Design and analysis of a hybrid radio frequency and visible light communication system. IEEE Trans. Commun. **65**(10), 4334–4347 (2017)
9. X. Wu, M. Safari, H. Haas, Joint optimisation of load balancing and handover for hybrid LiFi and WiFi networks, in *2017 IEEE Wireless Communications and Networking Conference (WCNC)* (2017), pp. 1–5
10. V.K. Papanikolaou, P.P. Bamidis, P.D. Diamantoulakis, G.K. Karagiannidis, Li-Fi and Wi-Fi with common backhaul: coordination and resource allocation, in *2018 IEEE Wireless Communications and Networking Conference (WCNC)* (2018)
11. H. Marshoud, V.M. Kapinas, G.K. Karagiannidis, S. Muhaidat, Non-orthogonal multiple access for visible light communications. IEEE Photon. Technol. Lett. **28**(1), 51–54 (2016)
12. Z. Ding, X. Lei, G.K. Karagiannidis, R. Schober, J. Yuan, V.K. Bhargava, A survey on non-orthogonal multiple access for 5G networks: research challenges and future trends. IEEE J. Sel. Areas Commun. **35**(10), 2181–2195 (2017)
13. Z. Ding, P. Fan, H.V. Poor, Impact of user pairing on 5G nonorthogonal multiple-access downlink transmissions. IEEE Trans. Veh. Technol. **65**(8), 6010–6023 (2016)
14. V.K. Papanikolaou, P.D. Diamantoulakis, G.K. Karagiannidis, User grouping for hybrid vlc/RF networks with noma: a coalitional game approach. IEEE Access **7**, 103299–103309 (2019)
15. P.D. Diamantoulakis, K.N. Pappi, Z. Ding, G.K. Karagiannidis, Wireless-powered communications with non-orthogonal multiple access. IEEE Trans. Wirel. Commun. **15**(12), 8422–8436 (2016)
16. J. Fakidis, S. Videv, S. Kucera, H. Claussen, H. Haas, Indoor optical wireless power transfer to small cells at nighttime. J. Lightw. Technol. **34**(13), 3236–3258 (2016)
17. P.D. Diamantoulakis, G.K. Karagiannidis, Z. Ding, Simultaneous lightwave information and power transfer (SLIPT). IEEE Trans. Green Commun. Netw. **2**, 1 (2018)
18. M.R. Zenaidi, Z. Rezki, M. Abdallah, K.A. Qaraqe, M.S. Alouini, Achievable rate-region of VLC/RF communications with an energy harvesting relay, in *Proceedings of the GLOBECOM 2017-2017 IEEE Global Communications Conference* (2017), pp. 1–7
19. H. Kazemi, M. Safari, H. Haas, A wireless optical backhaul solution for optical attocell networks. IEEE Trans. Wirel. Commun. **18**(2), 807–823 (2019)
20. T. Komine, M. Nakagawa, Fundamental analysis for visible-light communication system using LED lights. IEEE Trans. Consum. Electron. **50**(1), 100–107 (2004)
21. H. Ma, L. Lampe, S. Hranilovic, Coordinated broadcasting for multiuser indoor visible light communication systems. IEEE Trans. Commun. **63**(9), 3313–3324 (2015)
22. J.M. Kahn, J.R. Barry, Wireless infrared communications. Proc. IEEE **85**(2), 265–298 (1997)

23. Y. Xiao, P.D. Diamantoulakis, Z. Fang, Z. Ma, L. Hao, G.K. Karagiannidis, Hybrid lightwave/RF cooperative noma networks. IEEE Trans. Wirel. Commun. **19**(2), 1154–1166 (2020)
24. A.M. Abdelhady, O. Amin, A. Chaaban, M. Alouini, Downlink resource allocation for multichannel TDMA visible light communications, in *Proceedings of the IEEE Global Conference Signal and Information Processing (GlobalSIP)* (2016), pp. 1–5
25. J.B. Wang, Q.S. Hu, J. Wang, M. Chen, J.Y. Wang, Tight bounds on channel capacity for dimmable visible light communications. J. Lightw. Technol. **31**(23), 3771–3779 (2013)
26. X. Li, R. Zhang, L. Hanzo, Cooperative load balancing in hybrid visible light communications and WiFi. IEEE Trans. Commun. **63**(4), 1319–1329 (2015)
27. X. Wu, M. Safari, H. Haas, Access point selection for hybrid Li-Fi and Wi-Fi networks. IEEE Trans. Commun. **65**(12), 5375–5385 (2017)
28. K.R Apt, A. Witzel, A generic approach to coalition formation. Int. Game Theory Rev. **11**(03), 347–367 (2009)
29. A. Chaaban, Z. Rezki, M. Alouini, On the capacity of the intensity-modulation direct-detection optical broadcast channel. IEEE Trans. Wirel. Commun. **15**(5), 3114–3130 (2016)
30. Wi-Charge. Products. https://www.wi-charge.com/products [online] (Accessed on May 12 2021)
31. C. Chen, D.A. Basnayaka, H. Haas, Downlink performance of optical attocell networks. J. Lightwave Technol. **34**(1), 137–156 (2016)
32. H. Tran, G. Kaddoum, P.D. Diamantoulakis, C. Abou-Rjeily, G.K. Karagiannidis, Ultra-small cell networks with collaborative RF and lightwave power transfer. IEEE Trans. Commun. **67**(9), 6243–6255 (2019)
33. E. Boshkovska, D.W.K. Ng, N. Zlatanov, R. Schober. Practical non-linear energy harvesting model and resource allocation for swipt systems. IEEE Commun. Lett. **19**(12), 2082–2085 (2015)
34. E. Boshkovska, D.W.K. Ng, N. Zlatanov, A. Koelpin, R. Schober, Robust resource allocation for MIMO wireless powered communication networks based on a non-linear eh model. IEEE Trans. Commun. **65**(5), 1984–1999 (2017)

Chapter 10
Resource Allocation in 6G Optical Wireless Systems

Osama Zwaid Aletri, T. E. H. El-Gorashi, and Jaafar M. H. Elmirghani

Abstract The abundant optical spectrum is a promising part of the electromagnetic spectrum for 6G communication systems. The visible light spectrum which is a part of the optical spectrum, can be used to provide communication and illumination simultaneously. Visible light communication (VLC) systems have been widely researched, however, little work has focused on the area of multiple access. This chapter studies wavelength division multiple access (WDMA) techniques in VLC systems to support multiple users. In addition, the optimization of resource allocation is considered in this chapter by developing a mixed-integer linear programming (MILP) model that can be used to maximize the signal to noise and interference ratio (SINR) while supporting multiple users. The optimized resource allocation results in the best assignment of access points (APs) and wavelengths to users. Different indoor environments such as office, data center and aircraft cabins are evaluated in this chapter. A laser diode (LD) with four wavelengths (red, green, yellow and blue) is used to provide high bandwidth for communication and white light for illumination. Also, an angle diversity receiver (ADR) is utilized to receive signals and reduce noise and interference by exploiting the spatial domain.

1 Introduction

The number of users connected to the Internet has increased significantly and there is a growing demand for high data rates. The radio spectrum, which is the current spectrum that is broadly used in the indoor environment, faces various limitations,

The original version of this chapter was revised. The correction to this chapter is available at
https://doi.org/10.1007/978-3-030-72777-2_22

O. Z. Aletri (✉) · T. E. H. El-Gorashi · J. M. H. Elmirghani
Institute of Communication and Power Networks, School of Electronic and Electrical
Engineering, University of Leeds, Leeds, UK
e-mail: ml15ozma@leeds.ac.uk; t.e.h.elgorashi@leeds.ac.uk; j.m.h.elmirghani@leeds.ac.uk

Y. Wu et al. (eds.), *6G Mobile Wireless Networks*, Computer Communications
and Networks, https://doi.org/10.1007/978-3-030-72777-2_10

187

such as it being a scarce and congested spectrum resulting in limited channel capacity and low transmission rates. Thus, several techniques have been introduced to minimize such limitations in the spectrum, including advanced modulation, smart antennas and the concept of multiple-input and multiple-output (MIMO) systems [1, 2]. Achieving high data rates beyond 10 Gbps for each user is challenging when using the current radio spectrum, especially if the number of connected devices continues to increase. Cisco expects the increase in the number of connected devices from 2017 to 2021 to be approximately 27 times. Therefore, an alternative spectrum should be discovered to meet these demands. The optical spectrum is one such spectrum that promises to support multiple users at high data rates. Optical wireless communication (OWC) systems have been introduced by researchers as a potential solution. OW systems have important advantages, such as excellent channel characteristics, abundant bandwidth and low cost components compared to the current radio frequency (RF) wireless technology [3–10]. OWC systems can provide Tb/s aggregate data rates in an indoor environment for the sixth generation wireless communication. OWC data rates exceeding 25 Gbps per user have been demonstrated in downlink communication in indoor settings [9–20]. The uplink has been studied in different works [21, 22], also, energy efficiency is an area that must be given an attention [23]. In addition, various configurations of transmitter and receiver have been studied to reduce the delay spread [15, 24–31] while increasing the signal-to-noise ratio (SNR) [32–36]. Different multiple access methods used in radio frequency (RF) systems, such as sharing the time, frequency, wavelength or code domains, can be used in OWC systems to support multiple users. Utilizing resources efficiently is required to avoid signal quality degradation. Thus, attention has been given to methods that can be used to efficiently share the OWC resources including space, power, wavelength and time resources.

This chapter introduces an indoor OWC system that provides multiple access using wavelength division multiple access (WDMA). The optimization of resource allocation in terms wavelengths and access points to different users has been consider in this work by maximizing the sum over all users of Signal to Interference-plus-Noise Ratio (SINRs). A Mixed Integer Linear Programme (MILP) model has been developed to optimize the resource allocation. The rest of this chapter is organized as follows: Section 2 describes the transmitter and receiver design, Section 3 outlines the relevant multiple access techniques including WDMA and the developed MILP model. The simulation setup and results in different indoor environments are given in Sect. 4 and finally Sect. 5 presents the conclusions.

2 Transmitter and Receiver Design

2.1 Transmitter Design

In this work a visible light illumination source that utilizes red, yellow, green and blue (RYGB) laser diodes (LDs) is also used for as a transmitter (access point (AP)) for communication at the same time. RYGB LDs are used to provide four

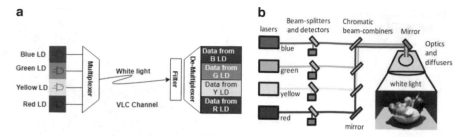

Fig. 10.1 WDM used in VLC systems using LDs

wavelengths, as shown in Fig. 10.1a, and each wavelength can carry a different data stream from to support multiple users. The LDs are used to offer high bandwidth to transfer data and support high data rates. RYGB LDs can be used safely in indoor environments to provide illumination, as stated in [37]. Four mixed LDs colours can be used to offer white illumination, as shown in Fig. 10.1b. The white colour source is generated by combining the four different LD colours utilizing chromatic beam-splitters. The combined beams are reflected using a mirror and subsequently pass through a diffuser to reduce speckle before the illumination is passed to the room environment.

2.2   Receiver Design

In this work, an angle diversity receiver (ADR) that consists of four branches (see Fig. 10.2) was used in the three different indoor environments to collect signals from different directions. Each of the four branches of the ADR has a narrow field of view (FOV) to reduce interference from other users and inter-symbol interference (ISI). The design of the ADR is similar to [9, 22, 38]. Each branch of the ADR covers different areas in the indoor environment by orienting the branch to that area based on Azimuth (Az) and Elevation (El) angles. These angles were set as follow: The Az angles of the four branches were set to be 45°, 135°, 225° and 315°, whereas, the El angles of these branches were set to 70°. The FOV of each branch of the receiver has been chosen to be 21° as a narrow FOV reduces both interference and ISI. Each branch has a detector area of 10 mm^2 and responsivity equal to 0.4, 0.35, 0.3 and 0.2 A/W for red, yellow, green and blue wavelengths respectively.

3   Multiple Access

3.1   Wavelength Division Multiple Access (WDMA)

Several Multiple access (MA) techniques that are utilized in RF systems can also be utilized in OWC systems. Wavelength division multiple access (WDMA) has

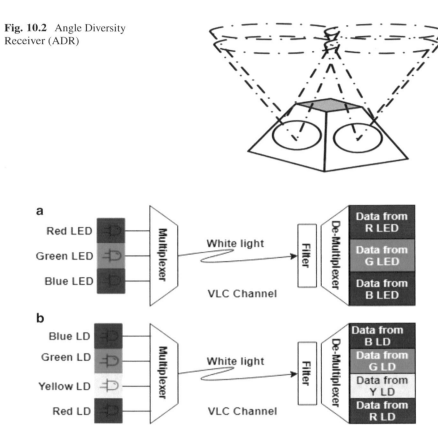

Fig. 10.2 Angle Diversity Receiver (ADR)

Fig. 10.3 WDM used in VLC systems: (**a**) LDs transmitter and (**b**) WDM receivers

been a subject of interest in many studies in OWC systems to support multiple users by sharing wavelengths among users [6, 38–41]. A multiplexer is used at the transmitter to aggregate wavelengths into a single optical beam. The receiver uses a de-multiplexer to separate the wavelengths. Two different light sources are utilized in OWC systems, namely RGB LEDs and RYGB LDs and both can support WDMA (see Fig. 10.3) [6, 42].

3.2 MILP Model

This section presents a MILP model developed to optimize resource allocation based on maximizing the sum of SINRs for all possible users [38]. This MILP model is used to assign users or devices to access points (APs) and wavelengths while maximizing the sum of SINRs over all users. The input data of the MILP model is the pre-calculated received power, receiver noise and background noise which can

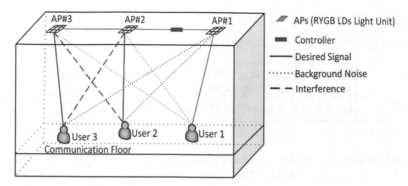

Fig. 10.4 Example of assignements in the MILP model in an indoor OWC environment

Table 10.1 Sets of the MILP model

US	Set of users in the indoor environment
AP	Set of access points
W	Set of wavelengths
B	Set of receiver branches

Table 10.2 Parameters of the MILP model

us, ui	us is desired user, ui refers to other users
ap, cp	ap is the allocated access point to user us, cp is the allocated access point to another users ui
λ	Refers to wavelength
b, f	Receiver branch, b for the desired user us and f for the other users ui
$P_{us,b}^{ap,\lambda}$	The electrical received power at the receiver from user us due to the received optical power from the access point ap using the wavelength λ at receiver branch b
$\sigma_{us,b}^{cp,\lambda}$	The shot noise at the receiver of user us due to the background unmodulated power from access point cp using wavelength λ at receiver branch b
σ_{Rx}	The receiver noise

then be used to measure all users' SINRs in each location of interest in the indoor environment. A simple example to show how the assignment is done is provided in Fig. 10.4, where three users, three APs and two wavelengths (Red and Blue) are used. User 1 is the only one assigned to the blue wavelength and only suffers from the background noise of the other APs. However, Users 2 and 3 share the same wavelength (Red) from different APs and they thus interfere with one another and receive background noise from AP 1.

The developed MILP model consists of sets, parameters and variables that are used to obtain the assignments.

Tables 10.1, 10.2 and 10.3 introduce these sets, parameters and variables respectively.

Table 10.3 Variables of the MILP model

$USINR_{us,b}^{ap,\lambda}$	SINR of user us allocated to access point ap using wavelength λ at receiver branch b
$S_{us,b}^{ap,\lambda}$	A binary selector function where 1 refers to the assignment of user us to access point ap using wavelength λ at receiver branch b
$\phi_{us,ui,b,f}^{ap,cp,\lambda}$	Linearization variable that is not negative value, $\phi_{us,ui,b,f}^{ap,cp,\lambda} = USINR_{us,b}^{ap,\lambda} S_{ui,f}^{cp,\lambda}$

The Objective of the MILP model is to maximize the sum of all users' SINRs which can be defined as follows:

$$Maximise \sum_{us \in \mathcal{US}} \sum_{ap \in \mathcal{AP}} \sum_{\lambda \in \mathcal{W}} \sum_{b \in \mathcal{B}} USINR_{us,b}^{ap,\lambda}, \tag{10.1}$$

The SINR of each user can be calculated as follows:

$$USINR_{us,b}^{ap,\lambda} = \frac{Signal}{Interference+Noise}$$

$$= \frac{P_{us,b}^{ap,\lambda} S_{us,b}^{ap,\lambda}}{\sum\limits_{\substack{cp \in \mathcal{AP} \\ cp \neq ap}} \sum\limits_{\substack{ui \in \mathcal{US} \\ ui \neq us}} \sum_{f \in \mathcal{B}} P_{us,b}^{cp,\lambda} S_{ui,f}^{cp,\lambda} + \sum\limits_{\substack{cp \in \mathcal{AP} \\ cp \neq ap}} \sigma_{us,b}^{cp,\lambda} \left[1 - \sum\limits_{\substack{ui \in \mathcal{US} \\ ui \neq us}} \sum_{f \in \mathcal{B}} S_{ui,f}^{cp,\lambda} \right] + \sigma_{Rx}} . \tag{10.2}$$

where the receiver noise σ_{Rx} can be given by:

$$\sigma_{Rx} = N_R B_E, \tag{10.3}$$

where N_R is the density of receiver noise measured in (A^2/Hz) and B_E is the electrical bandwidth.

The background noise $\sigma_{us,b}^{cp,\lambda}$ can be evaluated as:

$$\sigma_{us,b}^{cp,\lambda} = 2e \left(R \ PO_{us,b}^{cp,\lambda} \right) B_o B_e, \tag{10.4}$$

where R is the receiver responsivity, e is the electron charge, $PO_{us,b}^{cp,\lambda}$ is the received optical power from unmodulated APs and B_o is the optical bandwidth.

The electrical received power $P_{us,b}^{ap,\lambda}$ can be calculated as:

$$P_{us,b}^{ap,\lambda} = \left(R \ PO_{us,b}^{ap,\lambda} \right)^2, \tag{10.5}$$

where $PO_{us,b}^{ap,\lambda}$ is the received optical power.

The numerator in Eq. (10.2) is the received signal multiplied by the selector binary variable; while, the denominator consists of two parts: The first part is the sum of interferences coming from all APs that use the same wavelength for communication, and the second part is the sum of the background noise coming from all unmodulated APs using the same wavelength.

Rewriting Eq. (10.2) gives:

$$
\sum_{\substack{cp \in \mathcal{AP} \\ cp \neq ap}} \sum_{\substack{ui \in \mathcal{US} \\ ui \neq us}} \sum_{f \in \mathcal{B}} USINR_{us,b}^{ap,\lambda} \, P_{us,b}^{cp,\lambda} S_{ui,f}^{cp,\lambda}
$$

$$
+ \sum_{\substack{cp \in \mathcal{AP} \\ cp \neq ap}} USINR_{us,b}^{ap,\lambda} \, \sigma_{us,b}^{cp,\lambda} \left[1 - \sum_{\substack{ui \in \mathcal{US} \\ ui \neq us}} \sum_{f \in \mathcal{B}} S_{ui,f}^{cp,\lambda} \right] \tag{10.6}
$$

$$
+ USINR_{us,b}^{ap,\lambda} \, \sigma_{Rx} = P_{us,b}^{ap,\lambda} S_{us,b}^{ap,\lambda},
$$

Equation (10.6) can be rewritten as:

$$
\sum_{\substack{cp \in \mathcal{AP} \\ cp \neq ap}} \sum_{\substack{ui \in \mathcal{US} \\ ui \neq us}} \sum_{f \in \mathcal{B}} \left(P_{us,b}^{cp,\lambda} - \sigma_{u,f}^{b,\lambda} \right) USINR_{us,b}^{ap,\lambda} \, S_{ui,f}^{cp,\lambda}
$$

$$
+ \sum_{\substack{cp \in \mathcal{AP} \\ cp \neq ap}} USINR_{us,b}^{ap,\lambda} \, \sigma_{us,b}^{cp,\lambda} + USINR_{us,b}^{ap,\lambda} \, \sigma_{Rx} = P_{us,b}^{ap,\lambda} S_{us,b}^{ap,\lambda}, \tag{10.7}
$$

The first part of Eq. (10.7) consists of interference and background noise which is a nonlinear part composed of the multiplication of a continuous variable and a binary variable. To linearize this part, the same equations as [38, 43] were utilized which introduces a linearization variable that is non-negative; $\phi_{us,ui,b,f}^{ap,cp,\lambda} = \gamma_{us,b}^{ap,\lambda} \, S_{ui,f}^{cp,\lambda}$ as shown below:

$$
\phi_{us,ui,b,f}^{ap,cp,\lambda} \geq 0. \tag{10.8}
$$

$$
\phi_{us,ui,b,f}^{ap,cp,\lambda} \leq \alpha \, S_{ui,f}^{cp,\lambda}, \qquad \forall us, ui \in \mathcal{US}, \forall ap, cp \in \mathcal{AP}, \forall \lambda \in \mathcal{W}, \forall b, f \in \mathcal{B}
$$
$$
(us \neq ui, ap \neq cp) \tag{10.9}
$$

where α is a very large value, $\alpha \gg USINR$.

$$
\phi_{us,ui,b,f}^{ap,cp,\lambda} \leq USINR_{us,b}^{ap,\lambda}, \qquad \forall us, ui \in \mathcal{US}, \forall ap, cp \in \mathcal{AP}, \forall \lambda \in \mathcal{W}, \forall b, f \in \mathcal{B}
$$
$$
(us \neq ui, ap \neq cp) \tag{10.10}
$$

$$\phi_{us,ui,b,f}^{ap,cp,\lambda} \geq \alpha \ S_{ui,f}^{cp,\lambda} + USINR_{us,b}^{ap,\lambda} - \alpha, \quad \forall us, ui \in \mathcal{US}, \forall ap, cp \in \mathcal{AP},$$
$$\forall \lambda \in \mathcal{W}, \forall b, f \in \mathcal{B}$$
$$(us \neq ui, ap \neq cp)$$

$$(10.11)$$

Using linearization Eqs. (10.8)–(10.11) to replace the first part of Eq. (10.7), Eq. (10.7) can be re-written as follows:

$$\sum_{\substack{cp \in \mathcal{AP} \\ cp \neq ap}} \sum_{\substack{ui \in \mathcal{US} \\ ui \neq us}} \sum_{f \in \mathcal{B}} \left(P_{us,b}^{cp,\lambda} - \sigma_{u,f}^{b,\lambda} \right) \phi_{us,ui,b,f}^{ap,cp,\lambda}$$
$$+ \sum_{\substack{cp \in \mathcal{AP} \\ cp \neq ap}} USINR_{us,b}^{ap,\lambda} \ \sigma_{us,b}^{cp,\lambda} + USINR_{us,b}^{ap,\lambda} \ \sigma_{Rx} = P_{us,b}^{ap,\lambda} S_{us,b}^{ap,\lambda},$$

$$(10.12)$$

The MILP model is subject to three constraints as follows:

The first constraint ensures that a wavelength that belongs to an AP can only be allocated once and can be expressed as follows:

$$\sum_{us \in \mathcal{US}} \sum_{b \in \mathcal{B}} S_{us,b}^{ap,\lambda} \leq 1, \qquad\qquad \forall ap \in \mathcal{AP}, \forall \lambda \in \mathcal{W} \qquad (10.13)$$

The second constraint ensures that all users are assigned to one access point, one wavelength and one branch (due to the use of selection combining (SC) in the ADR receiver), i.e., one receiver branch is selected per user and can be written as follows:

$$\sum_{ap \in \mathcal{AP}} \sum_{\lambda \in \mathcal{W}} \sum_{b \in \mathcal{B}} S_{us,b}^{ap,\lambda} = 1 \ , \qquad\qquad \forall us \in \mathcal{US} \qquad (10.14)$$

The last constraint ensures that the SINR of each user should not go below 15.6 dB, which is the threshold, in order to provide a bit error rate (BER) 10^{-9} using on-off keying (OOK) modulation. This constraint can be expressed as follows:

$$USINR_{us,b}^{ap,\lambda} \geq 10^{\frac{15.6}{10}}, \qquad\qquad \forall us \in \mathcal{US}, \forall ap \in \mathcal{AP}, \forall \lambda \in \mathcal{W}, \forall b \in \mathcal{B}$$
$$(10.15)$$

The CPLEX solver was used to solve the MILP model.

4 Evaluation in Different Indoor Environments

Three different indoor environments were evaluated in this work; office, cabin and data centre. The optical channel is modelled using a ray tracing algorithm similar to [44, 45]. In addition, up to the second order reflection were modelled in this work due to the third and higher order reflections having a very low impact on the received optical power [44]. The office was divided into surfaces and one of these surfaces was divided into many small elements. These elements reflect signals following a Lambertian model and act as the secondary small emitters [46]. The area of the elements has a significant impact on the resolution of the results. Increasing the element's area, results in a reduction in the temporal resolution of the results. However, the simulation running time increases when the element's area is reduced.

4.1 Office

An empty office was examined in this work. The following sections introduce the system configuration, evaluation setup and results in a multi-user scenario.

4.1.1 Office OWC System Configuration

In this section, an unfurnished office without doors and windows is considered, as shown in Fig. 10.5. Table 10.4 introduces the OWC system parameters used.

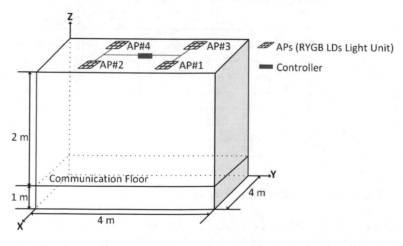

Fig. 10.5 Office dimensions

Table 10.4 The Office OWC parameters

Parameters	Configurations	
Office		
Walls and ceiling reflection coefficient	0.8 [46]	
Floor reflection coefficient	0.3 [46]	
Number of reflections	1	2
Area of reflection element	5 cm × 5 cm	20 cm × 20 cm
Order of Lambertian pattern, walls, floor and ceiling	1 [46]	
Semi-angle of reflection element at half power	60°	
Transmitters		
Number of transmitters' units	4	
Transmitters locations (x, y, z)	(1 m, 1 m, 3 m), (1 m, 3 m, 3 m), (3 m, 1 m, 3 m) and (3 m, 3 m, 3 m)	
Number of RYGB LDs per unit	9	
Transmitted optical power of red, yellow, green and blue LD	0.8, 0.5, 0.3 and 0.3 W	
Semi-angle at half power	60°	
Receiver		
Receiver noise current spectral density	4.47 pA/$\sqrt{\text{Hz}}$ [45]	
Receiver bandwidth	5 GHz	

Table 10.5 The optimized resource allocation of APs and wavelengths

User	Location (x, y, z)	Access point	Wavelength	Receiver branch
1	(0.5 m, 0.5 m, 1 m)	1	Yellow	1
2	(0.5 m, 2.5 m, 1 m)	2	Red	1
3	(1.5 m, 1.5 m, 1 m)	1	Red	3
4	(1.5 m, 3.5 m, 1 m)	2	Yellow	3
5	(2.5 m, 0.5 m, 1 m)	3	Red	1
6	(2.5 m, 2.5 m, 1 m)	4	Yellow	1
7	(3.5 m, 1.5 m, 1 m)	3	Yellow	3
8	(3.5 m, 3.5 m, 1 m)	4	Red	3

4.1.2 Office OWC System Setup and Results

A scenario of eight users was examined in this part where two users were placed under each AP. Table 10.5 shows the locations of the 8 users and the optimized resource allocation using the MILP model described in Sect. 3.2.

Figure 10.6 shows the channel bandwidth, SINR and data rate of the 8 user scenario. All user locations support a high channel bandwidth above 8 GHz. In

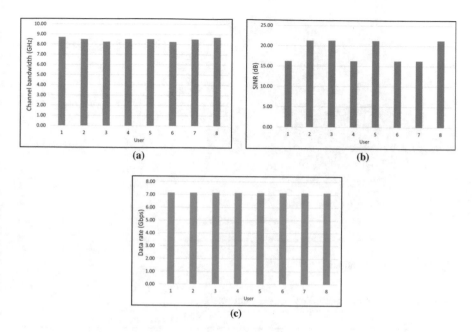

Fig. 10.6 (**a**) Channel bandwidth, (**b**) SINR and (**c**) data rate

addition, all user locations can provide high SINR above the threshold (15.6 dB). However, users assigned to the Yellow wavelength have a lower SINR compared to those assigned to Red wavelength. The reason for that is that the Yellow wavelength has a lower transmitted power (needed to ensure the version of white colour desired [38]) and receiver responsivity compared to the Red wavelength. In terms of data rate, all users can support a data rate above 7 Gbps.

4.2   Aircraft Cabin

An aircraft cabin downlink is evaluated in this section to support multiple users. OWC does not contribute further electromagnetic interference and is therefore attractive in this environment. The system configuration, including the type of aircraft, is shown in the next section. The scenario evaluated and the result are provided in the following section.

4.2.1   Aircraft Cabin OWC System Configuration

The type of aircraft examined as an indoor environment in this work is an Airbus A321neo (see Fig. 10.7) [47]. This aircraft has one class (economy) which consists

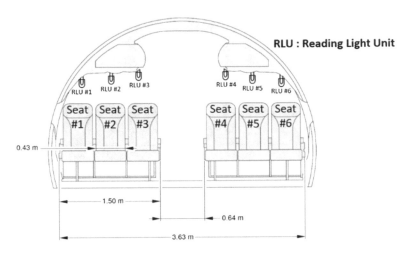

Fig. 10.7 Cabin dimentions

Table 10.6 The aircraft cabin OWC system parameters

Parameters	Configurations	
Cabin		
Walls and ceiling reflection coefficient	0.8 [46]	
Floor reflection coefficient	0.3 [46]	
Number of reflections	1	2
Area of reflection element	5 cm × 5 cm	20 cm × 20 cm
Order of Lambertian pattern, walls, floor and ceiling	1 [46]	
Semi-angle of reflection element at half power	60°	
Transmitters		
Number of transmitters' units	6	
Transmitters locations (RLU)	Above each seat	
Number of RYGB LDs per unit	3	
Transmitted optical power of red, yellow, green and blue LD	0.8, 0.5, 0.3 and 0.3 W	
Semi-angle at half power	19°	
Receiver		
Receiver noise current spectral density	4.47 pA/$\sqrt{Hz}$ [45]	
Receiver bandwidth	5 GHz	

of 202 passenger seats. The dimensions of the cabin are 36.85 m length × 3.63 m width. The cabin surfaces were divided into small sectors to carry out ray tracing to evaluate the channel impulse response, its delay spread and banwidth. Table 10.6 introduces the parameter used for cabin communication. Communication is blocked below the surface of the seats.

4.2.2   Aircraft Cabin OWC System Setup and Results

In this part, three devices per passenger have been assumed to examine the resource allocation problem for these devices. The MILP model introduced in Sect. 3.2 was used here to optimize the resource allocation. The locations of the three devices are assumed as follows: device 1 is placed at the centre of the seat, while devices 2 and 3 are located at a different corners of the seat. The optimized resource allocation of APs and wavelengths is shown in Table 10.7.

The channel bandwidth, SINR and data rate are shown in Fig. 10.8. Device number one in all seats, which is located at the centre of the seat has the best channel bandwidth and SINR compared to the other devices. All devices in all seats can support high SINR beyond the threshold and high data rate beyond 7 Gbps.

4.3   Data Centre

This section examines the resource allocation for the downlink OWC system in a data centre. The following sections discuss the data centre configuration and the results.

4.3.1   Data Centre OWC System Configuration

The proposed data centre is assumed to be divided into pods, with each pod having dimensions of: 5 m length × 6 m width × 3 m height as shown in Fig. 10.9. Each pod contains two rows and each row includes five racks. In this data centre, each rack has its own switch at the top of the rack. This switch works as a communication coordinator to connect the rack's servers and the racks inside the data centre. The racks' dimensions are shown in Fig. 10.9. One meter or more has been set between the two rows of racks, as well as between the walls and each row of racks for ventilation. Each rack has its own receiver placed at the centre of the top of the rack to receive signals, as illustrated in Fig. 10.9. Table 10.8 introduces the simulation parameters of the data centre.

4.3.2   Data Centre OWC Setup and Results

A data centre consisting of ten racks is examined in this section. Each rack has its own receiver located at the centre of the top of each rack. The MILP model in Sect. 3.2 was used here to assign racks to APs and wavelengths. Table 10.9 illustrates the optimized resource allocation of APs and wavelengths to each rack.

Table 10.7 The optimized resource allocation of APs and wavelengths

Device #	Reading light unit #	Branch #	Wavelength	Reading light unit #	Branch #	Wavelength	Reading light unit #	Branch #	Wavelength
Passenger #1				Passenger #2			Passenger #3		
1	1	4	Red	2	2	Red	3	3	Red
2	1	2	Green	2	2	Yellow	3	2	Green
3	1	3	Yellow	2	3	Green	3	3	Yellow
Passenger #4				Passenger #5			Passenger #6		
1	4	3	Red	5	2	Red	6	4	Red
2	4	3	Yellow	5	3	Green	6	3	Yellow
3	4	2	Green	5	2	Yellow	6	2	Green

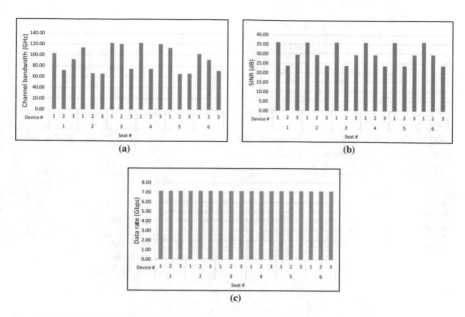

Fig. 10.8 (a) Channel bandwidth, (b) SINR and (c) data rate

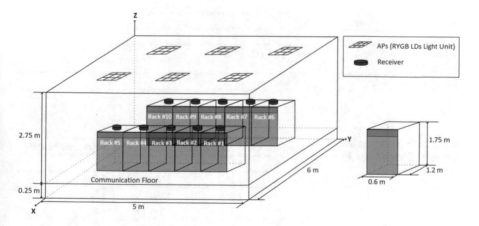

Fig. 10.9 Data center dimensions

Figure 10.10 illustrates the achieved channel bandwidth, SINR and data rate. All racks support a high channel bandwidth and high SINR above 15.6 dB which is the threshold. Racks assigned to the Red wavelength provide a higher SINR compared to the other racks. All racks can offer a high data rate beyond 7 Gbps.

Table 10.8 The data centre OWC parameters

Parameters	Configurations	
Data centre		
Walls and ceiling reflection coefficient	0.8 [46]	
Floor reflection coefficient	0.3 [46]	
Number of reflections	1	2
Area of reflection element	5 cm × 5 cm	20 cm × 20 cm
Order of Lambertian pattern, walls, floor and ceiling	1 [46]	
Semi-angle of reflection element at half power	60°	
Transmitters		
Number of transmitters' units	6	
Transmitters locations (RLU)	(1.6 m, 1.5 m, 3 m), (1.6 m, 2.5 m, 3 m), (1.6 m, 3.5 m, 3 m), (4.4 m, 1.5 m, 3 m), (4.4 m, 2.5 m, 3 m) and (4.4 m, 3.5 m, 3 m)	
Number of RYGB LDs per unit	9	
Transmitted optical power of red, yellow, green and blue LD	0.8, 0.5, 0.3 and 0.3 W	
Semi-angle at half power	60°	
Receiver		
Receiver noise current spectral density	4.47 pA/$\sqrt{\text{Hz}}$ [45]	
Receiver bandwidth	5 GHz	

Table 10.9 The optimized resource allocation of APs and wavelengths

Rack #	Access point	Wavelength	Receiver branch
1	1	Red	1
2	1	Yellow	4
3	2	Red	2
4	3	Yellow	2
5	3	Red	3
6	4	Red	2
7	4	Yellow	3
8	5	Red	1
9	6	Yellow	2
10	6	Red	4

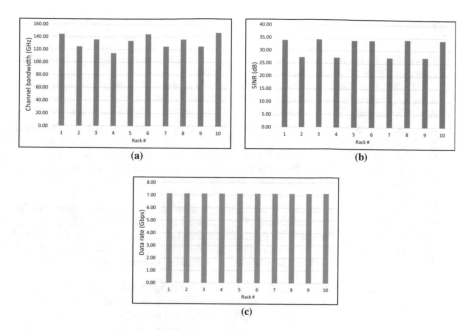

Fig. 10.10 (**a**) Channel bandwidth, (**b**) SINR and (**c**) data rate

5   Conclusions

This chapter introduced and discussed the optimization of resource allocation in an optical wireless communication (OWC) system, specifically, a visible light communication (VLC) system. A MILP model was developed to optimize the resource allocation. A wavelength division multiple access (WDMA) system was considered to support multiple users. The optimized resource allocation shows the best assignment of an access point (AP) and wavelength to users. A visible light illumination engine that uses red, green, yellow and blue (RGYB) laser diodes (LD) was utilized to offer high bandwidth for the communication purposes and white light for illumination. An angle diversity receiver (ADR) was utilized to receive signals and reduce noise and interference. Three different indoor environments (office, cabin and data centre) were evaluated using different scenario in this chapter. A scenario of eight users was examined in the office environment, while, in the cabin, three devices per passenger were evaluated. In the data centre, each rack connects individually to an AP. All scenarios in all environments can support a high data rate beyond 7 Gbps.

Acknowledgements The authors would like to acknowledge funding from the Engineering and Physical Sciences Research Council (EPSRC), INTERNET (EP/H040536/1), STAR (EP/K016873/1) and TOWS (EP/S016570/1). All data is provided in the results section of this paper. The first author would like to thank Umm Al-Qura University in the Kingdom of Saudi Arabia for funding his PhD scholarship.

References

1. A. Alexiou, M. Haardt, Smart antenna technologies for future wireless systems: Trends and challenges. IEEE Commun. Mag. **42**(9), 90–97 (2004)
2. A.J. Paulraj, D.A. Gore, R.U. Nabar, H. Bölcskei, An overview of MIMO communications – A key to gigabit wireless. Proc. IEEE, 198–217 (2004)
3. Z. Ghassemlooy, W. Popoola, S. Rajbhandari, *Optical Wireless Communications: System and Channel Modelling with Matlab*® [Internet]. Ieeexplore.Ieee.Org (2012), 513p. Available from: https://books.google.co.in/books?hl=en&oi=fnd&pg=PP1&dq=Optical+Wireless+Communications:+System+and+Channel+Modelling+with+MATLAB%EF%BF%BD&ots=Ya_CDECP3e&lr=&id=jpXGCN1qVQ4C&redir_esc=y#v=onepage&q=Optical%20Wireless%20Communications%3A%20System%20and%20C
4. F.E. Alsaadi, M.A. Alhartomi, J.M.H. Elmirghani, Fast and efficient adaptation algorithms for multi-gigabit wireless infrared systems. J. Light. Technol. **31**(23), 3735–3751 (2013)
5. A.T. Hussein, J.M.H. Elmirghani, 10 Gbps mobile visible light communication system employing angle diversity, imaging receivers, and relay nodes. J. Opt. Commun. Netw. [Internet] **7**(8), 718–735 (2015) Available from: https://www.osapublishing.org/abstract.cfm?URI=jocn-7-8-718
6. S.H. Younus, J.M.H. Elmirghani, in *WDM for High-Speed Indoor Visible Light Communication System*. International Conference on Transparent Optical Networks. Girona (2017), pp. 1–6
7. A.T. Hussein, M.T. Alresheedi, J.M.H. Elmirghani, 20 Gb/s mobile indoor visible light communication system employing beam steering and computer generated holograms. J. Light. Technol. **33**(24), 5242–5260 (2015)
8. A.T. Hussein, M.T. Alresheedi, J.M.H. Elmirghani, in *25 Gbps Mobile Visible Light Communication System Employing Fast Adaptation Techniques*. 2016 18th International Conference on Transparent Optical Networks (ICTON). Trento (2016)
9. O. Aletri, M.T. Alresheedi, J.M.H. Elmirghani, Transmitter Diversity with Beam Steering, in *2019 21st International Conference on Transparent Optical Networks (ICTON)*, (IEEE, Angers, France, 2019), pp. 1–5
10. O. Aletri, M.O.I. Musa, M.T. Alresheedi, J.M.H. Elmirghani, Visible Light Optical Data Centre Links, in *2019 21st International Conference on Transparent Optical Networks (ICTON)*, (IEEE, Angers, France, 2019), pp. 1–5
11. O. Aletri, M.T. Alresheedi, J.M.H. Elmirghani, Optical Wireless Cabin Communication System, in *2019 IEEE Conference on Standards for Communications and Networking (CSCN) [Internet]*, (IEEE, GRANADA, Spain, 2019), pp. 1–4. Available from: https://ieeexplore.ieee.org/document/8931345/
12. O. Aletri, M.O.I. Musa, M.T. Alresheedi, J.M.H. Elmirghani, Co-existence of Micro, Pico and Atto Cells in Optical Wireless Communication, in *2019 IEEE Conference on Standards for Communications and Networking (CSCN) [Internet]*, (IEEE, GRANADA, Spain, 2019), pp. 1–5. Available from: https://ieeexplore.ieee.org/document/8931323/
13. K.L. Sterckx, J.M.H. Elmirghani, R.A. Cryan, Sensitivity assessment of a three-segment pyrimadal fly-eye detector in a semi-disperse optical wireless communication link. IEE Proc. Optoelectron. **147**(4), 286–294 (2000)
14. A.G. Al-Ghamdi, J.M.H. Elmirghani, Performance evaluation of a triangular pyramidal fly-eye diversity detector for optical wireless communications. IEEE Commun. Mag. **41**(3), 80–86 (2003)
15. M.T. Alresheedi, J.M.H. Elmirghani, Hologram selection in realistic indoor optical wireless systems with angle diversity receivers. IEEE/OSA J. Opt. Commun. Netw. **7**(8), 797–813 (2015)
16. A.T. Hussein, M.T. Alresheedi, J.M.H. Elmirghani, Fast and efficient adaptation techniques for visible light communication systems. J. Opt. Commun. Netw. **8**(6), 382–397 (2016)

17. S.H. Younus, A.A. Al-Hameed, A.T. Hussein, M.T. Alresheedi, J.M.H. Elmirghani, Parallel data transmission in indoor visible light communication systems. IEEE Access [Internet] **7**, 1126–1138 (2019) Available from: https://ieeexplore.ieee.org/document/8576503/

18. A.G. Al-Ghamdi, M.H. Elmirghani, Optimization of a triangular PFDR antenna in a fully diffuse OW system influenced by background noise and multipath propagation. IEEE Trans. Commun. **51**(12), 2103–2114 (2003)

19. A.G. Al-Ghamdi, J.M.H. Elmirghani, in *Characterization of Mobile Spot Diffusing Optical Wireless Systems with Receiver Diversity*. ICC'04 IEEE International Conference on Communications (2004)

20. F.E. Alsaadi, J.M.H. Elmirghani, Performance evaluation of 2.5 Gbit/s and 5 Gbit/s optical wireless systems employing a two dimensional adaptive beam clustering method and imaging diversity detection. IEEE J. Sel. Areas Commun. **27**(8), 1507–1519 (2009)

21. M.T. Alresheedi, A.T. Hussein, J.M.H. Elmirghani, Uplink design in VLC systems with IR sources and beam steering. IET Commun. [Internet] **11**(3), 311–317 (2017) Available from: http://digital-library.theiet.org/content/journals/10.1049/iet-com.2016.0495

22. O. Aletri, M.T. Alresheedi, J.M.H. Elmirghani, Infrared Uplink Design for Visible Light Communication (VLC) Systems with Beam Steering, in *2019 IEEE International Conference on Computational Science and Engineering (CSE) and IEEE International Conference on Embedded and Ubiquitous Computing (EUC) [Internet]*, (IEEE, New York, NY, USA, 2019), pp. 57–60. Available from: http://arxiv.org/abs/1904.02828

23. J.M.H. Elmirghani, T. Klein, K. Hinton, L. Nonde, A.Q. Lawey, T.E.H. El-Gorashi, et al., GreenTouch GreenMeter core network energy-efficiency improvement measures and optimization [Invited]. J. Opt. Commun. Netw. **10**(2), 250–269 (2018)

24. M.T. Alresheedi, J.M.H. Elmirghani, 10 Gb/s indoor optical wireless systems employing beam delay, power, and angle adaptation methods with imaging detection. IEEE/OSA J. Light. Technol. **30**(12), 1843–1856 (2012)

25. K.L. Sterckx, J.M.H. Elmirghani, R.A. Cryan, Pyramidal fly-eye detection antenna for optical wireless systems. IEE Colloq. Opt. Wirel. Commun.. (Ref. No. 1999/128), 5/1–5/6 (1999)

26. F.E. Alsaadi, M. Nikkar, J.M.H. Elmirghani, Adaptive mobile optical wireless systems employing a beam clustering method, diversity detection, and relay nodes. IEEE Trans. Commun. **58**(3), 869–879 (2010)

27. F.E. Alsaadi, J.M.H. Elmirghani, Adaptive mobile line strip multibeam MC-CDMA optical wireless system employing imaging detection in a real indoor environment. IEEE J. Sel. Areas Commun. **27**(9), 1663–1675 (2009)

28. M.T. Alresheedi, J.M.H. Elmirghani, Performance evaluation of 5 Gbit/s and 10 Gbit/s mobile optical wireless systems employing beam angle and power adaptation with diversity receivers. IEEE J. Sel. Areas Commun. **29**(6), 1328–1340 (2011)

29. F.E. Alsaadi, J.M.H. Elmirghani, Mobile multi-gigabit indoor optical wireless systems employing multibeam power adaptation and imaging diversity receivers. IEEE/OSA J. Opt. Commun. Netw. **3**(1), 27–39 (2011)

30. A.G. Al-Ghamdi, J.M.H. Elmirghani, Line strip spot-diffusing transmitter configuration for optical wireless systems influenced by background noise and multipath dispersion. IEEE Trans. Commun. **52**(1), 37–45 (2004)

31. F.E. Alsaadi, J.M.H. Elmirghani, High-speed spot diffusing mobile optical wireless system employing beam angle and power adaptation and imaging receivers. J. Light. Technol. **28**(16), 2191–2206 (2010)

32. H.H. Chan, K.L. Sterckx, J.M.H. Elmirghani, R.A. Cryan, Performance of optical wireless OOK and PPM systems under the constraints of ambient noise and multipath dispersion. IEEE Commun. Mag. **36**(12), 83–87 (1998)

33. A.G. Al-Ghamdi, J.M.H. Elmirghani, Spot diffusing technique and angle diversity performance for high speed indoor diffuse infra-red wireless transmission. IEE Proc. Optoelectron. **151**(1), 46–52 (2004)

34. A.G. Al-Ghamdi, J.M.H. Elmirghani, Analysis of diffuse optical wireless channels employing spot-diffusing techniques, diversity receivers, and combining schemes. IEEE Trans. Commun. **52**(10), 1622–1631 (2004)

35. F.E. Alsaadi, J.M.H. Elmirghani, Adaptive mobile spot diffusing angle diversity MC-CDMA optical wireless system in a real indoor environment. IEEE Trans. Wirel. Commun. **8**(4), 2187–2192 (2009)

36. J.M.H. Elmirghani, R.A. Cryan, New PPM-CDMA hybrid for indoor diffuse infrared channels. Electron. Lett. **30**(20), 1646–1647 (1994)

37. A. Neumann, J.J. Wierer, W. Davis, Y. Ohno, S.R.J. Brueck, J.Y. Tsao, Four-color laser white illuminant demonstrating high color-rendering quality. Opt. Express [Internet] **19**(S4), A982 (2011) Available from: https://www.osapublishing.org/oe/abstract.cfm?uri=oe-19-S4-A982

38. O. Aletri, A.A. Alahmadi, S.O.M. Saeed, S.H. Mohamed, T.E.H. El-Gorashi, M.T. Alresheedi, et al., Optimum resource allocation in optical wireless systems with energy-efficient fog and cloud architectures. Philos. Trans. R. Soc. A: Math. Phys. Eng. Sci. [Internet] **378**(2169) (2020) Available from: https://royalsocietypublishing.org/doi/10.1098/rsta.2019.0188. Cited 2020 July 25

39. O. Aletri, A.A. Alahmadi, S.O.M. Saeed, S.H. Mohamed, T.E.H. El-Gorashi, M.T. Alresheedi, et al., Optimum Resource Allocation in 6G Optical Wireless Communication Systems, in *2020 2nd 6G Wireless Summit (6G SUMMIT) [Internet]*, (IEEE, Washington, DC, 2020), pp. 1–6. Available from: https://ieeexplore.ieee.org/document/9083828/. Cited 2020 May 8

40. G. Cossu, A.M. Khalid, P. Choudhury, R. Corsini, E. Ciaramella, 3.4 Gbit/s visible optical wireless transmission based on RGB LED. Opt. Express [Internet] **20**(26), B501–B506 (2012) Available from: http://www.ncbi.nlm.nih.gov/pubmed/23262894

41. Y. Wang, Y. Wang, N. Chi, J. Yu, H. Shang, Demonstration of 575-Mb/s downlink and 225-Mb/s uplink bi-directional SCM-WDM visible light communication using RGB LED and phosphor-based LED. Opt. Express [Internet] **21**(1), 1203 (2013) Available from: https://www.osapublishing.org/oe/abstract.cfm?uri=oe-21-1-1203

42. F.-M. Wu, C.-T. Lin, C.-C. Wei, C.-W. Chen, Z.-Y. Chen, K. Huang, in *3.22-Gb/s WDM Visible Light Communication of a Single RGB LED Employing Carrier-Less Amplitude and Phase Modulation*. Optical Fiber Communication Conference/National Fiber Optic Engineers Conference 2013 [Internet] (2013), p. OTh1G.4. Available from: https://www.osapublishing.org/abstract.cfm?uri=OFC-2013-OTh1G.4

43. M.S. Hadi, A.Q. Lawey, T.E.H. El-Gorashi, J.M.H. Elmirghani, Patient-centric cellular networks optimization using big data analytics. IEEE Access **7**, 49279–49296 (2019)

44. J.R. Barry, J.M. Kahn, W.J. Krause, E.A. Lee, D.G. Messerschmitt, Simulation of multipath impulse response for indoor wireless optical channels. IEEE J. Sel. Areas Commun. **11**(3), 367–379 (1993)

45. A.T. Hussein, J.M.H. Elmirghani, Mobile multi-gigabit visible light communication system in realistic indoor environment. J. Light. Technol. **33**(15), 3293–3307 (2015)

46. F.R. Gfeller, U. Bapst, Wireless in-house data communication via diffuse infrared radiation. Proc. IEEE **67**(11), 1474–1486 (1979)

47. *Aircraft Characteristics – Airport Operations and Technical Data – Airbus* [Internet]. Available from: https://www.airbus.com/aircraft/support-services/airport-operations-and-technical-data/aircraft-characteristics.html. Cited 2020 May 8

Chapter 11
Machine Type Communications in 6G

Tristan Braud, Dimitris Chatzopoulos, and Pan Hui

Abstract The sixth generation of mobile networks (6G) is expected to be deployed in the early 2030s. By this time, the density of autonomous Internet-connected machines is expected to explode up to hundreds of devices per cubic meter. These devices (1) generate voluminous multisensory data, (2) access sophisticated artificial intelligence-based services with high frequency, and (3) have widely diverse constraints in terms of latency, bandwidth, energy, and computation power. Such devices are not operated by humans, and communicate with each other and with remote servers located on the network core or edge: The wireless communications between these machines are called machine-type communications (MTC) and can either be between multiple machines that collectively collect and process multidimensional information or between machines that interact with services located on servers. Representative examples include autonomous driving, piloting crewless aerial vehicles, smart grid energy trading, and others. In this chapter, we define the following requirements for 6G, following the predicted density and heterogeneity of the autonomous connected device landscape: (1) ultra-dense wireless communication networks, (2) massive multi-access edge computing, (3) large-scale autonomous operation of devices with heterogeneous requirements and constraints. To address these requirements, 6G will enable a convergence of computing, energy, and communication for device- and application-aware communications. We discuss how 6G can achieve such convergence and highlight the future trends for MTC to ubiquitously integrate the computing landscape.

T. Braud (✉) · D. Chatzopoulos
Department of Computer Science and Engineering, Hong Kong University of Science and Technology, Hong Kong, China
e-mail: braudt@ust.hk; dcab@cse.ust.hk

P. Hui
Department of Computer Science and Engineering, Hong Kong University of Science and Technology, Hong Kong, China

Department of Computer Science, University of Helsinki, Helsinki, Finland
e-mail: panhui@ust.hk

207

1 Introduction

In recent years, the capabilities and intricacy of Internet-connected devices have increased dramatically, allowing to delegate more operations than before to automated systems [1, 2]. Nowadays, sensors, actuators, servers, and other connected physical machines can interact and operate without any human intervention.

Machine-type communications (MTC), also known as machine-to-machine (M2M) communications designates such as automated operation and communication between devices, through fixed or wireless networks. Use cases include connected vehicles and roadside infrastructure [3], smart grids [4], and factory automation [5]. In the upcoming years, the number of autonomous connected devices is expected to explode, progressively replacing human-operated devices as main users of the Internet [6].

MTC has widely different characteristics from human-type communications (HTC), where a human operator ultimately interacts with the devices. Human-type communication often revolves around interactivity on a human scale. As such, low latency and high bandwidth are often the two main parameters of HTC. Besides, most HTC applications have some degree of tolerance with variations in bandwidth and latency. For instance, cloud gaming can tolerate up to 160 ms [7] round trip latency, while the video quality can be temporarily decreased in case of bandwidth drops. On the other hand, MTC has widely different constraints, depending on the devices and the considered use cases. Devices participating in the MTC ecosystem need to account for limited computation capabilities, very long or very short transmission distance, ultra-high device density, and limited energy consumption while addressing constraints in terms of reliability, bandwidth, and latency that are strongly depending on the application [8]. As a result, multiple protocols were developed for long-distance communication (SigFox [9] and LoRA [10]), swarms of devices (802.15.4 [11]), or personal area networks (BLE[12]). The fifth generation of mobile networks (5G) aimed to bring some convergence in this fragmented landscape [13]. However, the task of bringing together the multiple use cases of MTC together with HTC within a single communication method is far from trivial. Humans have finite sensing capabilities, with a higher tolerance to variations in the network's core parameters. In 2020, machines have already surpassed human sensing capabilities, and some M2M applications present latency tolerance outside of typical human applications (see Table 11.1). Despite the recent deployment of 5G networks around the world, migrating from an HTC-centric network to an MTC-centric network remains an open question for researchers and industry alike.

The sixth generation of mobile networks (6G) is in its infancy. However, several researchers, companies, and organizations have released white papers on the topic [14–16], allowing us to refine a future vision of 6G and MTC. 6G will be driven by a significant increase in the number of connected autonomous machines [17]. Depending on the source, it is expected that by 2030, between 50 billion [6] and 500 billion [18] devices will be connected to the Internet, significantly exceeding the human population. As envisioned by Letaief et al. *"6G will go beyond mobile*

Table 11.1 Human and machine perception limitations. Typical values are given for devices in 2020

	Human	Machine
FOV	5° central (text character recognition)	360°
	120° azimuth peripheral	
	100° zenith peripheral	
Resolution	Very high in central vision	Unlimited, 360° (8–16 K for typical 360° cameras)
Audible range	20 Hz–20 kHz (often narrower)	Unlimited (20 Hz–20 kHz for a typical smartphone microphone)
Visible range	380–700 nm	Unlimited (360–1000 nm for a typical camera sensor)
Latency	<20 ms (AR/VR)	<1 ms (safety applications)
	<100 ms (cloud gaming)	>10 min (energy-constrained sensing)
	<1 s (web browsing)	

Internet and will be required to support ubiquitous AI services from the core to the end devices of the network. Meanwhile, AI will play a critical role in designing and optimizing 6G architectures, protocols, and operations" [14]. As a result, machines will be the main users of network and computation resources. Such a paradigm shift from human-centric communication networks to machine-centric communication networks will lead to significant challenges, as follows:

- *Performance:* 6G will accommodate hundreds of billions of devices, that continuously exchange information in real-time. 6G networks will thus require to support thousands of concurrent access to the medium for a single access point, with a bandwidth large enough for real-time transmission of large volumes of information, and sub-millisecond latency.
- *Network Architecture:* 6G cannot only rely on performance improvement to account for the migration from HTC to MTC. MTC brings much more diverse use cases driven by the constraints of the devices and applications. Some devices may require sub-millisecond latency for safety applications (e.g., connected vehicles), while others may only be able to communicate once every hour due to a limited energy consumption requirement. Such devices should not only be able to communicate, but the network should also provide strategies to accommodate both use cases. The increased complexity of the traffic will, therefore, require to significantly adapt the network architecture towards AI-operated networks that automatically optimize, adapt, and maintain wireless communication to autonomous devices.
- *Reliability:* With the development of highly sensitive applications where human lives are at stake, 6G networks will need to provide reliability guarantees, with robust fallback operations for the most critical applications.
- *Trustworthiness:* Devices in the 6G landscape will continually monitor, sense, and interpret the physical world and its actors. As such, data on 6G networks will

be sensitive by nature, and the privacy of the data subjects needs to be enforced. The strong reliance on AI will require specific mechanics of technical robustness, human agency, non-discrimination, transparency, and accountability.

In this chapter, we address these challenges following a bottom-up approach. We survey the current works to develop the idea of a future machine-centric 6G network that does not impede on HTC, from the physical to the application layer. The rest of this chapter is organized as follows. In Sect. 2, we summarize the devices and applications that will participate in the 6G MTC landscape. We then discuss the fundamental techniques enabling MTC at physical and access layers in Sects. 3.3 and 3.4. We then move on to the network and transport layers and display the technologies and strategies 6G should develop to achieve the requirements mentioned above in Sect. 4. Finally, we detail application-specific requirements and discuss how a cross-layer approach can bring convergence between these requirements Sect. 5.

2   MTC Applications and Devices

With the explosion of the IoT, more devices than ever are connected to the Internet. These devices are becoming part of entirely autonomous systems, that the human user only accesses through a single endpoint, abstracting the complexity of the underlying architecture. In this section, we establish the applications that will likely shape the landscape of MTC in 2030.

2.1   General, Non-critical IoT

Before introducing more specific applications, we expect the bulk of MTC applications to be composed of small to medium scale swarms of devices of IoT devices, each operating within a well-defined scope. These swarms of devices primarily sense their environment and periodically update a more powerful nearby or remote machine. This setting is characterized by proximity between the devices, low data rates, and limited mobility. Besides, applications are generally delay-tolerant (up to several minutes) and can afford some degree of information loss (with few explicit exceptions). Smart homes are an excellent example of such a setting. A multitude of smart devices is fixed in close proximity within a single location. These devices periodically report information related to temperature, humidity, motion, electric consumption, etc. to a central processing unit, located either within the same physical location or in the cloud. Oftentimes, each device reports a single parameter (e.g. temperature) and only sporadically sends small-sized data packets over the network, with a period on a scale of minutes. Some, if not most of these devices are either redundant or are targeting an application that can afford some degree of losses. For instance, in the case of automated heating, if a temperature sensor in a single bedroom fails to transmit a value, the application can reuse the previous value or infer from the temperature sensors located in nearby rooms with little impact on the end result. Few applications require extreme reliability. In a smart home, such applications are mainly safety-related (e.g., smoke detection or intrusion detection), and require real-time and reliable delivery of the information to the concerned party.

Although these applications generate primarily best-effort traffic that has a low footprint on the network, due to its magnitude, general IoT will still represent a significant challenge for 6G. Even with intermittent, low-data rate transmission, the IoT will result in hundreds of thousands if not millions of devices connecting concurrently to the same cell and generate considerable amounts of data in the core network. Besides, the diversity of applications and their associated constraints will significantly complicate network planning to maintain a high quality of service. Although solutions such as vertical network slicing may allow us to isolate such applications, they are neither convenient nor cost-effective to deploy at a smaller scale.

2.2 Connected Vehicles and Road Safety

In recent years, connected vehicles have become an increasingly active application of MTC. These vehicles are embedded with a wide range of sensors, computation units, and communication capabilities. Similarly, fixed roadside infrastructure is starting to provide sensing, computation, and communication capabilities. By facilitating the exchange of information between vehicles, roadside infrastructure, user devices, and remote servers, connected vehicles enable a multitude of applications, pertaining to both HTC and MTC. Such applications encompass in-car infotainment, driver assistance, emergency systems, and crash prevention, traffic optimization, and autonomous driving. In the scope of connected vehicles, MTC applications represent not only the applications that involve minimal human intervention (e.g., autonomous driving, automated braking, crash avoidance solutions), but also the systems that aggregate and interpret information to present it to the human driver in a condensed form. For instance, many driver assistance systems strongly rely on automated communication between vehicles and roadside units to detect or infer potential obstacles, only to display a small share of this information to the driver.

Connected vehicles are a use case of MTC in 6G that has both stringent requirements and strict constraints. Besides, in-car infotainment and basic driver assistance systems (e.g., traffic jam detection) that only require large bandwidth and a reasonable latency (1s), most applications can have direct consequences on human life. As such, the latency between an entity sensing information and a car taking a decision should be kept to an absolute minimum, ideally in the order of the millisecond. Furthermore, connected vehicles require extreme reliability of the network and computation infrastructure as information delay or loss can have dramatic consequences. Regarding bandwidth, current connected vehicles's sensors generate between 3 Gbps (10.8 Tb/h) and 40 Gbps (144 Tb/h) of data[19]. Exchanging these amounts of data with nearby vehicles or the infrastructure requires considerable bandwidth, as hundreds of vehicles may be connected to the same cell. For these reasons, contrary to most other MTC applications, connected vehicles integrate powerful computation units that allow them to process information on-board. This operation allows connected vehicles to perform emergency decisions without depending on the availability of the network or the external computation resources. However, many applications still require communication between vehicles, with the above constraints.

2.3 Smart Cities as Organic Systems

City-systems are becoming increasingly complex and distributed, encompassing both connected vehicles and general IoT as described above, but also a multitude of other applications on a large scale. These massive organic entities can be compared to computer systems, where information is exchanged between the various components and presented to the user through a single interface, at a much larger scale. By 2030, the scale and complexity of smart cities will be such that it will be impossible for humans to operate them directly. Facing such complexity, it is critical to developing abstraction layers to interact with future smart cities. As such, it is necessary to consider a smart city as a single massive distributed system. MTC is the enabler of such evolution. In traditional computer systems, buses interconnect the elements composing the system. Similarly, in smart cities, buses are replaced by wireless networks that interconnect the devices that autonomously sense, aggregate, interpret, and display urban data. Servers deployed at the edge and in the cloud assume the role of central processing units and storage devices, while end user equipment such as smartphones, smart wearables, and augmented reality headsets replace screens.

With almost unlimited sensing and processing capabilities, smart cities may encounter a bottleneck with the communication medium. As in computer systems, low performance at a single point can dramatically reduce the performance of the computer system. For instance, in a desktop computer, a low-speed hard-drive will impede the perceived operating speed of the system. In smart cities, low network bandwidth, high latency, or high jitter in a single area may degrade the performance of the entire city-system, and even have an impact on the human end-user [20].

As such, the challenges posed by smart cities are not only in terms of the resources available but also raise questions related to the actual network deployment. With the convergence of wireless technologies brought by 6G, future smart cities will be intrinsically correlated to future cellular network geographic deployments. In addition to purely engineering problems, city-systems will face issues related to urban planning.

2.4 Industry Automation

In recent years, the industry has gone through a complete digital transformation. An essential facet of Industry 4.0 is autonomous production methods, enabled by MTC. The increase in performance brought by 6G will be critical in establishing completely autonomous production lines. Besides, experts foresee that the next industrial revolution will be defined by the cooperation between humans and robots on production lines, introducing cognitive capabilities in production lines[21]. Such a requirement will increasingly leverage the capabilities of 6G and require flawless communication between the different systems composing the production line.

Most industrial applications rely on control loops that are updated with a high frequency and are characterized by an extremely reliable and stable environment. For instance, the closed control loop of a servo motor may update more than 100 times per second. As such, industry automation requires sub-millisecond latency (ideally 0.1 ms) with extreme reliability (loss rate below 10^{-9})[22] to accommodate the high frequency of the control loop updates. With such low latency targets, jitter is another critical parameter for industrial applications. On the other hand, such control applications require a lower bandwidth than the other applications mentioned above, with payloads in the order of 10–100 bytes.

2.5 Body Area Networks

A body area network is a wireless sensor network where the devices are placed over the body of the user. Devices participating in body area networks (BANs) may be embedded inside the body as implants, carried on a fixed position (e.g., smartwatch), or carried by the user at all times (e.g., smartphone). Such body area networks primarily target telemedicine, with applications as varied as active rehabilitation, patient monitoring, or assisted living [23], although novel applications have also been developed related to disaster management [24].

Applications may present various requirements, depending on how critical they are to the user. However, we can identify the following four critical requirements for BANs. First, human life may directly depend on the reliability of the system. As such, *high reliability* is required, both in terms of accuracy of the health data and in the transmission of such data. Similarly, some patient monitoring applications handle emergency data, that require ultra-fast response times. *Low-latency* is thus key in many applications. Body area networks are likely to track either medical information or personal information. *Privacy and Security* should be at the core of BAN systems, by design and by default. Finally, *Power Consumption* is a concern for all permanent or semi-permanent devices, such as sensors embedded as implants. These devices will rely on energy harvesting techniques and join the category of low-to-zero energy IoT.

Due to the physical location of devices participating in BANs, achieving such requirements may be difficult. Many independent devices may be located in the same location, interfering with each other. Besides, the human body attenuates radio frequencies and can act as a physical barrier between devices. Finally, the limited power consumption of some devices limits their computation and communication capabilities.

2.6 Low-to-Zero Energy IoT

The deployment of IoT devices, which are equipped with multiple sensors and communication interfaces, are able to form device-to-device networks, and performing complex tasks, in modern buildings that have access to the smart grid and renewable energy resources will make low-to-zero energy buildings feasible [25]. The cross-technology communication protocols for MTC that will be integrated into 6G will allow various types of sensors and actuators to utilize the collectively collected voluminous data and decide when it makes the most sense, both from a cost and efficiency standpoint, to adapt the energy consumption of a building. MTC-enabled applications in smart buildings for minimal energy consumption will be assisted by the AI-based components of 6G for predicting the energy needs as well as trading energy with surrounding entities in need.

Besides low-energy consumption, some devices may require to operate independently from the power grid. Typical examples of such devices include sparse deployments in rural environments such as forest fire detection, or, at the opposite end of the spectrum, ultra-dense urban deployments, where connecting the massive number of devices to the power grid would require significant modifications to the existing infrastructure. Energy harvesting will, therefore, be at the core of 6G, with the development of new techniques such as ambient backscatter communication.

3 Medium Access and Network Architecture for 6G MTC

With the explosion of devices connected to the Internet and the ever-increasing bandwidth and latency requirements of MTC applications, network congestion will be one of the major challenges of 6G MTC. Millions of devices will compete to access the wireless medium at the same time. Meanwhile, the core network will need to handle the additional load caused by billions of connections. In this section, we first review the current and future network access techniques before moving on to network architectures that will enable massive deployment of MTC while respecting the constraints of the applications mentioned in the previous section.

Table 11.2 Differences between HTC and MTC traffic

	HTC	MTC
Packet size	Large packets (1–100 KB)	Small packets (10–100 B)
Direction	Downlink	Uplink
Data rates	High (Mb/s)	Low (10–100 Kb/s)
Traffic patterns	Session-based	Sporadic and unpredictable
Concurrent users	Hundreds per cell	Millions per cell
Mobility	High	Fixed—100 km/h

3.1   MTC Traffic Characterization

Current cellular and core networks are traditionally HTC-centric. As such, they are designed to handle traffic patterns generated by human users. Such traffic is generally on the downlink direction, with large data packets and high data rates requirements. On the other hand, latency requirements are often relative to the human capacity to process information and rarely go below 20 ms. Most cellular network deployments are devised to handle at most hundreds of users per cell. Finally, human users are expected to be mobile, with frequent handovers.

MTC wireless access patterns are significantly different from HTC. In 2020, the goal fixed for 5G was in the hundreds of thousands of devices connected to a single cell [26]. By 2030, millions of devices are expected to connect to a single cell. However, contrary to human users, most of these devices communicate only sporadically, sending short bursts of data on the uplink. Besides, if we expect specific applications such as connected vehicles, most devices will remain in a fixed location, significantly simplifying network provisioning. We summarize the differences between HTC and MTC traffic in Table 11.2.

Nowadays, cellular networks are primarily targeting HTC, with occasional MTC uses. Most deployments of 4G, and even 5G, are involving asymmetric links with more capacity on the downlink than the uplink. MTC applications that present specific requirements in latency, bandwidth, range, or energy consumption that cannot be achieved through cellular networks use other technologies, such as 802.15.4, Sigfox, LoRA, or BLE. In the upcoming years, we will observe a shift from a majority of HTC traffic to a majority of MTC traffic, both in number and volume. It is thus critical to reconsider the paradigms on which 6G will be built to enable convergence between HTC and MTC within a single networking technology.

3.2   Service Classes

In 5G, traffic is divided into three main service classes: enhanced Mobile Broadband (eMMB) massive MTC (mMTC), Ultra-Reliable Low-Latency Communication (URLLC). These service classes can encompass the requirements of most MTC applications, whether it is general IoT (eMMB), massive deployments of MTC devices, as is the case in smart cities (mMTC), or critical MTC applications, including collision detection in connected vehicles (URLLC). However, 6G will be characterized by a significantly increased number of devices and applications with much more heterogeneous applications. As such, we foresee the following service classes in the future:

- **Critical and Dependable MTC:** This service class encompasses all MTC applications that require low latency and ultra-reliability to operate. It encompasses applications where human life may be at stakes such as connected and automated driving, or applications where latency may build up, and failure may result in significant losses such as automated production lines.
- **Ultra-dense MTC:** Some MTC deployments will be characterized by hundreds to thousands of devices being deployed in very close proximity. This service class considers the case where the number of devices surpasses the capacity of the medium by several orders of magnitude, while still requiring some degree of reliability and low-latency.
- **Scalable Massive MTC:** Opposite to Ultra-Dense MTC, Scalable Massive MTC considers the case of MTC networks spreading over wide areas, at city, region, or even country scale. In some areas, such networks may experience limited connectivity and high latency, and rely primarily on secondary communication methods, opportunistic networks, or satellite communication [27].
- **Hyper-Mobile Critical MTC:** Many critical MTC applications are characterized by high mobility, whether it is autonomous vehicles or health monitoring. One of the major impacts of mobility in current mobile networks is high variability in reliability and latency [28]. This service class will require prioritization and mobility prediction in order to plan ahead of the application's needs.
- **Low-to-Zero Energy MTC:** These applications will require to operate in low energy conditions, sometimes in batteryless scenarios exploiting energy harvesting. With such extreme energy constraints, this service class will require a global simplification of the techniques used for medium access control, congestion control, and error coding, to minimize the number of necessary computations and information transmission.
- **Best Effort MTC:** Finally, as described in Sect. 2, not all MTC applications have specific requirements, and some general IoT services can operate with a reasonable best-effort QoS.

Besides MTC-specific traffic classes, we foresee that HTC applications may also follow similar patterns, supported by a close collaboration between machines and humans. As such, we will observe convergence in the traffic patterns, characterized by the end application as opposed to the end-user.

3.3 Physical Layer

Multiple solutions have been proposed for addressing the extreme requirements of MTC at the physical layer. In this section, we focus on a selection of techniques that address specific aspects of MTC for 6G.

3.3.1 Transmission Medium

To enable massive MTC deployment, it will be necessary to densify the deployment of 6G wireless infrastructure significantly. Under the currently available frequency bands, this densification will inevitably lead to interference between the network cells and considerably degrade the quality of service. It is thus primordial to extend the range of available frequencies. By using millimetre waves, 5G aims to provide a dense, hyper-local infrastructure for line-of-sight communication, allowing a multitude of microstations to be deployed in urban environments. Most operators are migrating towards a three-tiered infrastructure, consisting of a low-band tier (600 MHz) as a "blanket spectrum" for most applications, a mid-band tier (sub-6 GHz) for urban areas, and a high-band tier (over 6 GHz) for ultra-dense, high-speed, and low-range communication [29]. 6G will need to pursue such effort, and more particularly, investigate the mostly unlicensed higher frequency bands (60–100 GHz), and even THz communication [30] for large-scale hyper-local MTC applications (dense smart homes, personal area networks). Similarly, the existing lower frequency bands will require to be extended to withstand the increased load for longer-range applications in dense urban areas.

In-line with the current widening of the frequency spectrum, visible light communication (VLC) has also been extensively studied for ultra-local communication. VLC can achieve high transmission rates with low latency by leveraging inexpensive LEDs and photodetectors. In the context of MTC, it is possible to reuse existing light sources such as the devices' status LEDs to transmit information. Besides, VLC requires near-line of sight conditions for optimal signal recovery, thus limiting interference both in the spatial and the frequency domain. Finally, hybrid visible light communication and radio frequency (VLC/RF) networks allow for combining the advantages of both technologies in indoor environments [31]. As such, VLC represents an ideal candidate in ultra-dense environments where devices periodically transmit a small amount of data (e.g., home IoT). In the same direction, protocols based on near field communication (NFC) can offer a low-interference alternative, especially in cases where fixed NFC-enabled devices are exchanging data on demand, in the MTC context, when NFC-readers appear nearby [32, 33].

Besides opening the frequency spectrum to address the various use-cases of MTC, 6G will require convergence in the transmission technologies. Until now, there has been a clear separation between broadband networks, wireless broadband networks, residential wireless networks, non-terrestrial internet access (satellites, UAVs, stratospheric balloons), and personal area networks and other specialized protocols. Increasing the number of available frequency bands will bring a convergence between wireless broadband networks, wireless residential networks, and wireless personal area networks by enabling the requirements of each application to be integrated within a single technology. Further down that line, satellite communication, and more particularly low-earth orbit satellites networks can be integrated within the 6G infrastructure to provide service to traditionally unreachable areas and alternative paths to disgorge potentially congested paths. Drones and stratospheric balloons can also provide similar functionalities, with the added advantage of fast and scalable deployment, allowing for backup links to be set up in a matter of minutes in case of significant traffic congestion, or other infrastructure perturbations (cell breakdown, natural disasters).

Overall, at the radio link level, 6G will require a significant convergence of technologies to provide an ultra-dense, ultra-reliable network technology that addresses the extreme bandwidth and latency requirements of MTC applications.

3.3.2 Modulation

Ambient Backscatter Communications (AmBC)
AmBC transforms existing wireless networks into sources of power for wireless devices. By modulating and reflecting existing RF signals such as TV broadcast signals or mobile communication, it enables battery-free devices to communicate without even generating wireless signals, as shown in Fig. 11.1. Such technology allows us to distribute sensors without requiring access to the power grid, or even

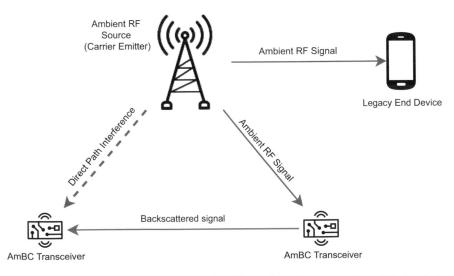

Fig. 11.1 Ambient backscatter communication. Transceivers harvest ambient RF signals to transmit information with low-power requirements

to a battery [34], while providing full Internet connectivity [35]. However, AmBC faces significant challenges to be deployed at large scale:

- **Interference between the carrier and the AmBC signal:** AmBC reuses signals that already contain encoded data. Recovering a signal encoded as small variations in an existing signal is not a trivial task. Ideally, AmBC requires the carrier signal to have a fixed amplitude for easy recovery. Unfortunately, RF signals in the wild rarely have a constant amplitude, significantly complicating the AmBC signal recovery. Besides, AmBC requires a good knowledge of the carrier signal's modulation scheme in order not to interfere with the original transceiver. If the carrier symbol and the AmBC symbol have a similar size, they may interfere with each other, and prevent not only recovery of the AmBC signal, but also the carrier signal in case of massive AmBC deployment [36].
- **Direct path interference:** The carrier signal is often several orders of magnitude higher than the AmBC signal. As such, the direct path signal is a direct source of the interference with the AmBC signal. Several techniques have been proposed for suppressing direct path interference for various wireless communication techniques [34, 37, 38]. However, these strategies often rely on high-resolution hardware, which is also energy-consuming.
- **Decentralized:** AmBC does not provide a centralized access controller that determines which device may access the medium at a given point in time. As such, accessing the medium in a setting where thousands of devices are expected to transmit concurrently requires careful planning. Although collision-avoidance mechanisms are possible, they rely on techniques such as random number generators that may be difficult to implement on ultra-low-power terminals. Similarly, common techniques to solve the hidden terminal problem may result in higher energy consumption.
- **Range:** AmBC suffers from substantial path loss limiting its application to a scale shorter than a meter. As such, its application is only possible in a few well-defined scenarios, and cannot address most settings requiring devices to operate without access to the power grid [39].

With these constraints, deploying AmBC at scale will represent one of the significant challenges for low-to-zero energy IoT. On the other hand, addressing these challenges will enable the development of devices that do not require access to the power grid, thus facilitating dense and large-scale deployment.

Beamforming

6G will leverage new high-frequency ranges over 100 GHz. These frequency ranges suffer from a high attenuation that considerably limits their application. On the other hand, one of the main benefits of such high frequencies is the small-size of the wavelength, below 1 mm. As such, it is possible to design a 10×10 antenna array in a surface of $1 \, cm^2$, thus enabling massive multiple-input and multiple-output (MIMO) systems at the scale of the device. Such MIMO systems can then direct the transmission of a signal by building constructive and destructive interference, with significant improvements over the omnidirectional transmission.

Beamforming allows us to introduce spatial multiplexing, allowing more devices to transmit simultaneously and without interference. Beam-division multiple-access (BDMA) has been suggested for 5G as it allows to transmit over several frequencies and thus adapt the capacity to the number of devices connected to the cell. 6G will face a never-before-seen increase in the volume and the complexity of the traffic, primarily caused by the massive deployment of MTC devices. To address such a challenge, the research community has been considering the following directions [40]:

- **XL-MIMO:** Extra-large MIMO considers the use case where thousands of antennas are used to serve a large number of users, such as MTC devices. XL-MIMO can take the form of either a massive array of antennas at the base station [41] or a distributed network of devices that intercommunicate with on-device antenna arrays [42]. XL-MIMO will enable high spatial resolution and considerably lower the necessary transmission power for energy-constrained devices.
- **Intelligent Beamforming:** With the increasing complexity of the network (massive number of devices, mobility), machine learning techniques can be applied to dynamically optimize the beamforming process, for instance, to predict user mobility and the resulting efficiency [43].
- **Communication and Sensing:** Large arrays of antennas have other uses than communication. Beamforming can typically be used in radar systems to generate high-resolution directional beams. As such, it is possible to combine beamforming for communication, and for sensing applications such as environment sensing or positioning. For instance, in vehicular networks, beamforming can be used to communicate with other vehicles and, at the same time, participate in collision avoidance mechanisms [44].
- **AmBC and Beamforming:** The directionality brought by beamforming makes AmBC more complex, as devices require location awareness to transmit signal using AmBC accurately. However, when the legacy receiver and the AmBC receiver are co-located, it is possible to exploit the directionality of the legacy transmitter to transmit information using AmBC with fewer interference [40].
- **Spatial non-orthogonal multiple access:** Beamforming allows to focus the transmission of a signal towards its receiver. However, in the case of massive MTC, the number of devices located in close proximity may be much larger than the resolution of the transmitter. It is possible to design non-orthogonal multiple access methods in the spatial domain that we will describe more in detail in the next section.

Non-orthogonal Multiple Access

Wireless networks allocate radio resources to users based on orthogonal multiple access. Such schemes have been present in every generation of mobile networks, from the frequency division multiple access (FDMA) of 1G to orthogonal frequency division multiple access (OFDMA) in 4G. With the exponentially increasing densification of connected devices, we expect traditional orthogonal multiple access

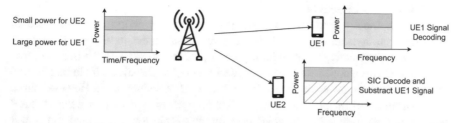

Fig. 11.2 Power-domain NOMA. Multiple signals are superposed in the power domain and transmitted over the same channel. At the receiver side, Successive Interference Cancellation decodes the strongest signal first, then substracts it from the received signal to obtain the desired signal

techniques to show their limits to accommodate a large number of devices. One potential solution to this issue consists in spreading communication over the spectrum by exploiting previously unused frequency bands such as THz communication [45], or visible light communication [46]. Another solution consists in developing new, non-orthogonal multiple access (NOMA) schemes [47] (Fig. 11.2).

NOMA allows serving users to share the same time or frequency resources in parallel. There are currently multiple techniques to perform NOMA, the most popular being in the code domain or the power domain. Code-domain NOMA multiplexes users by employing user-specific spreading sequences in the code-domain. These spreading sequences have sparse, low-density, and low inter-correlation properties [48]. Power-domain NOMA relies on Superposition Coding (SC) to allow more users to access the same resource and Successive Interference Cancellation (SIC) at the receiver side for recovering each communication. More recently, spatial domain NOMA has been proposed for millimetre Wave, exploiting spatial beams [49]. NOMA has been proposed as part of the multiple access strategies in 5G to accommodate the increasing number of users. However, most current techniques rely on the base station pre-allocating the spreading sequences or transmission powers. In the context of 6G and massive MTC, only a small (yet unknown) portion of the devices is sending at the same time. As such, these strategies result in a massive and unnecessary overhead. It is thus more convenient to develop grant-free NOMA schemes [50]. For instance, MC-NOMA exploits properties of Frame Theory to multiplex users over a finite number of orthogonal resources [51]. With such unpredictable network conditions, leveraging AI is an option to optimize latency and reliability [52]. The constraints of MTC devices in computing power and energy also require careful design of the transceiver architecture. Such transceivers should rely on low-complexity yet efficient multi-user detection algorithms [53]. Finally, NOMA can also be considered in the case of large-scale deployment of Ambient Backscatter communication [54].

NOMA enables new possibilities in the realm of collaborative and device-to-device communications by allowing the multiplication of concurrent transmission over a single channel.

Coding for Short Packets

Information theory predicts the existence of error-correcting codes capable of achieving the channel capacity [55]. As such, researchers have been thriving to achieve error-correcting codes with low error probability close to the Shanon limit. For instance, turbo codes are used in a wide variety of applications, including 4G and 5G, thanks to their low error probability within 1 dB of Shanon limit. However, turbo codes suffer from many disadvantages, including a high decoding delay, modest performance in high signal-to-noise environments, and low performance at lower bit error rates. As such, turbo codes are not an acceptable solution to address the diversity of traffic brought by MTC.

5G uses low-density parity-check (LDPC) and polar codes as error-correcting code for the data channel and the control channel, respectively. This transition was primarily brought by the need for high throughput in 5G networks. LDPC also supports a wide range of block lengths and coding rates, brings high-performance guarantees, allows parallelization of decoding, and brings a lower complexity than turbo codes. As such, LDPC has been selected as the error-correcting code for the enhanced mobile broadband (eMBB) service class, while the mMTC and URLLC (Ultra-Reliable Low-Latency Communication) service class is still under standardization. As we have seen in previous sections, multiple applications of MTC rely on very low size packets and thus require ultra-reliable coding codes for smaller block lengths. LDPC [56] shows critical limitations at lower block lengths and thus requires significant adaptations for addressing some of the constraints of mMTC.

Contrary to LDPC, Bose–Chaudhuri–Hocquenghem (BCH) codes have large minimum distances allowing for high reliability at low BER. However, BCH codes require the block length to follow specific rules, limiting its flexibility. Convolutional codes show good performance for short packets. In particular, tail-biting convolutional codes avoid the tail loss caused by zero-tail termination, increasing the reliability at the expense of a more complex decoding process. Finally, polar codes have already been selected for the control channel of eMBB, characterized by the transmission of short packets. Contrary to LDPC, they perform well without floor for shorter block length [57].

There is no one-size-fits-all solution that can tackle every requirement for every class, but the three classes defined by 5G (eMMB, mMTC, and URLLC) can already cover a vast portion of the future applications of MTC. The diversity of requirements and constraints of MTC applications envisioned for 6G will necessitate defining further traffic subclasses, onto which various error codes will be applied.

3.4 MAC Layer

One of the most crucial AI-assisted network functionalities is on radio resource management. More specifically, for the MAC layer, 6G will enable dynamic scheduling of resources using AI-based algorithms. In contrast to significant changes in most layers of the protocol stack between the different generations of mobile networks, the medium access control protocols in 5G are almost identical to the first ALOHA-like schemes [58]. Even worse, the current standard for 5G is connection-oriented, a very inefficient approach for MTC, primarily due to the vast increase in the density of IoT devices' deployment. Random access protocols in 6G will work in synergy with physical layer protocols, under the umbrella of intelligent radios, using data-driven methods in which receivers have access to information about the devices' activation patterns.

MAC Protocols in the THz Band and Beyond
The most crucial difference between traditional wireless communication systems and systems designed for the THz band is that the former utilize omnidirectional antennas and their mac protocols are based on carrier sensing while systems designed for the THz band utilize highly directional beams, and as a result employ MAC protocols that are based on handshakes. MAC protocols for communications in the THz band need to take into account the inherent spatial and spectral features in order to bypass problems like the deafness problem and the line of sight blockage problem [59].

Considering that THz communications can be part of a wide range of networks (nano networks, wireless networks on chip, personal area networks, local area networks, data center networks, and intersatelite networks) AI and context-based MAC-layer protocols are expected to be utilized for dealing the deafness and the line of sight blockage problem while guaranteeing Tbps-level links [60].

Beyond the THz band, VLC may also be considered for transmission of ultra-local information. However, VLC suffers from significant interference issues between the different transmitting entities. As such, collision detection and avoidance protocols are necessary to enable reliable information transmission in dense MTC deployments. Multiple studies were performed for sharing access on hybrid RF/VLC networks [31, 61], to provide load balancing [62], and even machine-learning enabled resource allocation [63]. Overall, there is no one-size-fits-all, and a multitude of PHY and MAC protocols are being deployed to adapt to specific uses of VLC [64].

Device-to-Device (D2D) Communication
D2D was planned to integrate 4G through LTE Direct [65]. This technology would have allowed for devices to communicate using LTE without the need to connect to the base station. Among other promises, it would have brought connectivity between up to a thousand devices and a lowered energy consumption, even outperforming traditional LTE [66]. However, it was never truly implemented. In the age of 5G, D2D communication was envisioned for peer-to-peer communication, an alternative communication method that allows relieving the burden at a congested base station by allowing devices to directly communicate with each other [67]. 6G will blur the line between user equipment and base stations, and assemble currently distinct communication methods within the same standardized technology [68]. As a result, D2D will be at the core of 6G, not only as a backup channel in case of congestion, but also to support some elements of the network infrastructure itself. As we have seen in Sect. 3.3.1, mobile swarms of UAvs and stratospheric balloons will become

an integral part of 6G. As such, it will be necessary to develop reliable protocols for efficiently routing information in core network within these self-assembling swarms.

Vehicular Networks

Vehicular networks, and more particularly vehicle-to-vehicle communication (V2V) will bring additional challenges for D2D communication in 6G. Vehicles are indeed highly mobile, and require to communicate with each others under low latency and high reliability constraints. Besides, vehicular communication often involve communication between vehicles (V2V), but also with the infrastructure (V2I), and with computing resources at the edge (V2E) and the cloud (V2C). Such a setting allows to combine the capabilities of vehicles towards a global task [69], but also raises multiple challenges.. Challenges include providing rapid and precise channel division in multi-radio environments, radio configuration in highly mobile conditions, and beamforming at high speed [70]. 6G will require a combination of intelligent radio, network intelligence, and on-board AI to optimize the communication and provide the tight requirements of vehicular networks.

4 Network Intelligence

The most significant difference between 6G and previous generations of wireless mobile communications is the fact that instead of connecting more "things" together, it will connect "intelligent entities" together. Artificial Intelligence (AI) has a crucial role in this world-scale integration of intelligent entities that can take nontrivial decisions. The design of 6G will go past traditional mobile Internet architectures and support ubiquitous AI services for devices located anywhere between the network core and the edge infrastructure. Moreover, AI techniques like machine and deep learning will be applied to the design and the optimization of 6G architectures, protocols, and operations. More specifically, 6G will employ AI techniques for its function and simultaneously, via the low-latency and high-bandwidth communication channels to specialized hardware, enable applications that operate based on computationally intensive and data-hungry AI models. Federated learning is probably the most representative example of an AI technique that will be widely used in MTC using the features that are missing from 5G and will be implemented in 6G. Federated learning, a recently emerged technique for training models across multiple machines, allows machines that generate and store data, to collectively train a machine learning model without requiring them to exchange their data but only the exchange a set of learned parameters [71]. Machines that have access to locally stored data communicate periodically with a dedicated machine, that is usually located in the cloud, and send their parameter updates. On each period, the dedicated server aggregates the collected updates, selects a subset of the machines using a sampling mechanism and sends the aggregates parameters to each of the sampled machines. After receiving the new parameters, a sampled machine trains a local model for a fixed period of time using the stored data and at the end sends back the calculated parameters. Considering that it is possible for a machine to not have the computational capabilities to train the model locally, even for a small number of iterations, it can employ cloud resources using split learning techniques [72].

Thus, 6G will require the support of new service types that will (1) back computation offloading between small devices with low capabilities and resourceful entities located in the network, (2) be more context-aware and adaptive to changes in link congestion, network topology and mobility, and (3) establish ultra-reliable and low-latency communications for events with spatially and temporally changing device densities, traffic patterns, and spectrum and infrastructure availability [14]. These services will handle the need for distributed computations, training machine learning models using the federated learning paradigm as well as edge inference. Representative examples are autonomous driving, piloting crewless aerial vehicles, smart grid energy trading, and others.

Besides supporting intelligent devices, 6G will require in-network intelligence to address the massive increase of scale and usages presented by MTC applications. 5G generalized the usage of software defined networking (SDN) and Network Function Virtualization (NFV) both in the user plane and in the backbone to provide programmable control and management of network resources. Among other advantages, such technologies allow to perform vertical network slicing, splitting a 5G network into multiple logical networks, each providing per-application guarantees [73]. Although providing significant advantages as compared to previous generations, we predict that by 2030, vertical slicing will not be sufficient to accommodate the large number of devices and the variety of applications. 6G will bring such network softwarization to the next level by enabling automated control of the network functions through AI. Network intelligence will provide dynamic and on-demand resources for each individual application. Similarly, AI can instrument Software Defined Radio (SDR) to handle in real-time the resource allocation for the multiple communication technologies and frequency bands [14].

5 Situation-Aware Applications

As presented in Sect. 2, future MTC applications will encompass a variety of QoS requirements. Many emerging applications consider high reliability and low latency as top priorities. For instance, connected vehicles require sub-millisecond latency and extreme reliability for collision avoidance systems. Similarly, in industrial automation, latency adds up on production chains, and a single failure of the system can lead to high monetary loss. On the other hand, more passive applications such as home IoT, or smart city sensing, can afford a more relaxed transmission environment, sometimes more than HTC applications. In order for these heterogeneous applications to cohabit together, exploiting network situation awareness is key in designing cooperative communication schemes that respect and enforce the various levels of QoS.

Graceful degradation consists of keeping minimal functionality, even when the network conditions are sub-optimal. Such a concept can be applied to critical systems in order to prevent catastrophic failure, but also to non-critical applications exploiting network situation awareness for detecting congestion and making space for applications with higher reliability and latency constraints. Although often presented in HTC [74, 75], graceful degradation will have significants applications in MTC [76]. Such a concept allows to provide a tradeoff between power, delay, and reliability in MTC networks while reducing the impact of transient peaks in-network utilization for critical applications. Even within a single MTC application, it is possible to reduce some information flows to favour the most essential ones. For instance, a connected vehicle encountering a congested communication network can suspend the infotainment service for passengers, reduce the frequency of updates for the navigation system, and keep transmitting the entirety of the collision avoidance system's information flow, thus ensuring the safety of the passengers at the cost of a slightly less comfortable ride.

In order to provide graceful degradation, it is necessary to adopt a cross-layer approach, where each layer garners information from the lower layers to adapt its network behaviour. In case of congestion on the link, it is necessary for the physical and datalink layer to inform the application of the limited access. As such, the application can decide to temporarily throttle its transmission rate by selecting the most essential information. Similarly, the application should work hand-in-hand with the network and transport layers to adapt the transmission to the global network state, and perform internal QoS to prioritize traffic in case of congestion. Although breaking the typical OSI model, such a convergence allows to develop applications that internally prioritize traffic depending on the network condition and allow for critical systems to operate even in degraded network conditions.

In centralized MTC deployments, an MTC controller can easily sense the network conditions and make decisions on whether to trigger the MTC devices under its supervision [77]. In general, such a controller embed a higher computing power than devices and can deploy advanced machine learning algorithms to take, learn, and improve decisions based on the network situation. However, as we have seen in previous sections, the current trend for MTC tends towards grant-free resource allocation schemes, where devices cooperate towards efficiently sharing the communication medium [78]. New mechanisms should be adapted for low energy and low complexity, on-device network sensing, and interpretation.

Network awareness for MTC has been discussed as early as 2014 for MTC applications not to jeopardize the access probability of HTC applications in LTE networks. Future generations of mobile networks will be increasingly facing the case of MTC applications interfering with other MTC applications. Solutions have been developed to provide congestion control by deferring some activated MTC devices transmitting their access requests [79]. Newer approaches use statistical provisioning [80], traffic prediction [81], and machine learning [82] for massive access optimization in cellular MTC applications.

6 Conclusion

MTC in 6G will be characterized by a massive number of devices deployed in close proximity. The heterogeneity of applications will result in a never-before-seen diversity of constraints and requirements. To address such challenges, new strategies will have to be applied at every layer of the transmission process, from the access medium itself to the application.

In this chapter, we summarized the most promising research tracks and provided a vision of how 6G can achieve massive heterogeneous MTC. We saw that expanding the spectrum at the physical layer can significantly improve the bitrates while reducing interference between devices by increasing the number of available channels while exploiting frequency ranges with high attenuation (and thus a low range). When combined with specific modulation techniques such as AmBC, beamforming, or NOMA, it is thus possible to accommodate a massive number of devices in a limited area. At the same time, newly designed MAC-layer AI-based protocols are required to deal with the deafness and the sight blockage problems in THz communications by considering their spectral and unique features. Moreover, we discussed how 6G differs from previous generations by not only connecting "things" together by also by adding "intelligent entities" in every layer of the protocol stack. Finally, situation-aware applications can gracefully degrade their operation to support the most critical applications, both internally and externally.

We hope to have provided a comprehensive overview of the most promising strategies for addressing the paradigm shift that the popularisation of MTC will present in the upcoming years. Future cellular networks will progressively migrate from HTC-centric to MTC-centric, leading to additional, yet less predictable stress on the network, that only a convergence between technologies will allow tackling.

References

1. L. Pu, X. Chen, J. Xu, X. Fu, D2D fogging: an energy-efficient and incentive-aware task offloading framework via network-assisted D2D collaboration. IEEE J. Selec. Areas Commun. **34**(12), 3887–3901 (2016)
2. D. Chatzopoulos, C. Bermejo, E.u. Haq, Y. Li, P. Hui, D2D task offloading: a dataset-based Q&A. IEEE Commun. Mag. **57**(2), 102–107 (2019)
3. S. Chen, J. Hu, Y. Shi, Y. Peng, J. Fang, R. Zhao, L. Zhao, Vehicle-to-everything (V2X) services supported by lte-based systems and 5G. IEEE Commun. Stand. Mag. **1**(2), 70–76 (2017)
4. X. Fang, S. Misra, G. Xue, and D. Yang, Smart grid–the new and improved power grid: a survey. IEEE Commun. Surveys Tutorials **14**(4), 944–980 (2011)
5. B. Holfeld, D. Wieruch, T. Wirth, L. Thiele, S.A. Ashraf, J. Huschke, I. Aktas, J. Ansari, Wireless communication for factory automation: an opportunity for lte and 5G systems. IEEE Commun. Mag. **54**(6), 36–43 (2016)
6. Statista, Number of internet of things (IOT) connected devices worldwide in 2018, 2025 and 2030 (2020). https://www.statista.com/statistics/802690/worldwide-connected-devices-by-access-technology/. Accessed 23 July 2019

7. M. Jarschel, D. Schlosser, S. Scheuring, and T. Hoßfeld, An evaluation of QoE in cloud gaming based on subjective tests, in *2011 Fifth International Conference on Innovative Mobile and Internet Services in Ubiquitous Computing* (2011), pp. 330–335

8. Z. Dawy, W. Saad, A. Ghosh, J.G. Andrews, E. Yaacoub, Toward massive machine type cellular communications. IEEE Wirel, Commun. **24**(1), 120–128 (2017)

9. Sigfox connected objects: Radio specifications - ref.: Ep-specs rev.: 1.5, SigFox (2020). Accessed 23 July 2019

10. N. Sornin, M. Luis, T. Eirich, T. Kramp, O. Hersent, Lorawan 1.1 specification. LoRa Alliance (2015). Accessed 23 July 2019

11. IEEE standard for local and metropolitan area networks–part 15.4: Low-rate wireless personal area networks (LR-WPANs). *IEEE Std 802.15.4-2011 (Revision of IEEE Std 802.15.4-2006)* (2011), pp. 1–314

12. Bluetooth core specification v 5.2, *Bluetooth*(2019). Accessed 23 July 2019

13. ETSI, ETSI - mobile technologies - 5g, 5g specs|future technology. https://www.etsi.org/technologies/5g. Accessed 23 July 2019

14. K.B. Letaief, W. Chen, Y. Shi, J. Zhang, Y.A. Zhang, The roadmap to 6G: Ai empowered wireless networks. IEEE Commun. Mag. **57**(8), 84–90 (2019)

15. 6G - the next hyper-connected experience for all. Samsung Research (2020) Accessed 23 July 2019

16. B. Aazhang, P. Ahokangas, H. Alves, M.-S. Alouini, J. Beek, H. Benn, M. Bennis, J. Belfiore, E. Strinati, F. Chen, K. Chang, F. Clazzer, S. Dizit, D. Kwon, M. Giordiani, W. Haselmayr, J. Haapola, E. Hardouin, E. Harjula, P. Zhu, Key drivers and research challenges for 6G ubiquitous wireless intelligence (white paper) (2019)

17. N.H. Mahmood, S. Böcker, A. Munari, F. Clazzer, I. Moerman, K. Mikhaylov, O. Lopez, O.-S. Park, E. Mercier, H. Bartz, R. Jäntti, R. Pragada, Y. Ma, E. Annanperä, C. Wietfeld, M. Andraud, G. Liva, Y. Chen, E. Garro, F. Burkhardt, H. Alves, C.-F. Liu, Y. Sadi, J.-B. Dore, E. Kim, J. Shin, G.-Y. Park, S.-K. Kim, C. Yoon, K. Anwar, P. Seppänen, White Paper on Critical and Massive Machine Type Communication Towards 6G (2020)

18. Cisco edge-to-enterprise IoT analytics for electric utilities solution overview. *Cisco* (2018). Accessed 23 July 2019

19. J. Zhang, K.B. Letaief, Mobile edge intelligence and computing for the internet of vehicles. Proc. IEEE **108**(2), 246–261 (2019)

20. L.H. Lee, T. Braud, S. Hosio, P. Hui, Towards augmented reality-driven human-city interaction: Current research and future challenges (2020). Preprint arXiv:2007.09207

21. V. Özdemir, N. Hekim, Birth of industry 5.0: Making sense of big data with artificial intelligence,"the internet of things" and next-generation technology policy. OMICS: J. Integr. Biol. **22**(1), 65–76 (2018)

22. G. Berardinelli, N.H. Mahmood, I. Rodriguez, P. Mogensen, Beyond 5G wireless irt for industry 4.0: Design principles and spectrum aspects, in *2018 IEEE Globecom Workshops (GC Wkshps)* (2018), pp. 1–6

23. R. Negra, I. Jemili, A. Belghith, Wireless body area networks: applications and technologies. Procedia Comput. Sci. **83**, 1274–1281 (2016)

24. M. Cicioğlu, A. Çalhan, IoT-based wireless body area networks for disaster cases. Int. J. Commun. Syst. **33**(13), e3864 (2020)

25. R. Jain, V. Goel, J.K. Rekhi, J.A. Alzubi, IoT-based green building: Towards an energy-efficient future, in *Green Building Management and Smart Automation* (IGI Global, Hershey, 2020), pp. 184–207

26. J. Sachs, P. Popovski, A. Höglund, D. Gozalvez-Serrano, P. Fertl, M. Dohler, T. Nakamura, *Machine-Type Communications* (Cambridge University Press, 2016), pp. 77–106

27. A. Aucinas, J. Crowcroft, P. Hui, Energy efficient mobile M2M communications. Proc. ExtremeCom **12**, 1–6 (2012)

28. T. Braud, T. Kämäräinen, M. Siekkinen, P. Hui, Multi-carrier measurement study of mobile network latency: The tale of hong kong and helsinki, in *2019 15th International Conference on Mobile Ad-Hoc and Sensor Networks (MSN)* (IEEE, Piscataway, 2019), pp. 1–6

29. J. Horwitz, The definitive guide to 5G low, mid, and high band speeds. *Venture Beat Online Magazine* (2019)
30. T. Kürner, S. Priebe, Towards THz communications-status in research, standardization and regulation. J. Infr. Millimeter Terahertz Waves **35**(1), 53–62 (2014)
31. M. Amjad, H.K. Qureshi, S.A. Hassan, A. Ahmad, S. Jangsher, Optimization of mac frame slots and power in hybrid VLC/RF networks. IEEE Access **8**, 21653–21664 (2020)
32. K. Sucipto, D. Chatzopoulos, S. Kosta, P. Hui, Keep your nice friends close, but your rich friends closer – computation offloading using NFC, in *IEEE INFOCOM 2017 - IEEE Conference on Computer Communications* (2017), pp. 1–9
33. D. Chatzopoulos, C. Bermejo, S. Kosta, P. Hui, Offloading computations to mobile devices and cloudlets via an upgraded NFC communication protocol. IEEE Trans. Mobile Comput. **19**(3), 640–653 (2020)
34. V. Liu, A. Parks, V. Talla, S. Gollakota, D. Wetherall, J.R. Smith, Ambient backscatter: Wireless communication out of thin air. ACM SIGCOMM Comput. Commun. Rev. **43**(4), 39–50 (2013)
35. B. Kellogg, A. Parks, S. Gollakota, J.R. Smith, D. Wetherall, Wi-Fi backscatter: Internet connectivity for RF-powered devices, in *Proceedings of the 2014 ACM Conference on SIGCOMM* (2014), pp. 607–618
36. K. Ruttik, R. Duan, R. Jäntti, Z. Han, Does ambient backscatter communication need additional regulations?, in *2018 IEEE International Symposium on Dynamic Spectrum Access Networks (DySPAN)* (2018), pp. 1–6
37. J. Qian, F. Gao, G. Wang, S. Jin, H. Zhu, Noncoherent detections for ambient backscatter system. IEEE Trans. Wirel. Commun. **16**(3), 1412–1422 (2017)
38. G. Yang, Y. Liang, R. Zhang, Y. Pei, Modulation in the air: Backscatter communication over ambient ofdm carrier. IEEE Trans. Commun. **66**(3), 1219–1233 (2018)
39. R. Duan, X. Wang, H. Yigitler, M.U. Sheikh, R. Jantti, Z. Han, Ambient backscatter communications for future ultra-low-power machine type communications: challenges, solutions, opportunities, and future research trends. IEEE Commun. Mag. **58**(2), 42–47 (2020)
40. S. Chen, S. Sun, G. Xu, X. Su, Y. Cai, Beam-space multiplexing: Practice, theory, and trends, from 4G TD-LTE, 5G, to 6G and beyond. IEEE Wirel. Commun. **27**(2), 162–172 (2020)
41. E.D. Carvalho, A. Ali, A. Amiri, M. Angjelichinoski, R.W. Heath, Non-stationarities in extra-large-scale massive mimo. IEEE Wirel. Commun. **27**(4), 74–80 (2020)
42. V.C. Rodrigues, A. Amiri, T. Abrão, E. de Carvalho, P. Popovski, Low-complexity distributed xl-mimo for multiuser detection, in *2020 IEEE International Conference on Communications Workshops (ICC Workshops)* (2020), pp. 1–6
43. T. Maksymyuk, J. Gazda, O. Yaremko, D. Nevinskiy, Deep learning based massive mimo beamforming for 5G mobile network, in *2018 IEEE 4th International Symposium on Wireless Systems within the International Conferences on Intelligent Data Acquisition and Advanced Computing Systems (IDAACS-SWS)* (2018), pp. 241–244
44. J.A. Zhang, X. Huang, Y.J. Guo, J. Yuan, R.W. Heath, Multibeam for joint communication and radar sensing using steerable analog antenna arrays. IEEE Trans. Vehic. Technol. **68**(1), 671–685 (2019)
45. A.A. Boulogeorgos, E.N. Papasotiriou, A. Alexiou, A distance and bandwidth dependent adaptive modulation scheme for thz communications, in *2018 IEEE 19th International Workshop on Signal Processing Advances in Wireless Communications (SPAWC)* (2018), pp. 1–5
46. A.U. Guler, T. Braud, P. Hui, Spatial interference detection for mobile visible light communication, in *2018 IEEE International Conference on Pervasive Computing and Communications (PerCom)* (IEEE, Piscataway, 2018), pp. 1–10
47. S. Islam, M. Zeng, O.A. Dobre, Noma in 5G systems: Exciting possibilities for enhancing spectral efficiency (2017). Preprint arXiv:1706.08215
48. D. Duchemin, J.-M. Gorce, C. Goursaud, Code domain non orthogonal multiple access versus aloha: A simulation based study, in *2018 25th International Conference on Telecommunications (ICT)* (IEEE, Piscataway, 2018), pp. 445–450

49. W. Shao, S. Zhang, H. Li, N. Zhao, O.A. Dobre, Angle-domain noma over multicell millimeter wave massive mimo networks. IEEE Trans. Commun. **68**(4), 2277–2292 (2020)
50. M.B. Shahab, R. Abbas, M. Shirvanimoghaddam, S.J. Johnson, Grant-free non-orthogonal multiple access for iot: A survey. IEEE Commun. Surv. Tutor. **22**(3), 1805–1838 (2020)
51. R. Stoica, G.T.F.d. Abreu, Massively concurrent noma: A frame-theoretic design for non-orthogonal multiple access, in *2018 52nd Asilomar Conference on Signals, Systems, and Computers* (2018), pp. 461–466
52. N. Ye, X. Li, H. Yu, A. Wang, W. Liu, X. Hou, Deep learning aided grant-free noma toward reliable low-latency access in tactile internet of things. IEEE Trans. Ind. Inf. **15**(5), 2995–3005 (2019)
53. Y. Chen, A. Bayesteh, Y. Wu, B. Ren, S. Kang, S. Sun, Q. Xiong, C. Qian, B. Yu, Z. Ding, et al., Toward the standardization of non-orthogonal multiple access for next generation wireless networks. IEEE Commun. Mag. **56**(3), 19–27 (2018)
54. Q. Zhang, L. Zhang, Y. Liang, P. Kam, Backscatter-noma: a symbiotic system of cellular and internet-of-things networks. IEEE Access **7**, 20000–20013 (2019)
55. C.E. Shannon, Probability of error for optimal codes in a gaussian channel. Bell Syst. Tech. J. **38**(3), 611–656 (1959)
56. T. Richardson, Error floors of LDPC codes, in *Proceedings of the Annual Allerton Conference on Communication Control and Computing*, vol. 41 (The University; 1998, 2003), pp. 1426–1435
57. M. Shirvanimoghaddam, M.S. Mohammadi, R. Abbas, A. Minja, C. Yue, B. Matuz, G. Han, Z. Lin, W. Liu, Y. Li, et al., Short block-length codes for ultra-reliable low latency communications. IEEE Commun. Mag. **57**(2), 130–137 (2018)
58. F. Clazzer, A. Munari, G. Liva, F. Lazaro, C. Stefanovic, P. Popovski, From 5G to 6G: Has the time for modern random access come? (2019). Preprint arXiv:1903.03063
59. C. Han, X. Zhang, X. Wang, On medium access control schemes for wireless networks in the millimeter-wave and terahertz bands. Nano Commun. Netw. **19**, 67–80 (2019)
60. I.F. Akyildiz, A. Kak, S. Nie, 6G and beyond: the future of wireless communications systems. IEEE Access **8**, 133995–134030 (2020)
61. V.K. Papanikolaou, P.D. Diamantoulakis, P.C. Sofotasios, S. Muhaidat, G.K. Karagiannidis, On optimal resource allocation for hybrid VLC/RF networks with common backhaul. IEEE Trans. Cognit. Commun. Netw. **6**(1), 352–365 (2020)
62. A. Adnan-Qidan, M. Morales-Céspedes, A.G. Armada, Load balancing in hybrid VLC and RF networks based on blind interference alignment. IEEE Access **8**, 72512–72527 (2020)
63. S. Shrivastava, B. Chen, C. Chen, H. Wang, M. Dai, Deep q-network learning based downlink resource allocation for hybrid RF/VLC systems. IEEE Access **8**, 149412–149434 (2020)
64. A.R. Ndjiongue, T.M.N. Ngatched, O.A. Dobre, A.G. Armada, VLC-based networking: Feasibility and challenges. IEEE Netw. **34**(4), 158–165 (2020)
65. 3GPP, Overview of 3GPP release 12 v0.2.0, Technical Report, 3GPP (2015). https://www.3gpp.org/ftp/Information/WORK_PLAN/Description_Releases/. Accessed 28 Aug 2020
66. M. Condoluci, L. Militano, A. Orsino, J. Alonso-Zarate, G. Araniti, LTE-direct vs. WiFi-direct for machine-type communications over LTE-a systems, in *2015 IEEE 26th Annual International Symposium on Personal, Indoor, and Mobile Radio Communications (PIMRC)* (2015), pp. 2298–2302
67. R.I. Ansari, C. Chrysostomou, S.A. Hassan, M. Guizani, S. Mumtaz, J. Rodriguez, J.J.P.C. Rodrigues, 5G D2D networks: Techniques, challenges, and future prospects. IEEE Sys. J. **12**(4), 3970–3984 (2018)
68. S. Zhang, J. Liu, H. Guo, M. Qi, N. Kato, Envisioning device-to-device communications in 6G. IEEE Netw. **34**(3), 86–91 (2020)
69. P. Zhou, T. Braud, A. Zavodovski, Z. Liu, X. Chen, P. Hui, J. Kangasharju, Edge-facilitated augmented vision in vehicle-to-everything networks. IEEE Trans. Vehic. Technol. **69**, 1–1 (2020)
70. F. Tang, Y. Kawamoto, N. Kato, J. Liu, Future intelligent and secure vehicular network toward 6G: machine-learning approaches. Proc. IEEE **108**(2), 292–307 (2020)

71. Q. Yang, Y. Liu, T. Chen, Y. Tong, Federated machine learning: Concept and applications. CoRR **abs/1902.04885** (2019)
72. A. Singh, P. Vepakomma, O. Gupta, R. Raskar, Detailed comparison of communication efficiency of split learning and federated learning (2019). Preprint arXiv:1909.09145
73. A.A. Barakabitze, A. Ahmad, R. Mijumbi, A. Hines, 5G network slicing using sdn and NFV: A survey of taxonomy, architectures and future challenges. Comput. Netw. **167**, 106984 (2020)
74. T. Braud, F.H. Bijarbooneh, D. Chatzopoulos, P. Hui, Future networking challenges: The case of mobile augmented reality, in *2017 IEEE 37th International Conference on Distributed Computing Systems (ICDCS)* (IEEE, Piscataway, 2017), pp. 1796–1807
75. T. Schierl, C. Hellge, S. Mirta, K. Gruneberg, T. Wiegand, Using h. 264/AVC-based scalable video coding (SVC) for real time streaming in wireless IP networks, in *2007 IEEE International Symposium on Circuits and Systems* (IEEE, Piscataway, 2007), pp. 3455–3458
76. M. Shehab, E. Dosti, H. Alves, M. Latva-aho, On the effective capacity of MTC networks in the finite blocklength regime, in *2017 European Conference on Networks and Communications (EuCNC)* (IEEE, Piscataway, 2017), pp. 1–5
77. W.U. Rehman, T. Salam, A. Almogren, K. Haseeb, I. Ud Din, S.H. Bouk, Improved resource allocation in 5G MTC networks. IEEE Access **8**, 49187–49197 (2020)
78. B. Han, V. Sciancalepore, O. Holland, M. Dohler, H.D. Schotten, D2D-based grouped random access to mitigate mobile access congestion in 5G sensor networks. IEEE Commun. Mag. **57**(9), 93–99 (2019)
79. S. Duan, V. Shah-Mansouri, Z. Wang, V.W.S. Wong, D-ACB: Adaptive congestion control algorithm for bursty M2M traffic in LTE networks. IEEE Trans. Vehic. Technol. **65**(12), 9847–9861 (2016)
80. M. Shehab, E. Dosti, H. Alves, M. Latva-aho, Statistical QOS provisioning for MTC networks under finite blocklength. EURASIP J. Wirel. Commun. Netw. **2018**(1), 1–14 (2018)
81. S. Ali, W. Saad, N. Rajatheva, A directed information learning framework for event-driven M2M traffic prediction. IEEE Commun. Lett. **22**(11), 2378–2381 (2018)
82. B. Sliwa, R. Falkenberg, T. Liebig, N. Piatkowski, C. Wietfeld, Boosting vehicle-to-cloud communication by machine learning-enabled context prediction. IEEE Trans. Intell. Transport. Syst. **21**, 3497–3512 (2019)

Chapter 12
Edge Intelligence in 6G Systems

Christina Chaccour and Walid Saad

Abstract In this chapter, we provide a vision for edge intelligence as a key building block of 6G wireless systems. As we evolve towards a new breed of wireless services, enabled by an Internet of Intelligence system, wireless networks cannot operate in a rigid, static, and conventional infrastructure. Instead, networks need to be capable of providing distributed decision making where and when needed. On the one hand, the foreseen 6G services require operational intelligence to orchestrate wireless functions and resources in the presence of small, but massive heterogeneous data dispersed across edge devices. On the other hand, they require service intelligence to execute complex applications such as high-precision manufacturing and holographic teleportation. Subsequently, artificial intelligence at the edge must manage large-scale, heterogeneous, massive small data to perform highly precise and accurate decision making, while linking the result to a contextual awareness. To break the heterogeneity in data and the intermittent nature of resources, we propose a framework deploying unsupervised learning mechanisms via meta-learning which enables clustering trends in separate tasks which are later meta-learned. Furthermore, we identify the need to provide a sustainable and proactive network design and optimization, by enabling an explainable edge intelligence backed by data-science and theory-based models.

1 Introduction

With the emergence of the Internet of Things (IoT) with 5G, the classical internet infrastructure as we know it has witnessed a paradigm shift necessitating a need

This research was supported by the Office of Naval Research (ONR) under MURI Grant N00014-19-1-2621.

C. Chaccour · W. Saad (✉)
Wireless@VT, Bradley Department of Electrical and Computer Engineering, Virginia Tech, Blacksburg, VA, USA
e-mail: christinac@vt.edu; walids@vt.edu

for novel cloud computing capabilities, moving part of its lengthy processing and storage closer to its connected devices. Thus, the concept of *edge computing* emerged as an effective approach to extend the capabilities of computation, network connection, and storage from a cloud server to the edge of the network [1]. To date, edge computing research has been focused on devising edge techniques that can reduce the overhead and latency associated with performing computing at a remote cloud, thus, providing, for example, ultra reliable low latency communications (URLLC) for critical IoT applications. Meanwhile, an unprecedented proliferation of new Internet of Intelligence (IoI) services have emerged. Examples include, eXtended reality (XR) (the spectrum of virtual reality (VR) to augmented reality (AR) services), connected holography, haptics, autonomous driving, and tactile internet. terabit per second (Tbps)-level data rates, near-zero end-to-end (E2E) latency, improved energy efficiency, and high E2E reliability are necessary breakthroughs to provide the IoI applications with real-time and immersive communications. Furthermore, in 6G [2–4], the network needs to provide these milestones to the user *instantaneously* and *continuously*, rather than on average; thus, making the processing and communication requirements more stringent. Moreover, IoI services are deployed on heterogeneous devices that are either completely human-centric and necessitate a high level of immersion, or device-centric with the goal of mimicking human behavior and intellect, e.g. robotics, driving, digital twins, etc. In other words, the requirements of such services are not only more stringent but are beyond the conventional wireless requirements, dictated by the timely arrival of transmitted packets. For example, VR calls for perceptual and haptic requirements that translate onto an instantaneous high-rate and high-reliability low latency communications (HRLLC) and low jitter [5, 6]. Violating such requirements could expose the VR user to motion sickness and dizziness. Hence, instead of merely raising the bar of wireless requirements, such services need to be governed by *experienced and human-like* artificial intelligence (AI) frontiers deployed at the network edge and capable of unleashing the potential of massive small heterogeneous data, services, and resources. Therefore, the concept of *edge computing* in 5G must evolve to *edge intelligence* in 6G to deliver an unprecedented and proactive network optimization.

Recently, a number of papers looked into the concept of edge computing [7–9]. The work in [7] studied the joint communication and computation resource allocation to minimize the weighted-sum delay of all devices in a cloud-edge collaboration system. The authors in [8] jointly investigated the problem of fog network[1] formation and task distribution while considering a hybrid fog-cloud architecture. In [9], a new low-latency and reliable communication-computing system design for enabling mission-critical applications is proposed. However, these works were mostly limited to IoT and 5G services, and did not encompass an

---

[1]The literature uses the terms fog and edge interchangeably, nevertheless, strictly speaking, it can be defined as a layer that consists of various networking equipment for data offloading at the network edge.

intelligently driven edge for complex 6G networks and applications. Meanwhile, a number of surveys and tutorials related to edge intelligence appeared in [10–13]. The works in [10–12] provided a thorough and comprehensive survey on the literature surrounding edge intelligence. In [13] the authors explored the key building blocks and theoretical principles of edge machine learning warranting a clean-slate design in terms of neural network architectures. However, the works in [10–13] did not discuss the 6G enabling decentralized learning mechanisms, nor the challenges of emerging 6G services necessitating a collaborative edge mechanism. Different from this state of art, in this chapter, we focus on how edge intelligence will be a key enabler for 6G applications. In particular, 6G services will be deployed in more complex environments dictated by higher mobility and dynamics, meanwhile fueled by channels that are sparser and driven by higher uncertainty and susceptibility. Under these conditions, such services require higher intellect capabilities and need to be driven by human driven behavior to provide participating users with an immersive experience in applications like XR and holographic teleportation, or need to mimic human intelligence in applications like autonomous driving and industrial automation. Moreover, this AI at the edge needs to be provided with unprecedented learning mechanisms capable of exploiting small heterogeneous data and resources in contrast with conventional big data needed for successful training processes. Additionally, such AI mechanisms need to be flexible, scalable and transparent migrating from rigid black box mechanisms to ones supported with evidence and explainability. As such, providing us with the tools to improve these mechanisms as well as providing the user audience with insights and trust.

The main goal of this chapter is to provide a holistic forward looking vision of edge intelligence in 6G that identifies the unprecedented high precision and human-mimicking wireless requirements along with their key enablers. This vision delineates the role and the challenges of edge intelligence in foreseen future IoI applications. Subsequently, it outlines concrete research approaches and frameworks needed to provide an effective deployment of IoI services and a collaborative edge intelligence. These approaches, entitle empowering the edge with proactive learning strategies capable of unleashing the potential of scarce, distributed, and heterogeneous data. Subsequently, our research road-map identifies the difference between AI at the edge and centralized AI. On the one hand, the road-map emphasizes the distinctive features of edge data and its established models and their need for breakthrough learning mechanisms such as meta-learning and theory-guided data science models. On the other hand, it characterizes the unprecedented edge training and inference mechanisms driven by specialization, generalization, and explainability.

The rest of this chapter is organized as follows. Section 2 looks onto the definition of the edge and why we need to move to it. Section 3 answers the question: "How is the edge intelligence different than centralized AI?". Summaries are stated in Sect. 4.

2   What Is the Edge and Why Do We Need It in 6G?

The edge was first thought to be a key enabler of closer service-specific processing and data storage to users physically and logically. Nevertheless, the drivers of 6G (e.g., densification, higher rates, and energy efficiency) [2, 4] require us to move to an intelligent edge acting as a new communication infrastructure to enable new services that cannot be attained in the development of 5G networks. These services push the wireless network requirements bar to a level that cannot be manually optimized or fulfilled without advanced and experienced intellect. For example holographic teleportation, an evolutionary version of ultimate XR, necessitates data rates in the Tbps range and an E2E latency less than a 1 ms simultaneously. Thus, enabling it to transport five dimensions (each featuring one the five senses) in contrast to conventional and simple packets transmission. Moreover, industrial automation services need to be governed by error-free and highly intellectual decision makers at the edge enabling extremely high reliability in the order of 10^{-9}. As a matter of fact, infringing the reliability and the performance metrics in such mission-critical services can lead to high risk errors. For example, as shown in [14], outdated information in AR can pose threats to the livelihoods of people when used in surgical operations. Furthermore, in some scenarios, given that the control is performed on private user data, the governing edge needs to enable links based on trust rather than conventional hierarchical models. Hence, posing many questions on the network architecture and the learning mechanisms needed to safeguard the personal data.

Given the aforementioned 6G challenges, edge AI will bring in unprecedented distributed intelligence as close as possible to the user. Edge AI will act as the intelligent backbone of the network enabling synchronized and instantaneous streams of communications for highly demanding 6G applications. Furthermore, moving to the edge will establish network connections based on trust instead of depending on a central cloud intelligence. Thus, safeguarding the user raw data, and enabling device-to-device communication in stringent latency and privacy settings. Edge's distinctive intelligence will enable networks to have contextual awareness, flexible and highly accurate and precise decision making. Indeed, 6G edge intelligence contrasts itself with conventional edge computing in 5G where the main objective was to merely provide information with less overhead and lower latencies. Next, we will first elaborate on the wireless requirements and the challenges posed by the network that concretely lead to a migration towards the edge. Subsequently, we will emphasize on mechanisms enabling a collaborative and intelligent edge encompassing a spectrum of structures and a range of proximities to the end user.

2.1 The Wireless Migration Towards the Edge

2.1.1 Towards Massive Small Data and Denser Cells

Large scale IoT networks connected a plethora of heterogeneous wireless devices that require storage, processing, and computing of data without relying on a remote cloud. As such, edge computing in 5G needed to handle task distribution, offloading, caching, etc. Nonetheless, with the migration towards high frequency (HF) (i.e. millimeter wave (mmWave) and terahertz (THz)), cells need to shrink in size, the deployment of base stations (BSs) needs to be denser, and the coherence time of channels will become very small. Therefore, the data transmitted by such devices and their resources are only valid in a very limited spatial-temporal zone. In other words, the resources managed have a more intermittent and varying nature, i.e. linked to a very small cell, a narrow beam, and a limited temporal availability as a result of sparser channels and more complex environments in 6G applications. For example, the concept of *ephemeral edge computing* was introduced by [1], to capture cases in which edge computing for IoT devices occurs in a relatively short time period. Hence, IoI devices go through more ephemeral experience given the smaller coherence time, the more prominent effect of blockage, and the shorter the communication range. Moreover, heterogeneity in IoI is not limited to devices, but it is also characteristic of the services offered, as well as the radio, computation and energy resources; thus, leading to a scattered and skewed distribution of massive *small data* all over the network. As such, massive small heterogeneous data will pose challenges to the training processes of network learning, which perform ideally when accompanied with big data. Subsequently, this requires to move to edge intelligence which will leverage new proactive strategies using unprecedented learning mechanisms capable of making accurate and critical decision-making despite the scarcity of central big data and the intermittent nature of resources.

2.1.2 The Need for High-Rate and High-Reliability Low Latency Communications (HRLLC)

5G predominantly supports three generic services: enhanced mobile broadband (eMBB), massive machine type communications (mMTC), and URLLC. In other words, 5G applications focused either on standalone high rate services or short packet URLLC services. With the sixth generation, a new breed of applications such as ultimate XR, holography, and connected robotics and autonomous systems (CRAS) services [2] require simultaneously a high data rate and a *continuous* low latency, i.e., a blend of eMBB and URLLC with more stringent technological requirements. Consequently, providing a network that can meet the *reliability, low latency, and high data rate* requirement can naturally be addressed by the use of abundant bandwidth, available at THz and mmWave. While providing promising high data rates, such channels are characterized by their high susceptibility to blockages, short communication range, and the molecular absorption for THz. As such, driven by this uncertainty, an unprecedented edge predictive mechanism is needed to characterize the fundamental tradeoffs that need to be addressed on the network architecture and resources, finally guaranteeing a *seamless* experience to the user. In other words, AI at the edge will be the key enabler of such services, breaking the uncertainty barrier and soliciting users' five senses. For example, ultimate XR services operating at the THz band, require a continuous pencil narrow beam alignment between the users and the BS antennas, in the light of the narrowness of the beams, a minute misalignment will lead to a disruption in the experience. As a matter of fact, the edge needs to consider all the angles pertaining to the optimization with the exact increased precision, i.e., interference mitigation in a highly dense environment, user association in short communication range, etc. Indeed, a fundamental need for an intelligent way to orchestrate the radio and application resources is needed. Hence, the governing predictive mechanism needs to be driven by *experience*, i.e., the network management learned for a set of resources should be adaptable for a different type of radio resources in different environments. Only a transferable learning mechanism, i.e. a learning mechanism that can transfer knowledge from a previously learned task to another closely related to it, can optimize the network performance and provide the user *instantaneous* guarantees in such highly varying channels. Hence, on the one hand, the intelligent edge needs to fueled with a *service intelligence* to provide all the applications with robustness with respect to the user experience, i.e., granting industrial automation and robotics systems with human behavioral intelligence leads to precise machinery and innovations. On the other hand, edge *operational intelligence* provides the network with self-sustaining capabilities, allowing it to utilize its resources scalably, energy efficiently, and finally overcoming the highly dynamic and complex environments stemming from 6G applications [2].

2.1.3 Towards Privacy-Preserving Intelligence

An effective deployment of IoI services requires gathering massive small and ephemeral data from edge devices. As such, everything from self-driving cars to wearables in foreseen 6G services will now have to deal with personal data that can locate the activity, location, and preferences of the user. Even though the data is ephemeral, collecting it in a large scale fashion, and supplementing it with inference insights can provide entire timelines about users private life. Henceforth, to enforce a strong privacy policy, edge intelligence allows *raw data* to be processed at the vicinity of its source, thus enabling a trust bound. Moreover, given that edge intelligence relies on an inherent distributed learning architecture, raw datasets will be shared in a decentralized fashion preserving their privacy. For example, raw datasets gathered from wireless brain-computer interactions [2] not only jeopardize personal lives, but also compromise access over our emotions, moods and thoughts. Thus, making it completely prohibitive to store, process or train such datasets or their associated models on a centralized cloud entity. As such, one technique that enables learning decentralized data is federated learning (FL). In fact, FL is a distributed implementation of machine learning (ML) using which edge devices can perform on-device ML model training while only exchanging ML model parameters with a central controller to collaboratively find a shared optimal ML model [15, 16]. Nevertheless, while FL preserves the raw datasets of users, several unique challenges arise to provide a complete private system. Such challenges are dictated by a tradeoff between privacy and accuracy, resulting from the fact that injecting noise provides a differential privacy [17]. Subsequently, the most prominent threats to privacy in FL [18] occur throughout an inference during the learning process or an inference over the outputs. The latter implies a leakage of data outputs from one participant to another, whereas the former implies any participant in the federation inferring information about another participant's private dataset. In fact, the work in [18] proposes a scalable approach that protects against inference threats and produces models with high accuracy. Furthermore, decentralized authentication is another issue of prime importance for IoI devices. Here, distributed ledger technologies exploiting blockchain systems will be a key enabler for distributed and decentralized authentication. This discussion remains outside the scope of this chapter. Next, we will discuss the different collaborative mechanisms that empower the edge with more flexibility based on the network's specific needs.

2.2   How to Enable Collaborative Edge Intelligence?

Edge intelligence aims to enhance the benefits provided with the deployment of edge computing in 5G which include URLLC, traffic optimization, and energy savings. In its primitive form in 5G, edge computing mainly served to push cloud services closer to the end user, namely to the *edge*. Nevertheless, as the devices are becoming increasingly more heterogeneous with a different ranges of memory, processing units, and capabilities. It is important to address this heterogeneity in all network design and optimization approaches instead of adopting a one size fits them all technique. Hence, in beyond 5G and 6G, *edge* is not a specific entity anymore, instead it is a range of proximities that enable us to empower the network with an intelligent foundation enabling reliable and immersive experiences through its evolutionary applications. The computing capability of devices ranges from a low processing power for IoT sensors and wearables, to high processing capability devices like GPU powered workstations that can easily process and train complex deep neural network (DNN). Given this wide range of capabilities, the edge needs to encompass a spectrum of ranges vis-à-vis the service and the device needs as shown in Fig. 12.1. Henceforth, the computational burden can be weighted out between the *edge device* and the *edge server*, these weights are assessed based on the latency requirements of the application and its computational efficiency. This

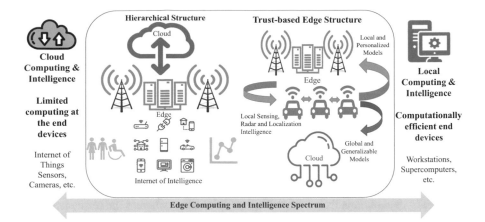

Fig. 12.1 Illustrative example of edge computing and intelligence spectrum in 6G

further leads to a tradeoff between the privacy, latency, and the accuracy of the model learned aiming to optimize the performance. A similar concept of levels of edge was discussed in [12], where they categorize edge intelligence into six levels, based on the amount and path length of data offloading. Here, level 0 relates to a pure cloud intelligence (training and inference on the cloud) and the highest level relates to an all on device intelligence. All the levels in between share the training and the inference among the cloud, edge and the device in collaboration based on the level.

It is important to note that, as also shown in Fig. 12.1, hierarchical structures were predominant in edge computing when deployed in 5G. While they still make sense for a large range of applications in 6G, trust-based structures break the ladder between the edge device and the cloud, and establish physical and logical links that act in the best interest of privacy of raw data, low latency and accuracy. The dynamics vary also based on the application and on the degree of privacy and latency that needs to be established. One example of non hierarchical models is the collaborative FL mechanism suggested in [15], where devices that cannot connect to the BS directly can associate with neighboring users to successfully engage in FL. Henceforth, enabling a wide range of levels of communication and inference, as well as different edge structures (hierarchical or trust-based) tailored to the needs of the application empowers the edge with a robust collaborative mechanism. Similar to the shift in the topology and structure of the edge, centralized learning mechanisms will fail to provide an accelerated learning process with a scaled algorithm given the massively distributed small data, and the aforementioned communication constraints such as privacy, security, power, and range limitations. Next, we will discuss the evolutionary modifications required on AI systems to provide the promised 6G performance.

3   How Is Edge Intelligence Different than Centralized AI?

The three pillars of every ML system are: data, training/model, and test-ing/inference. In contrast to conventional AI mechanisms adopted in previous generations of wireless communications and in generic ML applications, fueling the edge with intelligence transforms the three pillars of the ML system in a drastic way leading to significant changes in the design and deployment of ML. In what follows, we will emphasize how the changes in each of the pillars reflect novel deployment and design challenges. Also, we will shed light on some prospective solutions while emphasizing the caveats and benefits of each on the network as a whole.

3.1   Edge Data and Established Models

The proliferation of ML systems was mainly dictated by the availability of big data. In an ideal scenario, big data free from biases and noise can lead to a substantially successful training and prediction mechanism. Nevertheless, when intelligence is operated at the edge, injecting our training algorithm with immense data sets is not a luxury that is available. In fact, even if available, such lengthy training periods are not acceptable given that the end goal is to operate in a low latency and reliable realm. For example, the concept of big data has been transformed to *scattered small data* that is distributed among different entities in the network. This is the reason why, edge intelligence demands distributed and multi-agent learning. Moreover, such small data can vary from highly non i.i.d. to i.i.d. based on the level of heterogeneity in the resources and connectivity. Furthermore, small data can be aggregated in massive amounts and starts to mimic big data, or can suffer from a scarcity despite the aggregation due its highly ephemeral nature. Hence, the novel characteristics pertaining to small data can be addressed as follows.

3.1.1   Addressing the Heterogeneity and Non-i.i.d. Data

Having heterogeneity in computer resources and devices transforms the network management task, from a single task process to multiple tasks. In fact, having distributed small data in the network, i.e. having data points from ephemeral heterogeneous experiences, closely looks like multiple tasks data-sets in multi-task learning and meta-learning. When the number of data points of every task is very minimal, the learning mechanism is known as *few shot learning*. In fact, solving such problems with single task learning will make the agent incapable of generalizing the model or inferring the right decisions given the highly varying dynamics. Here, the agent at testing time will consider the network in unlearned state. Even with abundant data, acquiring the agent to learn the network at the infinitely many dynamic conditions is not a feasible technique. Hence, meta-learning not only provides the agent with learning skills in the face of data scarcity but also fuels the agent with more generalization in the face of sparse, highly uncertain and varying channels. Nevertheless, the meta-learning paradigm is still in its nascent stage and has been successful so far for labeled tasks a-priori or deterministic reinforcement learning (RL) mechanisms where a task can be easily identified. Handling problems such as channel parameters estimation, resource management, and network slicing is subject to a highly stochastic and uncertain conditions especially at HF bands. This leads to three main problems:

- Obtaining categorized datasets according to tasks in such problems is not a trivial task, given that many dynamics are intertwined; as such, the learning agent cannot explicitly define the tasks which consists of the input for the training process
- To account for the instantaneous behavior of the network, *extreme events* need to be taken into account to guarantee high reliability and a high quality of physical experience (QoPE). Such events will make the anomaly detection process in the preprocessing stage demanding given that true extreme data points might be confused with anomalous and noisy points infused with biases.
- In multi task scenarios, task datasets might not be readily available from users due to the highly dynamic and ephemeral edge network managements, whereas in single task scenarios, datasets are also not readily available from users to achieve the accuracy needed in the predictive end-result.

Defining Tasks for Operational Edge Intelligence

Given that the wireless setting does not have predefined tasks when coming across problems like resource allocation, channel estimation, and edge caching; it is important for us to preprocess the raw data from edge users and edge servers, enabling us to transform them to predefined tasks. The term *task* in the meta-learning community is different than its semantic meaning in the English language.

For a supervised learning setting, when learning multiple tasks, a task is defined as a [19]:

$$\mathcal{T}_i^S \triangleq \{p_i(\mathbf{x}), p_i(\mathbf{y} \mid \mathbf{x}), \mathcal{L}_i\} \tag{12.1}$$

where $\mathbf{x}$ is the feature vector, $\mathbf{y}$ is its label, $p_i(\mathbf{x})$ and $p_i(\mathbf{y} \mid \mathbf{x})^2$ are the probability distribution functions of the features and the class labels vector given the feature label vector respectively. $\mathcal{L}_i$ is the loss function. Moreover in an RL setting, a task is defined as a Markov decision process (MDP) as follows [19]:

---

[2] While we might not have access to these functions, they allow the data scientist categorize what constitutes a task in the scope of the meta-learning problem faced.

$$\mathscr{T}_i^R \triangleq \left\{ \mathcal{S}_i, \mathscr{A}_i, p_i\left(\mathbf{s}_1\right), p_i\left(\mathbf{s}' \mid \mathbf{s}, \mathbf{a}\right), r_i(\mathbf{s}, \mathbf{a}) \right\} \qquad (12.2)$$

where $\mathcal{S}_i$ is the state space, $\mathscr{A}_i$ is the action space, $p_i\left(\mathbf{s}_1\right)$ is the initial state distribution, $p_i\left(\mathbf{s}' \mid \mathbf{s}, \mathbf{a}\right)$ is the dynamics distribution, and $r_i(\mathbf{s}, \mathbf{a})$ is the reward function.

In fact, RL problem settings are more common in wireless networks given that a set of predefined and label tasks is not readily available. In fact, the environment feeding back the learning agent a reward is the fundamental teaching mechanism that allows to user to learn the learning concept. Indeed, establishing an accurate reward function is an important factor in any RL setting and even more importantly in meta-learning, nevertheless, such a discussion is outside the scope of this chapter.

Pertaining to edge learning, the learnt tasks mainly reflect a specific *trend* among users to break the inherent heterogeneity in the data. One way to obtain these tasks is to benefit from the concept of unsupervised learning and cluster the data points in clusters of trends allowing us to form clusters of quasi-homogeneous trends. In fact, such a paradigm has been suggested to tackle deterministic RL problems via unsupervised learning in [20]. Henceforth, as shown in Fig. 12.2 we further extend [20] to look into stochastic and highly dynamic environments, where the unsupervised learning tries to cluster the data in trends. If meaningful trends are found, the meta-RL procedure is carried out after proposing the cluster of tasks. Otherwise, distributed learning mechanisms like FL and democratized learning need to be established based on the problem and the service being analyzed.

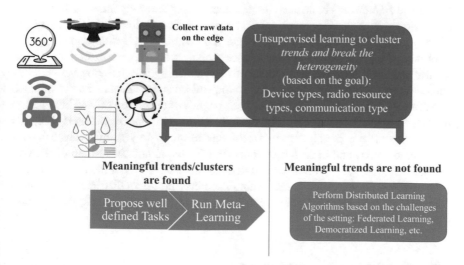

Fig. 12.2 Approach to break the heterogeneity in data

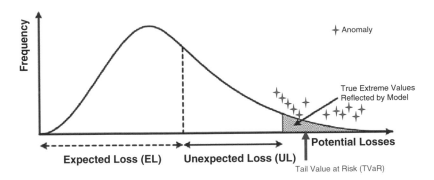

Fig. 12.3 Model of a loss function with true extreme values and anomalies

Distinguishing Extreme Behaviors from Anomalies

Prior to the emergence of powerful processing units, most of ML models were mere theoretical frameworks. As such, most of optimization methods prior to ML depended on pure theory-based models. Nevertheless, as the complexity of problems grew and the processing units evolved immensely, engineers were more likely to reach out to data science based models given that they rely on training machines based on *true data*. Henceforth, given that wireless communications operate in a scientific framework that is not completely arbitrary, instead of abandoning theoretical frameworks built, such models can be used to *orient, initialize and bound* data science models. The work in [21] was the first to look into the paradigm of theory-guided data science to leverage the wealth of scientific knowledge for improving the effectiveness of data science model. Moreover, we extend that vision further and look into to the problem of distinguishing true extreme values from anomalous points. As such, theoretical frameworks can provide us with the variance of extreme values, and thus provide us with bounds of the region shaded in Fig.12.3. In fact, to quantify accurately the shaded region, *risk and ruin theory* [22, 23] provide powerful economics frameworks that allow us to scrutinize the behavior of extreme values and their distributions. In [5], the authors quantified of the risk for an unreliable VR performance and derived the distribution of the tail of the E2E delay using such frameworks. Integrating such insights about distributional models with data science models finally contributes to overall solid models and learning mechanisms.

Addressing Scarcity in Dataset Points

Meta-learning and multi-task learning indeed do not require huge datasets, nevertheless, requisite large datasets of *tasks*. While we were capable of isolating trends and heterogeneity based on unsupervised learning and clustering techniques, these trends might not reflect *all possible trends* that allow agents to generalize to *new unseen trends and instances.* Similarly and more simply, even at the edge, single task learning is inevitable if edge servers are in the quest of network management and optimization vis-à-vis homogeneous resources. In such a case, the agent might also suffer from scarcity in data points. One promising solution for this challenge is to preprocess the raw data points with generative adversarial networks (GANs). The authors in [24] considered this approach for single task learning where a GAN was used to pre-train the deep-RL framework using a mix of real and synthetic data, finally contributing to a deep-RL framework that has been exposed to a broad range of network conditions. As a matter of fact, extending this framework to generate synthetic datasets of tasks, provides the agent with powerful inference tactics, finally leading to a generalizable edge learning mechanism immune of long trial and error periods.

3.1.2   Leveraging Edge Data for an HRLLC Realm

Given that data is heterogeneous, non i.i.d., and is distributed across edge servers and devices, preprocessing data indeed is a fundamental task in edge intelligence as pointed out previously. Having that in mind, such data should additionally be manipulated to allow the training algorithms to operate in a HRLLC realm. As a matter of fact, in mission critical services failing to obey the low latency and reliability constraints by a minute amount is as severe as failing to learn the trend of interest. Subsequently, we should benefit from the fact that *all data is not created equal.* For instance, in VR content, the background is more predictable but has a higher rendering load given the rich details complex textures; whereas the foreground environment is less predictable but has a much lighter computational complexity [25]. Such characteristics about the data can be used to the benefit of the network, for example for the background environment one solution would be to leverage the edge servers' computing capability to render the image, whereas the learning (since it is predictable and does not require a lot of rounds) is done on the edge device containing the preferences of the user and its raw data.

Furthermore, there is a big debate whether large models are inherently needed for a high accuracy. Indeed, large models are required for *robust* training, but *can we actually do more with less?* The concept of knowledge distillation and compression was suggested in [26] where compound compression was performed in terms of depth by distilling a shallower student network with a similar accuracy from a deeper teacher network. Incorporating such mechanisms to edge learning are necessary to finally operate in an HRLLC realm.

3.2   Edge Training and Prediction Mechanisms

3.2.1   Towards Generalizable and Personalized Learning

Given that the user is at the heart of 6G applications, the learned insights are not merely an image classification or a simple automated spam email checker. Ideally speaking, the behavior and interactions of edge intelligence should *mimic human intelligence*. As a matter of fact, the data learned is common and personalized exactly like people are. While some behavior tends to be common across most people, other characteristics tend to be very personal. Starting from here, this phenomenon says a lot about the data learned and the heterogeneity that we have previously discussed. Intelligence at the edge needs to be able to learn tasks with a high generalization and personalization margin. In other words, while the learned trends tend to be different across resources or devices, a common denominator shared should be always taken into the equation. Thus, parameters related to the common behavior shared across learned tasks needs to be improved and ought to be progressively learned based on shared experiences among the distributed agents learning. For example, *common* parameters can be reused by new agents initializing their parameters instead of learning from scratch, hence contributing for a faster convergence. A recent framework that acknowledges specialization and generalization known by *democratized learning* [27, 28] has emerged, it is one of the emerging distributed learning frameworks along FL. Here, the agents according to their different characteristics form appropriate groups that are tailored for a specific specialization. Such groups are self-organized in a hierarchical structure where the biggest group shares the most common knowledge across all agents, then groups start to shrink in size with more specialized skills. Subsequently, a learning mechanism driven by *specialization and generalization* not only mimics human behavior but also tends to be fairer since it can reduce bias significantly.

Furthermore, to provide a single agent with the ability to deliver goals that sustain diversity and personalization while considering heterogeneous context, the paradigm of Bayesian clustering can be leveraged. In [29], such a framework was introduced in order to collaboratively provide edge online caching. In particular, caching is considered a service intelligence rather than an operational intelligence given that it targets the popularity of contents, and thus optimizes the service experience rather than the wireless experience. Also, in [29] the authors propose a dynamic clustering policy that presents two policies: a diversity policy where one allows to identify the heterogeneity by treating clusters separately, and a second uniformity policy adopted within each cluster to speed up the learning process. Indeed, such frameworks need to be more elaborated to take part in large scaled distributed learning paradigms governing the overall structure of edge learning.

3.2.2 Towards a Robust and Transparent Learning Mechanism

With artificial neural networks (ANNs) constituting the key building block to most ML mechanisms being universal function approximators, the complexity of ML models is increasingly becoming higher. While such complexity might lead to accurate decision making in some instances, higher uncertainty is associated with the understanding, the interpretability, and the explainability of such models. Such uncertainty associated with black box models raises a lot of questions, especially when applied in mission critical applications. In other words, in applications such

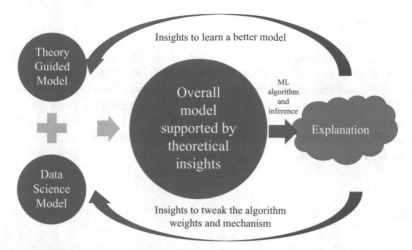

Fig. 12.4 Illustrative example of the general recipe for a transparent AI mechanism

as AR used in factories and manufacturing plants, a single mistake can lead to substantial hazards, and hence necessitates the need for a transparency to foster confidence in such situations. Moreover, such transparency should transform learning mechanisms from simple binary decision makers onto ones that provide supportive information to claim such decisions. Furthermore, beside the transparency for the sake of the audience or persona involved in the situation, explainability and interpretability is needed for experts building these mechanisms. As shown in Fig. 12.4, such explanation can put us one step further in terms of the success of the model and the algorithm, leading further to more accurate and robust inference. A number of works looked into the paradigm of explainable AI [30–32], nevertheless such works have not emphasized how explainable AI will be a key enabler in edge intelligence, wireless networks, or 6G applications.

Another problem faced by current ML techniques is the inability to learn complex trends (without infinitely running the algorithm). In here, an important question to ask is whether a smarter structure than ANN can be implemented for current cumbersome learning tasks. One potential solution for this question is quantum ML. Inspired by quantum mechanics, which is well known to produce complex and atypical patterns in data [33], an intuition suggests that small quantum information processors can easily learn and infer computationally difficult tasks for classical computers. As a matter of fact, the benefits of parallelism and superposition can be used to represent eigenstates in quantum powered RL algorithms [34]. Speedups are not limited to RL algorithms, but also provide exponential benefits to generative models as well. Certainly, quantum ML is a potential 6G key enabler that should be further researched.

4 Summary

In this chapter, we have provided an overview on edge intelligence and how it is a key enabler for 6G technologies. Specifically, we looked into the wireless requirements and transformations seen in this new breed of services that necessitate an intelligent backbone supporting their network interactively far from the conventional role of edge computing in 5G. Subsequently, we looked into the necessary changes that need to occur on the AI mechanism governing the edge, in order to successfully deliver complex 6G applications with their promised performance.

References

1. G. Lee, W. Saad, M. Bennis, An online framework for ephemeral edge computing in the internet of things (2020). Preprint. arXiv: 2004.08640
2. W. Saad, M. Bennis, M. Chen, A vision of 6G wireless systems: applications, trends, technologies, and open research problems. IEEE Netw. **34**(3), 134–142 (2020)
3. M. Mozaffari, A.T.Z. Kasgari, W. Saad, M. Bennis, M. Debbah, Beyond 5G with UAVs: foundations of a 3d wireless cellular network. IEEE Trans. Wirel. Commun. **18**(1), 357–372 (2018)
4. I.F. Akyildiz, A. Kak, S. Nie, 6G and beyond: the future of wireless communications systems. IEEE Access **8**, 133995–134030 (2020)
5. C. Chaccour, M.N. Soorki, W. Saad, M. Bennis, P. Popovski, Can terahertz provide high-rate reliable low latency communications for wireless VR? (2020). Preprint. arXiv: 2005.00536
6. F. Bonato, A. Bubka, S. Palmisano, D. Phillip, G. Moreno, Vection change exacerbates simulator sickness in virtual environments. Presence Teleoperat. Virt. Environ. **17**(3), 283–292 (2008)
7. J. Ren, G. Yu, Y. He, G.Y. Li, Collaborative cloud and edge computing for latency minimization. IEEE Trans. Vehic. Technol. **68**(5), 5031–5044 (2019)
8. G. Lee, W. Saad, M. Bennis, An online optimization framework for distributed fog network formation with minimal latency. IEEE Trans. Wirel. Commun. **18**(4), 2244–2258 (2019)
9. M.S. Elbamby, C. Perfecto, C.-F. Liu, J. Park, S. Samarakoon, X. Chen, M. Bennis, Wireless edge computing with latency and reliability guarantees. Proc. IEEE **107**(8), 1717–1737 (2019)
10. Y. Shi, K. Yang, T. Jiang, J. Zhang, K.B. Letaief, Communication-efficient edge AI: algorithms and systems (2020). Preprint. arXiv: 2002.09668
11. D. Xu, T. Li, Y. Li, X. Su, S. Tarkoma, P. Hui, A survey on edge intelligence (2020). Preprint. arXiv: 2003.12172
12. Z. Zhou, X. Chen, E. Li, L. Zeng, K. Luo, J. Zhang, Edge intelligence: paving the last mile of artificial intelligence with edge computing. Proc. IEEE **107**(8), 1738–1762 (2019)
13. J. Park, S. Samarakoon, M. Bennis, M. Debbah, Wireless network intelligence at the edge. Proc. IEEE **107**(11), 2204–2239 (2019)
14. C. Chaccour, W. Saad, On the ruin of age of information in augmented reality over wireless terahertz (THz) networks, in *Proceedings of IEEE Global Communications Conference (Globecom)*, Taipei, December 2020
15. M. Chen, H.V. Poor, W. Saad, S. Cui, Wireless communications for collaborative federated learning in the internet of things (2020). Preprint. arXiv: 2006.02499
16. M. Chen, Z. Yang, W. Saad, C. Yin, H.V. Poor, S. Cui, A joint learning and communications framework for federated learning over wireless networks (2019). Preprint. arXiv: 1909.07972

17. H.B. McMahan, D. Ramage, K. Talwar, L. Zhang, Learning differentially private recurrent language models (2017). Preprint. arXiv: 1710.06963
18. S. Truex, N. Baracaldo, A. Anwar, T. Steinke, H. Ludwig, R. Zhang, Y. Zhou, A hybrid approach to privacy-preserving federated learning, in *Proceedings of the 12th ACM Workshop on Artificial Intelligence and Security* (2019), pp. 1–11
19. C. Finn, P. Abbeel, S. Levine, Model-agnostic meta-learning for fast adaptation of deep networks, in *Proceedings of 2017 International Conference on Machine Learning*, Sydney, August 2017
20. K. Hsu, S. Levine, C. Finn, Unsupervised learning via meta-learning (2018). Preprint. arXiv: 1810.02334
21. A. Karpatne, G. Atluri, J.H. Faghmous, M. Steinbach, A. Banerjee, A. Ganguly, S. Shekhar, N. Samatova, V. Kumar, Theory-guided data science: a new paradigm for scientific discovery from data. IEEE Trans. Knowl. Data Eng. **29**(10), 2318–2331 (2017)
22. A.J. McNeil, Extreme value theory for risk managers. Departement Mathematik ETH Zentrum **12**(5), 217–37 (1999)
23. D.C.M. Dickson, *Insurance Risk and Ruin* (Cambridge University Press, Cambridge, 2016)
24. A.T.Z. Kasgari, W. Saad, M. Mozaffari, H.V. Poor, Experienced deep reinforcement learning with generative adversarial networks (GANs) for model-free ultra reliable low latency communication (2019). Preprint. arXiv: 1911.03264
25. F. Guo, R. Yu, H. Zhang, H. Ji, V.C.M. Leung, X. Li, An adaptive wireless virtual reality framework in future wireless networks: a distributed learning approach. IEEE Trans. Vehic. Technol. **69**(8), 8514–8528 (2020)
26. A. Polino, R. Pascanu, D. Alistarh, Model compression via distillation and quantization (2018). Preprint. arXiv: 1802.05668
27. M.N.H. Nguyen, S.R. Pandey, T.N. Dang, E.-N. Huh, C.S. Hong, N.H. Tran, W. Saad, Self-organizing democratized learning: towards large-scale distributed learning systems (2020). Preprint. arXiv: 2007.03278
28. M.N.H. Nguyen, S.R. Pandey, K. Thar, N.H. Tran, M. Chen, W. Saad, C.S. Hong, Distributed and democratized learning: philosophy and research challenges (2020). Preprint. arXiv: 2003.09301
29. J. Liu, D. Li, Y. Xu, Collaborative online edge caching with bayesian clustering in wireless networks. IEEE Internet of Things J. **7**(2), 1548–1560 (2019)
30. S. Chari, D.M. Gruen, O. Seneviratne, D.L. McGuinness, Foundations of explainable knowledge-enabled systems (2020). Preprint. arXiv: 2003.07520
31. W. Samek, K.-R. Müller, Towards explainable artificial intelligence, in *Explainable AI: Interpreting, Explaining and Visualizing Deep Learning* (Springer, New York, 2019), pp. 5–22
32. A.B. Arrieta, N. Díaz-Rodríguez, J.D. Ser, A. Bennetot, S. Tabik, A. Barbado, S. García, S. Gil-López, D. Molina, R. Benjamins, et al., Explainable Artificial Intelligence (XAI): concepts, taxonomies, opportunities and challenges toward responsible AI. Inf. Fusion **58**, 82–115 (2020)
33. J. Biamonte, P. Wittek, N. Pancotti, P. Rebentrost, N. Wiebe, S. Lloyd, Quantum machine learning. Nature **549**(7671), 195–202 (2017)
34. S.J. Nawaz, S.K. Sharma, S. Wyne, M.N. Patwary, M. Asaduzzaman, Quantum machine learning for 6G communication networks: state-of-the-art and vision for the future. IEEE Access **7**, 46317–46350 (2019)

Chapter 13
6G CloudNet: Towards a Distributed, Autonomous, and Federated AI-Enabled Cloud and Edge Computing

Isiaka A. Alimi, Romil K. Patel, Aziza Zaouga, Nelson J. Muga, Armando N. Pinto, António L. Teixeira, and Paulo P. Monteiro

Abstract The current 5G network deployment is anticipated to support a massive number of connections and offer a very high-data-rate to ensure enhanced mobile broadband with ultra-reliable and low-latency services. However, the full connectivity requirements by the future digital society may be challenging for 5G. Consequently, for effective support of future demands, research attention is towards beyond 5G wireless communications. This chapter presents in-depth studies on the envisage use cases, network architectures, deployment scenarios, and technology-driven paradigm shifts for the sixth generation (6G) networks. Beyond 5G connectivity will be based on digital twin worlds for accurate representation and unification of the physical and biological worlds. To realize this, innovative ideas will emerge to define 6G system requirements and technologies. Besides, the swift advancement in artificial intelligence makes it one of the viable solutions for 6G networks. Moreover, key technical areas and technology advances such as new spectrum bands, cognitive spectrum sharings, photonics-based cognitive radio, innovative architecture models, terahertz communications, holographic radio, advanced duplex, and advanced modulation schemes are envisaged to define the 6G radio access network (RAN). Consequently, the 6G RAN is anticipated to be an exceptionally dynamic, inherently intelligent ultra-dense heterogeneous network with effective support for all things.

I. A. Alimi (✉) · R. K. Patel · A. Zaouga · N. J. Muga
Instituto de Telecomunicações, Universidade de Aveiro, Aveiro, Portugal
e-mail: iaalimi@ua.pt; romilkumar@ua.pt; aziza.zaouga@av.it.pt; muga@ua.pt

A. N. Pinto · A. L. Teixeira · P. P. Monteiro
Instituto de Telecomunicações, Departamento de Electrónica, Telecomunicações e Informática, Universidade de Aveiro, Aveiro, Portugal
e-mail: anp@ua.pt; teixeira@ua.pt; paulo.monteiro@ua.pt

© The Author(s), under exclusive license to Springer Nature Switzerland AG 2021
Y. Wu et al. (eds.), *6G Mobile Wireless Networks*, Computer Communications and Networks, https://doi.org/10.1007/978-3-030-72777-2_13

1 Introduction

The fifth-generation (5G) networks have been progressively deployed for commercial use across various markets. The 5G network aims at delivering a number of novel services and vertical applications with ultra-low-latency, high-bit rates, high-capacity, and ubiquitous access to the subscribers [1]. In this context, it is envisioned that the 5G system may considerably transform the mobile networks and significantly enhance the existing fourth-generation (4G) network performance. Besides, it is anticipated to support various advanced technologies like Information-Centric Networking (ICN) for effective management of the network traffic; network slicing for enabling the swift deployment of various services; software-defined networking (SDN) for ensuring adequate flexibility in the network; as well as millimeter-wave and massive multi-input-multi-output (MIMO) for enhancing the system achievable information rate (AIR) [2]. Besides the aforementioned performance enhancement regarding the latency and data rates, massive connection density and other system improvements are anticipated [3, 4]. Based on this, the International Telecommunication Union (ITU) has defined different 5G network use cases such as massive Machine-Type Communications (mMTC), enhanced Mobile BroadBand (eMBB), as well as Ultra-Reliable and Low-Latency Communication (URLLC) [5].

Moreover, the related applications have stringent specifications regarding critical system parameters such as power consumption, number of connected devices, latency, bit-rate, and availability. It is noteworthy that the use cases are generic and each of them is mainly focusing on the optimization of at least one key design parameter that could be the bit-rate, the latency, or the number of connected devices [6]. Besides, there are ongoing efforts on the sixth-generation (6G) system to establish what the 6G will be and the potential enabling technologies and use cases. The 6G mobile network is envisioned to be an exceptionally dynamic, inherently intelligent, extremely heterogeneous, and ultra-dense network that can support all things with ultra-low-latency (i.e. 1 ms) and ultra-high-speed (i.e. 1 Tbps) transmission [7]. The real design of 6G is expected to depend on the 5G developments and performance limits that will be created by the existing networks. In this regard, as the 5G truly manifests, the areas where it underperformed will be part of what the 6G will be addressing [8].

This chapter presents in-depth studies on the envisaged use cases, network architectures, deployment scenarios, and technology-driven paradigm shifts for the 6G networks. Besides, it presents AI as an efficient platform for realizing high levels of automation that are essential for the optimization and management of the current and future complex networks. Moreover, key technical areas and technology advances such as cognitive spectrum sharings, photonics-based cognitive radio, innovative architecture models, terahertz communications, holographic radio, and advanced modulation schemes are presented for the 5G and beyond networks.

> **! Attention**
>
> It should be noted that the main drivers of 6G will not only be owing to the performance limits and challenges that will be unfolding in the 5G era but also based on the unrelenting network evolution and technology-driven paradigm shift trends that characterized the wireless networks.

2   5G and Beyond Networks

In this section, we present different RAN architectures and discuss the associated concepts. Also, we discuss the main requirements, enabling technologies, related technical challenges, and proffer potential solutions on the approaches that can be adopted to realize an efficient system.

2.1   *Cloud Radio Access Networks*

The 5G network is envisaged to deliver relatively better performance compared to its preceding networks. In this context, it is expected to deliver exceptional user experience regarding high-data-rate, ultra-low-latency, and ubiquitous access compared to 4G networks. Moreover, there are a remarkable number of machine-type communications (MTC) that support fully automatic data generation, processing, exchange, and actuation between intelligent machines [9, 10]. 5G is expected to support massive MTC (mMTC) with a huge number of mobile devices and applications. This will help in enhancing the user experience by offering eMBB services and in delivering uRLLC for critical communication and control services [10, 11]. Although a lot of applications such as video-driven machine-human interaction, real-time closed-loop robotic control, virtual reality (VR), augmented reality (AR), and mixed reality (MR) based on high definition 360° video streaming with low-latency and high-bit-rate requirements can be supported by the emerging 5G networks, it is envisaged that more challenging applications such as real-time sharing and updating of high-resolution maps for control of autonomous systems will be supported by the evolving 5G networks [6]. Also, apart from the enhanced mobile broadband, 5G is anticipated to enable the Fourth Industrial Revolution (Industry 4.0) through digitalization and connectivity of all things. Besides, digital twins of a number of objects that will be established at the edge clouds will be an indispensable basis for the next-generation digital world [12].

Moreover, in the next-generation, digital twin worlds of both physical and biological entities with a huge amount of capacity at ultra-low-latency will be the fundamental platform for supporting the new digital services. Also, digitalization is expected to initiate the creation of innovative virtual worlds. In this perception, based on a range of degrees, digital representations of imaginary objects will be integrated with the digital twin world to bring about an MR and a super-physical world. For instance, as smart devices such as watches and heart rate monitors are transforming into wearables, skin patchable, body implant, ingestible, and brain activity detectors, human biology will accurately be mapped and integrated into digital and virtual worlds. This will facilitate new super-human potentials. Also, efficient and intuitive human control of the entire worlds (physical, virtual, or biological) will be ensured by the AR user interfaces [12].

Furthermore, for efficient support of the 5G use cases, deployment scenarios, and the associated requirements, there is a growing consensus of opinion that all the existing network architectures have to be restructured [11, 13]. In this sense, advancements in the radio access networks (RANs) and in the core networks are highly imperative for the achievement of the envisaged goals and objectives of the 5G networks. There are a number of viable RAN architectures that have been considered for the 5G. One of such is the cloud RAN (C-RAN) that has been consented as a promising architecture. Apart from the fact that C-RAN implementation can speed up the network deployment, it can also enhance the power efficiency and save costs (i.e. capital and operation expenditures) [14]. This has been demonstrated by several operators both in different field trials and commercial networks that are based on the 4G systems. With evolution towards 5G, there is a considerable advancement in the C-RAN as well. There are many additional technologies such as SDN, virtualization, and novel FH solutions that have been proposed and adopted in the C-RAN [11].

2.2 C-RAN Architecture

A traditional base station (BS) comprises a baseband unit (BBU) and a remote radio unit (RRU) as the logical network entities [11]. Moreover, the 4G network deployments have been leveraging two major forms of network architectures which are the Distributed RANs (D-RANs) and the C-RANs. In the former, the whole BS stack is incorporated in each of the cell sites. In the latter, each of the cell sites just contains the radiating elements and the remote radio heads (RRHs), while their BBUs are normally deployed and centralized in a remote location to form a BBU pool. Along the progress line in the C-RAN architecture, the BBU can be accomplished on a virtualization platform for further flexibility [11, 15]. It is noteworthy that C-RAN is an application model of a cloud computing-based RAN (CC-RAN). Other prevalent models are the heterogeneous cloud radio access network (H-CRAN) and fog computing-based RAN (F-RAN) [14].

As aforementioned, the C-RAN architecture offers comparative advantages such as better energy efficiency and enhanced network capacity based on inter-cell coordination. Nevertheless, very stringent requirements are imposed on the common public radio interface (CPRI) based FH transport networks. The mobile FH connects the RRHs to the centralized BBUs [16]. On the other hand, in the D-RAN, IP packets are conveyed between the cell site and the core network by means of the backhaul transport network. It is noteworthy that the FH and backhaul networks are implemented completely as distinct entities based on various technologies in the existing deployments [15]. For instance, in the C-RAN architecture, digital radio over fiber-based CPRI has been adopted for digitized radio samples transmission between the network entities. However, it is challenging for the CPRI-based FH networks to scale to 5G RANs [15, 17]. To address this, different innovative solutions have been presented. Apart from the H-CRAN and F-RAN that have been presented to alleviate the issue, other viable solutions like the RAN functional splits have been presented to split the RAN functionalities between the centralized location and the cell site. Nevertheless, this approach presents a tradeoff between the centralization gains and the transport network requirements. Similarly, the 5G RAN architecture in which next-generation NodeB (gNB) incorporates three modules, namely, the distributed unit (DU), the centralized unit (CU), and the radio unit (RU) has been adopted by the 3GPP to support multiple functional splits with flexible location [15, 17, 18]. In the same context, the CPRI Cooperation has presented an eCPRI[1] that ensures a reduction in the data rate demands between the enhanced radio equipment controller (eREC—i.e. CU) and enhanced radio equipment (eRE—i.e. DU) through a flexible functional decomposition, while limiting the eRE complexity [15, 23]. Besides, traffic transmission schemes such as edge cache and mobile data offloading can be exploited to alleviate the stringent requirements and the effect of bandwidth-intensive services delivery over the mobile networks [14].

2.3 5G RAN Architecture

Based on the Release 15 specification, various terminologies in conjunction with interfaces and functional modules are presented by the 3GPP 5G RAN architecture known as NG-RAN. The 3GPP NG-RAN contains a set of radio base stations (BSs) that are known as gNodeBs (gNBs) [24]. Unlike the 4G RAN architecture illustrated in Fig. 13.1a, the 5G gNB in Fig. 13.1b comprises three logical network entities. These network entities are the CU, DU, and RU. It is noteworthy that those entities could be deployed in multiple combinations within the network environment as illustrated in Fig. 13.2. With the 3GPP NG-RAN architecture, a gNB can have a CU as well as several associated DUs. Therefore, a typical gNB architecture can be considered as a Mini-C-RAN.

Figure 13.3 depicts a typical 5G C-RAN architecture with centralized multiple CUs for the effective coordination of the connected DUs. Besides, as illustrated in Fig. 13.3, 5G C-RAN architecture supports two sets of transport networks which are the FH and midhaul links [16, 24]. The FH links connect the DU to one or

---

[1]The eCPRI is an enhancement to run CPRI over switched Ethernet (packet-based). In the literature, it is called Ethernet-based CPRI [19], enhanced CPRI [20, 21], and evolved CPRI [22].

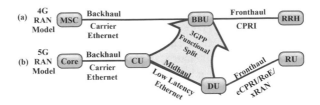

Fig. 13.1 4G and 5G RAN architectural contrast. MSC: Mobile switching center; DU: Distributed unit; CU: Centralized unit; RU: Radio unit; CPRI: Common public radio interface; eCPRI: enhanced CPRI; RAN: Radio access network; xRAN: Extensible radio access network; RoE: Radio over Ethernet; RRH: Remote radio head

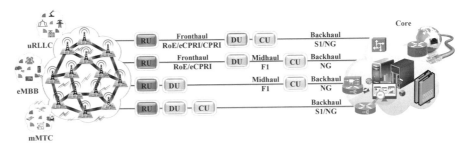

Fig. 13.2 Flexible distribution of RAN entities and functionalities

more RUs, while the midhaul links connect the CU and the associated DUs [11]. In addition, it should be noted that the emerging 5G use cases like mMTC, eMBB, and uRLLC require an extensive range of quality-of-service (QoS) as regards the latency, bandwidth, and reliability. Moreover, the traditional fully centralized C-RAN mainly centered on a "one size fits all" approach. However, this approach is inefficient for addressing the wide QoS requirements by different traffic and the associated use cases. Also, the implementation of the approach imposes very stringent requirements on the FH links [25]. Consequently, the possibility of supporting such potentially wide-ranging QoS requirements has led to a number of concerns regarding latency and bandwidth requirements.

Moreover, there are various viable methods in which different use cases can be efficiently and cost-effectively supported over a shared single network infrastructure. One of such approaches is network slicing [25, 26]. Network slicing presents flexible-based system implementations by creating several virtual (logical) slices (i.e. sub-networks) on an integrated physical infrastructure. Usually, this approach leverages schemes like cloud computing, SDN, and network function virtualization (NFV) [25–28]. Moreover, by exploiting network slicing, highly diverse services with a wide range of requirements can be supported on a shared infrastructure. This can be achieved by means of an independent configuration of each network slice [25, 28]. Based on the resources, the respective slice is an end-to-end virtualized network instance that is well-designed to meet the service requirements [25].

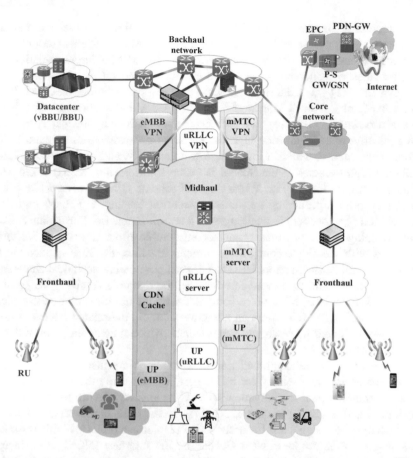

Fig. 13.3 A typical 5G C-RAN with network slicing architecture. CP: Control plane; UP: User plane; UE: User equipment; CDN: Content delivery networks; mMTC: Massive machine-type communications; uRLLC: Ultrareliable low-latency communications; eMBB: enhanced mobile broadband; RRU: Remote radio unit; VPN: Virtual private network; BBU: Baseband units; vBBU: virtual BBU; PDN-GW/P-GW: Packet data network gateway; EPC: Evolved packet core; SGSN: Serving GPRS Support Node; GPRS: General packet radio service

In addition, NFV is an advanced technology that has been leveraged in the network slicing to achieve a virtualized network and to attain infrastructure virtualization as well. In this regard, with network slicing implementation, the underlying physical infrastructure and the related network functions can be appropriately instantiated, connected, and incorporated over 5G networks [25, 28]. Besides, it is noteworthy that network slicing is different from a virtual private network (VPN) that is usually employed for isolating different networks [28]. It goes beyond the VPN by offering both computing and storage resources for a variety of network slices [25].

Furthermore, innovative technologies such as millimeter-wave and multiple antennas technology that will be employed in 5G will impose stringent requirements on transport networks. For instance, with massive MIMO implementation, the FH data rates will increase to hundreds of Gbps based on the traditional CPRI employment [11]. To address this, various transport solutions such as Radio over Fiber (RoF), Optical Transport Network (OTN), and Multicore Fiber (MCF) transmission have been presented for 5G FH. Although these high-capacity transport solutions offer notable advantages, their practical implementations demand further efforts. For instance, RoF is a low-cost, low-complexity, and future-proof solution while the high-capacity of the Spatial Division Multiplexing (SDM)-based MCF solution is highly fascinating. When the RoF scheme is implemented along with MCF, they offer a promising transmission solution for the 5G FH. Nevertheless, this solution presents certain challenges in a real-life scenario. For instance, MCF is an emerging technology that demands considerable effort for commercialization compared with the single-core SMF solutions. Besides, the RoF application for uplink (UL) transmission (from RRU to CU) can pose certain technical challenges. For instance, in a bid to prevent interference, the dynamic receiving range of an amplifier is lower in an MCF when compared with the traditional wireless systems [11]. Also, although MCF offers an attractive solution, the related effects of inter-core crosstalk hinder its reach and performance, which may be an issue in the 5G C-RAN system [11, 29, 30].

In general, the aforementioned transport solutions are just trying to increase the transport network capacity for better performance while the primary concern regarding the FH signal is left unattended to. Consequently, the offered solutions are provisional and might not be able to stand the test of time. In this regard, a holistic approach to the FH signal and the associated FH Interface (FHI) is highly desirable. One of such is the proposed Next Generation FH Interface (NGFI). In principle, NGFI presents an FH interface in which the data rate is antenna-independent but traffic-dependent. The traffic-dependent attribute implies an increase in the traffic results in a corresponding increase in the FH rate. Therefore, unlike the traditional CPRI fixed FH rate, NGFI offers a dynamic increase or decrease in the FH rate in accordance with the actual network traffic. This helps significantly in improving transport efficiency. Moreover, based on the antenna-independent feature, the influence of the number of antennas on the FH can be minimized. This facilitates improved support for massive MIMO by the FH transport networks [11].

Moreover, the advent of NGFI sheds light on the need for efficient FH transport networks for the 5G and beyond systems. In this context, some organizations have been working relentlessly on different architectures and design requirements for the next-generation FH networks. For instance, a working group known as the NGFI 1914 has been instituted by the IEEE to develop 5G FH standards. Likewise, the 1914.1 project is based on architecture design and requirement development while the 1914.3 project focuses on the Ethernet encapsulation format definition for radio signal [11]. Moreover, a virtualized radio access network (vRAN) is a promising evolution of mobile networks due to its ability to significantly reduce costs, logically boost capacity, and enhance the customer experience. It is noteworthy that the main enabling scheme for vRAN is NFV. It helps in transforming a typical network architecture from hardware-based to software-based[2] and the concept of proprietary hardware-based BSs into a more agile, flexible, and cost-effective system.[3] However, vRAN deployment has been deemed to be much slower than anticipated compared with other network elements. Apart from the FH issue, another main hurdle to large-scale deployment of vRAN is vendor hostility. Operators have tried to develop an open interface between the BBU and RRU in the 4G. However, the transition from the CPRI to the eCPRI appears not to be given significant attention to ensure an open interface. Consequently, novel 6G network architecture is required to ensure effective support of control, communication, sensing, and computing convergence that is based on ultra-low-latency and ultra-high-data rates system with open interfaces [7, 8].

2.4   Future Mobile Systems: Beyond 5G

The long-distance propagations are usually based on a low-frequency spectrum band that is capable of supporting a wide coverage. Nevertheless, due to its comparatively narrow bandwidth, it is more effective for low transmission rate applications. Similarly, as a result of the considerable increase in the traffic as well as the related network requirements for the existing and the emerging use cases, it is imperative to move beyond the low-frequency bands (sub-6 GHz), toward higher carrier frequencies such as millimeter-wave (mm-wave) band and optical frequencies [6, 31]. In this regard, the mm-wave band has been adopted for the 5G applications. The mm-wave band is capable of delivering bandwidths around some gigahertz. Nonetheless, based on the current and the anticipated traffic increase, it might be challenging to meet the bandwidth requirements of the *Network 2030* (6G and Beyond) efficiently with mm-wave band exploitation [31]. Besides, it should be noted that the conventional abstraction of channel states based on less feedback and simplified modeling may be insufficient, specifically at higher carrier frequencies. This is due to the incapability of generalized models to emulate unknown channels. Subsequently, a paradigm shift will be imperative for more robust models. Besides, since the conventional interference-cancellation schemes are not optimal, advanced techniques of interference utilization are emerging. With the heterogeneous nature and envisaged massive connections, tighter cooperation, as well as more aggressive resource sharing, will be some of the enabling features of the future wireless networks. However, this will result in a growing challenge regarding interference management. In this regard, a concerted research effort is required to realize more

---

[2]Network virtualization involves the evolution from custom-built network nodes to network functionality that is implemented in software. Usually, the software is running on a generic hardware computing platform.

[3]A vendor-agnostic hardware aids revolution across a variety of software ecosystems.

efficient radio resource utilization. For instance, with interference exploitation, the system performance can be improved significantly [8].

Furthermore, a Focus Group for Network 2030 has been established by the ITU in July 2018. This Group aims to guide the global information and communications technology (ICT) community in the exploration of the potential system technologies towards the development of 2030 and beyond networks. In this regard, extensive research efforts are ongoing regarding the 6G network [8, 31, 32]. According to the perceptions of the Group, 6G will span over various new concepts regarding services, holographic media, Internet Protocol (IP), and network architecture. Moreover, relevant research areas have been considered by other groups in several organizations and countries. Some of the works are focusing on artificial intelligence (AI) and distributed computing; near-instant, reliable, unlimited wireless connectivity; as well as antennas and materials for supporting future circuits and devices [8, 32]. Besides, based on the existing research developments, 6G network architecture is expected to leverage various schemes such as AI, higher frequency bands, green energy, super IoT, mobile ultra-broadband, as well as symbiotic radio and satellite-assisted communications (e.g. space-air-terrestrial-sea integrated (SATSI)) [7].

Additionally, with symbiotic radio and satellite-assisted communications, efficient, flexible, and ubiquitous connections can be facilitated for the envisaged massive things (i.e. for supporting super IoT) [7]. Besides, as will be discussed in Sect. 3, Machine Learning (ML) algorithms are expected to be promising solutions for wireless AI networks [31]. Also, beyond 5G networks are envisaged to offer a considerable improvement by employing AI techniques [6]. Thus, operators have been employing ML/AI not only for enhancing the network performance but also to curtail deployment costs. Besides, AI presents an efficient platform for realizing high levels of automation that are essential for the optimization and management of the current and future complex networks. Based on the offered opportunities by the ML/AI, providers can seamlessly and swiftly shift from network to service management [8]. In this context, the fundamental target will be to improve the entire 5G applications and vertical use cases based on cognition capability. This can be effectively achieved by exploiting cognitive computing systems that leverage the ML algorithms and the available huge amounts of data [6]. Likewise, to ensure that the uMUB is well-supported with ultra-high-data-rate and ultra-low latency, 6G will exploit broader frequency bands such as visible light communication (VLC), terahertz (THz), and sub-THz bands. However, the 6G network has to be an ultra-dense system to address the associated high path-loss at higher frequency bands. Furthermore, low-latency and high-speed device-to-device (D2D) communication as well as ultra-massive MIMO communication will be good solutions for enhancing the system performance. Hence, one of the main aspects of 6G is the ultra-dense heterogeneous network [7].

In addition, the 6G system is expected to support higher carrier frequencies for a relatively smaller number of antennas. Also, for increased resolution, it will have broadened bandwidth. However, one of the major challenges will be on how RF signals can be analyzed and processed promptly over a very broad bandwidth in real-time and without any prior knowledge of the carrier frequency, signals, and modulation format. In this context, a photonics-defined[4] system offers full-spectrum capacity and huge bandwidth. Therefore, it will be an ideal platform for supporting the future 6G system [8].

Although potential improvement by the AI to the physical layer in the development of 6G technologies demands further consideration, its implementation results in a flexible and agile air interface, with enhanced efficiency. Moreover, as a consequence of the constrained flexibility and bandwidth of the hardware platform, conventional AI-based cognitive radio encounters a substantial gap between flexible modules and mobile network deployment. In the light of this, integration of ML and photonics-defined radio[5] will be one of the promising key evolutions of AI in 6G. Besides, in cognitive radio applications, neuromorphic photonic[6] ideas have been employed for RF fingerprinting of crowded and complex environments. Its implementation at the front-end of RF transceivers can help in the offloading of complex signal processing operations to the photonic chip, thereby, addressing the latency and bandwidth limitations of the existing DSP solutions. Consequently, a combination of photonics technology and AI in 6G infrastructures can offer high-reliability, low-latency, and scalable AI [8].

Furthermore, regarding the bandwidth improvement and data rate enhancement, the THz band to be utilized by the 6G can theoretically offer three orders of magnitude greater than the achievable one in the mm-wave band [31]. Consequently, 6G aims at achieving ubiquitous connectivity through the integration of satellite communication networks as well as underwater communications to deliver global coverage. Besides, intelligent driving and industry revolutions will be the core requirements of 6G. So, new service classes such as ultrahigh-speed-with-low-latency communications (uHSLLC), ubiquitous mobile ultra-broadband (uMUB), and ultra-high data density (uHDD) have been described for the 6G networks [2, 8]. For instance, future intelligent factories will be characterized by densely concentrated intelligent mobile robots that will be accessing high-performance computing resources wirelessly. The setup will be based on a distributed intelligent network with terabytes computing capacity. In this scenario, the robots have to respond swiftly to changing conditions to satisfy the requirements of various applications. To achieve this, the robots will require enormous computing capacity to process huge amounts of data. In this regard, optical wireless and THz wireless are promising solutions with inherently huge bandwidth that can support the required data density. Consequently, uHDD will be essential to offer enormous wireless capacity that

---

[4]Photonics-defined system is a scheme that focuses on an extension of microwave photonics by leveraging various areas such as digital photonics (optical computing and photonic DSP), coherent optics, and photonic analog-to-digital/digital-to-analog conversion (ADC/DAC).

[5]Photonics-defined radio is the convergence of integrated microwave photonics, integrated coherent optics, and photonic digital signal processing (DSP) [8].

[6]Neuromorphic photonics is an evolving field with a tight strategic nexus between photonics and neuroscience. It merges the benefits of electronics and optics to produce systems with superior features regarding interconnectivity, efficiency, and information density [8, 33].

can be efficiently supported by uHSLLC [8]. In this context, a dedicated study group has been formed in the IEEE 802.15 for the THz spectral allocations as well as standardizations. Similarly, several companies such as Intel and Huawei have been performing experiments in these bands [31]. Moreover, further information regarding the requirements, technologies, challenges, and solutions for the 5G and beyond networks can be found in [14] and the references therein. Moreover, with the envisaged complex networks, automation is indispensable for effective optimization and management. In this context, Sect. 3 discusses trends towards intelligent-based network optimization.

> **! Attention**
>
> In general, the 6G mobile network has to be defined implicitly as regards the system architecture, fundamental technology, and mobile terminal (MT) types. For instance, the 6G MT type has to evolve beyond the current 4G/5G smartphone. Also, there will be intelligent and notable evolution in the fundamental technology, as a result, software-defined radio will just be a complementing technology. Besides, regarding the network architecture, a paradigm shift from the CloudRAN to a distributed and autonomous AI-enabled edge computing that will leverage schemes such as RF holography and photonic technology is envisaged. Furthermore, smart driving (smart vehicles) and smart industry (smart manufacturing/smart mobile robots) are expected to be some of the main focus of the 6G services.

3 Trends Towards Intelligent-Based Optimized Networks

There have been unprecedented challenges regarding network design, scalability, and optimization. For effective deployment and implementation of different use cases, mainly, the high-stake applications in which low-latency and reliable communication are required, the trend is to shift from the conventional cloud-based schemes to distributed and collaborative AI-enabled approaches as depicted in Fig. 13.4. In this section, we present the trends towards intelligent networks.

3.1 AI-Enabled Cloud and Edge Computing

Cloud computing is a paradigm that facilitates convenient, ubiquitous, on-demand network access. It can be implemented over a shared pool of computing resources that feature programmable options. Also, one of the fundamental features of cloud computing is the establishment of a platform that seems to offer *infinite* computing resources that are on-demand. Nevertheless, the resources supported by a single cloud are mostly limited. Besides, a single cloud could be unable to handle the

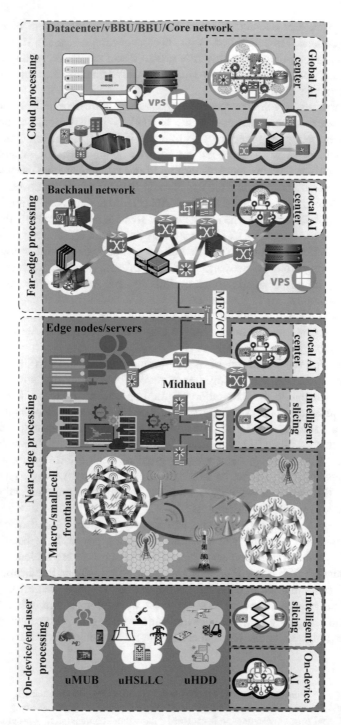

Fig. 13.4 A typical intelligence-native (AI-enabled) 6G architecture in which local and/or global processing and service provisioning can be supported based on the requirements of the underlying use cases. uHSLLC: Ultrahigh-speed-with-low-latency communications; uMUB: Ubiquitous mobile ultra-broadband; and uHDD: Ultra-high data density

abrupt surge in user demands. The aforementioned challenges can be addressed by the Intercloud. An Intercloud is an interconnection of clouds that facilitates cooperation among them, resulting in a global *cloud of clouds*. In this context, a cloud without adequate resources to satisfy the user demands can exploit resources from other clouds [34].

The exceptional revolutions regarding the number of devices in the network have caused an unprecedented increase in the bandwidth-intensive mobile applications and services being supported by the Internet [10, 35]. Based on this, there is a growing increase in data processing devices at the edge [6]. Moreover, the anticipated stringent requirements regarding ultra-low-latency, high user experience continuity, and high reliability for the 5G and beyond wireless networks demand additional localized services within the RANs. This service localization has to be, as much as possible, close to the mobile users. To attend to this demand, the concept of mobile edge computing (MEC) has been presented. MEC aims to unify the telco, information technology (IT), and cloud computing to facilitate direct cloud services delivery from the network edge. This offers advantages such as extended reachability, improved cloud service availability, and minimized network latency [14]. Conceptually, MEC focuses on ways of extending cloud capabilities like computing, storage, and caching to the edge with the intention of enhancing the broadband experience and reduce end-to-end latency over the RAN to/from IoT devices and mobile users [6, 14].

Furthermore, the advancement in the mobile user equipment (UE) has paved the way for local edge computing. For instance, there is growing advancement in the mobile UE due to the integration of multiple sensors, quasi-sufficient storage resources, extended battery lifespan, as well as high-performance and high-end processors. Based on this, UE can be considered as a personal mobile workstation platforms that facilitate light-level on-device AI processing for future intelligent networking. Consequently, there are concerted research efforts on the utilization of on-device AI owing to its capability for dynamic local network environment sensing. In this regard, network parameters can be optimally and efficiently adjusted to meet various network requirements [7].

Moreover, the advances in the ML, mainly deep-learning (DL), driven primarily by the availability of additional data and computing power, have been changing means by which services and applications have been delivered. Nonetheless, the conventional ML/AI[7] concept is based on a centralized scheme that is implemented using a cloud computing model. In this approach, the global dataset is presented a priori and employed in the training stage by exploiting the offered massive storage and computing power. However, with emerging novel types of intelligent devices and high-stake use cases such as robotic assemblers, remotely controlled

---

[7] AI is an advanced concept that is based on the development of automated and intelligent machines with identical fundamental characteristics as human intelligence, while, ML is an application or a subset of AI that permits machines to learn from data without being explicitly programmed. The ML uses algorithms to parse data and then learn from it so as to predict and decide on events.

drones, autonomous vehicles, and AR/VR applications, the cloud-based ML might be insufficient. Because the aforementioned are real-time applications that demand not only for high-reliability but also for a millisecond of latency may not be tolerated [6, 36].

3.2 Distributed and Federated AI

Interclouds can be broadly categorized into federated clouds and multi-clouds. In the former, network infrastructures are readily interconnected by the providers to support the exchange and sharing of resources among themselves. Moreover, federated clouds are grouped into centralized[8] and peer-to-peer[9] approaches. Also, clouds that interconnect at the same layer and different layers are known as the horizontal federation and vertical federation, respectively. Furthermore, in a multi-cloud, there is no voluntary interconnection and sharing of infrastructures among providers. Consequently, consumers are responsible for resource management across multiple clouds. An *agent*[10] -*based cloud computing* that exploits intelligent agents can be employed for boosting cloud resource trading. The agents can be used for bolstering discovery, matching, selection, composition, negotiation, scheduling, workflow, and monitoring of Intercloud resources [34, 37].

Furthermore, the envisaged requirements of the aforementioned applications demand a novel system design that will mainly be focusing on the introduction of intelligence at the edge devices. Consequently, the applications have been attracting considerable attention in the low-latency, distributed, collaborative, and reliable ML. In this context, the focus has been shifting from the cloud-based platform, with centralized training and inference, to an innovative edge AI-enabled platform. The platform is anticipated to offer intelligent services over the mobile or fixed access network to the edge devices. It is noteworthy that the wireless AI-enabled platform for the edge is capable of addressing the associated challenges of the cloud-computing paradigm that may render it inappropriate for edge computing[11] [6, 36].

Moreover, a typical edge AI-enabled architecture is illustrated in Fig. 13.4. In the edge AI-enabled architectures, to effectively support the network requirements, the training data have to be randomly distributed across a substantial number of edge devices.[12] Besides, in the architecture, apart from the access to a fraction of the

---

[8]In this, resource allocation is accomplished by a central entity.

[9]There is no central authority.

[10]An agent is a computer system that can make decisions independently, carry out actions autonomously, and interact with other agents by means of cooperation, coordination, and negotiation.

[11]Some of the challenges are the demand for large training datasets, difficulty in the effective handling of dynamically changing environments, and huge computational costs.

[12]In this context, the edge devices could be network base stations (BSs) and/or mobile devices.

data by each edge device, training and inference are collaboratively implemented based on the accessible computation and storage power. Likewise, the architecture facilitates effective communication and exchange locally trained models instead of private data between the edge devices. Based on the massive amount of the envisaged edge devices, it is imperative to ensure data dimensionality reduction, cleansing, and abstraction[13] [6, 36].

Moreover, edge AI-enabled platform offers various benefits such as low-latency and reduced cost. In this regard, it has been recognized as a promising system that not only focuses on reliable and secure communications but also on related to on-device resource constraints regarding memory, energy, and computing power. Also, the offered platform can effectively support the deployment of beyond 5G networks that can function as distributed processing elements with sufficient capacity for AI at the edge. Besides, edge AI-enabled platform presents collaborative computing that is within very close proximity to the distributed end-user devices. The network interconnection could be wireless and/or optical links (optical fiber and free-space optics) that can be efficiently managed over the metro and/or access network infrastructure. For instance, mobile autonomous air-ground systems are anticipated to have a profound impact on our everyday life and will be common devices and platforms for distributed AI applications such as area surveying, disaster recovery, and autonomous driving. It should be noted that the unprecedented autonomous cooperating systems will pose new challenges in various aspects such as distributed systems, networking, and resource management [38]. For instance, implementation of a wireless AI at the network edge presents certain research challenges regarding communication, jointly optimized training, and control considering hardware requirements, reliability, end-to-end (E2E) latency, privacy, and security [6, 36].

In addition, the associated network constraints can be effectively managed by smartly deploying/offloading certain computational tasks in proximity to the edge nodes. Based on this approach, the core-cloud will be engaged as the last resort for processing [14, 38]. So, the conventional approach of traversing the RAN, core network (CN), and the Internet for connection can be prevented [14]. This helps in reducing the associated latency, enhancing global intelligence, and improving the global system performance by leveraging local edge applications and resources [14, 38].

In general, distributed AI applications demand seamless, effective, and efficient communication with appropriate computation mechanisms over the entire computing spectrum cloud, fog, and edge [38]. Besides, since resource allocation may frequently fluctuate during the execution,[14] a wireless edge AI-enabled platform demands the development of efficient algorithms for distributed optimization

---

[13]This is due to the storage challenges of the related huge amount of monitored data.

[14]There can be abrupt changes in the computing load and the availability of the services both locally and globally. In some cases, this may be disruptive and destructive particularly to the central processing units (CPUs) and graphics processing units (GPUs). Similarly, it is noteworthy that the communications become more intermittent and unreliable, the more we move towards the edge [38].

between a massive number of the envisaged edge devices that will be sharing a model to be trained. The training can be realized through distributed data in a huge number of interconnected edge devices. An instance of such a decentralized ML training scheme is the federated learning (FL) approach.[15] However, since FL is in its developmental stage, there is a need for a concerted effort on means of attending to certain basic challenges such as the co-design of ML communication, considering wireless channel characteristics and on-device constraints [6, 36].

3.3 *Potential Synergism Between the ML and Communications*

As aforementioned, communication systems with salient features such as low-latency, high-speed, and high-reliability are crucial to effectively support the ML/AI at the edge. This initiates research in the area of utilization of communications for ML (CML). Likewise, ML for Communications (MLC) exploits edge ML for improving communications. For instance, edge ML/AI implementations have to be optimized by considering the channel dynamics, communication overheads, and on-device constraints such as memory, energy, and computing [6, 36].

Furthermore, the performance of RANs can be enhanced and optimized by exploiting edge ML. This opens up various research directions that are novel for the MLC. For instance, it is envisaged that AI-enabled next-generation of networks such as the 6G will optimally exploit both the physical and computing resources at the network edge. This will facilitate content-aware intelligent service delivery. Besides, AI-enabled algorithms are envisioned to aid in the software-defined control and management plane of the RAN. This can help in determining the required resources in accordance with inherent network parameters or external factors. Besides, it can assist in the identification of potential device malfunctions, physical failures, as well as malicious attacks [6, 36]. In general, MLC and CML are anticipated to have a significant influence on the intelligent next-generation systems as will be discussed in Sect. 4.

4 Potential Technology Transformation for 6G

The 6G RAN connectivity will be about ensuring seamless integration of various worlds as discussed in Sect. 2 and depicted in Fig. 13.5. The integration will help in the realization of ubiquitous universal computing that is appropriately distributed among multiple local devices, cloud, and Intercloud. Besides, it will facilitate end devices to evolve beyond being single entities to a set of multiple local entities

---

[15]The FL is an edge ML training scheme that supports the periodical exchange of neural network (NN) weights and gradients of mobile devices during local training.

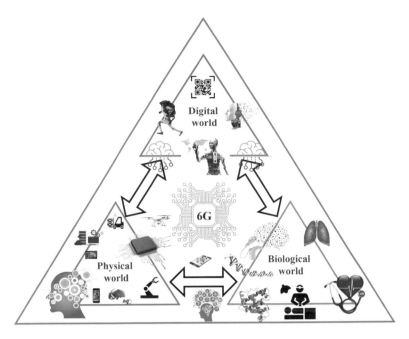

Fig. 13.5 Biological, physical, and digital worlds integration by 6G.

with well-synchronized functions, for establishing the new man-machine interface. Likewise, knowledge-based (agent-based) systems that are capable of processing, storing, and converting data into actionable knowledge will be supported. Similarly, precision sensing and actuation for controlling the physical world to realize a unified experience will be featured [12]. In this section, we discuss a variety of potential technology transformation that is anticipated to influence the 6G system.

4.1 Artificial Intelligence Exploitation

As aforementioned, AI and ML algorithms, particularly DL, have been experiencing a considerable advancement, making them attractive solutions for a number of fields such as computer vision, image classification, network security, and social networks. ML/AI concept is normally implemented in scenarios where there are readily available substantial amounts of data for training. For instance, reinforcement learning (RL) has been gaining considerable applications in a range of robotic control. Besides, DL is one of the innovative techniques that have been significantly penetrating wireless systems for performance enhancement. In this context, it is envisaged that 5G and beyond systems deployment will be based on distributed ML/AI techniques that will be embedded in various network nodes. This will facilitate prompt adaptation through automation to the network conditions that inherently fluctuate. Consequently, ML/AI can be an attractive alternative to various model-based Layer 1 and Layer 2 algorithms like channel estimation, equalization, preamble detection, and user scheduling [12].

Moreover, as 5G is to be ultra-dense network (UDN) deployments, beyond 5G network deployments are expected to range from a super-dense network (SDN) to a tremendously-dense network (TDN) [14]. However, it should be noted that the relatively huge network densification in 6G will pose various challenges such as high energy consumption, immense signaling, high-cost, rigorous interference, and very complex resource management. Based on this, ML/AI techniques will be good tools in intelligent automated system operation, optimization, maintenance, and management [7, 12]. For instance, the techniques can be exploited in the configuration of an optimum subset of beams that can effectively illuminate the coverage area, considering the cell traffic patterns. Besides, with UDN-TDN for the massive number of envisaged devices, during deployment, the number of parameters to be configured will be huge. The associated challenges can be addressed by the ML/AI techniques through automation-based end devices localization and network optimization [12]. In general, through sensing and network big data training, complex 6G networks will be more intelligent through the ML/AI techniques implementation. Consequently, their implementation will offer a number of salient features such as status prediction, dynamic optimization, self-healing, proactive configuration, and network environment sensing [7].

Apart from the anticipated applications of ML/AI techniques in the 5G and beyond RANs, these techniques will also play vital roles in the end-to-end network automation. This will help in addressing the complexity of related orchestration across multiple layers and network domains. Besides, it will facilitate the dynamic adaptation of edge and cloud resources based on the inherent nature of the network demands, service deployment, and system reliability. In the following subsection, we discuss the envisaged potential ML/AI techniques implementation for the 6G systems.

4.1.1 Quasi-Autonomous Mobile Devices with AI-Enabled D2D Communication

As discussed in Sect. 3.1, the advancement in mobile devices facilitates their ability to perform comparably to mobile workstations. Actually, future mobile will be equipped with sensors, sufficient storage resources, and powerful processors that are sufficient enough to perform comparatively special functions. Based on this, the trend will be towards handheld mobile devices that are capable of performing different functions such as wireless reality sensing, environment sensing, AR/VR cloud gaming, ML, photonic computing, and holography processing. Despite the intense processing that they will be executing, their portability will still be maintained. Moreover, based on the envisaged resources that they will be equipped with, mobile devices will evolve from being smart to AI-driven. In this context, they will be able to support on-device AI through local data training. Also, with on-device ML, they will be able to predict common interest content, preempt moving trajectory, and sense the local environment. In general, AI-driven/AI-enabled mobile devices will facilitate automated network configuration and preemptive network management in the 6G networks, resulting in the system performance enhancement. Nevertheless, on-device AI alone will not be sufficient for preempting and managing the envisaged huge and dynamic networks. As a result, based on the network demand, hierarchical AI (global, local, and on-device) have to be exploited for network optimization [7].

As aforementioned, hierarchical AI will be exploited in the 6G communication systems. For instance, shared information from the on-device and local AI will be leveraged not only to extend mobile devices' field of view in D2D communication but also to increase their perception regarding the local network environment. In addition, the unprecedented data traffic and computation increase impose a massive workload on the cloud and MEC servers. So, their joint implementation may not be sufficient to fulfill the required QoE by mobile devices. In this regard, edge computing can be adopted to ensure effective D2D communication. Based on this, to enhance the system performance, multiple UE can collaborate so that a set can address the needs of a cooperating entity.[16] Moreover, for optimal allocation and management of the associated network resources for D2D-enhanced MEC in the 6G networks, ML/AI techniques will be leveraged to ensure intelligent optimization and allocation based on the computing capability and network requirements. Besides, hierarchical AI will be exploited for intelligent resource sensation, prediction, configuration, aggregation, evaluation, and monitoring. This will help in the estab- lishment of logical interconnection for dynamic resources [7].

Furthermore, they are expected to be embedded with multiple radio interfaces that have the potential for full frequency radio access. This will help in the connection with several systems with diverse QoS requirements. Besides, with the implementation of higher frequency bands in 6G, multiple small size antennas can be embedded in mobile devices. The antenna system can exploit large-scale antenna technology such as massive MIMO. Moreover, multiple domain sensing applications such as gesture detection, healthcare monitoring, security sensing, body scanning, as well as 3D imaging and mapping will be enabled by THz. Besides, mobile devices will be able to offer cognitive ability high-speed data transmission, full-duplex transmission, MIMO communication, and considerable data caching [7].

4.1.2 Self-Optimizing RAN Transceivers

The envisaged THz communication in 6G ultra-dense heterogeneous networks will present very strict directional characteristics. For example, to deliver the information successfully, the transmitter beam has to be pointed perfectly to the receiver antenna. This requirement renders network management to be a very demanding

---

[16]The capacity of a single UE is comparatively weak. So, for edge computing, computationally intensive tasks can be assigned to the collaborating clusters.

task given the complexity and dynamism of the 6G network compared with the prior generations. Based on this, the conventional network management approaches will be inefficient, making innovative methods indispensable in the envisioned 6G ultra-dense heterogeneous networks. Therefore, the trend will be towards intelligent network optimization and management, for satisfying diverse QoS requirements. In this context, some ML algorithms, such as unsupervised learning, supervised learning, and reinforcement learning have been implemented to facilitate intelligent network optimization, network design, resource allocation, and UAVs or small BSs deployment [7].

In the future, it is deemed that AI-enabled CN and RAN will enable next-generation networks, not only to sense and learn from their environment, but also to make decisions by means of big data training. This will help in automatic prediction and adaptation to the network change and subsequently facilitate the realization of optimal performance through self-configuration [7]. For example, there are various research efforts on a DL system that have demonstrated its capability to learn and communicate over quasi-static links. In this regard, DL performs more efficiently compared with a model-based system [39, 40]. It demands no explicit design for parameters such as constellations, waveforms, or reference signals. The DL leverages extensive training between the transmitter and receiver [41]. Through the training, the transceiver learns to select the optimal design parameters. It is noteworthy that the required end-to-end learning technique may be impractical in complex multi-user environments that change dynamically. So, the anticipated design criteria for the 6G communication will facilitate effective learning for certain design options. So, this is expected to support air-interface optimization regarding the choice of environment, spectrum, hardware, and application requirements. Besides, it is anticipated that the air-interface will also adapt to the hardware capabilities [12, 42].

In general, it is envisioned that the global intelligent management of 6G networks will be aided by multi-level and distributed AI. Precisely, the core network will support the global AI center, while the local AI center will be embedded in the conventional mobile BSs or MEC servers. In this regard, UE can as well offer prospects for on-device data training that is capable of sensing and learning not only from the local channel pattern but also from traffic patterns and moving trajectory. This can help in understanding the various features regarding user behavior. It can also aid in the network status prediction. Therefore, 6G networks are expected to offer a platform that can support global, local, and on-device AI. This multi-level AI will be cooperating and leveraging one another's benefits to ensure that the network operation and optimization are well-maintained.

4.1.3 Cognition- and D2D-Aided Cooperative NOMA

User fairness, spectrum efficiency, and connectivity density can be enhanced with the integration of cognitive radio networks (CRNs) and non-orthogonal multiple access (NOMA). NOMA implementation can aid secondary UE to facilitate efficient cooperation with primary UE to achieve spectrum access opportunities. Besides, the joint implementation of D2D and NOMA can allow communication between one D2D transmitter and several D2D receivers. This NOMA-based communication can ensure enhanced D2D performance gains [7].

4.1.4 Photonics-Based Cognitive Radio and Cognitive Spectrum Management

As aforementioned in Sect. 2.4, a combination of photonic technologies and ML will offer improved performance for 6G and facilitate a photonics-based cognitive radio system with various notable advantages. It worth mentioning that the introduced AI in 5G mainly focuses on network operation, maintenance, and management. This is to ensure smart management and maintenance. Unlike the 5G, the 6G network is expected to be based on intelligent system architecture as well as a real cognitive radio network (CRN). Based on this, to be truly intelligent, AI will be supported across the protocol stack of each network node [8].

Furthermore, the envisaged photonics-based cognitive radio will leverage heterogeneous and hierarchical AI architectures. A typical framework will comprise all-photonic entities, a photonic neural network, and a spectrum computing unit in which graphics processing unit (GPU)-based neural network accelerators will be exploited. It should be noted that the neural network processing is usually based on a conventional GPU. Employment of such equipment for 6G, with an anticipated immense signal processing, will make it to be very energy inefficient. In the light of this, photonics-based neural networks are promising solutions that are capable of executing computations in parallel with comparatively lesser energy than GPUs. Consequently, photonics-based neural networks are energy-efficient and offer the platform for scalable AI systems [8].

Moreover, the 6G photonics-based cognitive radio system is anticipated to support multipurpose and multiband signals. In this system, a photonic neural network will be exploited to process, identify, and split different types of signals such as mmwave, radar, FSO, THz, and LIDAR. Besides, since the training step happens to be a very computationally-intensive aspect of the neural network application, optical domain implementation will help in enhancing the speed, efficiency, and power consumption of artificial neural networks. Nevertheless, the employment of GPUs for training normally results in a more accurate process compared with photonic counterparts. At large, a heterogeneous, hierarchical, and hybrid AI cognitive radio architecture leads to a flexible and agile air interface [8].

In addition, as aforementioned in Sect. 2.4, the low-frequency spectrum exploitation will continue to dominate wide-area coverage. This is owing to the offered better propagation properties in the NLOS compared with the higher-frequency counterparts. Moreover, with 5G evolution, an additional spectrum has to be allocated to support several use cases. This act can eventually cause spectrum shortages in the bands below 6 GHz. Consequently, novel spectrum-use techniques are demanded in the 6G to ensure improved local access, coexistence, and spectrum sharing among multiple operators. This will enable multiple generation networks and related innovative technologies to be supported not only to coexist but also to share spectrum. Based on the advances in radio technology for supporting multi-band operation as well as learning techniques like deep RLs, the main spectrum sharing challenges and management can be alleviated with efficient autonomous sharing systems [43]. For instance, network densification demands the implementation of advanced beamforming techniques. In this regard, spectrum utilization is increasingly becoming local. This will help in ensuring more cost-effective spectrum reuse and will also facilitate notable coexistence among cognitive sharing systems [12].

4.1.5   Context-Aware Implementation

As aforementioned, 6G will leverage ML/AI techniques for system performance enhancement using innovative data acquisition and processing schemes. For instance, it is envisaged that communication system optimization will be based on context-aware implementation. In this regard, awareness of various factors such as the traffic patterns, environment, location, and mobility patterns will be seamlessly integrated into the optimization schemes by exploiting the ML/AI techniques. Based on this, the variation in different factors can be acquired, predicted, processed in real-time, and subsequently employed for communication system optimization using deep-learning networks [12].

In addition, as massive network elements will be supporting various use cases, the knowledge on the use of communication can be learned from device characteristics and different emanating traffic patterns. This can facilitate accurate service personalization to ensure automatic and appropriate service delivery. In this context, a number of learning techniques such as transfer learning and FL will be employed in the performance optimization process. For instance, FL can play crucial roles in ensuring that network infrastructure and devices co-learn to facilitate effective end-to-end operations. Besides, various types of RL will also be exploited across different nodes and for different purposes such as resource allocation optimization and system parameter control [12].

4.2   New Spectrum Band Exploitation

The growing demand for higher capacity and peak rates has brought about the implementation of mm-wave and massive antenna arrays coupled with beamforming techniques for 5G networks. To ensure continuous support of the increasing system demand, 6G systems are anticipated to exploit sub-THz bands from 114 GHz to 300 GHz. Sub-THz spectrum can be used in the cellular systems to support different hauls such as FH, midhaul, and backhaul links. With the current evolution, these links are expected to be integrated for effective support in the future. In addition to the potential support for the hauls, sub-THz spectrum can be employed in short-range communications like the high-speed optical interconnect for data centers and rack-to-rack communications for edge data centers [12].

Besides the sub-THz spectrum bands exploitation, cost-effective massive MIMO schemes will facilitate more efficient use of the spectrum in mm-wave and centimeter wave (cm-wave) bands. Furthermore, with an increase in the network density and more cost-effective massive MIMO technologies, the trend will be towards the implementation of multi-user MIMO in the mm-wave bands. This will facilitate the available spectrum exploitation based on the potential for massive-scale implementation. It is noteworthy that for the 6G system, at the lower cm-wave band, there is a limitation on the application of massive MIMO. This is owing to the antenna element size that could be relatively large. Also, at sub-gigahertz frequencies, material penetration properties and basic path loss will be relatively better. Likewise, in the 6G networks, wide-area coverage will also be based on the lower-frequency bands. Besides, there will be notable evolution from the static spectrum assignment to dynamic ones that will leverage AI techniques [12].

Moreover, to facilitate spectrum exploitation, innovative radio frequency integrated circuits (RFICs) that are based on-board or on-chip antenna arrays coupled with phase shifters will be employed. Besides, the circuits should be able to offer narrow beams. Also, to realize a huge capacity through either single-user or multiple-user MIMO, hybrid beamforming will be an enabling scheme. This will help in evolving the initial mm-wave systems that are mainly based on analog beamforming and with a limited number of users that can be concurrently supported on a single panel. Other envisaged innovative schemes for the 6G RAN are ML/AI-based power consumption reduction based on new waveforms design, advanced signal processing, enhanced channel measurements, and related channel models [12].

4.3 Network Architectures and Technologies

There are a number of potential solutions that can be exploited in the technological path of the 6G mobile communication system. Also, based on the existing research developments, 6G network architecture is expected to leverage various schemes as depicted in Fig. 13.6. In this subsection, we present various potential key 6G technologies and system architectures.

4.3.1 Ultra-Dense Heterogeneous Networks with Sub-networks

The 5G has been intently designed to meet the challenging requirements by supporting innovative schemes like time-sensitive networking (TSN) bridge functionality. This trend will be exploited and build on by the 6G. This will enable 6G to offer wire-grade connectivity and reliability for a range of applications and services. In

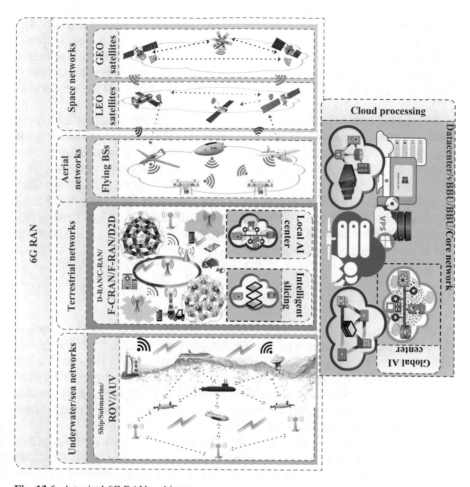

Fig. 13.6 A typical 6G RAN architecture

this context, both static and mobile devices, whether isolated, sparsely, or densely deployed should be effectively supported by the 6G system. To achieve this, 6G networks are envisaged to be highly densified with autonomous sub-networks. This will help in ensuring that the most crucial services in the sub-network stay uninterrupted in spite of no or poor connectivity to the holistic network [12]. Consequently, for the ultra-reliability, multiple path connectivity with infrastructure as well as opportunistic D2D connections will be demanded [7, 12]. This can probably result in cell-less architectures.[17] Also, considerable evolution is anticipated in the Integrated

---

[17]Seamless mobility support can be ensured by the cell-less network without the usual overhead that is owing to handovers, especially at THz frequencies.

Access and Backhaul (IAB) architecture[18] for effective support of the 6G ultra-dense heterogeneous wireless systems. In general, sub-network supports by the 6G will offer various benefits such as extremely low-latency, high-data rates, enhanced reliability, dynamic service execution, robust security, and improved resilience. These salient features will cut across the edge cloud and devices in the sub-network [12].

Furthermore, based on the envisaged extreme densification, there will be much shorter wireless transmission distance in 6G. In this regard, there will be a notable evolution of D2D communication into the 6G ultra-dense heterogeneous wireless systems for efficient support of the anticipated huge network devices with a diverse set of applications. Also, ultra-dense networks may result in performance degradation mainly owing to severe interference. This may subsequently bring about high energy consumption and excessive handover, especially for high-speed mobile devices. These challenges can be attended to with the future evolution of D2D-based and NOMA-based communication systems [7].

4.3.2   Innovative Network Slicing

As discussed in Sect. 2.3, apart from a variety of sub-networks that will be supported by the 6G and beyond networks, the network slicing and virtualization in the 5G will evolve for effective support of 6G innovative use cases. In this context, it is envisaged that an extremely intelligent and specialized network slicing that will be based on various software stacks will be employed for supporting each network slice. When optimized by AI, the flows can be well-automated to meet diverse services with a variety of requirements [12, 25].

Furthermore, D2D clusters with dynamic features are capable of offering virtual and/or physical resources such as network, storage, memory, and computation. This facilitates resource proliferation at the network edge. Besides, network resources can be efficiently integrated and federated at the edge by the network operators with D2D-enabled intelligent network slicing implementation. Also, it is imperative to exploit hierarchical AI to discover and manage a huge number of D2D clusters. This will help in ensuring a real-time-based dynamic network slicing. Similarly, based on the hierarchical AI, intelligent orchestration and management of virtual resources can be realized through the underlying resources and network slices status evaluation and monitoring. Besides, through virtual network functions (VNFs), instances of federated network slice can be effectively generated and/or dynamically adjusted to fulfill diverse service requirements without negatively affecting other slice instances [7].

---

[18]The required backhaul capacity of an ultra-dense system is very huge and the associated fiber optic-based infrastructure is not only costly but also time-consuming. So, IAB-based nodes can support both access and backhaul traffics with wireless links [44].

Moreover, there will be considerable evolution in the RAN virtualization to ensure further split/disaggregation of the associated RAN functions. In this perspective, the functions will be dispersed into modular micro-services. Based on this, they can be adaptively organized into RAN implementations that are slice-specific. The flexible slice-specific functionality is envisaged to be implemented across various network entities such as the relays, edges, cell sites, gateway devices, and clouds based on the requirements of the specific slice. It is noteworthy that such distinctive slices demand considerable advances in the network orchestration and service management [12].

4.3.3   Converged Network

6G is expected to be a converged system to ensure network simplification and improve system performance. This subsection focuses on the 6G network convergence.

RAN-Core Convergence

As discussed in Sect. 2.3, to facilitate network flexibility and enhance the system performance, 5G BS has been compartmentalized into the CU and DU. The physical Layer 1 and real-time Layer 2 of the protocol stack are implemented on the DU, while the non-real-time Layer 2 and Layer 3 functions are supported by the CU. Also, as aforementioned, the CU is usually based on a virtualized function and can be deployed in the edge or metro cloud to serve several DUs. Similarly, the core functions have also been gaining significant virtualization. The functionality can be at the regional or metro clouds. Besides, for low-latency services, it can be implemented at the cloud edges. Considering the significant increase in the higher-layer RAN functions centralization, as well as the core functions distribution, a significant simplification can be realized through the combination of certain functions of the RAN and core. Consequently, certain 5G RAN and core functional blocks are envisaged to be integrated. This can pave the way for a coreless RAN [12].

Space-Terrestrial Integrated Network

The growing increase in the traffic and the dispersed nature of the UE will make it very challenging for the existing terrestrial networks to be able to offer ubiquitous connectivity and support global coverage. Thus, the 6G network is expected to offer ubiquitous connections so that the anticipated massive devices and applications can be well-supported. To ensure this, it should be a converged network. For instance, 6G is expected to go beyond the 5G space–air–terrestrial networks by integrating underwater/sea networks, resulting in a 6G SATSI network. This will help in ensuring that ubiquitous mobile ultra-broadband is globally maintained [7, 8].

Furthermore, in such a network, the space-based comprise a set of interconnected satellites that are in the same or different orbit planes. Similarly, the aerial network entails flying base stations (BSs), low altitude unmanned aerial vehicles (UAVs), and high altitude platforms (e.g. airship and airplane). Likewise, the ground-based (terrestrial) network consists of cellular mobile networks (M2M communication, edge, cloud, and core networks), mobile satellite terminals, and satellite ground stations [45, 46]. Also, the sea-based network comprises an underwater acoustic network, ultrashort-wave communication, shortwave communication, and submarine optical network [7]. With the convergence of space−terrestrial network, a seamless high-speed mobile coverage in global 3D space can be accomplished by exploiting various associated features such as wide line-of-sight, broad space coverage, and low-loss transmission. For instance, a 5G-defined space−terrestrial integrated network is based on a high-throughput satellite system in which mm-wave or microwave bands are employed for delivering 100 Mb/s broadband services [8].

Moreover, for the 6G uMUB use case, a new architecture will be adopted for the SATSI integrated network. Meanwhile, free-space coherent/self-coherent optical communication systems offer huge capacity that can support communication among aerial-space, space-space, space-terrestrial, and terrestrial-sea nodes [47]. Consequently, with integrated hybrid optics-mm-wave convergence, more than 100 Gb/s space-terrestrial integration network can be realized. Since mm-wave is susceptible to rain fading but not cloud and fog fading, while free-space optics is prone to fog and cloud fading but not rain fading, they can be employed to complement each other. Also, it is imperative to employ diversity schemes to enhance the system performance [48, 49]. Based on this implementation, various links (optical, mm-wave, and hybrid links) can be adopted dynamically by the access network in line with the channel conditions and system requirements [8].

4.3.4 Holographic Radio

In modern wireless communication systems, interference is one of the main impairments that normally hinder the QoS. Many approaches mainly focus on interference elimination, minimization, and avoidance. In contrast to the conventional opinion that deems unwanted signals to be a harmful phenomenon, in the 6G system, interference will be exploited in the development of holographic communication systems that are highly precise and energy-efficient. Moreover, it is noteworthy that the entire-space RF holography or computational holographic radio is the technology with the highest potential and utmost level of interference exploitation. Besides the precise control of the entire space, holographic radio can realize the full closed-loop of the electromagnetic field by employing spatial wave field synthesis and spatial-spectral holography. Hence, not only the spectrum efficiency can be greatly improved but also the network capacity. Moreover, integration of imaging and wireless communication in addition to the uHDD 6G services can be realized [8].

4.3.5 All-Photonic RANs

The envisaged 6G use cases such as uHDD, uMUB, and uHSLLC demand ultra-broadband and ultra-low-latency, concurrently. These requirements are challenging for 5G to fulfill due to the usually employed schemes and the associated huge processing that cause an increase in the latency. Meanwhile, it has been observed that an all-optical network is capable of offering outstanding performance regarding latency. In view of this, the associated ultrafast signal processing of all-optical integrated devices (passive, active, or both) has been deemed as a promising solution for enhancing the next-generation RAN performance, especially for time-critical applications and services. However, it is noteworthy that an all-optical/photonic RAN implementation is still challenging owing to the need for a high-power RF photonic front-end. Despite this, it is envisaged that the network entities of the 6G will be all-photonic based. In the optical domain, this will offer a platform for supporting the enormous broadband signal generation and processing [8].

4.4 New Security, Trust, and Privacy Models

There has been a considerable increase in the rate of interconnection not only among the people but also among devices. This is facilitated by billions of IoT devices like sensors. However, an increase in the interconnection also brings about the vulnerability of every individual and society as a whole. For instance, different sensitive information such as financial and health data are transmitted routinely through the network. When the transmitted signal is tapped, data of millions of users and billions of applications are exposed to manipulation and theft, making physical security alone insufficient. Therefore, communication security and privacy measures in each generation of the network are of paramount importance. Also, the critical network infrastructure and IoT devices have to be safeguarded from data leakage and unexpected service outages [50]. 6G networks are expected to be robust to different likely network threats. This will be achieved through advanced authorization policy across the trust boundaries at different levels [12]. This can be achieved by FL with the capability to keep data training at each mobile device (e.g. UE). Consequently, FL will be leveraged to learn a shared global model from mobile/edge devices [7].

Furthermore, the current cryptographic algorithms being employed for securing communication in the IoT can be broken easily by means of quantum computers. In this context, the trend is towards quantum-resistant cryptosystems for securing the IoT. Besides, the cryptographic schemes must not be vulnerable to attacks from both classical and quantum computers. Nonetheless, for the cryptographic schemes to be effective solutions, there are various challenges such as secret key rates, cost, co-existence, and secure multiparty computation (SMC) to be addressed [51–53].

5 Conclusion

6G and beyond networks are anticipated to integrate intelligence for seamless evolution from the conventional cloud-based RAN schemes to distributed and collaborative edge AI-enabled paradigms. The edge AI-enabled architecture is expected to offer a significant evolution toward intelligent and converged 6G networks. This will facilitate the envisioned Internet of Everything (IoE) and make deployment of various applications/services such as the Industry 4.0, holographic communications, smart city, autonomous driving, sustainable development, enhanced energy efficiency, and AR/VR/MR, not only a reality but also deeper and broader. In this context, the convergence of control, communication, sensing, and computing functionalities that has not been receiving much attention in the 5G will be required in the 6G IoE applications. However, the edge AI-enabled approach presents a number of novel challenges due to the associated applications that have to operate in dynamic and real-time environments. The related scenarios have to meet strict end-to-end latency, bit-rate, on-device constraints, reliability, and security requirements. In this context, to satisfy the required QoS for various use cases, ML/AI techniques are essential tools for automated system operation, optimization, maintenance, and management in the 6G ultra-dense heterogeneous networks. Furthermore, with photonics-based cognitive radio, the air interface can be more agile and flexible, resulting in improved efficiency. Besides, computational holographic radio can enhance network capacity and spectrum efficiency significantly. Also, it can help in imaging and wireless communication integration in 6G ultra-dense heterogeneous networks.

Acknowledgments This work was supported in part by Fundação para a Ciência e a Tecnologia (FCT) through national funds, by the European Regional Development Fund (FEDER), through the Competitiveness and Internationalization Operational Programme (COMPETE 2020) of the Portugal 2020 framework, under the project DSPMetroNet (reference POCI-01-0145-FEDER-029405) and project UIDB/50008/2020-UIDP/50008/2020 (actions DigCORE and QUESTS). It was also supported in part by the ORCIP (CENTRO-01-0145-FEDER- 022141), SOCA (CENTRO-01-0145-FEDER-000010) and Project 5G (POCI-01-0247-FEDER-024539), RETIOT, POCI-01-0145-FEDER-016432.

References

1. J. Zhu, M. Zhao, S. Zhang, W. Zhou, Exploring the road to 6G: ABC – foundation for intelligent mobile networks. China Commun. **17**(6), 51–67 (2020). https://doi.org/10.23919/JCC.2020.06.005
2. T. Huang, W. Yang, J. Wu, J. Ma, X. Zhang, D. Zhang, A survey on green 6G network: architecture and technologies. IEEE Access **7**(175), 758–175, 768 (2019)
3. R. Katti, S. Prince, A survey on role of photonic technologies in 5G communication systems. Photon. Netw. Commun. **38**, 85–205 (2019). https://doi.org/10.1007/s11107-019-00856-w

4. P. Chanclou, L.A. Neto, K. Grzybowski, Z. Tayq, F. Saliou, N. Genay, Mobile fronthaul architecture and technologies: a RAN equipment assessment [invited]. IEEE/OSA J. Opt. Commun. Netw. 10(1), A1–A7 (2018). https://doi.org/10.1364/JOCN.10.0000A1

5. A. Liberato, M. Martinello, R.L. Gomes, A.F. Beldachi, E. Salas, R. Villaca, M.R.N. Ribeiro, K. Kondepu, G. Kanellos, R. Nejabati, A. Gorodnik, D. Simeonidou, RDNA: residue-defined networking architecture enabling ultra-reliable low-latency datacenters. IEEE Trans. Netw. Serv. Manage. 15(4), 1473–1487 (2018). 10.1109/TNSM.2018.2876845

6. I. Tomkos, D. Klonidis, E. Pikasis, S. Theodoridis, Toward the 6G network era: opportunities and challenges. IT Professional 22(1), 34–38 (2020). https://doi.org/10.1109/MITP.2019.2963491

7. S. Zhang, J. Liu, H. Guo, M. Qi, N. Kato, Envisioning device-to-device communications in 6G. IEEE Netw. 34(3), 86–91 (2020). https://doi.org/10.1109/MNET.001.1900652

8. B. Zong, C. Fan, X. Wang, X. Duan, B. Wang, J. Wang, 6G technologies: key drivers, core requirements, system architectures, and enabling technologies. IEEE Vehic. Technol. Mag. 14(3), 18–27 (2019). 10.1109/MVT.2019.2921398

9. E. Dutkiewicz, X. Costa-Perez, I.Z. Kovacs, M. Mueck, Massive machine-type communications. IEEE Netw. 31(6), 6–7 (2017). https://doi.org/10.1109/MNET.2017.8120237

10. I.A. Alimi, A. Tavares, C. Pinho, A.M. Abdalla, P.P. Monteiro, A.L. Teixeira, *Enabling Optical Wired and Wireless Technologies for 5G and Beyond Networks*, Chap. 8 (IntechOpen, London, 2019), pp. 177–199. https://doi.org/10.5772/intechopen.85858

11. C. I, H. Li, J. Korhonen, J. Huang, L. Han, RAN revolution with NGFI (xHaul) for 5G. J. Lightwave Technol. 36(2), 541–550 (2018). https://doi.org/10.1109/JLT.2017.2764924

12. H. Viswanathan, P.E. Mogensen, Communications in the 6G era. IEEE Access 8, 57,063–57,074 (2020). https://doi.org/10.1109/ACCESS.2020.2981745

13. C. I, C. Rowell, S. Han, Z. Xu, G. Li, Z. Pan, Toward green and soft: a 5G perspective. IEEE Commun. Mag. 52(2), 66–73 (2014). https://doi.org/10.1109/MCOM.2014.6736745

14. A.I. Alimi, A.L. Teixeira, P.P. Monteiro, Toward an efficient C-RAN optical Fronthaul for the future networks: a tutorial on technologies, requirements, challenges, and solutions. IEEE Commun. Surv. Tutor. 20(1), 708–769 (2018). https://doi.org/10.1109/COMST.2017.2773462

15. D. Camps-Mur, J. Gutierrez, E. Grass, A. Tzanakaki, P. Flegkas, K. Choumas, D. Giatsios, A.F. Beldachi, T. Diallo, J. Zou, P. Legg, J. Bartelt, J.K. Chaudhary, A. Betzler, J.J. Aleixendri, R. Gonzalez, D. Simeonidou, 5G-XHaul: a novel wireless-optical SDN transport network to support joint 5G Backhaul and Fronthaul services. IEEE Commun. Mag. 57(7), 99–105 (2019). https://doi.org/10.1109/MCOM.2019.1800836

16. H. Adams, Mobile transport for 5G networks. iHS markit white paper, IHS Markit (2018)

17. I.A. Alimi, P.P. Monteiro, Functional split perspectives: a disruptive approach to RAN performance improvement. Wirel. Pers. Commun. 106(1), 205–218 (2019). https://doi.org/10.1007/s11277-019-06272-7

18. P. Sehier, P. Chanclou, N. Benzaoui, D. Chen, K. Kettunen, M. Lemke, Y. Pointurier, P. Dom, Transport evolution for the RAN of the future [invited]. IEEE/OSA J. Opt. Commun. Netw. 11(4), B97–B108 (2019). https://doi.org/10.1364/JOCN.11.000B97

19. L. Li, M. Bi, H. Xin, Y. Zhang, Y. Fu, X. Miao, A.M. Mikaeil, W. Hu, Enabling flexible link capacity for eCPRI-based Fronthaul with load-adaptive quantization resolution. IEEE Access 7, 102,174–102,185 (2019). https://doi.org/10.1109/ACCESS.2019.2930214

20. G. Otero Pérez, D. Larrabeiti López, J.A. Hernández, 5G new radio Fronthaul network design for eCPRI-IEEE 802.1CM and extreme latency percentiles. IEEE Access 7, 82,218–82,230 (2019). https://doi.org/10.1109/ACCESS.2019.2923020

21. S.T. Le, T. Drenski, A. Hills, M. King, K. Kim, Y. Matsui, T. Sizer, 400 Gb/s real-time transmission supporting CPRI and eCPRI traffic for hybrid LTE-5G networks, in *2020 Optical Fiber Communications Conference and Exhibition (OFC)* (2020), pp. 1–3

22. G. Kalfas, M. Agus, A. Pagano, L.A. Neto, A. Mesodiakaki, C. Vagionas, J. Vardakas, E. Datsika, C. Verikoukis, N. Pleros, Converged analog fiber-wireless point-to-multipoint architecture for eCPRI 5G Fronthaul networks, in *2019 IEEE Global Communications Conference (GLOBECOM)* (2019), pp. 1–6. https://doi.org/10.1109/GLOBECOM38437.2019. 9013123

23. Common Public Radio Interface: eCPRI Interface Specification (2018). eCPRI specification v1.1, CPRI. http://www.cpri.info/spec.html

24. G. Brown, New Transport Network Architectures for 5G RAN. Heavy reading white paper, Fujitsu (2018)

25. C. Song, M. Zhang, Y. Zhan, D. Wang, L. Guan, W. Liu, L. Zhang, S. Xu, Hierarchical edge cloud enabling network slicing for 5G optical Fronthaul. IEEE/OSA J. Opt. Commun. Netw. **11**(4), B60–B70 (2019). https://doi.org/10.1364/JOCN.11.000B60

26. L. Ma, X. Wen, L. Wang, Z. Lu, R. Knopp, An SDN/NFV based framework for management and deployment of service based 5G core network. China Commun. **15**(10), 86–98 (2018). https://doi.org/10.1109/CC.2018.8485472

27. G. Wang, G. Feng, T.Q.S. Quek, S. Qin, R. Wen, W. Tan, Reconfiguration in network slicing-optimizing the profit and performance. IEEE Trans. Netw. Serv. Manage. **16**(2), 591–605 (2019). https://doi.org/10.1109/TNSM.2019.2899609

28. J. Ordonez-Lucena, P. Ameigeiras, D. Lopez, J.J. Ramos-Munoz, J. Lorca, J. Folgueira, Network slicing for 5G with SDN/NFV: concepts, architectures, and challenges. IEEE Commun. Mag. **55**(5), 80–87 (2017). https://doi.org/10.1109/MCOM.2017.1600935

29. M. Morant, A. Macho, R. Llorente, On the suitability of multicore fiber for LTE-advanced MIMO optical Fronthaul systems. J. Lightwave Technol. **34**(2), 676–682 (2016). https://doi. org/10.1109/JLT.2015.2507137

30. J.L. Rebola, A.V.T. Cartaxo, T.M.F. Alves, A.S. Marques, Outage probability due to intercore crosstalk in dual-core fiber links with direct-detection. IEEE Photon. Technol. Lett. **31**(14), 1195–1198 (2019). https://doi.org/10.1109/LPT.2019.2921934

31. L. Zhang, Y. Liang, D. Niyato, 6G visions: mobile ultra-broadband, super internet-of-things, and artificial intelligence. China Commun. **16**(8), 1–14 (2019)

32. Y. Yuan, B. Zong, S. Parolari, Y. Zhao, Potential key technologies for 6G mobile communications. Sci. China Inf. Sci. **63**(8), 183,301 (2020). http://engine.scichina.com/ publisher/ScienceChinaPress/journal/SCIENCECHINAInformationSciences/63/8/10.1007/ s11432-019-2789-y,doi=

33. M.A. Nahmias, B.J. Shastri, A.N. Tait, T.F. de Lima, P.R. Prucnal, Neuromorphic photonics. Opt. Photon News **29**(1), 34–41 (2018). https://doi.org/10.1364/OPN.29.1.000034. http:// www.osa-opn.org/abstract.cfm?URI=opn-29-1-34

34. K.M. Sim, Agent-based approaches for intelligent intercloud resource allocation. IEEE Trans. Cloud Comput. **7**(2), 442–455 (2019). https://doi.org/10.1109/TCC.2016.2628375

35. I. Alimi, A. Shahpari, A. Sousa, R. Ferreira , P. Monteiro, A. Teixeira, Challenges and opportunities of optical wireless communication technologies, in *Optical Communication Technology*, Chap. 2, ed. by P. Pinho (IntechOpen, Rijeka, 2017). https://doi.org/10.5772/ intechopen.69113

36. J. Park, S. Samarakoon, M. Bennis, M. Debbah, Wireless network intelligence at the edge. Proc. IEEE **107**(11), 2204–2239 (2019). 10.1109/JPROC.2019.2941458

37. K.M. Sim, Agent-based cloud commerce, in *2009 IEEE International Conference on Industrial Engineering and Engineering Management* (2009), pp. 717–721. https://doi.org/10.1109/ IEEM.2009.5373228

38. D. Aguiari, A. Ferlini, J. Cao, S. Guo, G. Pau, Poster abstract: C-Continuum: edge-to-cloud computing for distributed AI, in *IEEE INFOCOM 2019 - IEEE Conference on Computer Communications Workshops (INFOCOM WKSHPS)* (2019), pp. 1053–1054. https://doi.org/ 10.1109/INFCOMW.2019.8845170

39. S. Dörner, S. Cammerer, J. Hoydis, S.T. Brink, Deep learning based communication over the air. IEEE J. Select. Top. Signal Process. **12**(1), 132–143 (2018). https://doi.org/10.1109/JSTSP. 2017.2784180

40. T.J. O'Shea, T. Roy, T.C. Clancy, Over-the-air deep learning based radio signal classification. IEEE J. Select. Top. Signal Process. **12**(1), 168–179 (2018). https://doi.org/10.1109/JSTSP. 2018.2797022

41. M.S. Sim, Y. Lim, S.H. Park, L. Dai, C. Chae, Deep learning-based mmWave beam selection for 5G NR/6G with sub-6 GHz channel information: algorithms and prototype validation. IEEE Access **8**, 51,634–51,646 (2020). https://doi.org/10.1109/ACCESS.2020.2980285

42. V. Raj, S. Kalyani, Backpropagating through the air: deep learning at physical layer without channel models. IEEE Commun. Lett. **22**(11), 2278–2281 (2018). https://doi.org/10.1109/ LCOMM.2018.2868103

43. F. Zhou, G. Lu, M. Wen, Y. Liang, Z. Chu, Y. Wang, Dynamic spectrum management via machine learning: state of the art, taxonomy, challenges, and open research issues. IEEE Netw. **33**(4), 54–62 (2019). https://doi.org/10.1109/MNET.2019.1800439

44. J.Y. Lai, W. Wu, Y.T. Su, Resource allocation and node placement in multi-hop heterogeneous integrated-access-and-Backhaul networks. IEEE Access **8**, 122,937–122,958 (2020). https:// doi.org/10.1109/ACCESS.2020.3007501

45. I.A. Alimi, A.L. Teixeira, P.P. Monteiro, Effects of correlated multivariate FSO channel on outage performance of space–air–ground integrated network (SAGIN). Wirel. Pers. Commun. **106**(1), 7–25 (2019). https://doi.org/10.1007/s11277-019-06271-8

46. I.A. Alimi, A.O. Mufutau, A.L. Teixeira, P.P. Monteiro, Performance analysis of space-air-ground integrated network (SAGIN) over an arbitrarily correlated multivariate FSO channel. Wirel. Pers. Commun. **100**(1), 47–66 (2018). https://doi.org/10.1007/s11277-018-5620-x

47. R.K. Patel, I.A. Alimi, N.J. Muga, A.N. Pinto, Optical signal phase retrieval with low complexity DC-value method. J. Lightwave Technol. **38**(16), 4205–4212 (2020). https://doi. org/10.1109/JLT.2020.2986392

48. I.A. Alimi, P.P. Monteiro, A.L. Teixeira, Analysis of multiuser mixed RF/FSO relay networks for performance improvements in Cloud Computing-Based Radio Access Networks (CC-RANs). Opt. Commun. **402**, 653–661 (2017). https://doi.org/10.1016/j.optcom.2017.06.097. http://www.sciencedirect.com/science/article/pii/S0030401817305734

49. I.A. Alimi, P.P. Monteiro, A.L. Teixeira, Outage probability of multiuser mixed RF/FSO relay schemes for Heterogeneous Cloud Radio Access Networks (H-CRANs). Wirel. Pers. Commun. **95**(1), 27–41 (2017). https://doi.org/10.1007/s11277-017-4413-y

50. G. Brown, Quantum-safe security white paper: why quantum technologies matter in critical infrastructure and IoT. White paper, v1.0, ID Quantique SA (2017). https:// marketing.idquantique.com/acton/media/11868/why-quantum-technologies-matter-in-critical-infrastructure-and-iot

51. C. Cheng, R. Lu, A. Petzoldt, T. Takagi, Securing the Internet of Things in a quantum world. IEEE Commun. Mag. **55**(2), 116–120 (2017). https://doi.org/10.1109/MCOM.2017. 1600522CM

52. A.N. Pinto, N.A. Silva, A.J. Almeida, N.J. Muga, Using quantum technologies to improve fiber optic communication systems. IEEE Commun. Mag. **51**(8), 42–48 (2013). https://doi.org/10. 1109/MCOM.2013.6576337

53. M. Lemus, M. Ramos, P. Yadav, N. Silva, N. Muga, A. Souto, N. Paunkoviè, P. Mateus, A. Pinto, Generation and distribution of quantum oblivious keys for secure multiparty computation. Appl. Sci. **10**, 4080 (2020)

Chapter 14
Cloud Fog Architectures in 6G Networks

Barzan A. Yosuf, Amal A. Alahmadi, T. E. H. El-Gorashi, and
Jaafar M. H. Elmirghani

Abstract Prior to the advent of the cloud, storage and processing services were
accommodated by specialized hardware, however, this approach introduced a
number of challenges in terms of scalability, energy efficiency, and cost. Then came
the concept of cloud computing, where to some extent, the issue of massive storage
and computation was dealt with by centralized data centers that are accessed via the
core network. The cloud has remained with us thus far, however, this has introduced
further challenges among which, latency and energy efficiency are of the pinnacle.
With the increase in embedded devices' intelligence came the concept of the Fog.
At the edge of the network, large numbers of storage and computational devices
exist, where some are owned and deployed by the end-users themselves but most
by service operators, such devices are called fog nodes. This means that cloud
services are pushed further out from the core towards the edge of the network, hence
reduced latency is achieved. Fog nodes are massively distributed in the network,
some benefit from wired connections, and others are connected via wireless links.
The question of where to allocate services remains an important task and requires
extensive attention. This chapter introduces and evaluates cloud fog architectures in
6G networks paying special attention to latency, energy efficiency, scalability, and
the trade-offs between distributed and centralized processing resources.

The original version of this chapter was revised. The correction to this chapter is available at
https://doi.org/10.1007/978-3-030-72777-2_22

B. A. Yosuf (✉) · A. A. Alahmadi · T. E. H. El-Gorashi · J. M. H. Elmirghani
Institute of Communication and Power Networks, School of Electronic and Electrical
Engineering, University of Leeds, Leeds, UK
e-mail: b.a.yosuf@leeds.ac.uk; aaalahmadi@iau.edu.sa; T.E.H.Elgorashi@leeds.ac.uk;
J.M.H.Elmirghani@leeds.ac.uk

Y. Wu et al. (eds.), *6G Mobile Wireless Networks*, Computer Communications
and Networks, https://doi.org/10.1007/978-3-030-72777-2_14

1 Introduction

During the past several decades, computing paradigms have evolved from distributed models that included dedicated hardware such workstations to a more centralized model that is widely referred to as cloud computing. Cloud computing data centers are typically accessed via the Internet as they are attached to the core network [1]. This remote processing using a centralized cloud approach would not have been possible had it not been for the great advancement in both wired and wireless communication networks in terms of the speed at which data could be transmitted [2]. To some extent, cloud computing resolved the issue of massive storage and computation requirements of many applications. Cloud computing has two important traits [3]. First, centralization facilitates economies of scale through minimizing the cost of administration and operations. Second, speeding up innovations as individuals and organizations can avoid the operational and capital expenditure associated with owning a data center [3]. However, despite the cloud's on-demand and scalability merits, accessing its resources requires traversing through the access, metro and core network layers that can be prohibitively costly. This cost can be in terms of the communication latency due to the distance between data source nodes and the cloud and the high power consumption induced due to the transport network [4].

As a result, a new model of computing was proposed by Cisco in 2012, which is widely known as fog computing [2]. The term "fog" is used metaphorically to differ from the "cloud" as it is near to the ground [5]. In the same way, the main goal of fog computing is to extend the cloud services from the core to the edge of the network [6]. Thus, fog resources (i.e. computing and storage) are geographically distributed in the network through an *N-tier* deployment whereby heterogeneous fog resources are provided at different hierarchical levels [1]. According to the OpenFog consortium, any device can act as a fog node, be it embedded type CPUs onboard smart IoT devices or servers co-located with ISP's regional offices to processing servers located at local offices and/ or customer premises [7].

Driven by the emergence of the Internet of Things (IoT), the number of connected devices is expected to grow exponentially. Estimates have reported these devices to range between 25 and 50 billion, generating around 79.4 zettabytes of data [8, 9]. Evidently, given the rate at which connected devices grow, engineers from both industry and academia have already begun research into 6G networks, while 5G is currently being rolled out commercially. 6G networks are expected to support a new breed of next-generation applications (e.g. augmented reality, remote surgery, etc.) and an abundance of connectivity for the massive number of connected devices also referred to as supper IoT [10, 11]. One of the major aspects of 6G networks is integrating machine learning (ML) tools such as artificial intelligence (AI) for data analytics in order to move away from manual configurations/optimizations to a more intelligent network in the future [10].

As stated by one of the first white papers on 6G, a latency threshold of 0.1 ms and an improvement of $10\times$ in energy efficiency are one of the most important key performance indicators (KPI) [12]. Therefore, investigating cloud fog architectures in future 6G networks is imperative, because if all the collected raw data, which is generated at the edge network was to be processed centrally by the cloud will lead to slow decision making due to latency as well as increased power consumptions due to the transport network [13]. Hence, the interplay between the fog and the cloud in terms of energy efficiency and latency will become an important aspect of 6G networks due to the increased use of processing and data analytics [14]. The centralized cloud data center (DC) can provide increased processing capabilities and sophistication but may result in higher power consumption and increased latency. Thus, resource allocation problems will become vital where specialized compute and storage hardware must be accessed either in the distant cloud or in the edge processing fog nodes. It is anticipated that through the complementary features and the cooperation between the fog and the cloud, a more efficient and greener network can be achieved [15].

Different from the works that address the resource provisioning problem in cloud fog architectures only on individual layers, this chapter introduces a comprehensive optimization model based on Mixed Integer Linear Programming (MILP) and extends on the work in [16] by paying special attention to latency as well as energy consumption. The optimization model in this chapter considers (1) elements in the IoT, access, metro, and core layers, (2) simultaneous task requests generated from multiple IoT groups to capture the trade-offs between local and remote and/ or centralized processing, (3) a generic MILP model that is independent of the type of processing and/or networking technologies which allows for a holistic focus on energy-efficiency with a global perspective. Also, the work in this chapter benefits from our previous contributions in improving energy efficiency of cloud DCs [17–21], big data analytics [22–25], network coding in core networks [26, 27], energy efficient optical core networks [13, 28–34], and virtualization in the core and IoT networks [35–37].

2 The Proposed Cloud Fog Architecture

Devices at the edge network such as surveillance cameras, smart phones, IoT objects are expected to possess low-power embedded and specialized CPUs, which can collectively provide enormous amounts of computational resources due to their massive numbers and very low latencies due their distributed nature and proximity to end-users. The IoT and fog nodes are highly heterogeneous in terms of their processing resources and efficiencies, which poses a number of challenges in the resource allocation problem in future 6G networks. Therefore, the proposed cloud fog architecture shown in Fig. 14.1 comprises of five distinct layers of processing, which are the IoT, CPE, Access Fog, Metro Fog, and Cloud DC layers. A generated task emanates from the IoT layer and may be hosted by any processing devices in the aforementioned layers given that they have enough computational resources.

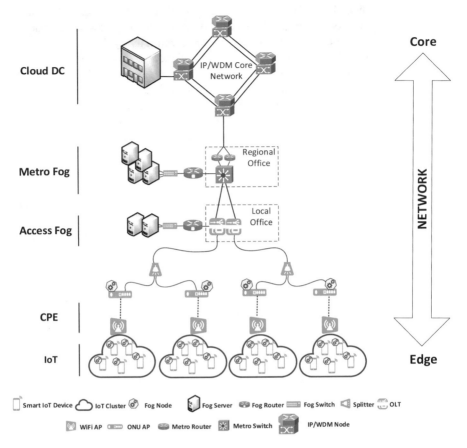

Fig. 14.1 A multi-layer cloud fog architecture supported by a PON access network

2.1 IoT Layer

This is the bottom-most layer and comprises of all the generic smart IoT nodes such as tablets, phones, vehicles etc. Two types of entities are defined in this layer, which are called source nodes and IoT nodes. The sensor nodes are a subset of the IoT nodes, the only difference between them is that the former is generating task requests while the latter remains idle. A real-world example can include a smart surveillance system whereby one or more cameras actively send video streams while the rest of the cameras remain idle due to little or no motion detected by their mounted passive infrared (PIR) sensors. The IoT nodes are wirelessly connected to the wireless access points (APs) in their respective zones and the generated tasks from the source nodes are offloaded to the next layer(s) for data analysis, if local resources are insufficient. A Wi-Fi link is considered between the IoT and APs as it is an ideal choice for data-intensive applications compared with other wireless counterparts such as Bluetooth, Zigbee, LoRa, etc. In indoor environments, this link may also be replaced with a visible indoor light link in order to support very high data rates per user terminal [38].

2.2 CPE Layer

This layer comprises of the customer premises equipment (CPE) such as ONUs and Wi-Fi Access Points (APs). These devices are typically situated in proximity to the IoT nodes. The ONU, since it has multiple Ethernet ports and acts as a switch can be equipped with embedded type CPUs that have similar and/or higher capacity than the IoT nodes [39]. Small organizations or even end-users can deploy their own CPE nodes at locations such as APs, routers, gateways, etc. In this chapter, Optical Networking Units (ONUs) represent the CPE nodes and they are part of the Passive Optical Network (PON) deployment [40]. PON technologies have promising potentials as they offer high data rates for data-intensive applications, at relatively low cost and PONs are particularly suitable due to their high scalability [41]. The main role of the CPE nodes is to act as controllers that collect and coordinate the allocation of the generated tasks by the source nodes. Thus, these CPE nodes are assumed to be fully aware of the available processing resource across all layer of the architecture shown in Fig. 14.1. As can be seen, each CPE node is connected to a separate IoT group which represents a geographical area. The CPE nodes due to their coordination and allocation roles can communicate with other group(s) through the PON access network. Thus, the tasks generated from one group can also be allocated to other zones [42]. The access part of the PON deployment will be explained next.

2.3 Access Fog Layer

This layer comprises of multiple Optical Line Terminals (OLTs) that are responsible for aggregating data traffic from the connected ONU devices. A single fiber link can be split in the ratio of 1:N and with next generation PONS (NG-PONs) a splitting ratio of 1:256 can be achieved [43]. This is particularly suitable for 6G networks as it is expected that there will be hundred(s) of devices per cubic meter [12]. The processing capability available to this layer is higher than that of the IoT and the CPE as several high-end servers are used to form a fog collocated with the OLT [44]. However, the number of servers that can be deployed at this layer is still finite, which can be due to space limitations as OLT devices are usually installed in small local offices and/or enclosed in street cabinets. Thus, more intensive tasks will need to be offloaded to the next layer for processing.

2.4   Metro Fog Layer

This layer comprises of multiple edge routers and a single ethernet switch that acts as the entry point to the metro and edge network as can be seen in Fig. 14.1. The Ethernet switch is mainly used to provide access to public clouds and is also used for traffic aggregation from one or more access networks (OLT devices in this chapter). The main role of the edge routers is to perform traffic management and authentication and usually multiple edge routers are used for redundancy purposes [41]. The computational resources available to the metro fog layer are typically significantly higher than that of the IoT and lower fog layers due to the number of users it supports, however the resources are still insignificant compared to the cloud DC [45].

2.5   Cloud DC Layer

This layer comprises of large data centers that are attached to the core network. High capacity IP/WDM fibre links are used to interconnect the core nodes. The IP/WDM core network consists of an optical layer and an IP layer. In the IP layer, IP core routers are deployed at each node to aggregate traffic and/or route traffic. In the optical layer, optical cross connects are used to establish the physical network links between the IP core routers. WDM fiber links utilize EDFAs, transponders and regenerators as part of the IP/WDM setup. The processing resources of the cloud DC are virtually infinite in comparison to the rest of the aforementioned processing layers. This is understandable since cloud DCs are not restricted by space, they are deployed to support vast number of applications and service [46].

3   MILP Model

The proposed cloud fog architecture is shown in Fig. 14.1. The optimization model minimizes the joint networking and processing power consumption resulting from IoT processing placement [42, 47]. Each task request comprises of CPU and traffic requirement. CPU is the amount of processing required in Million Instructions Per Second (MIPS) and traffic is the amount of data required to be transported in the network in Mbps. The optimization model considers the network topology in Fig. 14.1 as a bi-directional graph $G(N, L)$, where N is the set of all nodes, and L is the set of links connecting those nodes. A processing node's computational capacity is measured in MIPS, whilst the link's network capacity is given in Mbps.

The definitions of the sets, parameters, and variables used in the MILP model are as follows:

Sets:	
$\mathbb{N}$	Set of all nodes in the proposed architecture shown in Fig. 14.1
$\mathbb{N}_m$	Set of all neighbor nodes of node m in the proposed architecture shown in Fig. 14.1
$\mathbb{C}$	Set of IP/WDM core nodes, where $\mathbb{C} \subset \mathbb{N}$
$\mathbb{A}$	Set of Wi-fi access points (APs), where $\mathbb{A} \subset \mathbb{N}$
$\mathbb{O}$	Set of ONU devices in the PON, where $\mathbb{O} \subset \mathbb{N}$
$\mathbb{OT}$	Set of OLT devices in the PON, where $\mathbb{OT} \subset \mathbb{N}$
$\mathbb{MS}$	Set of Ethernet switches in the metro, where $\mathbb{MS} \subset \mathbb{N}$
$\mathbb{DC}$	Set of cloud data centers (DCs), where $\mathbb{DC} \subset \mathbb{N}$
$\mathbb{I}$	Set of generic IoT devices, where $\mathbb{I} \subset \mathbb{N}$
$\mathbb{P}$	Set of nodes with processing capability, where $\mathbb{P} \subset \mathbb{N}$ and $\mathbb{P} = \mathbb{I} \bigcup \mathbb{O} \bigcup \mathbb{OT} \bigcup \mathbb{MS} \bigcup \mathbb{DC}$
$\mathbb{S}$	Set of all IoT devices generating task requests, where $\mathbb{S} \subset \mathbb{I}$
IP/WDM Core network parameters:	
Pr	Maximum power consumption of an IP router port
Pt	Maximum power consumption of a transponder
Pe	Maximum power consumption of an EDFA
Po	Maximum power consumption of an optical switch
Prg	Maximum power consumption of a regenerator
Ir	Idle power consumption of an IP router port
It	Idle power consumption of a transponder
Ie	Idle power consumption of an EDFA
Io	Idle power consumption of an optical switch
Irg	Idle power consumption of a regenerator
B	Maximum data rate of a single wavelength
W	Number of wavelengths in a fibre
$\epsilon^{(r)}$	Energy per bit of a router port, where $\epsilon^{(t)} = \left(\frac{Pt - It}{B} \right)$
$\epsilon^{(t)}$	Energy per bit of a transponder, where $\epsilon^{(r)} = \left(\frac{Pr - Ir}{B} \right)$
$\epsilon^{(e)}$	Energy per bit of the EDFAs, where $\epsilon^{(e)} = \left(\frac{Pe - Ie}{B} \right)$
$\epsilon^{(o)}$	Energy per bit of the optical switches, where $\epsilon^{(o)} = \left(\frac{Po - Io}{B} \right)$
$\epsilon^{(rg)}$	Energy per bit of the regenerators, where $\epsilon^{(rg)} = \left(\frac{Prg - Irg}{B} \right)$
D_{mn}	Distance between core node m and core node n, where $m, n \in \mathbb{C}$
Se	Span distance between neighboring EDFAs
Sg	Span distance between two neighboring regenerators
A_{mn}	Number of EDFAs utilized on each fiber in the core network from node $m \in \mathbb{C}$ to $n \in \mathbb{C}$, where $A_{mn} = \left\lvert \left(\left(\frac{D_{mn}}{Se} \right) - 1 \right) \right\rvert + 2$
R_{mn}	Number of regenerators utilized between core node $m \in \mathbb{C}$ and core node $n \in \mathbb{C}$, $R_{mn} = \left\lvert \left(\frac{D_{mn}}{Sg} \right) - 1 \right\rvert$
PUE_C	Power usage effectiveness of IP/WDM core network node
Cloud data center parameters:	
$P^{(DS)}$	Maximum power consumption of cloud DC switch

(continued)

$I^{(DS)}$	Idle power consumption of cloud DC switch
$B^{(DS)}$	Data rate of cloud DC switch
$\epsilon^{(DS)}$	Cloud DC switch energy per bit, where $\epsilon^{(DS)} = \left(\frac{P^{(DS)} - I^{(DS)}}{B^{(DS)}} \right)$
$P^{(DR)}$	Maximum power consumption of cloud DC router
$I^{(DR)}$	Idle power consumption of cloud DC router
B^{DR}	Cloud DC router data rate
$\epsilon^{(DR)}$	Energy per bit of a cloud DC router, where $\epsilon^{(DR)} = \left(\frac{P^{(DR)} - I^{(DR)}}{B^{(DR)}} \right)$
PUE_DC	Power usage effectiveness of DC node, for processing and networking

Metro network and fog parameters:

$P^{(MS)}$	Maximum power consumption of a metro switch
$I^{(MS)}$	Idle power consumption of a metro switch
B^{MS}	Bit rate of a metro switch
$\epsilon^{(MS)}$	Metro switch energy per bit, where $\epsilon^{(MS)} = \left(\frac{P^{(MS)} - I^{(MS)}}{B^{(MS)}} \right)$
$P^{(MfS)}$	Maximum power consumption of a metro fog switch
$I^{(MfS)}$	Idle power consumption of a metro fog switch
B^{MfS}	Bit rate of a metro fog switch
$\epsilon^{(MfS)}$	Metro fog switch energy per bit, where $\epsilon^{(MfS)} = \left(\frac{P^{(MfS)} - I^{(MfS)}}{B^{(MfS)}} \right)$
$P^{(MR)}$	Maximum power consumption of a metro router
$I^{(MR)}$	Idle power consumption of a metro router
$B^{(MR)}$	Bit rate of a metro router
$\epsilon^{(MR)}$	Metro router energy per bit, where $\epsilon^{(MR)} = \left(\frac{P^{(MR)} - I^{(MR)}}{B^{(MR)}} \right)$
$P^{(MfR)}$	Maximum power consumption of a metro fog router
$I^{(MfR)}$	Idle power consumption of a metro fog router
$B^{(MfR)}$	Bit rate of a metro fog router
$\epsilon^{(MfR)}$	Metro fog router energy per bit, where $\epsilon^{(MfR)} = \left(\frac{P^{(MfR)} - I^{(MfR)}}{B^{(MfR)}} \right)$
PUE_M	Power usage effectiveness of a metro node, for processing and networking
$\mathcal{R}$	Metro router port redundancy

Access network and fog parameters:

$P^{(OT)}$	Maximum power consumption of OLT in the PON network
$I^{(OT)}$	Idle power consumption of OLT in the PON network
$B^{(OT)}$	Bit rate of OLT in the PON network
$\epsilon^{(OT)}$	OLT router energy per bit, where $\epsilon^{(OT)} = \left(\frac{P^{(OT)} - I^{(OT)}}{B^{(OT)}} \right)$
$P^{(O)}$	Maximum power consumption of an ONU in the PON network
$I^{(O)}$	Idle power consumption of an ONU in the PON network
$B^{(O)}$	Data rate of the Wi-fi interface of an ONU device in the PON network
$\epsilon^{(O)}$	ONU energy per bit, where $\epsilon^{(O)} = \left(\frac{P^{(O)} - I^{(O)}}{B^{(O)}} \right)$
$P^{(AfS)}$	Maximum power consumption of an access fog switch
$I^{(AfS)}$	Idle power consumption of an access fog switch
B^{AfS}	Bit rate of an access fog switch
$\epsilon^{(AfS)}$	Access fog switch energy per bit, where $\epsilon^{(AfS)} = \left(\frac{P^{(AfS)} - I^{(AfS)}}{B^{(AfS)}} \right)$
$P^{(AfR)}$	Maximum power consumption of an access fog router
$I^{(AfR)}$	Idle power consumption of an access fog router
$B^{(AfR)}$	Bit rate of an access fog router

(continued)

$\epsilon^{(AfR)}$	Access fog router energy per bit, where $\epsilon^{(AfR)} = \left(\frac{P^{(AfR)} - I^{(AfR)}}{B^{(AfR)}} \right)$
$P^{(CfR)}$	Maximum power consumption of CPE fog switch
$I^{(CfR)}$	Idle power consumption of an CPE fog switch
$B^{(CfR)}$	Bit rate of a CPE fog switch
$\epsilon^{(CfR)}$	CPE fog switch energy per bit, where $\epsilon^{(CfR)} = \left(\frac{P^{(CfR)} - I^{(CfR)}}{B^{(CfR)}} \right)$
$P^{(ap)}$	Maximum power consumption of an AP
$I^{(ap)}$	Idle power consumption of an AP
$B^{(ap)}$	Data rate of the AP
$\epsilon^{(ap)}$	AP Wi-fi interface energy per bit, where $\epsilon^{(ap)} = \left(\frac{P^{(ap)} - I^{(ap)}}{B^{(ap)}} \right)$
PUE_A	Power usage effectiveness of an access fog node, for processing and networking
Parameters of IoT devices:	
$P^{(iot)}$	Maximum power consumption of an IoT transceiver
$I^{(iot)}$	Idle power consumption of an IoT transceiver
$B^{(iot)}$	Data rate of the Wi-fi interface of an IoT device
$\epsilon^{(iot)}$	IoT Wi-fi interface energy per bit, where $\epsilon^{(iot)} = \left(\frac{P^{(iot)} - I^{(iot)}}{B^{(iot)}} \right)$
Parameters of processing devices:	
$P_d^{(cpu)}$	Maximum power consumption of processing device $d \in \mathbb{P}$, in Watts
$I_d^{(cpu)}$	Idle power consumption of processing device $d \in \mathbb{P}$, in Watts
$C_d^{(cpu)}$	Maximum capacity of processing device $d \in \mathbb{P}$ in million instructions per second (MIPS)
$E^{(mips)}$	Energy per instruction of processing device $d \in \mathbb{P}$, where $E^{(mips)} = \left(\frac{P^{(cpu)} - I^{(cpu)}}{C^{(cpu)}} \right)$
Application parameters:	
$D_s^{(cpu)}$	Processing task in MIPS requested by source node $s \in \$$
$T_s^{(cpu)}$	Data rate traffic in Mbps requested by source node $s \in \$$
C_{mn}	Capacity of link (m, n), where $m \in N$ and $n \in \mathbb{N}_m$
δ	Portion of the idle power of equipment attributed to the use case
Δ	Number of MIPS required to process 1 Mb of traffic
M	Large enough number
Variables:	
L^{sd}	Traffic demand between source node $s \in \$$ and processing device $d \in \mathbb{P}$
L_{mn}^{sd}	Traffic flow between source node $s \in \$$ and processing device $d \in \mathbb{P}$, traversing node $m \in N$ and $n \in \mathbb{N}_m$
L_d	Volume of aggregated traffic by node $d \in N$
$\mathcal{B}_m$	$\mathcal{B}_m = 1$, if network node $m \in N$ is activated, otherwise $\mathcal{B}_m = 0$
θ_d	Traffic in node $d \in \mathbb{P}$ for processing, where $\theta_d = \lambda_d \Omega_d$
Γ_{mn}	If $\Gamma_{mn} = 1$, link (m, n) in the core network, where $m \in \mathbb{C}, n \in (\mathbb{N}_m \cap \mathbb{C})$ is used, otherwise $\Gamma_{mn} = 0$
ρ^{sd}	Processing task of source node $s \in \$$ allocated to processing device $d \in \mathbb{P}$
Ω^{sd}	$\Omega^{sd} = 1$, if processing task of source node $s \in \$$ is allocated to destination node $d \in \mathbb{P}$, otherwise $\Omega^{sd} = 0$
Ω^d	$\Omega^d = 1$, if processing node $d \in \mathbb{P}$ is turned ON, otherwise $\Omega^d = 0$
$\mathcal{N}_d$	Number of processing servers used at node $d \in \mathbb{P}$
W_{mn}	Number of wavelengths used in fiber link (m, n), where nodes $m, n \in \mathbb{C}$
F_{mn}	Number of fibers used on link $(m, n) \in \mathbb{C}$
Ag_m	Number of core router aggregation ports activated at node $m \in \mathbb{C}$

(continued)

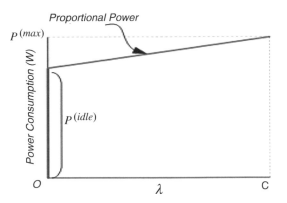

Proportional Power

Fig. 14.2 The adopted power
profile with two parts; (a)
proportional power
consumption and (b) idle
power consumption

In this chapter, we adopt the power profile depicted in Fig. 14.2, which consists of
an idle and proportional section. The idle section is consumed as soon as the device
is turned ON, regardless of the load (MIPS or traffic). Whereas the proportional
section is dependent on the amount of workload (processing and/or traffic) that is
allocated to the device. Almost all devices adopt a linear power profile similar to
the one shown in Fig. 14.2 [48]. Hence, in practical settings, idle power represents a
large proportion of the maximum power of a device (networking or processing) and
therefore cannot be ignored.

The architecture considered spans across multiple layers of processing and
networking. Therefore, it becomes a necessity to fairly represent the utilization
characteristics of both the networking and processing devices. In the literature,
when idle, servers are reported to consume around 60% of their maximum power
consumption, whilst networking nodes are reported to consume around 90% of their
maximum power consumption [49]. In this chapter, both ratios are assumed for the
idle power consumption of both networking and processing elements. However,
large networking equipment, such as those found in the access to the core layer,
are assumed to consume a portion (3% based on [50]) of the total idle power
consumption. This is a reasonable assumption since such devices can be shared
by many applications due to the number of users that can be connected to them. The
total power consumption given the power profile in Fig. 14.2 is calculated using Eq.
(14.1):

$$Total\ Power\ Consumption = \left(\frac{P^{(\text{max})} - P^{(idle)}}{C} \right) \lambda + P^{(idle)} \qquad (14.1)$$

where $P^{(max)}$ is the maximum power consumption of the device (networking or
processing) which is consumed as soon as the device is activated regardless
of the load λ and (*Pmax*) is the maximum power consumption of the device,
when it is utilized at full capacity C. The linear curve represents the proportional

power consumption. For networking devices, this is expressed as energy per bit and likewise, for processing, it is expressed as energy per instruction.

The total power consumption of the proposed cloud fog architecture is composed of the power consumption in processing nodes and the power consumption in the network. The processing power consumption also includes the power consumption of the networking elements needed for intra processing node's communication.

3.1   Network Power Consumption (net _ pc)

The total power consumption in the core network, under the non-bypass light path approach [51] is composed of:

The power consumption of core router ports:

$$PUE\_C\left[\sum_{m\in C}\left(\epsilon^{(r)}L_m\right) + \sum_{m\in C}\left(\delta Ir\left(Ag_m + \sum_{n\in(N_m\cap C)}W_{mn}\right)\right)\right] \qquad (14.2)$$

The power consumption of transponders:

$$PUE\_C\left[\sum_{m\in C}\left(\epsilon^{(t)}L_m\right) + \sum_{m\in C}\sum_{n\in(N_m\cap C)}\left(\delta It\,W_{mn}\right)\right] \qquad (14.3)$$

The power consumption of EDFAs:

$$PUE\_C\left[\sum_{m\in C}\left(\epsilon^{(t)}L_m A_{mn}\right) + \sum_{m\in C}\sum_{n\in(N_m\cap C)}\left(\delta Ie A_{mn}F_{mn}\right)\right] \qquad (14.4)$$

The power consumption of optical switches:

$$PUE\_C\left[\sum_{m\in C}\left(\epsilon^{(o)}L_m\right) + \sum_{m\in C}\left(\delta Io B_m\right)\right] \qquad (14.5)$$

The power consumption of regenerators:

$$PUE_C\left[\sum_{m\in C}\left(\epsilon^{(rg)}L_m Rg_{mn}W_{mn}\right) + \sum_{m\in C}\sum_{n\in(N_m\cap C)}\left(\delta Irg\,Rg_{mn}W_{mn}\right)\right]$$
$$(14.6)$$

The metro network power consumption consists of the power consumption of metro routers and switches, which is given as:

$$
PUE_M \left[\mathcal{R} \sum_{m \in \mathbb{MR}} \left(\epsilon^{(MR)} L_m \right) + \mathcal{R} \sum_{m \in \mathbb{MR}} \left(\delta I^{(MR)} \mathcal{B}_m \right) \right.
$$
$$
\left. + \sum_{m \in \mathbb{MS}} \left(\epsilon^{(MS)} L_m \right) + \sum_{m \in \mathbb{MS}} \left(\delta I^{(MS)} \mathcal{B}_m \right) \right]
$$

(14.7)

The power consumption of the PON access network comprises of the power consumption of OLT and ONU devices:

$$
PUE_A \left[\sum_{m \in \mathbb{OT}} \left(\epsilon^{(OT)} L_m \right) + \sum_{m \in \mathbb{OT}} \left(\delta I^{(OT)} \mathcal{B}_m \right) + \sum_{m \in \mathbb{O}} \left(\epsilon^{(O)} L_m \right) \right.
$$
$$
\left. + \sum_{m \in \mathbb{O}} \left(\delta I^{(O)} \mathcal{B}_m \right) \right]
$$

(14.8)

The Wi-Fi AP's power consumption is given as:

$$
\sum_{m \in \mathbb{AP}} \left(\epsilon^{(ap)} L_m \right) + \sum_{m \in \mathbb{AP}} \left(\delta I^{(ap)} \mathcal{B}_m \right)
$$

(14.9)

The IoT devices' transceivers power consumption is given as:

$$
\sum_{m \in \mathbb{I}} \left(\epsilon^{(iot)} L_m \right) + \sum_{m \in \mathbb{I}} \left(\delta I^{(iot)} \mathcal{B}_m \right)
$$

(14.10)

3.2 Processing Power Consumption (pr _ pc)

The total power consumption of the processing nodes is composed of:
The processing power consumption of IoT devices:

$$
\sum_{s \in \mathbb{S}} \sum_{d \in \mathbb{I}} \left(E_d^{(i)} \rho^{sd} \right) + \sum_{d \in \mathbb{I}} \left(I^{(pr)} \mathcal{N}_d \right)
$$

(14.11)

The processing power consumption of CPE fog servers:

$$\sum_{s\in\mathbb{S}}\sum_{d\in\mathbb{O}}\left(E_d^{(i)}\rho^{sd}\right)+\sum_{d\in\mathbb{O}}\left(I_d^{(pr)}\mathcal{N}_d\right) \tag{14.12}$$

The processing power consumption of access fog servers:

$$PUE\_A\left[\sum_{s\in\mathbb{S}}\sum_{d\in\mathbb{O}\mathbb{T}}\left(E_d^{(i)}\rho^{sd}\right)+\sum_{d\in\mathbb{O}\mathbb{T}}I_d^{(pr)}\mathcal{N}_d\right] \tag{14.13}$$

The processing power consumption of metro fog servers:

$$PUE\_M\left[\sum_{s\in\mathbb{S}}\sum_{d\in\mathbb{M}\mathbb{S}}\left(E_d^{(i)}\rho^{sd}\right)+\sum_{d\in\mathbb{M}\mathbb{S}}\left(I_d^{(pr)}\mathcal{N}_d\right)\right] \tag{14.14}$$

The processing power consumption of cloud data center (DC) servers:

$$PUE\_DC\left[\sum_{s\in\mathbb{S}}\sum_{d\in\mathbb{D}\mathbb{C}}\left(E_d^{(i)}\rho^{sd}\right)+\sum_{d\in\mathbb{D}\mathbb{C}}\left(I_d^{(pr)}\mathcal{N}_d\right)\right] \tag{14.15}$$

The intra cloud DC power consumption is composed of the power consumption of the cloud LAN, which consist of a router and a switch:

$$PUE\_DC\left[\sum_{d\in\mathbb{D}\mathbb{C}}\left(\epsilon^{(DR)}\theta_d\right)+\sum_{d\in\mathbb{D}\mathbb{C}}\left(\delta I^{(DR)}\Omega^d\right)+\sum_{d\in\mathbb{D}\mathbb{C}}\left(\epsilon^{(DS)}\theta_d\right)\right.$$
$$\left.+\sum_{d\in\mathbb{D}\mathbb{C}}\left(\delta I^{(DS)}\Omega^d\right)\right] \tag{14.16}$$

The intra metro fog power consumption consists of power consumption of metro fog routers and switches:

$$PUE\_M\left[\sum_{d\in\mathbb{M}\mathbb{S}}\left(\epsilon^{(MfR)}\theta_d\right)+\sum_{d\in\mathbb{M}\mathbb{S}}\left(\delta\epsilon^{(MfR)}\Omega^d\right)+\sum_{d\in\mathbb{M}\mathbb{S}}\left(\epsilon^{(MfS)}\theta_d\right)\right.$$
$$\left.+\sum_{d\in\mathbb{M}\mathbb{S}}\left(\delta I^{(MfS)}\Omega^d\right)\right] \tag{14.17}$$

The MILP model's objective is to minimize the total power consumption as follows:

Minimize: $net\_pc + pr\_pc$
Subject to the following constraints:

$$\sum_{n\in\mathbb{N}_m} L_{mn}^{sd} - \sum_{n\in\mathbb{N}_m} L_{nm}^{sd} = \begin{cases} L_{sd} & m=s \\ -L_{sd} & m=d \\ 0 & otherwise \end{cases} \quad \forall s\in\mathbb{S}, d\in\mathbb{P}, m\in\mathbb{N}: s \neq d.$$

(14.18)

Constraint (14.18) is the flow conservation constraint. It ensures that the total incoming traffic at a node is equal to the total outgoing traffic of that node; if the node is not a source or a destination node.

$$\sum_{d\in\mathbb{P}} \rho^{sd} = D_s^{(cpu)} \quad \forall s \in \mathbb{S}$$

(14.19)

Constraint (14.19) ensures that processing service demand per IoT source node $s \in S$ is met at a given destination node.

$$\rho^{sd} \geq \Omega^{sd} \quad \forall s \in \mathbb{S}, d \in \mathbb{P}$$

(14.20)

$$\rho^{sd} \leq M\Omega^{sd} \quad \forall s \in \mathbb{S}, d \in \mathbb{P}$$

(14.21)

Constraints (14.20) and (14.21) are used to ensure that the binary variable $\rho^{sd} = 1$ if destination node $d \in P$ is activated to host the processing demand of source node $s \in S$.

$$\sum_{d\in\mathbb{P}} \Omega^{sd} \leq K \quad \forall s \in \mathbb{S}$$

(14.22)

Constraint (14.22) ensures that the number of sub-services a processing demand can be divided into is less than or equal to K, hence $K = 1$ implies no service splitting is allowed.

$$\mathcal{N}_d \leq \mathcal{V}_d \forall d \in \mathbb{P}$$

(14.23)

Constraint (14.23) ensures that the number of servers activated at a processing node $d \in \mathbb{P}$, does not exceed the maximum available number of servers in that node.

$$\mathcal{N}_d \geq \frac{\sum_{s\in\mathbb{S}} \sum_{d\in\mathbb{P}} \rho^{sd}}{C^{cpu}}$$

(14.24)

$$\sum_{s\in\mathbb{I}} \Omega^{sd} \geq \Omega^d \quad \forall d \in \mathbb{P} \tag{14.25}$$

$$\sum_{s\in\mathbb{I}} \Omega^{sd} \leq M\Omega^d \quad \forall d \in \mathbb{P} \tag{14.26}$$

Constraints (14.25) and (14.26) are used to ensure that, the binary variable $\Omega^d = 1$ if processing node $d \in \mathbb{P}$ is activated, otherwise $\Omega^d = 0$.

$$\begin{aligned}
\lambda_m = &\sum_{\substack{s \in \mathbb{S}: \\ m = s}} \sum_{d\in\mathbb{P}} \sum_{n\in\mathbb{N}_m} L_{mn}^{sd} \\
&+ \sum_{\substack{s \in \mathbb{S}: \\ m \neq s}} \sum_{\substack{d \in \mathbb{P}: \\ s \neq d}} \sum_{n\in\mathbb{N}_m} L_{nm}^{sd} \qquad \forall m \in S
\end{aligned} \tag{14.27}$$

$$L_m = \sum_{\substack{s \in \mathbb{S}: \\ m \neq s}} \sum_{\substack{d \in \mathbb{P}: \\ s \neq d}} \sum_{n\in\mathbb{N}_m} L_{nm}^{sd} \qquad \forall m \in (\mathbb{I}\cup\mathbb{A}\mathbb{P}\cup\mathbb{O}\cup\mathbb{O}\mathbb{T}\cup\mathbb{M}\mathbb{S}\cup\mathbb{M}\mathbb{R}\cup\mathbb{D}\mathbb{C})$$

$$\tag{14.28}$$

$$L_m = \sum_{\substack{s\in\mathbb{S} \\ s \neq d}} \sum_{\substack{d \in \mathbb{P}: \\ }} \sum_{\substack{n \in \mathbb{N}_m: \\ n \in (\mathbb{N}_m \cap \mathbb{C})}} L_{mn}^{sd} \qquad \forall m \in \mathbb{C} \tag{14.29}$$

Constraint (14.27) gives the traffic generated or received by an IoT node with the first term representing its role as a source and the second term representing IoT node serving demands of other IoT nodes. Constraint (14.28) gives the traffic traversing/received by a node of the access, metro, and cloud network. Constraint (14.29) gives the traffic traversing the core nodes.

$$\theta_d \leq M\Omega^d \quad \forall d \in \mathbb{P} \tag{14.30}$$

$$\theta_d \leq L_d \quad \forall d \in \mathbb{P} \tag{14.31}$$

$$\theta_d \geq \lambda_d - \left(1 - \Omega^d\right)M \quad \forall d \in \mathbb{P} \tag{14.32}$$

Constraints (14.30), (14.31) and (14.32) are used to linearize the non-linear equation $\lambda_d \Omega_d$, where $d \in P$. This ensures that traffic on a processing node $d \in P$ is only accounted for if it is destined to that node for processing.

$$L_m \geq \mathcal{B}_m \quad \forall m \in \mathbb{N} \tag{14.33}$$

$$L_m \leq M\mathcal{B}_m \quad \forall m \in \mathbb{N} \tag{14.34}$$

Constraints (14.33) and (14.34) are used to ensure that, the binary variable $\mathcal{B}_m = 1$ if network node $m \in N$ is activated, otherwise $\mathcal{B}_m = 0$.

$$L^{sd} = T^{(DR)}\Omega_{sd} \quad \forall s \in \mathbb{S}, d \in \mathbb{P} \tag{14.35}$$

Constraint (14.35) ensures that traffic is only directed to the destination node that is hosting a processing service.

$$\sum_{s\in\mathbb{S}} \sum_{\substack{d \in \mathbb{P}: \\ s \neq d}} L^{sd}_{mn} \leq C_{mn} \quad \forall m \in (\mathbb{I} \cup \mathbb{AP} \cup \mathbb{O} \cup \mathbb{OT} \cup \mathbb{MS} \cup \mathbb{MR} \cup \mathbb{DC}): n \in \mathbb{N}_m$$

$$\tag{14.36}$$

Constraint (14.36) ensures that the total traffic carried on link m, n, in all layers except the core does not exceed the link capacity.

$$Ag_m \geq \frac{L_m}{B} \quad \forall m \in \mathbb{C} \tag{14.37}$$

Constraint (14.37) gives the number of aggregation router ports at each IP/WDM node.

$$\sum_{s\in\mathbb{S}} \sum_{\substack{d \in \mathbb{P}: \\ s \neq d}} L^{sd}_{mn} \leq W_{mn} B \quad \forall m \in \mathbb{C}: n \in (\mathbb{C} \cap \mathbb{N}_m) \tag{14.38}$$

$$W_{mn} \leq W F_{mn} \forall m \in \mathbb{C}: n \in (\mathbb{C} \cap \mathbb{N}_m) \tag{14.39}$$

Constraints (14.38) and (14.39) represent the physical link capacity of the IP/WDM optical links. Constraint (14.38) ensures that the total traffic on a link does not exceed the capacity of a single wavelength while constraint (14.39) ensures the total number of wavelength channels does not exceed the capacity of a single fiber link.

4   Input Data for the MILP Model

4.1   Processing and Data Rates

In this chapter, we assume that processing task requirement is proportional to data rate, such that, for every bit of traffic, 1000 MIPS is required for processing. This assumption is based on the work in [52] where for a file of 10 kB, 69.23 MIPS are required for processing for visual processing applications. Thus, through we derive how many MIPS are required (Δ) to process a Mb of traffic using (14.40):

$$\Delta = \frac{69.23}{0.08} \cong 865.4. \tag{14.40}$$

For the sake of simplicity, we assume that a Mb of traffic requires approximately 1000 MIPS for processing. As for the bandwidth requirement, an online tool is used to estimate the required data rates for different video resolutions and this was estimated to be between 1–10 Mbps, which covers video resolutions between 1024×720 to 1600×1200 at 30 frames per second [53]. The CPU workload intensity is then calculated by multiplying the Δ by the amount of traffic. Thus, this makes the CPU demand proportional to the size of the traffic due to the assumption that the higher the traffic, the more features a video file will hold.

4.2   Power Consumption Data

The data for all the network devices in the network (except the core as it is shown in separate table) is shown in Table 14.1. We have use of manufacturer's and equipment datasheet where possible in order to represent a practical scenario. The idle power consumption of high capacity networking equipment is reported to consume up to 90% of equipment's maximum power consumption [40]. Since high capacity networking equipment is shared by many users and applications, we assume that the IoT application under consideration only consumes 3% (δ) of the equipments' maximum idle power consumption. This is based on Cisco's visual networking index for the years 2017–2022. It is reported that, globally, 3% of all video traffic on the Internet is due to surveillance applications [50]. As for processing device's idle power consumption, based on [49], we assume it is 60% of the maximum power consumption of the CPU. The processing devices' input data are summarized in Table 14.2. We estimate the processing capacity of processing nodes (in MIPS) using a technical benchmark published in [66]. It is reported that high-end CPUs process four instructions per cycle (I/C). Thus, to determine the maximum capacity of a processing device we have used the following equation

$$MIPS = clock \times I/C \tag{14.41}$$

Table 14.1 Data for all networking devices in the network except the core layer

Node	Maximum power (W)	Idle power (W)	δ	Data rate (Gb/s)
IoT (WiFi)	0.56 [54]	0.34 [55]	–	0.1 [54]
ONU (WiFi)	15 [56]	9 [56]	–	0.3 [56]
OLT	1940 [57]	60 [57]	3%	8600 [57]
Metro Router Port	30 [58]	27	3%	40 [58]
Metro Ethernet Switch	470 [59]	423	3%	600 [59]
Metro Router Redundancy (R)	2 [41]			

Table 14.2 Data of all processing devices

Node	Device	Maximum power (W)	Idle power (W)	GHz	k MIPS	Watts/ MIPS	Instruction per cycle
GP-DC server	Intel Xeon E5-2680	130 [60]	78	2.7 [60]	108	481μ	5
Metro Server	Intel X5675	95 [61]	57	3.06 [61]	73.44	517μ	4
Access Server	Intel Xeon E5-2420	95 [62]	57	1.9 [62]	34.2	1111μ	3
CPE Server	RPi 3 Model B	12.5 [63]	2	1.2 [64]	2.4	4375μ	2
IoT Server	RPi Zero W	3.96 [63]	0.5	1 [65]	1	3460μ	1

where I/C is the number of instructions a CPU can execute per clock cycle in GHz. To differentiate between the different types of CPUs and their efficiencies, we set the I/C of the Metro Fog server as a reference point. The efficiency of the processing decreases as one moves down the network hierarchy (from the core to the edge) [40]. At those layers where multiple servers can be deployed, networking infrastructure becomes a necessity to establish a LAN network between multiple active servers. Hence, we have used routers and switches accordingly to achieve this and Table 14.3 contains the data of all the devices utilized for this purpose. For the lower layers of the cloud fog architecture such as IoT and CPE, we assume embedded type processors such as Raspberry Pi (RPi) Zero W and Raspberry Pi (RPi) 3 Model B, respectively. We assume the cloud DC node is a single hop away from the aggregated traffic and the average distance is also assumed to span 2010 km on average, which is estimated using google maps for AT&T US network topology [67]. The power consumption of the IP/WDM core network is consistent with our previous work in [13] and all the parameters are summarized in Table 14.4.

Table 14.3 Data for networking devices used inside (intra-processing) Access Fog, Metro Fog and Data Centre processing units

Device	Maximum power (W)	Idle power (W)	Data rate (Gb/s)	Energy per bit (W/Gb/s)
Access Fog Router	13 W [58]	11.7	40 [58]	0.03
Access Fog Switch	210 W [59]	189	240 [59]	0.08
Metro Fog Router	13 W [58]	11.7	40 [58]	0.03
Metro Fog Switch	210 W [59]	189	600 [59]	0.04
DC LAN Router	30 [58]	27	40 [58]	0.08
DC LAN Switch	470 [59]	423	600 [59]	0.08

Table 14.4 Data for IP/WDM core network

Distance between two neighbouring EDFAs Se	80 (km) [13]
Number of wavelengths in a fibre (W)	32 [13]
Bitrate of a wavelength (B)	40 Gb/s
Distance between two neighbouring core nodes D_{mn}	509 km
Maximum power consumption of a router port Pr	638 (W) [13]
Idle power consumption of a router port Ir	574.2 (W)
Energy per bit of a router port $\epsilon^{(r)}$	1.6 W/Gb/s
Maximum power consumption of a transponder Pt	129 (W) [13]
Idle power consumption of a transponder It	116 (W)
Energy per bit of a transponder $\epsilon^{(t)}$	0.32 (W/Gb/s)
Maximum power consumption of an optical switch Po	85 (W) [13]
Idle power consumption of an optical switch Io	76.5 (W)
Energy per bit of an optical switch $\epsilon^{(o)}$	0.2 (W/Gb/s)
Maximum power consumption of a regenerator that reaches 2500 km Prg	71.4 (W) [13]
Idle power consumption of a regenerator Irg	64 (W)
Energy per bit of a regenerator $\epsilon^{(rg)}$	0.19 (W/Gb/s)

4.3 Power Usage Effectiveness (PUE)

The power usage effectiveness (PUE) is a ratio that is used to measure the efficiency of a facility such as DCs, ISP networks, etc. PUE is defined as the ratio of the total power consumed by a facility to the total power consumed by the communications

Table 14.5 PUE values used in the MILP model

Node	PUE
IoT	1
CPE	1
Access Fog (PUE_A)	1.5
Metro Fog (PUE_M)	1.4
Cloud DC (PUE_DC)	1.12 [35]
IP/WDM Core (PUE_C)	1.5 [21]

and processing elements within the facility. In DCs, Google reported that one of its DC has a PUE of 1.15 in 2018. In this chapter, we estimate the value of PUE on "space type", such that the value of PUE decreases with the increase in "space" [68]. Similarly, we increase PUE progressively in the proposed network architecture since the largest "Space Type" is generally occupied by cloud DCs connected to the core network. We assume that at the access and metro layers, processing and networking equipment have the same PUE as these two types of elements can be collocated in the same office/ building. The PUE value of the core network is consistent with one of our previous works, which is 1.5 [21]. Table 14.5 is a summary of the PUE values used in this chapter.

5 Scenarios and Processing Placement Results

5.1 Energy-Aware Processing Placement

In this subsection, we consider a capacitated case where extra processing capacity cannot be added to the processing nodes in the cloud fog architecture. Such design problems are faced in the short-term when the network is already designed, and the processing nodes have been put in place. It is important to note that the cloud DC offers unlimited processing capacity and hence always enough to host all the tasks. We evaluate the performance of the proposed cloud fog architecture given a processing placement problem, and 20 generic IoT devices that are divided into 4 groups uniformly. The total number of IoT devices in each group is based on a representative home LAN network which typically connects a single to few users [69]. The performance of the cloud fog approach is benchmarked against the baseline approach in which all of the task requests are simultaneously processed by the cloud DC and the types of scenarios evaluated in this sub-section are shown in Table 14.6.

Table 14.6 Types of scenarios evaluated in this sub-section

Use case	Source node distribution	Total # of source nodes	Total requested MIPS	
			Min	Max
Scenario #One	A single task request generated in any random IoT group	1	1k	10k
Scenario #Two	Five task requests generated in the same IoT group	5	5k	100k
Scenario #Three	Four task requests generated per IoT group	4	4k	80k
Scenario #Four	Five task requests generated per IoT group	20	20k	200k

5.2   Scenario One

In this scenario, it is assumed that at any given time, a single source node is generating task requests from any IoT group randomly. Figure 14.3 shows the total power consumption of the cloud fog approach versus the baseline solution. As was expected, during low workloads such as at 1000 MIPS, processing tasks are allocated to source nodes themselves and when processing capacity runs out at this layer, the model utilizes the closest processing layer to the source nodes, which is the CPE layer. This is justifiable due to the low-power embedded type CPUs onboard these devices and their close proximity to the source nodes, hence substantially lower networking and processing power consumption as can be seen in Fig. 14.4a. However, once the processing resources run out at the two aforementioned layers, the model utilizes the large servers located at the Access Fog. As shown in Fig. 14.3, allocating processing tasks to nodes with low-power embedded CPUs introduces substantial power savings of up to 98% and up to 46% when tasks are processed by larger fog servers. The allocation of the total processing tasks (in %) in the different processing layers is shown in Fig. 14.5. Although this result is optimal in this situation, it may not be so optimal when the processing capacity of the CPE layer can be expanded (i.e. evaluations done in an un-capacitated setting). The reason behind this is that CPE nodes consume considerably lower power compared with the servers at the Access Fog and they do not have any associated overheads such as PUE due to cooling requirements. Also, it is worthy of noting that the location of the source in the network highly influences the allocation of the processing tasks due to the power consumption of the network in order to access certain processing layer. It is for this reason we evaluate different scenarios in which source nodes are in different parts of the network. Also, the reason why the baseline curve is flat is because the DC's processing efficiency is substantially higher than the rest of the processing locations, hence the proportional power consumption increases in very

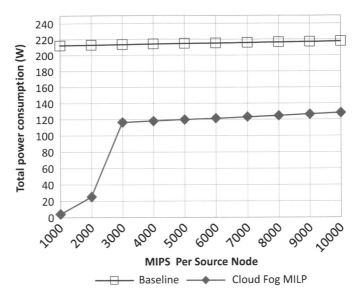

Fig. 14.3 Total power consumption in Scenario one of the cloud fog approach versus the baseline approach

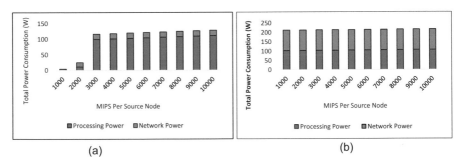

Fig. 14.4 Total power consumption in Scenario one broken down into network and processing power consumption in (**a**) cloud fog approach and (**b**) baseline approach

small steps. We would see a staircase curve should the number of servers increase due to the idle power consumption [35].

5.3 Scenario Two

In this scenario, we begin to observe the utilization of the Metro Fog node instead; and the Access Fog node has limited role to play in this case. This was anticipated because the Access Fog has a lower processing efficiency and a higher PUE value compared with the Metro Fog node. As shown in Fig. 14.7a, the Access Fog node

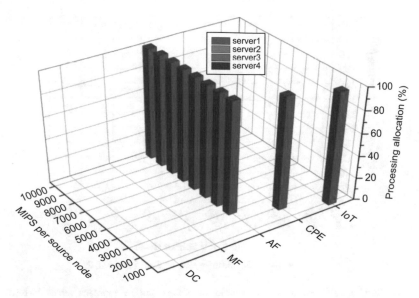

Fig. 14.5 Processing task allocation

is chosen to process the tasks at 2000 MIPS only because the network power consumption to access the Metro Fog node overrides the processing efficiency and lower PUE advantage of the Metro Fog node. However, as the workload increases (at and beyond 3000 MIPS), we can observe that the Metro Fog's efficiency compensates for its networking overhead, hence all tasks are processed at this layer. In Fig. 14.6, the cloud fog approach still produces substantial power savings despite the activation of larger fog servers to process the increased demands such as those in the Access Fog and Metro Fog layers. At a workload of 2000 MIPS, we can observe the impact of the single allocation constraint (i.e. task request per source node cannot be split) on the total power savings. Although the CPE nodes had enough capacity to host the majority of the tasks (9600 MIPS out of 10,000 MIPS), the optimization model is forced to allocate all the tasks to a larger server with sufficient capacity such as the Access Fog node in this case. Had the model considered the prospect of task splitting and/or adding further processing capabilities to the CPE layer, the results would have been different. These dimensions are thoroughly investigated in our previous work in [47]. Task splitting can help realize server utilization improvements and can help better pack the lower processing and networking layers of the cloud fog architecture as we know from the previous scenario that the IoT and CPE layers produce substantial savings (Figs. 14.7 and 14.8).

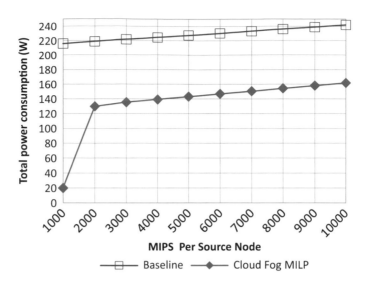

Fig. 14.6 Total power consumption in Scenario two of the cloud fog approach versus the baseline approach

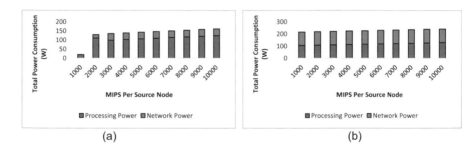

Fig. 14.7 Total power consumption in Scenario two broken down into network and processing power consumption in (**a**) cloud fog approach and (**b**) baseline approach

5.4 Scenario Three

As shown in Fig. 14.12, the trends in this scenario are similar to those observed in Scenario Two, except for the case at 2000 MIPS where instead of the Access Fog server, the model allocates the total demands to all of nodes in the CPE layer. This is primarily due to the geographical distribution of the source nodes in this scenario. Each CPE Fog server has enough processing capacity to process the task of the closest source node and the total demands happen to match the total processing capacity offered by the CPE layer. Hence, the model activates multiple low-power CPE servers in order to avoid the high idle power and associated PUE overheads of the higher fog layers such as the Access Fog and the Metro Fog nodes, as can be

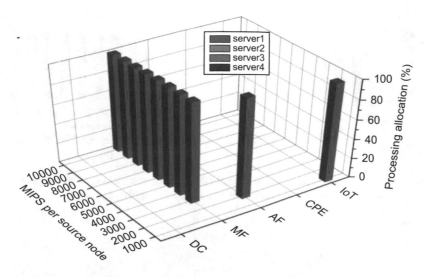

Fig. 14.8 Processing task allocation

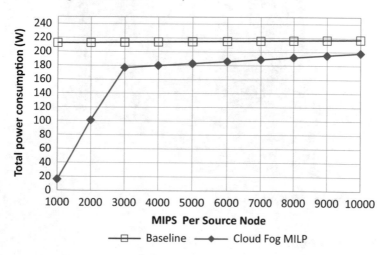

Fig. 14.9 Total power consumption in Scenario Three of the cloud fog approach versus the baseline approach

seen in Fig. 14.10a. A total power saving of 66% is achieved at 2000 MIPS and up to 55% power savings at workloads beyond 2000 MIPS, as shown in Fig. 14.9 (Figs. 14.10 and 14.11).

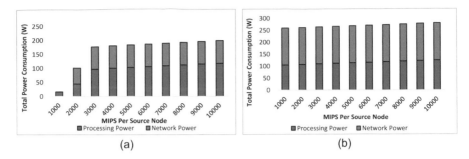

(a)   (b)

Fig. 14.10 Total power consumption in Scenario three broken down into network and processing power consumption in (**a**) cloud fog approach and (**b**) baseline approach

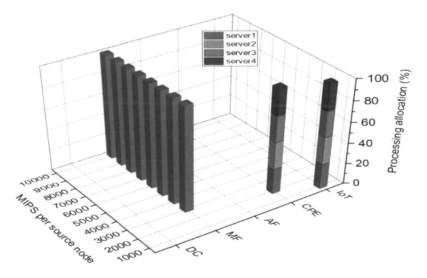

Fig. 14.11 Processing task allocation

5.5 Scenario Four

In this scenario, all the of the source nodes generate task requests, hence the total workload volume has increased substantially. We begin to observe similar trends as in previous scenarios. At very low load workloads, processing locally on source nodes is still the optimal choice in terms of total energy consumption as shown in Fig. 14.12. The allocation decisions shown in Fig. 14.14 confirm the superiority of the Metro Fog server over the Access Fog server for high workloads, as the Access Fog is never utilized. The model chooses to utilize the Metro Fog layer at four different workloads due to the processing and networking trade-off shown in Fig. 14.13a and b. At 4000 MIPS and 5000 MIPS, processing all the tasks in the Metro Fog layer results in activating two servers, therefore trading off the high network

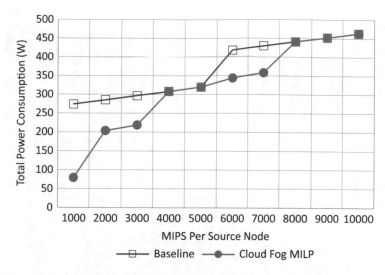

Fig. 14.12 Total power consumption in Scenario three of the cloud fog approach vs. the baseline approach

power consumption from accessing the cloud DC produces more power savings than choosing the Metro Fog. This is because the cloud DC has enough processing resources to server all the tasks on a single server, hence resulting in lower total server idle power consumption compared to the Metro Fog layer. However, at 6000 MIPS and 7000 MIPS, processing all the requests in the cloud DC results in the activation of two servers, hence the idle power consumption coupled with the core network power consumption renders the cloud DC solution no longer favourable and as a result the Metro Fog is chosen as the optimal location for processing all the tasks. At 8000 MIPS and beyond, the cloud DC solution due to its superiority in terms of processing capacity produces more power savings than the Metro Fog layer, hence it is chosen as the optimal allocation. Figure 14.12 shows power savings of up to 29% with the cloud fog approach. In this scenario, we show that the cloud fog approach does not replace the cloud DC, but instead the complementary features of both of these paradigms helps to achieve a better and more greener processing platform in the upcoming 6G networks (Fig. 14.14).

5.6 Energy and Delay Aware Processing Placement

In this section, we study the trade-off between power consumption and delay. We consider propagation delay between network nodes and queuing delay at different network nodes. We optimize the allocation of the processing resources in a multi-objective MILP optimization model to minimize the power consumption and delay equally.

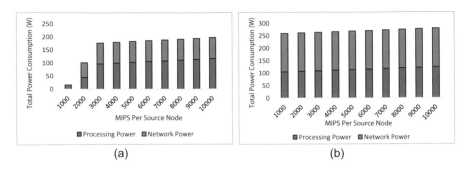

Fig. 14.13 Total power consumption in Scenario four broken down into network and processing power consumption in (**a**) cloud fog approach and (**b**) baseline approach

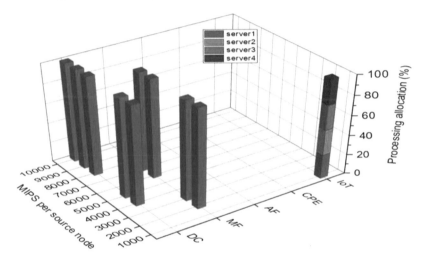

Fig. 14.14 Processing task allocation

The propagation delay is based on the distances between network nodes as this network covers a large geographical area, and is calculated using the following equation:

$$Propagation\ Delay = \frac{D}{C} \tag{14.42}$$

where D represents the distance, and C is the speed of light.

The values of distance (D) between each two nodes are based on the following assumptions:

1. The distance between the AP and surrounding IoT nodes is set to the standard coverage range of WLAN, (100 m) [70].

2. The distance between the AP and ONU is estimated based on an assumption that one ONU can connect to multiple APs in a typical LAN (100 m) [71].
3. The distance between the ONU and OLT is based on typical PON designs. We considered a design where the OLT is located in the telecom main office in the centre of the city. ONUs usually represent devices located at the end-users location (i.e., at home); usually such distance are around 5–20 km [72] so we assumed an average distance equal to (10 km).
4. The distance between the OLT and metro node (router and switch) was estimated based on the metro network design. The metro network usually has a radius of 20–120 km [73]. The OLT can be either collocated with the metro node in the same telecom office or located somewhere else in the local area of the metro node (1–10 km away). We based our estimation on the latter scenario with an approximate distance equal to (5 km) between the OLT and metro node.
5. The distance between the metro node and the core node (including the associated data centre), is given as the distance between two large cities, assuming the current city does not have a large central cloud. An example of such a distance is taken as the distance between Leeds and a large data centre in London, a (300 km) distance.

The queuing delay was modelled for each networking node as an M/M/1 queue with one server, where arrivals follow a Poisson process and the service rate is negative exponentially distributed, summarized in Fig. 14.15. The queueing delay was calculated based on aggregated traffic at each node (arrival rate) and the maximum capacity of the nodes (service rate), as given below

$$Queuing\ Delay = \frac{1}{\mu - \lambda} \tag{14.43}$$

where μ is the service rate, and λ is the arrival rate. We have considered in this work delay at the packet level. We used the Ethernet maximum packet size of 1500 bytes and therefore, expressed the arrival data rates as packets per second and expressed the service rates (transmission rates) in packets per second.

Three different service rate values are considered. We assumed that the AP works based on the wireless medium interface capacity (with 1 Gb/s service rate). Moreover, the core node, with the associated data centre, was assumed to work at 40 Gb/s, as they are part of the IP/WDM network. Other network devices were assumed to have a 10 Gb/s service rate based on GPON.

Fig. 14.15 M/M/1 Queueing model

The MILP model introduced in the previous sub-section was extended to jointly minimize power consumption and delay. To continue to use linear programming, Eq. (14.43) was converted to a linear form based on a pre-defined lookup table. This table includes all the possible generated traffic combinations (arrival rates indicator), indexed with the calculated queuing delay based on a fixed service rate. As we have three different service rate values in our designed network, three lookup tables were defined. Based on this arrival rate indicator, the queuing delay for a node was given as the value corresponding to the indicator in the lookup table.

The modified MILP defines the following additional sets, parameters, and variables:

Sets:	
$\mathbb{AR}$	Set of arrival rates
$\mathbb{SR}$	Set of service rates
Parameters:	
η_{as}	Queuing delay at arrival rate $a \in \mathbb{AR}$ and service rate $S \in \mathbb{SR}$, in the lookup table
$G1$	Large enough number with units of Mb/s
$G2$	Large enough number with units of ms
D_{mn}	Distance between any two nodes (m, n), where $m \in \mathbb{N}, n \in \mathbb{N}_m$
C	Speed of light, $C = 299, 792 \frac{km}{s}$
ΔRI	Refractive index of fibre, which is defined as the ratio of the speed of light in fibre to speed of light in free space; $\Delta RI = \frac{2}{3}$
Variable:	
ζ_{mn}^{sd}	Binary variable $\zeta_{mn}^{sd} = 1$ if the traffic flow sent from source node s to processing node d traverses physical link (m, n), where $s \in \mathbb{S}, d \in \mathbb{P}$, and $m, n \in \mathbb{N}$
Q_{mn}^{sd}	Queuing delay at node j experienced by the traffic from source node s to processing node d traversing physical link (m, n), where $s \in \mathbb{S}, d \in \mathbb{P}$ and $m, n \in \mathbb{N}$
Q_i	Queuing delay experienced by traffic aggregated at node $i \in \mathbb{N}$
Q_{sd}	Queuing delay of the traffic sent from source node $s \in \mathbb{S}$ to processing node $d \in \mathbb{P}$
Q	Total queuing delay of the network
R_{sd}	Propagation delay of the traffic sent from source node $s \in \mathbb{S}$ to processing node $d \in \mathbb{P}$
R	Total propagation delay of the network
λ_i	Arrival rate (total traffic) at each node $i \in \mathbb{N}$
H_{mn}	Arrival rate indicator for node $m \in \mathbb{N}$, $\sigma_{mn} = 1$ if the arrival rate of node m matches rate $n \in \mathbb{AR}$, it is 0 otherwise

All the power consumption equations in the previous sub-section were considered in this model. The total power consumed in is calculated as follows:

$$\text{Total Power Consumption} = \text{netpc} + \text{prpc} \qquad (14.44)$$

Additionally, the following equations are used to calculate the propagation and queuing delay for the network.

(1) The total propagation delay (R), is calculated based on the propagation delay between all source node and processing node pairs and is given as

$$R = \sum_{s \in \mathbb{S}} \sum_{d \in \mathbb{P}} R_{sd} \qquad \forall \ s \in \mathbb{S}, d \in \mathbb{P} \qquad (14.45)$$

where R_{sd} is is the propagation delay of the path traversed by traffic sent from each sourcenode $s \in \mathbb{S}$ to the processing node $d \in \mathbb{P}$, and is calculated as follows;

$$R_{sd} = \sum_{\substack{m \in \mathbb{N} \\ m \notin \mathbb{I}}} \sum_{n \in \mathbb{N}_m} \zeta_{mn}^{sd} \ \frac{D_{mn}}{\Delta RIC} \qquad \forall \ s \in \mathbb{S}, d \in \mathbb{P} \qquad (14.46)$$

$$R_{sd} = \sum_{\substack{i \in \mathbb{N} \\ i \in \mathbb{I}}} \sum_{j \in \mathbb{N}_m} \zeta_{mn}^{sd} \ \frac{D_{mn}}{\mathbb{C}} \qquad \forall \ s \in \mathbb{S}, d \in \mathbb{P} \qquad (14.47)$$

Equation (14.46) and (14.47) calculate the propagation delay for the traffic sent to the processing nodes via fibre or wireless links, respectively. A refractive index ΔRI with the value of $\frac{2}{3}$ is added to Eq. (14.46) to define the ratio of the speed of light in fibre to the speed of light in free space.

(2) The total queuing delay (Q), which is calculated based on the queuing delay experienced by traffic between all the source node and processing node pairs, and is given as:

$$Q = \sum_{s \in \mathbb{S}} \sum_{d \in \mathbb{P}} Q_{sd} \qquad \forall \ s \in \mathbb{S}, d \in \mathbb{P} \qquad (14.48)$$

where Q_{sd} is the queuing delay of the path traversed by traffic sent from each source node $s \in \mathbb{S}$ to processing node $d \in \mathbb{P}$, and is calculated as:

$$Q_{sd} = \sum_{m \in \mathbb{N}} \sum_{j \in \mathbb{N}_m} Q_{mn}^{sd} \qquad \forall \ s \in \mathbb{S}, d \in \mathbb{P} \qquad (14.49)$$

Equation (14.47) calculates the queuing delay for a traffic demand by summing the queuing delay experinced by the demand at each node.

The joint objective is defined as:
Minimize

$$\alpha\, P \;+\; \beta\, R \;+\; \gamma\, Q \tag{14.50}$$

where α, β, and γ are weight factors used for the following purposes :
(1) to scale the terms so that they are comparable in magnitude; (2) to emphasise
and de$-$emphasise terms (power, queuing delay and propagation delay) ; and (3) to
accommodate the units in the objective function. Therefore, α is a unitless factor,
and β & γ have units of $\frac{Watt}{sec}$.

In addition to the constraints in the previous sub-section, the model is subject to
the following additional constraints:

(1) The traffic estimation at each node:

$$\sum_{s\in\mathbb{S}}\sum_{d\in\mathbb{P}}\sum_{m\in\mathbb{N}m} \lambda_{nm}^{sd} = \lambda_m \qquad \forall\;\; m\in\mathbb{N}, m\notin\mathbb{S} \tag{14.51}$$

Constraint (14.51) calculates the traffic arrival at each node in the.
(2) The arrival rate indicator:

$$\sum_{n\in\mathbb{AR}} H_{mn}\cdot n = \lambda_m \qquad \forall\;\; m\in\mathbb{N}, m\notin\mathbb{S} \tag{14.52}$$

Constraint (14.52) creates indicators of the arrival rate for each node. This is
equal to 1 if the arrival rate is equal to n:

$$\sum_{n\in\mathbb{AR}} H_{mn} \leq 1 \qquad \forall\;\; m\in\mathbb{N}, m\notin\mathbb{S} \tag{14.53}$$

Constraint (14.53) ensures that each node has no more than one arrival rate
indicator for a given service rate.

(3) Queuing delay estimation:

$$\sum_{n\in\mathbb{AR}} H_{mn}\cdot \eta_{ns} = Q_m \qquad \forall\;\; m\in\mathbb{C}\cup\mathbb{DC}\;, \quad s = 40Gb/s \tag{14.54}$$

$$\sum_{n\in\mathbb{AR}} H_{mn}\cdot \eta_{ns} = Q_m \qquad \forall\;\; m\in\mathbb{N}, m\notin\mathbb{I}\cup\mathbb{C}\cup\mathbb{DC}\cup\mathbb{A}, \quad s = 10Gb/s \tag{14.55}$$

$$\sum_{n \in \mathbb{AR}} H_{mn} \cdot \eta_{ns} = Q_m \qquad \forall\ m \in \mathbb{A}\ ,\ s = 1Gb/s$$

$$(14.56)$$

Constraints (14.54)–(14.56) estimate the traffic delay for each node that operates at 40, 10, or 1 Gb/s, respectively.

$$L_{mn}^{sd} \geq \zeta_{mn}^{sd} \qquad \forall s \in \mathbb{S}, d \in \mathbb{P}, m\ and\ n \in \mathbb{N} \tag{14.57}$$

$$L_{mn}^{sd} \leq G1\ \zeta_{mn}^{sd} \qquad \forall\ s \in \mathbb{S}, n \in \mathbb{P}, i\ and\ j \in \mathbb{N} \tag{14.58}$$

Constraints (14.57) and (14.58) set $\zeta_{mn}^{sd} = 1$ if the traffic demand between the source node and the processing node is routed throughlink (m, n).

$$Q_{mn}^{sd} = Q_n\ \zeta_{mn}^{sd} \qquad \forall\ s \in \mathbb{S}, n \in \mathbb{P}, m\ and\ n \in \mathbb{N} \tag{14.59}$$

$$Q_{mn}^{sd} \leq G2\ \zeta_{mn}^{sd} \tag{14.60}$$

$$Q_{mn}^{sd} \leq Q_n \tag{14.61}$$

$$Q_{mn}^{sd} \geq Q_n - G2\left(1 - \zeta_{mn}^{sd}\right) \tag{14.62}$$

Equation (14.59) calculates the queuing delay at node m for the traffic sent from source node s to processing node d. As Eq. (14.59) involves the multiplication of two variables,Q_{mn}^{sd} and Q_m, constraints (14.60)–(14.62) are used to remove the nonlinearity of Eq. (14.59) and replace the relationship with an equivalent linear relationship.

6   Scenarios and Results

The model presented in the previous section is considered with the following variations of the objective function:

1. Minimizing the total power consumption only
2. Minimizing the traffic propagation delay only
3. Minimizing the power consumption and traffic propagation delay jointly.
4. Minimizing the traffic queuing delay
5. Minimizing the power consumption and traffic queuing delay jointly.
6. Minimizing the power consumption, traffic propagation and queuing delay jointly.

The previously described objective functions were combined into four evaluations that highlight the individual effects of the propagation and queuing delay, combined with the power consumption on the processing allocation decision, and both power and delay values. All the evaluations consider Scenario Two described in the previous sub-section, where we considered a cloud-fog-VEC allocation (CFVA) with low-density VNs (8VNs) and single five tasks were generated from the same IoT group. Each source node in group 1 generates a task with an increasing preocessing requirement (1000–10,000 MIPS) and a propotional data rate ranging from 1 to 10 Mbps per task.

6.1   Evaluation One: Power and Propagation Delay Minimization

In this evaluation, we study the joint minimization of the power consumption and the propagation delay (objective function case 3), and compare the results to the two cases where only the power (objective function case 1) or propagation delay (objective function case 2) are minimized. Figures 14.16 and 14.17 illustrate the total power consumption and the average propagation delay for the three cases considered in the objective function versus the traffic generated per demand, (1–10) Mbps. These results are reflected by the processing allocation illustrated in Fig. 14.18.

Figure 14.16 shows that the three objective functions have achieved the minimum power consumption when processing tasks locally in the IoT nodes. Local processing is confirmed to be the most energy efficient strategy (per the results generated from the previous sub-section). Moreover, it is confirmed that local processing can achieve minimize propagation delay as multiple hops and distances are avoided, as shown in Fig. 14.18. The propagation delay minimized case, in Fig. 14.17 produced the highest power consumption. The jumps in the curve (at 2 Mbps) are due to moving the allocation to a less efficient processing node that can support the demand. Moreover, activating two processing nodes, as shown in Fig. 14.18, causes

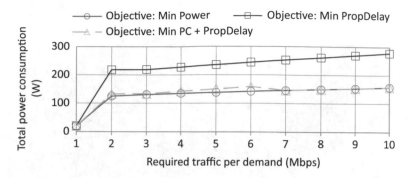

Fig. 14.16 Total power consumption (power and propagation delay minimization)

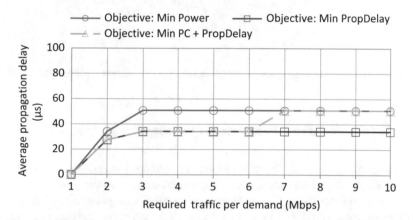

Fig. 14.17 Average propagation delay (power and propagation delay minimization)

an increase in the power consumption compared to the power minimization case. This increase continues as all tasks are allocated to the access fog by activating all its servers compared to one metro fog server in the power minimization case. With the joint minimization of the power and propagation delay, results led to a lower power consumption, by an average of 40%, compared to the delay minimization objective, as shown in Fig. 14.17. However, the propagation delay, in Fig. 14.18, increases at high traffic because the model activated a metro server and access server instead of two access servers in the delay minimization case. This is to achieve a balance between the power consumption and propagation delay. Activating an access server decreases the propagation delay but metro servers minimize the processing power consumption and therefore the total power consumption, as seen in Fig. 14.17.

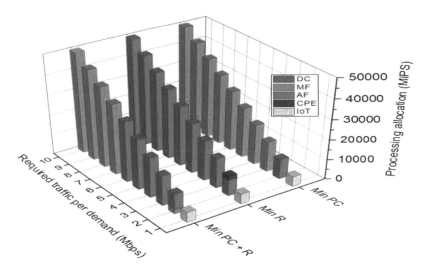

Fig. 14.18 Processing allocation at each processing node (power and propagation delay minimization)

6.2 Evaluation Two: Power and Queuing Delay Minimization

In this evaluation, we study the joint minimization of the power consumption and the queuing delay (objective function case 5), and compare it to the power minimized case (objective function case 1), and the queuing delay minimized objective function (objective function case 4). Figure 14.19 shows the total power consumption for the three cases. Similar to the propagation delay minimization case, minimizing queuing delay consumes the highest power consumption. This is due to allocating tasks to the processing nodes that guarantee the minimum hops and therefore minimum queuing delay experienced by each networking node. For example, as seen in Fig. 14.21, tasks are allocated to the CPE whenever it is sufficient. Subsequently, all tasks are allocated to the AF by activating all the servers in the access fog. On the other hand, relatively comparable power consumption results can be observed for the power minimized case (blue curve) and power and delay minimized case (yellow curve), in Fig. 14.19. The latter case causes more power consumption. With 3–4 Mbps generated traffic allocating low traffic tasks to AF achieves a balance between the power consumption and queuing delay. Figure 14.19 shows comparable average queuing delay for the three minimization cases. This is due to the fact that all the networking nodes in the access and metro layers operate with the same service rate (Fig. 14.20). The increase in the queuing delay in the power minimized cases is caused by the extra hop the traffic travels through when allocating the tasks to the metro layer, as shown in Fig. 14.21.

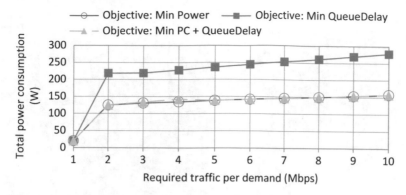

Fig. 14.19 Total power consumption (power and queuing delay minimization)

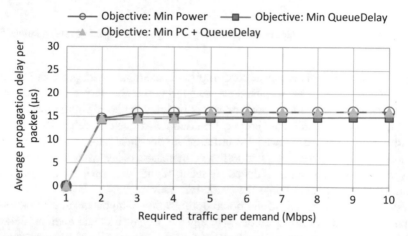

Fig. 14.20 Average queuing delay per packet (power and queuing delay minimization)

7   Conclusions and Future Work

In this chapter, we have evaluated a cloud fog architecture for future 6G networks paying special attention to energy efficiency and latency. We developed a Mixed Integer Linear Programming (MILP) model that is generic and independent of technology and application. We used the resultant model to investigate the processing task allocation problem in a representative IoT application. The results showed that in the cloud fog approach despite its limitations such as processing resources and distributed nature, substantial amounts of power savings can be achieved. The results also showed that regardless of how efficient edge processing is, the cloud DC will always remain relevant due to its abundance of processing capacity and efficiency. We have also investigated the joint optimization of power consumption, propagation delay and queueing delay when allocating processing tasks to the available servers. Three evaluations were considered with different

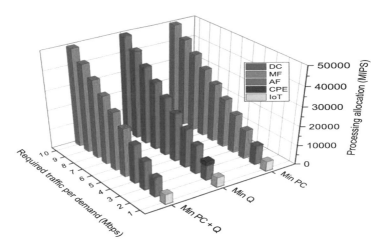

Fig. 14.21 Processing allocation at each processing node (power and queuing delay minimization)

objective functions where power consumption is examined with propagation delay and queueing delay. Our results show that the closer the server is to the source nodes, the lower the propagation and queuing delay achieved, as the distance and the number of hops affect the propagation delay and queuing delay. However, the queuing delay can be reduced by utilizing higher data rates in the networking devices. Therefore, the effects in terms of increased delay when allocating tasks to a remote location can be reduced by using higher data rates on the route, and minimizing the number of hops the traffic traverses.

Future work can introduce additional optimization components to the delay such as processing and transmission delay alongside propagation and queuing delay. It can also consider queuing at the wireless devices where IoT nodes are allowed to communicate with each other. Future studies can also include developing heuristic algorithms to mimic the behaviour of the MILP models as well as accounting for the dynamic nature of demands in uncertain network settings. It is also worth looking into duty cycling schemes (shallow and/or deep sleep) of both networking and processing equipment, which can result in further power savings. Also, it has to be observed that shutting down a network or processing element completely poses latency challenges in the start-up phase.

References

1. L. Baresi, D.F. Mendonca, in *Towards a Serverless Platform for Edge Computing*. Proceedings – 2019 IEEE International Conference on Fog Computing, ICFC 2019, June 2019, pp. 1–10. https://doi.org/10.1109/ICFC.2019.00008
2. Q.-V. Pham et al., A survey of multi-access edge computing in 5G and beyond: Fundamentals, technology integration, and state-of-the-art. IEEE Access, 1–1 (2020). https://doi.org/10.1109/ACCESS.2020.3001277

3. M. Satyanarayanan, The emergence of edge computing. Computer (Long. Beach. Calif.). **50**(1), 30–39 (2017). https://doi.org/10.1109/MC.2017.9
4. G. Premsankar, M. Di Francesco, T. Taleb, Edge computing for the internet of things: A case study. IEEE Internet Things J. **5**(2), 1275–1284 (2018). https://doi.org/10.1109/JIOT.2018.2805263
5. Fog | Definition of Fog by Merriam-Webster. https://www.merriam-webster.com/dictionary/fog. Accessed 24 July 2020
6. N. Abbas, Y. Zhang, A. Taherkordi, T. Skeie, Mobile edge computing: A survey. IEEE Internet Things J. **5**(1), 450–465 (2018). https://doi.org/10.1109/JIOT.2017.2750180
7. Introduction and Overview at W3C Open Day OpenFog Consortium (2017)
8. Q. Qi, X. Chen, C. Zhong, Z. Zhang, Integration of energy, computation and communication in 6G cellular internet of things. IEEE Commun. Lett. **24**(6), 1333–1337 (Jun. 2020). https://doi.org/10.1109/LCOMM.2020.2982151
9. M. Silverio-Fernández, S. Renukappa, S. Suresh, What is a smart device? – A conceptualisation within the paradigm of the internet of things. Vis. Eng. **6**, 3 (2018). https://doi.org/10.1186/s40327-018-0063-8
10. L. Zhang, Y.C. Liang, D. Niyato, 6G Visions: Mobile ultra-broadband, super internet-of-things, and artificial intelligence. China Commun. **16**(8), 1–14 (2019). https://doi.org/10.23919/JCC.2019.08.001
11. H. Yang, A. Alphones, Z. Xiong, D. Niyato, J. Zhao, K. Wu, Artificial intelligence-enabled intelligent 6G networks. arXiv Prepr. arXiv1912.05744, vol. 639798 (2019)
12. Key Drivers and Research Challenges for 6G Ubiquitous Wireless Intelligence | 1 (2019)
13. J.M.H. Elmirghani et al., GreenTouch GreenMeter core network energy efficiency improvement measures and optimization [Invited]. IEEE/OSA J. Opt. Commun. Netw. **10**(2), A250–A269 (2018). https://doi.org/10.1364/JOCN.10.00A250
14. 6G Channel – White Paper on 6G Networking. https://www.6gchannel.com/portfolio-posts/6g-white-paper-networking/. Accessed 25 July 2020
15. B.A. Yosuf, M. Musa, T. Elgorashi, J.M.H. Elmirghani, Impact of distributed processing on power consumption for IoT based surveillance applications. Int. Conf. Transp. Opt. Netw. **2019**, 1–5 (2019). https://doi.org/10.1109/ICTON.2019.8840023
16. B.A. Yosuf, M. Musa, T. Elgorashi, J. Elmirghani, Energy efficient distributed processing for IoT. arXiv Prepr. arXiv2001.02974 (2020)
17. H.M.M. Ali, T.E.H. El-Gorashi, A.Q. Lawey, J.M.H. Elmirghani, Future energy efficient data centers with disaggregated servers. J. Light. Technol. **35**(24), 5361–5380 (Dec. 2017). https://doi.org/10.1109/JLT.2017.2767574
18. X. Dong, T. El-Gorashi, J.M.H. Elmirghani, Green IP over WDM networks with data centers. J. Light. Technol. **29**(12), 1861–1880 (2011). https://doi.org/10.1109/JLT.2011.2148093
19. N.I. Osman, T. El-Gorashi, L. Krug, J.M.H. Elmirghani, Energy-efficient future high-definition TV. J. Light. Technol. **32**(13), 2364–2381 (2014). https://doi.org/10.1109/JLT.2014.2324634
20. A.Q. Lawey, T.E.H. El-Gorashi, J.M.H. Elmirghani, BitTorrent content distribution in optical networks. J. Light. Technol. **32**(21), 4209–4225 (2014). https://doi.org/10.1109/JLT.2014.2351074
21. A.Q. Lawey, T.E.H. El-Gorashi, J.M.H. Elmirghani, Distributed energy efficient clouds over core networks. J. Light. Technol. **32**(7), 1261–1281 (2014). https://doi.org/10.1109/JLT.2014.2301450
22. A.M. Al-Salim, A.Q. Lawey, T.E.H. El-Gorashi, J.M.H. Elmirghani, Energy efficient big data networks: Impact of volume and variety. IEEE Trans. Netw. Serv. Manag. **15**(1), 458–474 (Mar. 2018). https://doi.org/10.1109/TNSM.2017.2787624
23. A.M. Al-Salim, T.E.H. El-Gorashi, A.Q. Lawey, J.M.H. Elmirghani, Greening big data networks: Velocity impact. IET Optoelectron. **12**(3), 126–135 (Jun. 2018). https://doi.org/10.1049/iet-opt.2016.0165
24. M.S. Hadi, A.Q. Lawey, T.E.H. El-Gorashi, J.M.H. Elmirghani, Patient-centric cellular networks optimization using big data analytics. IEEE Access **7**, 49279–49296 (2019). https://doi.org/10.1109/ACCESS.2019.2910224

25. M.S. Hadi, A.Q. Lawey, T.E.H. El-Gorashi, J.M.H. Elmirghani, Big data analytics for wireless and wired network design: A survey. Comput. Netw. **132**, 180–199 (2018). https://doi.org/10.1016/j.comnet.2018.01.016

26. M. Musa, T. Elgorashi, J. Elmirghani, Bounds for energy-efficient survivable IP over WDM-networks with network coding. J. Opt. Commun. Netw. **10**(5), 471–481 (2018). https://doi.org/10.1364/JOCN.10.000471

27. M. Musa, T. Elgorashi, J. Elmirghani, Energy efficient survivable IP-over-WDM networks with network coding. J. Opt. Commun. Netw. **9**(3), 207–217 (2017). https://doi.org/10.1364/JOCN.9.000207

28. M.O.I. Musa, T.E.H. El-Gorashi, J.M.H. Elmirghani, Bounds on GreenTouch GreenMeter network energy efficiency. J. Light. Technol. **36**(23), 5395–5405 (2018). https://doi.org/10.1109/JLT.2018.2871602

29. X. Dong, T.E.H. El-Gorashi, J.M.H. Elmirghani, On the energy efficiency of physical topology design for IP over WDM networks. J. Light. Technol. **30**(12), 1931–1942 (2012). https://doi.org/10.1109/JLT.2012.2186557

30. B.G. Bathula, M. Alresheedi, J.M.H. Elmirghani, in *Energy Efficient Architectures for Optical Networks*. Proceedings of IEEE London Communications Symposium, London, September (2009)

31. B.G. Bathula, J.M.H. Elmirghani, in *Energy Efficient Optical Burst Switched (OBS) Networks*. 2009 IEEE Globecom Workshops, Gc Workshops (2009). https://doi.org/10.1109/GLOCOMW.2009.5360734

32. T.E.H. El-Gorashi, X. Dong, J.M.H. Elmirghani, Green optical orthogonal frequency-division multiplexing networks. IET Optoelectron. **8**(3), 137–148 (2014). https://doi.org/10.1049/iet-opt.2013.0046

33. X. Dong, T. El-Gorashi, J.M.H. Elmirghani, IP over WDM networks employing renewable energy sources. J. Light. Technol. **29**(1), 3–14 (2011). https://doi.org/10.1109/JLT.2010.2086434

34. X. Dong, A. Lawey, T.E.H. El-Gorashi, J.M.H. Elmirghani, in *Energy-Efficient Core Networks*. 2012 16th International Conference on Optical Networking Design and Modelling, ONDM 2012 (2012). https://doi.org/10.1109/ONDM.2012.6210196

35. L. Nonde, T.E.H. El-Gorashi, J.M.H. Elmirghani, Energy efficient virtual network embedding for cloud networks. J. Light. Technol. **33**(9), 1828–1849 (2015). https://doi.org/10.1109/JLT.2014.2380777

36. A.N. Al-Quzweeni, A.Q. Lawey, T.E.H. Elgorashi, J.M.H. Elmirghani, Optimized energy aware 5G network function virtualization. IEEE Access **7**, 44939–44958 (2019). https://doi.org/10.1109/ACCESS.2019.2907798

37. Z.T. Al-Azez, A.Q. Lawey, T.E.H. El-Gorashi, J.M.H. Elmirghani, Energy efficient IoT virtualization framework with peer to peer networking and processing. IEEE Access **7**, 50697–50709 (2019). https://doi.org/10.1109/ACCESS.2019.2911117

38. O. Aletri, A. Alahmadi, S.O.M. Saeed, S.H. Mohamed, T.E.H. El-Gorashi, M.T. Alresheedi, J.M.H. Elmirghani, Optimum resource allocation in optical wireless systems with energy-efficient fog and cloud architectures. R. Soc. Open Sci., 1–34 (2019). https://doi.org/10.1098/rsta.2019.0188

39. R. Deng, R. Lu, C. Lai, T.H. Luan, Towards power consumption-delay tradeoff by workload allocation in cloud-fog computing. IEEE Int. Conf. Commun., 3909–3914 (2015, 2015). https://doi.org/10.1109/ICC.2015.7248934

40. F. Jalali, K. Hinton, R. Ayre, T. Alpcan, R.S. Tucker, Fog computing may help to save energy in cloud computing. IEEE J. Sel. Areas Commun. **34**(5), 1728–1739 (2016). https://doi.org/10.1109/JSAC.2016.2545559

41. B.J. Baliga, R.W.A. Ayre, K. Hinton, R.S. Tucker, F. Ieee, *Green Cloud Computing: Balancing Energy in Processing , Storage , and Transport* (ScienceOpen, Inc., Burlington, MA, 2011)

42. A.A. Alahmadi, T. El-gorashi, J. Elmirghani, Energy efficient processing allocation in opportunistic cloud-fog-vehicular edge cloud architectures. arXiv:2006.14659 [cs.NI], June, (2020)

43. M. Radivojević, P. Matavulj, M. Radivojević, P. Matavulj, PON evolution. Emerg. WDM EPON, 67–99 (2017). https://doi.org/10.1007/978-3-319-54224-9_3
44. M. Taheri, N. Ansari, A feasible solution to provide cloud computing over optical networks. IEEE Netw. **27**(6), 31–35 (2013). https://doi.org/10.1109/MNET.2013.6678924
45. F. Jalali, S. Khodadustan, C. Gray, K. Hinton, F. Suits, in *Greening IoT with Fog: A Survey*. Proceedings – 2017 IEEE 1st International Conference on Edge Computing, EDGE 2017, September, pp. 25–31 (2017). https://doi.org/10.1109/IEEE.EDGE.2017.13
46. W. Yu et al., A survey on the edge computing for the internet of things. IEEE Access **6**, 6900–6919 (2017). https://doi.org/10.1109/ACCESS.2017.2778504
47. B.A. Yosuf, M.O.I. Musa, T.E.H. El-Gorashi, J.M.H. Elmirghani, Energy efficient distributed processing for IoT. arXiv:2001.02974 [cs.NI] (2020)
48. P. Chołda, P. Jaglarz, Optimization/simulation-based risk mitigation in resilient green communication networks. J. Netw. Comput. Appl. **59**, 134–157 (2016). https://doi.org/10.1016/J.JNCA.2015.07.009
49. D. Meisner, B.T. Gold, T.F. Wenisch, PowerNap: Eliminating server idle power
50. Cisco visual networking index: Forecast and trends, 2017–2022 White Paper – Cisco. https://www.cisco.com/c/en/us/solutions/collateral/service-provider/visual-networking-index-vni/white-paper-c11-741490.html. Accessed 26 Oct 2019
51. G. Shen, R.S. Tucker, Energy-minimized design for IP over WDM networks. IEEE/OSA J. Opt. Commun. Netw. **1**(1), 176–186 (2009)
52. C. Delgado, J.R. Gállego, M. Canales, J. Ortín, S. Bousnina, M. Cesana, On optimal resource allocation in virtual sensor networks. Ad Hoc Netw. **50**, 23–40 (2016). https://doi.org/10.1016/j.adhoc.2016.04.004
53. M. Kašpar, Bandwidth calculator. https://www.cctvcalculator.net/en/calculations/bandwidth-calculator/. Accessed 20 Mar 2019
54. Thepihut, USB Wifi adapter for the Raspberry Pi. https://thepihut.com/products/raspberry-pi-zero-w. Accessed 21 Mar 2019
55. RasPi.TV, How much power does Pi Zero W use? http://raspi.tv/2017/how-much-power-does-pi-zero-w-use. Accessed 21 Mar 2019
56. Cisco ME 4600 series optical network terminal data sheet – Cisco. https://www.cisco.com/c/en/us/products/collateral/switches/me-4600-series-multiservice-optical-access-platform/datasheet-c78-730446.html. Accessed 16 Mar 2018
57. Cisco ME 4600 series optical line terminal data sheet – Cisco. https://www.cisco.com/c/en/us/products/collateral/switches/me-4600-series-multiservice-optical-access-platform/datasheet-c78-730445.html. Accessed 20 Nov 2019
58. Cisco network convergence system 5500 series modular chassis data sheet – Cisco. https://www.cisco.com/c/en/us/products/collateral/routers/network-convergence-system-5500-series/datasheet-c78-736270.html. Accessed 26 Oct 2019
59. Cisco Nexus 9300-FX series switches data sheet – Cisco. https://www.cisco.com/c/en/us/products/collateral/switches/nexus-9000-series-switches/datasheet-c78-742284.html. Accessed 26 Oct 2019
60. Intel® Xeon® Processor E5-2680 (20M Cache, 2.70 GHz, 8.00 GT/s Intel® QPI) Product Specifications. https://ark.intel.com/content/www/us/en/ark/products/64583/intel-xeon-processor-e5-2680-20m-cache-2-70-ghz-8-00-gt-s-intel-qpi.html. Accessed 26 Oct 2019
61. Intel® Xeon® Processor X5675 (12M Cache, 3.06 GHz, 6.40 GT/s Intel® QPI) Product Specifications. https://ark.intel.com/content/www/us/en/ark/products/52577/intel-xeon-processor-x5675-12m-cache-3-06-ghz-6-40-gt-s-intel-qpi.html. Accessed 26 Oct 2019
62. Intel® Xeon® Processor E5–2420 (15M Cache, 1.90 GHz, 7.20 GT/s Intel® QPI) Product Specifications. https://ark.intel.com/content/www/us/en/ark/products/64617/intel-xeon-processor-e5-2420-15m-cache-1-90-ghz-7-20-gt-s-intel-qpi.html. Accessed 26 Oct 2019
63. FAQs – Raspberry Pi Documentation. https://www.raspberrypi.org/documentation/faqs/. Accessed 26 Oct 2019

64. Raspberry Pi 3: Specs, benchmarks & testing—The MagPi magazine. https://magpi.raspberrypi.org/articles/raspberry-pi-3-specs-benchmarks. Accessed 26 Oct 2019

65. Raspberry Pi Zero W (Wireless) | The Pi Hut. https://thepihut.com/products/raspberry-pi-zero-w. Accessed 26 Oct 2019

66. Cisco Industrial Benchmark (2016). https://www.cisco.com/c/dam/global/da_dk/assets/docs/presentations/vBootcamp_Performance_Benchmark.pdf. Accessed 16 Mar 2018

67. America – Google Maps. https://www.google.com/maps/search/america/@42.593777,-113.8013893,5z. Accessed 24 Dec 2019

68. A. Shehabi et al., *United States Data Center Energy Usage Report*. Lawrence Berkeley Natl. Lab. Berkeley, CA, Tech. Rep., No. June (2016), pp. 1–66

69. C. Gray, R. Ayre, K. Hinton, R.S. Tucker, in *Power Consumption of IoT Access Network Technologies*. 2015 IEEE Int. Conf. Commun. Work., pp. 2818–2823 (2015), https://doi.org/10.1109/ICCW.2015.7247606

70. S. Banerji, R.S. Chowdhury, On IEEE 802.11: Wireless LAN technology. Orig. Publ. Int. J. Mob. Netw. Commun. Telemat. **3**(4) (2013). https://doi.org/10.5121/ijmnct.2013.3405

71. Data Communications and Computer Networks: A Business User's Approach – Curt White – Google Books. https://books.google.co.uk/books/about/Data_Communications_and_Computer_Network.html?id=FjV-BAAAQBAJ&redir_esc=y. Accessed 05 Aug 2020

72. G. Kramer, B. Mukherjee, G. Pesavento, Ethernet PON (ePON): Design and analysis of an optical access network (2000)

73. P.P. Iannone, et al., in *A 160-km Transparent Metro WDM Ring Network Featuring Cascaded Erbium-Doped Waveguide Amplifiers*. OFC 2001. Optical Fiber Communication Conference and Exhibit. Technical Digest Postconference Edition (IEEE Cat. 01CH37171), vol. 3, pp. WBB3-W1-3 (2001) https://doi.org/10.1109/OFC.2001.928443

Chapter 15
Towards a Fully Virtualized, Cloudified, and Slicing-Aware RAN for 6G Mobile Networks

Mohammad Asif Habibi, Bin Han, Meysam Nasimi, Nandish P. Kuruvatti, Amina Fellan, and Hans D. Schotten

Abstract The fifth generation (5G) system was marketed as a true enabler of latency-aware ultra-reliable low latency communications (uRLLC), bandwidth-devouring enhanced mobile broadband (eMBB), and unlimited-things-centric massive machine type communication (mMTC) services for vertical industries and internet-of-everything (IoE). Among other promises, the eMBB services have gained significant momentum in various 5G deployments, such as using millimeter wave (mmWave) frequency band in the next-generation radio access network (NG-RAN). The early rollouts of the uRLLC and mMTC services show that the 5G system supports their basic features, however, supporting advanced and complicated use-cases of such services remains uncertain. These drawbacks are due to, but not limited to, the existing limitations in the adaptation of current technologies, such as multi-access cloud computing, network function virtualization, software-defined networking (SDN), network slicing, etc. in NG-RAN. In this regard, we aim to (i) provide a comprehensive overview on key concepts towards the RAN of sixth generation (6G) system; (ii) study various legacy RAN implementations, reported in literature, from different perspectives, and the motivation of their reconstruction and redesign with respect to the requirements of their forth-coming generations; (iii) present an exclusive review of the NG-RAN with special emphasize on the cloudification and virtualization of resources and services, and on the management and orchestration of RAN slices; (iv) identity the key drivers that leverage advanced mMTC, uRLLC, and eMBB services and applications; and (v) address key challenges and future research arising from the deployment of NG-RAN.

M. A. Habibi (✉) · B. Han · M. Nasimi · N. P. Kuruvatti · A. Fellan
Technische Universität Kaiserslautern, Kaiserslautern, Germany
e-mail: asif@eit.uni-kl.de; binhan@eit.uni-kl.de; nasimi@eit.uni-kl.de; Kuruvatti@eit.uni-kl.de; fellan@eit.uni-kl.de

H. D. Schotten
Technische Universität Kaiserslautern, Kaiserslautern, Germany
German Research Center for Artificial Intelligence (DFKI GmbH), Kaiserslautern, Germany
e-mail: schotten@eit.uni-kl.de; hans_dieter.schotten@dfki.de

Y. Wu et al. (eds.), *6G Mobile Wireless Networks*, Computer Communications and Networks, https://doi.org/10.1007/978-3-030-72777-2_15

1 Background on Key Concepts Towards RAN Architecture in 6G Wireless Network

Different visions about the 6G have been proposed since 2019 [6, 15, 28, 62, 66], generally sharing the same ambition of fulfilling even more extreme performance requirements than fifth generation (5G), e.g., higher throughput, larger capacity, improved reliability, lower latency, and reduced power consumption [25]. Yet, it is highly challenging, or even physically infeasible, to fulfill these extreme performance requirements simultaneously in a one-size-fits-all radio access network, nevertheless, there is no use-case scenario that requires to do so. Instead, the network shall be flexibly specialized with different specifications, each satisfying a particular use-case with individual performance requirements. Such approach of use-case driven network orchestration has been enabled by network slicing and was introduced in 5G. Now this approach shall be pushed to the next level in the future 6G era.

Existing studies on the outlook of 6G have also suggested numerous types of radio access technologies (RATs) to pave the way towards 6G. Most of them are counting machine learning (ML), artificial intelligence (AI), new spectrum covering terahertz (THz), visible light communication (VLC), and intelligent reflecting surfaces (IRS) as the most characterizing and essential 6G-enabling RATs. More specifically, the broadband new spectrum at higher frequencies is required to extend the channel capacity, so as to achieve the targeted radio performance. Since the radio channel at such high frequencies generically suffers from high propagation loss and blockage, 6G has to rely on new beamforming technologies such as ultra massive multiple input multiple output (MIMO) and IRS to maintain a reasonable channel budget [63, 67].

While ultra massive MIMO and IRS are promising to enable the exploitation of 6G's new spectrum, advanced cognitive online solutions are needed to adapt the antenna arrays and programmable surfaces, which is computationally challenging. In addition to that, the resource management in virtualized 6G network also requires a huge computational effort to achieve a satisfying resource efficiency. For solving complex optimization problems, ML and AI have been widely recognized as the most (if not the only) promising approaches. The future 6G network, therefore, is universally believed to be AI-empowered [36, 57].

Currently, most ML/AI solutions for networking systems have been deployed in centralized cloud servers, where data obtained all over the network are aggregated and processed. However, such a topology implies a long training period, and a slow reaction to the dynamics of the local environment, which may lead to a violation of the latency/tolerance constraints of future 6G applications. Additionally, privacy concerns may also be raised by the aggregated learning process. To cope with these issues, 6G is supposed to sink and distribute the AI algorithms to its edge, aiming at a fully decentralized edge intelligence with high security and capability of real-time learning [55, 68].

Along with the history of wireless networking so far, associated with the birth of every generation of wireless networks, there has been an architectural evolution in the RAN domain, to support the deployment of corresponding enabling technologies. More specifically: it was since the second generation (2G) that digital radio signals are used instead of analog signals; the third generation (3G) introduced the hierarchical cell structure that has been used till today; the fourth generation (4G) completely abandoned circuit switching technologies to become a pure IP-based system; the 5G leverages software-defined networking (SDN) and network function virtualization (NFV) technologies to organize the network infrastructure—including both the core network and the RAN—in a virtualized and cloudified fashion. Now, gazing from afar into the horizon of 2030, the community is amused with a curiosity about the roadmap towards 6G RAN. It is the mainstream opinion that 6G will also evolve the current RAN architecture as its predecessors have done. Some proposals can sound radical, e.g., the conception of a revolutionary serverless cellular network [23]. On the other hand, there have also been alternative or opposite points of view such as [20], which advocates that the latest 5G network architecture, with potential mild extensions in the future, shall be sufficient to fulfill the increasing wireless service demand over the next decade.

Despite the lack of definition of technological advancements, researchers in both academia and industry are working on various aspects of the 6G system. The current work focuses more on the requirements and emerging technologies. The requirements may be identified, once the key enabling technologies are defined and accepted by the standardization community. Among others, one of the key issues is to define the requirements, key enabling technologies, and architecture for 6G RAN. This study is thus the first attempt to review the current and legacy cloudified RAN architectures, discuss their capability to support future 6G technologies and the potentially upcoming 6G RAN evolution.

The remainder of this chapter is structured as follows: In Sect. 2 we review some legacy cloudified RAN architectures, including the C-RAN, the heterogeneous cloud RAN (H-CRAN), the V-CRAN with 5G virtualization technologies, and the fog RAN (F-RAN) with fog computing. In Sect. 3, the latest standardized 5G RAN, namely NG-RAN, is described. We then briefly introduce some of the most important 6G enabling technologies in Sect. 4, while discussing the impacts they may have on the future 6G RAN architecture, and the research problems arising from their implementation. Before closing this chapter, we provide our conclusions in Sect. 5.

2 Legacy RAN Architectures

2.1 C-RAN

In a 2G mobile system, the RAN architecture comprises of entire radio and base-band processing functions incorporated into the base-station (BS). On the other hand, 3G and 4G mobile networks deploy a distributed RAN (D-RAN) which splits radio and base-band processing functions into two separate nodes, namely the radio resource head (RRH) and the base-band unit (BBU). Due to the increase in user-data volume with diverse quality of service (QoS) requirements, network operators are obligated to satisfy these demands employing centralization and cloudification of BBUs and their corresponding RRHs. Such a centralized and cloudified RAN with a centralized BBU pool, where network resources are pooled, is called a C-RAN [11, 31].

The initial proposal for a C-RAN was done by IBM, which was called a wireless network cloud (WNC) [40]. Further descriptions and details were added by China Mobile Research Institute (CMRI) in a white paper [11]. Figure 15.1 depicts the key approach of the C-RAN, which is to segregate all BBUs from their corresponding

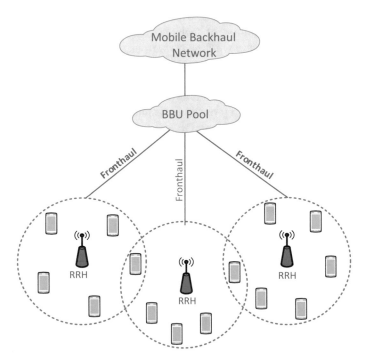

Fig. 15.1 The C-RAN architecture. The figure illustrates that RRHs and BBU are separated, however, the RRHs are connected to a shared and centralized baseband processing unit in a virtualized BBU pool through a reliable fronthaul transportation link

RRHs and to pool them into a centralized, cloudified, shared, and virtualized BBU pool. Each RRH unit is connected to its respective BBU pool via a fronthaul link. Up to tens of RRHs could be supported by each BBU pool, which is connected to the core network through a backhaul link. The C-RAN architecture leads to reduced capital expenditure (CAPEX) and operating expenditure (OPEX) of the mobile network operator, lower energy consumption, simplified network management and maintenance, increased network scalability, improved spectral efficiency, higher network throughput, and easier load balancing.

The C-RAN incorporates cloud computing into the 5G RAN architecture [11] based on two key principles; centralization and virtualization of baseband processing. The main goals of centralization are to enhance network performance, reduce energy consumption, and improve spectral efficiency. Whereas the target of virtualization of baseband processing is to reduce the CAPEX and OPEX of 5G mobile networks.

2.1.1 Types of C-RAN

C-RAN can be classified into two types based on the functions' split between RRH and BBU, namely fully centralized C-RAN and partially centralized C-RAN.

- In case of a fully centralized C-RAN, each function in Layer 1 (e.g., modulation, sampling, quantization, antenna, and resource block mapping), Layer 2 (e.g., transport-media access control) and Layer 3 (e.g., radio resource control) is present in the BBU (as depicted in Fig. 15.2). Few major improvements brought to the 5G mobile network due to the fully centralized C-RAN are a simpler expansion of network coverage, easier upgrade of network capacity, assistance to multi-standard operation, improved network resource sharing, and reinforced multi-cell collaborative signal processing. Even with the aforementioned advantages of the fully centralized C-RAN, there exists two key issues: the necessity of high bandwidth, and the transmission of baseband I/Q signals between RRH and BBU [11]. The use of a fully centralized C-RAN solution along with open platforms helps in the development and implementation of software-defined radio (SDR), leading to an upgrade of the air interface through software. This facilitates the convenient upgrading of RAN architecture while supporting a multi-standard operation.
- In the case of a partially centralized C-RAN, radio and baseband processing functions are present in the RRH, and the rest of the higher layers' functions are kept in the BBU (depicted in Fig. 15.3). Thus, it is made clear that all Layer 1 related functions are present in RRH, whereas BBU contains functions related to Layer 2 and Layer 3. The transmission bandwidth required between RRH and BBU is small in a partially centralized C-RAN, due to the shift of baseband processing from BBU to RRH. Nonetheless, it poses a problem of lower flexibility in network upgrades and makes multi-cell collaborative signal processing less convenient.

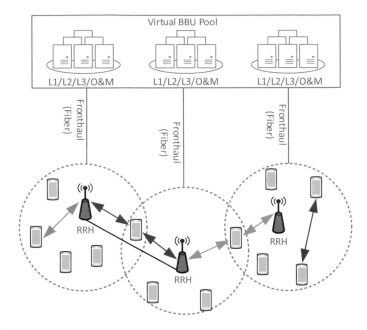

Fig. 15.2 The fully centralized C-RAN architectural framework. In this solution, all Layer 1, Layer 2, and Layer 3 functions are located in the BBU

To fulfill the diverse requirements of 5G, both fully and partially centralized C-RANs have been inspected and developed from several perspectives. Both types of C-RAN deployments rely on network characteristics. In either case, if the service provider aims to extend the coverage of the network or to split a cell with a motivation of increasing the capacity, it becomes simpler to deploy new RRHs and to eventually connect them with the BBU pool. On the other hand, if an increase in network load is detected by the service provider, then it simply requires a hardware upgrade in the BBU pool to enhance the processing capacity. It is important to note that the static coupling between RRH and BBU is relaxed with the deployment of C-RAN. The RRH is not anymore rigidly coupled with a dedicated physical BBU. Rather, a real-time virtualization technology enables each RRH to be connected to a virtual BS present in the BBU pool.

2.2 H-CRAN

2.2.1 Heterogeneous Network

When compared to the RAN architectures of long-term evolution (LTE)/LTE advanced (LTE-A) networks, the RAN architecture of the 5G mobile network is more heterogeneous. The authors in [21] anticipate the density of BSs in 5G RAN

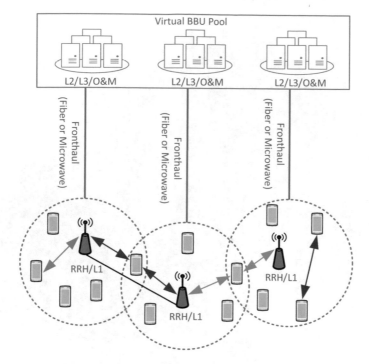

Fig. 15.3 The partially centralized C-RAN architectural framework. In this solution, Layer 1 functions are integrated into RRH, whereas Layer 2 and Layer 3 functions are located in BBU

to become as high as 40–50 BSs/Km². Furthermore, cellular networks are steadily experiencing increased data traffic demand. Thus, enhancements in system capacity and spectral efficiency are essential to satiate the end-user demands beyond 2020. One of the feasible ways to meet this requirement is through the deployment of small cells on the top of the underlying macrocellular layout leading to a so-called heterogeneous cellular layout. This idea is also supplemented by the well-established hypothesis of Shannon in 15.1. As it could be seen, N denotes the set of all heterogeneous RATs, and I_n indicates the set of all available subchannels in RAT n, total capacity of a system is approximately C_{sum}, where the terms B_i, S_i and N_i represent the bandwidth, the signal power, and the noise power in channel i, respectively. It is evident that C_{sum} is equivalent to the aggregate capacity of all subchannels. In order to increase C_{sum}, we need to deploy more macrocells and small cells, leading to a heterogeneous cellular environment (see Fig. 15.4). The macrocells in a heterogeneous cellular layout offer extensive coverage and seamless mobility over a larger area, while small cells supplement coverage and deliver higher capacity by transferring computation and communication nodes near to the end-users.

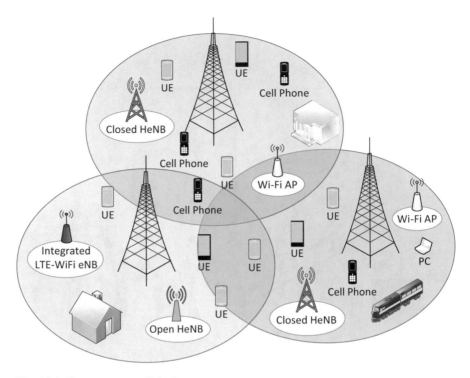

Fig. 15.4 Heterogeneous cellular layout

$$C_{\text{sum}} \approx \sum_{n \in \mathcal{N}} \sum_{i \in \mathcal{I}_n} B_i \log_2 \left(1 + \frac{S_i}{N_i}\right) \qquad (15.1)$$

A wide variety of cell types, e.g., femtocell, picocell, and microcell are included in the class of small cells or low-power nodes [22]. Such small cells are installed in diverse environments, such as homes, enterprise environments, hot spots, shopping malls, stadiums, train stations, and other smaller geographical areas, aiming to increase the overall network capacity and coverage, decrease network cost, and enhance spectral efficiency. Figure 15.4 shows several types of small cells that are deployed vigorously to tackle the demands of capacity and coverage in future 5G networks.

Installation of small cells is possible in both indoor and outdoor environments. The small cells which are deployed outdoor need low power (ranging from 250 mW to approximately 2 W). The cost of small cells is also lesser compared to the conventional macro BSs, which also require more power (ranging from 5 to 40 W) [14]. The small cells that are installed indoor require even lesser power (100 mW or less). Furthermore, the small cell has a significantly lesser coverage area than the macrocells as a result of lower transmit power, which can also restrict the data volume [51].

During the configuration of Indoor femtocells, access restrictions could be set flexibly. A femtocell can be configured as a closed femtocell (closed home enabled Node B (HeNB)) which allows access only to particular users. On the contrary, a femtocell can be configured as an open femtocell (open HeNB), allowing access to any user located in its coverage. Furthermore, another configuration called a hybrid femtocell is possible, where access is allowed to unsubscribed users but with an upper bound on accessible resources. The closed HeNB configures its restriction by maintaining a closed subscriber group (CSG) [14].

In the literature, extensive studies exist regarding the small cells and heterogeneous cellular networks. Recently, Cambridge University Press published a comprehensive book giving insight into small cells in the 5G mobile network [5]. This source discusses all the major features of small cells including their design, deployment, and optimization. It covers basic as well as advanced concepts related to small cells. Besides, it details emerging trends, research challenges, performance analysis, deployment strategies, standardization activities, environmental concerns along with energy efficiency, resource and mobility management in small cell heterogeneous networks. Reference [10], discusses the design and deployment of small cells along with respective technical challenges. The authors study key technical attributes namely; mobility management, interference management, coverage and capacity optimization, energy efficiency, backhaul, deployment planning, frequency assignment and access methods, and heterogeneous network management.

There exists a wide variety of studies on femtocells deployment. The majority of these works, discuss administration, operations, management, access, interference management, local internet protocol access (LIPA), and architecture [7, 44, 54, 64]. One of the major driving forces behind the deployment of a femtocell is its low OPEX/CAPEX. Reference [9] considers a blend of open HeNB (random deployment by end-users) and macro eNB (planned deployment by an operator) in a dense urban area, which shows only 70% of the total annual network cost when compared to a pure macrocellular layout. The works in [32, 52] evaluate the performance of heterogeneous network comprised of macro eNB, pico eNB, and open HeNB. Both studies conclude that limitations in the performance of the heterogeneous network are mainly due to limited coverage of small cells.

2.2.2 H-CRAN Architecture

A new RAN architecture has been recently proposed and is called the H-CRAN. The key goal of H-CRAN is to decouple the control and user planes. This is done to improve the functionalities and performance of the C-RAN architecture by implementing control plane functions only in the macro BSs. H-CRAN combines the advantages from C-RAN as well as heterogeneous networks leading to higher spectral efficiency, lesser energy consumption, and enhanced data rates. The H-CRAN architecture comprises of two cellular layouts [37, 61], the macro BSs (high power nodes) cellular layout, and the small BSs or RRHs cellular layout. The deployment of high power nodes (HPNs) is to improve network coverage and control network signaling. On the other hand, the small cells and RRHs are used to enhance network capacity and satisfy the diverse QoS requirements of end-users. Figure 15.5 shows the system architecture of H-CRAN that consists of three important functional modules [37]:

- **Enhanced-Cloud and Real-Time Virtualized BBU Pool:** This functional module integrates all the BBUs scattered to different cells into a BBU pool. Robust virtualization methods and strong cloud-computing are employed in building the BBU pool. Additionally, a connection exists between the BBU pool and high power nodes (HPNs) to facilitate the coordination of inter-tier interfaces between HPNs and the RRHs.

- **Extremely Reliable Transport Network:** Fig. 15.5 depicts the connection of all RRHs to their respective BBUs located in the BBU pool. Low latency, high bandwidth fronthaul link (e.g., optical fiber) interconnects the RRHs and BBUs. The data and control interfaces between BBU and macrocell base-stations (macrocell base-stations (MBSs)) (HPNs) are represented by S1 and X2, respectively.

- **High Number of Macro BSs, Small BSs, and RRHs:** Several types of cells coexist with each other in H-CRAN architecture (e.g., macro base stations, small base stations, RRHs). The network control, mobility management, and performance improvement are taken care of by the macro base stations, while small cells and RRHs aim at reducing transmission power and enhancing system capacity. The symbol processing and radio frequency functionalities are integrated into the RRHs, whereas, the BBU pool is incorporated with all other baseband physical processing functionalities of upper layers. In contrast, all functionalities from the physical layer to the network layer are integrated into the high power nodes.

In H-CRAN, enhanced cloud computing enables the centralized integration of all BBUs as well as assists in the segregation of functions among RRHs and BBUs. Furthermore, the decoupled control and user planes lead to the simpler and efficient management of heterogeneous mobile networks. Thus, when there is a need for expansion of network coverage and/or improvement in system capacity, then the mobile operators are only required to deploy new RRHs close to the end-users and just connect them to the BBU pool. Furthermore, it is easy for the implementation of agile software solutions. For instance, if the network operator needs to upgrade RANs and provide support for multi-standard operations, then it could be carried out by deploying SDRs and simply updating software.

There exists two key similarities between traditional C-RAN and H-CRAN from the architectural point of view (shown in Fig. 15.5): (a) A large number of RRHs are connected to a centralized BBU pool to achieve high cooperative gain and increased energy efficiency. (b) RRHs carry out RF and simple symbol processing functionalities, whereas, all the upper layer functionalities are executed in the BBU

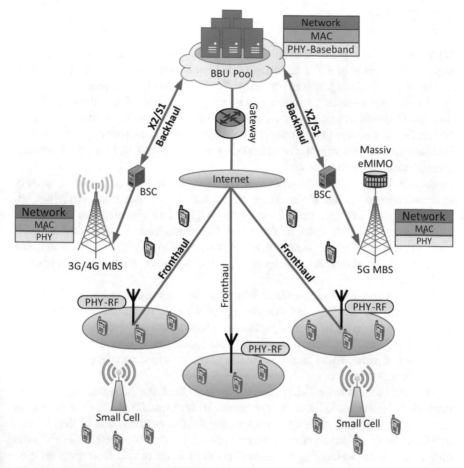

Fig. 15.5 System architecture of H-CRAN

pool. Although there are certain similarities between traditional C-RAN and H-CRAN, there still exist certain differences such as: (a) In the H-CRAN, the BBU pool and HPNs are connected via S1 and X2 interfaces (as shown in Fig. 15.5). This reduces the cross-tier interference between RRHs and HPNs that are coexisting in the same geographical area. (b) In the presence of HPNs, the demand on the fronthaul is reduced in the H-CRAN. This is because the HPNs deliver system broadcasting data and control signaling to the user equipments (user equipments (UEs)) leading to improved capacity and reduced time delay in fronthaul links. (c) In the H-CRAN, an RRH goes into a sleeping mode when it experiences a low traffic load. Thus energy efficiency is improved. In such a scenario, all the RRHs in sleeping mode are managed by the BBU pool.

2.3   V-CRAN

NFV has received great attention from both industry and academia, given the significant benefits of NFV in terms of efficient resource sharing and enhanced flexibility of scheduling. In the telecommunication networks, virtualization has been applied initially to the core network and is being further extended to the radio access domain. However, it is still in an early stage of development. Besides, the salient features of a wireless communication system such as user mobility, attenuation, interference, time-varying and broadcast channels make the task of wireless network virtualization more complicated.

Several levels of virtualization are possible in a cellular network, namely spectrum virtualization, air-interface virtualization, infrastructure virtualization, virtualization of multiple radio access technologies, and virtualization of computing resources. Virtualization in the cellular network leads to improved network performance, efficient resource utilization, reduction in CAPEX/OPEX, increased revenue, simplified migration to novel technologies, and creation of new businesses in the market.

Even though virtualization offers several advantages, three key issues still need attention: (a) An efficient and fair sharing of wireless resources and distribution among various virtual network operators needs to be considered. (b) The interferences arising due to resource utilization should be accurately considered. (c) Several technical and managerial issues need to be examined before executing virtualization in the wireless network.

To carry out end-to-end virtualization in the 5G cellular network, both the core network (CN) and RAN architecture needs to be virtualized. The virtualization concepts for the wired network can be reused for the virtualization of the CN in a mobile network. However, certain features of the wireless access network need to be considered while performing the virtualization of RAN or BS. New virtualization solutions are required compared to the ones used for wired networks due to the dynamic sets of end-users, their mobility, and varying channel conditions. Thus, the virtualization process becomes challenging for the mobile operator due to these attributes.

Due to the deployment of C-RAN, network resources are present nearer to the end-user. Thus, it is feasible to use a cloud data center to incorporate all core network-related functions and applications. To achieve lower latency and improved performance, mobile edge cloud (MEC) is being linked to the data centers by mobile operators. However, 5G mobile network is expected to fulfill diverse requirements originating from a massive number of user devices, low latency services, and demand of higher capacity from users and vertical industries. A potential way to satisfy these demands is to deploy NFV and SDN in the C-RAN, thereby virtualizing all the functions and resources as well as decoupling the data and control planes. This method of virtualization towards the access network gives rise to a new type of C-RAN known as V-CRAN.

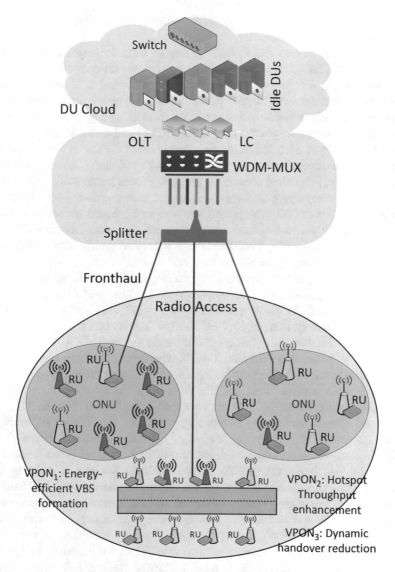

Fig. 15.6 System architecture of V-CRAN

2.3.1   System Architecture of V-CRAN

An example of V-CRAN architecture is depicted in Fig. 15.6, consisting of components such as digital unit cloud (DU cloud), time-wavelength-division multiplexed passive optical network (TWDM-PON), fronthaul, and virtualized base-station (V-BS) [59]. All the commercial servers offering baseband processing as well as layer 2 and layer 3 related functions are incorporated in the DU cloud. A layer 2 switch facilitates the interconnection among all distributed units (DUs), thereby allowing the exchange of data and signaling.

For the transfer of data and signaling between radio units (RUs) and DUs, the V-CRAN needs a fronthaul network with larger bandwidth and lower latency. One of the most feasible solutions to meet such a requirement of fronthaul is to use optical channels instead of a single fiber. The fronthaul makes use of TWDM-PON technology to allocate high-capacity optical channels among DUs. The TWDM-PON comprises a single optical line terminal (OLT) and multiple optical network units (ONUs). Figure 15.6 illustrates the connection of OLT with the DU cloud. It manages the provision of an optical transceiver and a line card (LC) to every DU for performing optical-electrical conversion and subsequent traffic delivery. Every LC associated with every single OLT is connected to a wavelength-division multiplexed multiplixer (WDM-MUX). This segregates the traffic based on wavelength. An ONU is placed at the end of each optical channel, far away from the DU cloud to expand the coverage of a TWDM-PON. It is located jointly with an RU and is provisioned with a reconfigurable transceiver.

One or more wavelengths are shared by the ONUs in the access network, leading to a virtualized passive optical network (V-PON). The V-PON is a virtualized channel that can link RUs with the DU. By using LC, each DU in the DU cloud is assigned exclusively to a single V-PON. This denotes that a single DU present in the DU cloud dynamically groups and controls the multiple RUs of a V-PON. Each cell is allocated to a dedicated V-PON whenever a V-BS experiences a high load situation (e.g., during the busy hours). Nevertheless, in the non-busy hours, the cells could be clustered to form a V-PON. As a result, there is a reduction in energy consumption as well as OPEX, and network resource utilization is efficient.

A novel idea of virtualizing computing resources of a BS in V-CRAN was recently proposed [59]. The virtualization of a BS is carried out at two discrete levels: a) hardware level (dedicated spectrum). b) flow level (shared spectrum) [12]. At the hardware level, a V-BS shares the radio equipment, while multiple protocol stacks of a BS run in the form of software. The standardization of hardware virtualization is completed and conventional mobile operators are already using this method to reduce OPEX and enhance energy efficiency. In the context of spectrum sharing-based models, virtualization at higher levels (e.g., flow level at the V-BS) is required for efficient multiplexing of the resources. Also, the spectrum sharing-based models assist in deployment scenarios, where mobile virtual network operators do not own the spectrum.

2.4   Fog-RAN

It is estimated that the data volume, diversity, and speed in the internet of things (IoT) and networks of the future will be increasing at an unprecedented rate. Furthermore, projections by the International Data Corporation (IDC) show that by the year 2020, the number of sensor-enabled objects connected to the Internet will be around 30 billion, count of connected cars will be approximately 110 million (equipped with 5.5 billion sensors), and the number of connected homes will be 1.2 million-amounting to the usage of roughly 200 million sensors [30]. One of the most suitable solutions to tackle these projections is to use cloud computing, which enables operations such as processing, storing, managing, and analyzing the data originating from the IoT devices, on remote servers present on the Internet. This supports the companies to avoid expenditure towards design, deployment, and maintenance of network infrastructure [53].

There are several limitations in the contemporary cloud computing methods namely, high end-to-end latency, significant communication charges, traffic overloading, and the need for bulk data processing. Hence, a new computing framework known as fog computing [48] was presented to process, store, manage, analyze, and act on network data. Cisco coined the term "Fog Computing" [49] based on the interpretation that "fog is a cloud close to the ground" and similarly, fog computing extends cloud computing to the edge of the network. Fog computing moves the tasks of communication, computation, control, and decision making towards the edge of the network. This provisions analysis of latency-critical data near to the IoT device, quick operations on data (in the span of milliseconds) and transfer of specific data to the data center in the cloud for statistical analysis or long term storage.

Fog computing enables a massive amount of storage, processing, control, communication, configuration, measurement, and management functions to be allocated at the edge of the mobile network [50]. In fog computing, the BBU pool can carry out the collaboration radio signal processing (CRSP), which can also be executed at the RRH or even further at the UE level (e.g., wearable smart UEs). To effectively support and integrate new types of UEs, the on-device processing and cooperative radio resource management (CRRM) with less distributed storing needs to be exploited. However, from the perspective of a mobile application, a UE doesn't need to be connected to the BBU pool for downloading the packets, as long as they are available locally and stored in the nearest RRHs.

After taking all the benefits offered by fog computing into account and addressing some critical issues of the H-CRAN, a new RAN architecture relying on fog computing known as F-RAN is proposed. Here, a computation task is split into the fog computing part and the cloud computing part [38]. This section deals with the physical architecture and literature review of the F-RAN. Therefore, we omit comprehensive discussions about applications, advantages, limitations, and system architecture of fog computing. Nevertheless, we refer the readers to [48–50, 53], in case they want to explore different aspects of fog computing and study comparisons between fog and cloud computing.

The F-RAN was devised to combine benefits from fog computing as well as C-RAN, which becomes instrumental in delivering better QoS to the end-users and handling extreme traffic demands. There are two types of F-RAN, namely the distributed F-RAN and the centralized F-RAN [38]. In a distributed F-RAN, certain functions like resource management, computation, and storage are moved from the BBU to the RRHs and user devices. However, the centralized F-RAN uses SDN and NFV to allow centralized control plan, simpler management, and resource allocation.

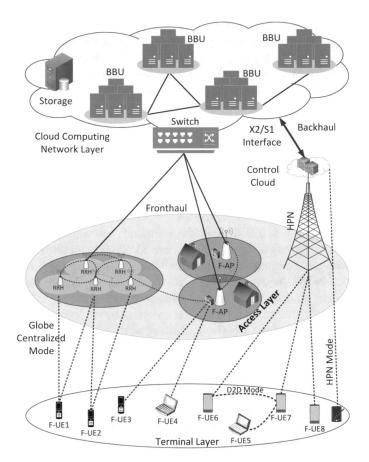

Fig. 15.7 System architecture of F-RAN

2.4.1 System Architecture

Figure 15.7 shows an exemplary system architecture of F-RAN, consisting of a terminal layer, a network access layer, and a cloud computing layer [56].

- The fog user equipments (F-UEs) belonging to the terminal layer along with the fog access points (F-APs) in the network access layer form the mobile fog computing layer. The F-UEs in the terminal layer access the HPN for obtaining system signaling related information. Additionally, F-UEs that are in proximity to each other can communicate using device-to-device (D2D) communication. The terminal layer of Fig. 15.7 illustrates an instance of such F-UE-based relay mode, where F-UE7 acts as a mobile relay and facilitates the communication between F-UE5 and F-UE6. In such cases, relevant F-UEs (in this instance F-UE5 and F-UE6) can transfer data between each other without HPN.

- The network access layer comprises of HPNs and F-APs. The task of HPNs is to distribute signaling related system information to every F-UE in its coverage, whereas the F-APs task is to process the data from F-UEs and pass it on. F-APs and HPNs are connected to the BBU pool in the cloud computing layer using fronthaul and backhaul links, respectively.
- The BBU pool in the cloud computing layer is similar to that of H-CRAN. Furthermore, the centralized chasing could also be found in this layer.

3 Next-Generation RAN

The NG-RAN is the latest RAN architecture specified by the Third Generation Partnership Project (3GPP) in Release 15 in conjunction with the 5G mobile communication network. It introduces new interfaces, functional components, functional split options, terminologies, etc. in the 5G network. The NG-RAN has been further developed in many aspects and its latest specifications have been released by the 3GPP in Release 16 [1]. The NG-RAN supports the provisioning of 5G services through the new radio (NR) air interface to the UE of the 5G network. Besides, the NG-RAN also supports the evolved universal terrestrial radio access network (E-UTRAN) services and provides it to the UE using LTE-A services.

Most of the functionalities of the NG-RAN in comparison with the previous RAN architectures are similar. However, there are some major differences between the functionalities of their corresponding components. For example, the NG-RAN should be slice-aware in order to provide differentiated QoS requirements to the enhanced mobile broadband (eMBB), ultra-reliable low latency communication (URLLC), and massive machine-type communication (mMTC) service types UEs. Since such a shift in service type was not introduced in previous RAN architectures, the functionalities and QoS related aspects should be designed in an efficient away in the NG-RAN components.

In NG-RAN, some functions and components are virtualized in four levels, namely the application level, the cloud level, the spectrum level, and the cooperation level. The main objective of virtualization in the NG-RAN is to map and implement the required services on the NG-RAN level in an efficient manner from the perspectives of energy, cost, radio resources, spectrum, and security; while guaranteeing the agreed service level agreement for each of the required services.

3.1 The NG-RAN Architecture

In Fig. 15.8, we illustrate the architecture of the NG-RAN [1]. As depicted, the NG-RAN consists of many next generation Node Bs (gNBs) and next generation evolved Node Bs (ng-eNBs). The gNBs and ng-eNBs are connected with the 5G core (5GC), mainly to the access and mobility management function (AMF) and

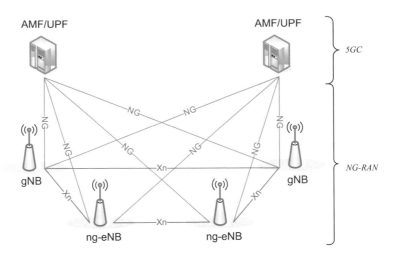

Fig. 15.8 The architectural framework of NG-RAN [1]

user-plane function (UPF), over NG interface. Besides, the gNBs and ng-eNBs are also interconnected with each other using the Xn interfaces (see Fig. 15.8).

As further illustrated in Fig. 15.10, the gNB contains three main functional blocks; namely the centralized unit (CU), the DU, and the RU. Depending on the service deployment scenarios and types, these functional modules can be deployed in multiple combinations. Each CU consists of multiple DUs. Each DU consists of multiple RUs. The CU and the DU are assumed to be fully virtualized. They are connected using the F1 interface. The RU is the antenna and radio frequency, therefore, is assumed to be a physical unit.

In NG-RAN, the gNBs and ng-eNB are interconnected among each other over Xn interface. Moreover, the CU and the DU are connected over the F1 interface. The F1 interface is responsible for the data transmission and signaling exchange of at least two endpoints. It separates the radio network and the transport network layers. Moreover, the F1 interface exchanges the UE-associated and non-UE-associated signaling. The latency and bandwidth requirements over the F1 interface are significantly important for the network to support the requested services of the UE. Therefore, the F1 and backhaul interfaces, the time-division multiplexing passive optical network (TDM-PON) with data rates over 10 Gb/s, must be sufficient to meet those requirements. The functions of the F1 interface are divided into two parts: the F1-Control Plane (F1-C) related functions and the F1-User Plane (F1-U) related functions. In the functions of a gNB (which is beyond the scope of this chapter), there are eight split options (option-1–option-8). In CU, DU, and RU these options are split into various combinations. Most commonly, option-2 is considered as a functional split between the CU and the DU, whereas option-7 is considered between the DU and the RU.

The virtualized resources (i.e., storage, computation, and communication) and physical resources (i.e., power, spectrum, hardware, etc.) are shared dynamically

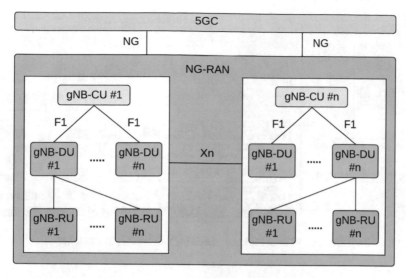

Fig. 15.9 The interconnection between gNBs and their corresponding components

among the gNBs and the components of each of the gNBs according to the requirements of the users or the tenant. The components of each of the gNB and their required resources are shared among n number of RAN network slice instances. The CU, DU, and RU of a gNB are supporting n number of RAN slices and their computational resources are shared among the RAN slices, according to vendor-specific algorithms (Fig. 15.9).

3.2   The Management and Orchestration of RAN Slices in NG-RAN

One of the emerging approaches in the NG-RAN architecture is RAN Slicing, which utilizes the SDN and NFV technologies. The former decouples the control and the user planes. The latter virtualizes the infrastructure, resources, and functions of a gNB into eMBB, mMTC, and URLLC RAN network slice subnets.

Each of the aforementioned RAN slice subnets (hereinafter RAN slices) has customized performance, functional, and operational requirements and thus requires its functions and resources to be well-isolated. The required resources of a RAN slice are divided into physical and virtual resources. The physical resources are managed by the 3GPP network slicing management system whereas the virtual resources are managed by the European Telecommunications Standards Institute (ETSI) NFV-management and orchestration (MANO).

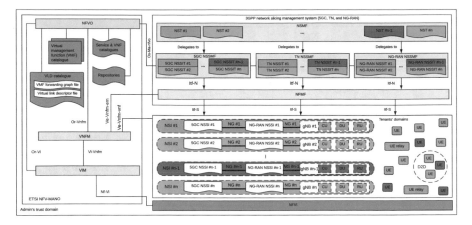

Fig. 15.10 The 3GPP and ETSI unified framework for RAN slicing in NG-RAN

In the beginning, the network slice management function (NSMF) receives slice related requirements from the communication service management function (CSMF) to allocate the planned virtual/physical resources of a network slice instance (NSI). The NSMF splits the slice related requirements into the transport network (TN), RAN, and CN slice subnet related requirements. To manage the life-cycle and required resources of three of the subnets, the NSMF delegates these requirements to their corresponding network slice subnet management function (NSSMF), namely the CN, TN, and RAN NSSMF (see Fig. 15.10).

The NSMF requests the 5GC NSSMF, the TN NSSMF, and the NG-RAN NSSMF to create the needed 5GC, TN, and NG-RAN NSSs, respectively. Every component in each of the subdomains has to be managed by its network function management function (NFMF). In Fig. 15.10, the NG-RAN subdomain is illustrated, where each of its components (namely the CU, DU, and the RU) are managed by its NFMF. The NFMF, the NSSMF, and the NSMF are the management entities of the 3GPP network slicing management system, which manage the physical resources of 3GPP networks.

However, an NSI or in particular the RAN slice is not only composed of physical resources but of virtual resources as well. The management and orchestration of virtual resources are beyond the scope of the 3GPP but fall within the scope of the ETSI NFV. The ETSI NFV introduced MANO that is responsible for the management and orchestration of virtual resources. It consists of three main functional blocks: the network function virtualization orchestration (NFVO), the virtual network function manager (VNFM), and the virtualized infrastructure manager (VIM). These functional blocks are interconnected using standard interfaces, and are described as follows:

• NFVO is responsible for the management and orchestration of network function virtualized infrastructure (NFVI), virtual resources, and realizing network services on the NFVI. The NFVO is also in-charge of the life-cycle management and on-boarding of virtual network functions (VNFs) and network services in its corresponding domain, i.e., the NG-RAN in this section.

- VNFM is responsible for the controlling, managing, and monitoring of the life-cycles of the VNFs and network services. The VNFM also controls the element management system (EMS) and the network management system (NMS).
- VIM is responsible for the management of the virtualized infrastructure and resources that are utilized by an NSI or network subnet slice instance (NSSI). It also allocates, upgrades, and releases the virtual functions and resources for an NSI or NSSI.

To allocate the virtualzed resources, the NSSMF of the RAN subnet triggers the NFV-MANO components in the NG-RAN to instantiate and configure the required VNFs and resources for the requested NSSI. The configuration of virtual resources shall take the shared and dedicated VNFs and resources into consideration. The RAN NSSMF shall also configure the required network management system components that are used by the RAN NSSI. Once the NSSMF complete the tasks, a confirmation is sent to the NSMF. The NSMF acknowledges and verifies the allocated virtual resources.

This joint framework and unified integration of both ETSI and 3GPP management systems lead to the effective management of an NSI or RAN slice in the NG-RAN. If we consider the RAN slice, it is obvious that the virtualized resources of the CU, DU, and RU are managed by the ETSI NFV-MANO; however, the physical resources are managed by the 3GPP slicing management system.

4 Key Enabling Technologies Towards 6G RAN Architecture

Based on the current progress in the 5G RAN architectures, the RAN architecture of the 6G wireless network is expected to consist of more than a single technology, architecture, access method, etc. It is predicted that recently introduced technologies in telecommunication, such as THz, network slicing, AI, ML, etc. will play a significant role in the 6G wireless networks. Moreover, the virtualiztion of the resources and infrastructure is also a major key enabling technology that will further expand into the 6G RAN.

In this section, we take a detailed look into the major key enabling technologies that are expected to be employed in the 6G RAN architecture. These key enabling technologies are expected to enhance spectrum efficiency, decrease CAPEX/OPEX, increase performance, and fulfill diversified requirements of end-users and vertical industries in comparison to those of 5G wireless network.

4.1 Virtualization, Cloudification, and Slicing

Some of the radio processing functions of NG-RAN are accommodated as VNFs over CU, DU, and RU, whereas others are distributed as physical network functions

(PNFs). The VNFs are running on top of VNF point of presence (PoP)s. The PoPs are virtualized cloud sites that are providing virtualized resources to VNFs. Most of the industrial partners and also those in academia assume the CU and DU as fully VNFs in the NG-RAN. However, there are still some of the PNFs in the RU that could be implemented as VNFs in the future.

The full virtualization of CU and DU and the partial virtualization of RU could increase the performance of the RAN architecture, decrease CAPEX/OPEX, simplify the network operation, etc. It will further improve the deployment of RAN slicing. We, therefore, believe that further research is required to virtualize some of the functions of RU in the RAN architecture towards 6G mobile networks.

4.2   Use-Case Specific RAN Slices

The NG-RAN supports a large amount of RAN slices. Each RAN slice provides communication services to a single use-case of a single tenant. However, there is a significant number of vertical industries that consist of multiple use-cases. Each use-case of each vertical industry; such as automobile, manufacturing, power grid, etc. is characterized by diverging service, network, and connectivity requirements. Such heterogeneity of a vertical industry cannot be effectively managed and efficiently mapped onto a single type of RAN slices.

Providing RAN slicing solutions to such types of verticals is critical in the NG-RAN. Therefore, further research is required in the NG-RAN to support RAN slices for such types of verticals. Among others, one of the crucial and essential research problem is to design a comprehensive framework for 6G RAN to support the RAN slices of multi-use-case verticals. This management and orchestration framework shall effectively manage the PNFs and VNFs of per-vertical per-use-case RAN slices in 6G RAN.

4.3   Slice-Aware Functional Splits and Function Placement

In NG-RAN, the current distribution of radio processing functions is done without considering the type of services. The one-size-fits-all functional split architecture is not efficient while supporting a large amount of eMBB, URLLC, and mMTC types of RAN slices. We believe that customized distribution of radio processing functions is needed to satisfy the requirements of the above-mentioned types of RAN slices. Therefore, in future research towards 6G RAN architecture, one of the interesting topics to be examined is the design of a gNB architecture, where all radio processing functions are distributed according to the type of RAN slices.

The customized functional split in gNB components enhances the performance of the RAN slices and improves the utilization of virtualized and physical resources. Moreover, it maintains a significant level of isolation among the RAN slices of different types while considering the metrics of service level agreements [24].

4.4   New Spectrum

The millimeter-wave (mmWave) has been introduced by the 5G NR, which significantly broadens the available bandwidth with new carrier frequencies up to 100 GHz. Nevertheless, it did not meet the demand for higher bandwidth. Looking forward to the 6G era, THz and VLC technologies are expected to play an important role in the NG-RAN to provide extremely high bandwidth [66].

Similar to mmWave, THz and light waves also suffer from high propagation loss and therefore rely highly on directive antennas and line-of-sight (LOS) channels while providing very limited coverage. However, when a satisfactory LOS link is available, the high carrier frequency offers a bandwidth that is significantly higher than any legacy technology, which makes it possible to simultaneously meet the extreme requirements in throughput, latency, and reliability [26, 33]. This makes THz and VLC technologies good supplementary solutions in addition to the mainstream radio frequency technologies for some specific use-cases, such as indoor and industrial scenarios.

NG-RAN provides a good capability of heterogeneous radio access technology (RAT), where the legacy RAT with low radio frequencies and the LOS-dependent RATs (THz and VLC) can co-exist well. THz and VLC may construct a new layer in the hierarchical RAN architecture (e.g., picocells), where heterogeneous cells with different RATs are overlaying on each other. The approach is similar to the introduction of mmWave in the 5G networks.

4.5   Intelligent Reflecting Surfaces

The high propagation loss and low diffraction in the high-frequency bands over 10 GHz lead to a substantial demand for advanced beamforming. While massive MIMO has proven to be an effective solution in the mmWave range, its capability can be challenged by the future 6G new spectrum. As an emerging and promising enhancement to traditional active antenna-based beamforming, the technology of IRS enables converting environmental objects with large surfaces into smart programmable reflectors, so that the environment can be exploited to create passive beamforming [39, 60], which significantly improves the channel state information (CSI) and enables physical layer security, at a low cost of energy consumption in comparison to active MIMO antenna arrays.

While IRS will be greatly contributing to 6G with its support to the new spectrum deployment, it relies on the external assessments that do not belong to the MNOs. Therefore, frameworks, interfaces, agreements, and protocols shall be designed and standardized, so that 6G operators become capable of widely accessing and exploiting IRS-equipped objects in different regions all over the world.

4.6   Coordinated Multi-Point

The concept of coordinated multi-point (CoMP) refers to a class of technologies, which lets multiple access points jointly serve multiple terminals, so as to build a network layer MIMO that brings more spatial diversity on top of classical physical layer MIMO approaches based on antenna arrays [45]. CoMP can be implemented in both downlink and uplink, for better spectral efficiency and fairness. In the downlink, inter-cell interference can be effectively jointly mitigated; in the uplink, multiple users can be jointly detected by multiple base stations.

In the 6G new spectrum that highly relies on the LOS link, the antenna level spatial diversity can be minimal, so the CoMP technologies that rely on base station level diversity [42] become important enablers to overcome challenges in these high-frequency bands, such as the blockage phenomena.

It shall be noted that CoMP is generally suggesting every UE to simultaneously hold multiple links to different access points (even when they are of the same RAT), which is out of classical cellular network design. Recent studies are even proposing serverless networking that completely relies on CoMP of RUs that are individually connected through fronthauls directly to the regional CU [23]. This calls for a complete revolution in the RAN architecture.

4.7   Non-orthogonal Multiple Access

The 3GPP LTE-A networks are implemented based on orthogonal frequency division multiple access (OFDMA), which is a typical instance of orthogonal multiple access (OMA) technologies prohibiting physical resource block (PRB) sharing by multiple users. In contrast to OMA, non-orthogonal multiple access (NOMA) technologies allow multiple users to reuse the same PRB. NOMA approaches can be generally divided into two categories, namely the power-domain (PD) NOMA and the code-domain NOMA. While PD-NOMA has been recently proposed and is attracting a lot of research interest, code-domain NOMA has a longer history in legacy systems (e.g., code division multiple access (CDMA) in 3G networks), and provides an alternative to PD with numerous variations such as trellis-coded multiple access (TCMA), interleave division multiple access (IDMA), multi-user shared access (MUSA), pattern-division multiple access (PDMA), and sparse-code multiple access (SCMA) [41].

Since the beyond 5G and the 6G networks are expected to simultaneously manage a massive amount of links, e.g., in the mMTC scenario and its future extensions, NOMA solutions appear to be promising since they provide higher bandwidth efficiency than OMA approaches. Recent studies have also demonstrated that NOMA can be effectively exploited in new spectrum, including mmWave [69], THz [65], and VLC [46]. Additionally, when deployed together with CoMP, NOMA has been proven to be capable of outperforming CoMP-OMA in both power efficiency and spectral efficiency [3, 4].

Being completely based on successive interference cancellation, NOMA requires cooperative decoding across different UEs. Specific D2D interfaces must be reserved for this functionality, and security/trust concerns shall be taken into account, to enable the deployment of NOMA in 6G.

4.8 Wireless Energy Harvesting and Wireless Power Transmission

In the past years, the rapid development of smart devices and IoT has motivated research in 5G to look for low power consumption solutions, which allow low-cost user devices such as networked sensors to achieve a long battery life up to 10 years. Beyond that stage, the 6G ambition towards a ubiquitous cyber-physical system (CPS) connecting everything, a.k.a. the internet of everything (IoE), is demanding a significantly denser deployment of communication devices, and therefore requires more sustainable solutions that can free mobile devices from limited battery life. To cope with this issue, the emerging technologies of wireless energy harvesting and wireless power transmission play an important role in 6G as key enablers [63].

If we expect the joint transmission of power and information to become a mainstream wireless solution in the next decade, it reasonably follows that the power transmission mechanism is integrated into the RAN design and standard protocols.

4.9 Non-terrestrial Communication

So far, all legacy and existing cellular systems have been designed to substantially rely on terrestrial base stations. Aiming at a better coverage rate, deployment of non-terrestrial infrastructures as part of the 6G network is being treated as an emerging topic, known as the integrated space and terrestrial network (ISTN). An ISTN is expected to consist of three layers: the ground-based layer constructed by terrestrial base stations, the airborne layer empowered by unmanned aerial vehicles (UAVs), and the spaceborne layer implemented by satellites [29].

Over the past years, it has been widely discussed to use UAVs in cellular networks as mobile access points or relays. As a flexible mobile supplement to the fixed-installed terrestrial gNBs, UAVs make it possible to dynamically plan the RAN, where the network capacity in different areas can be flexibly adjusted by deploying the UAVs to different locations. UAVs also provide a low-cost solution to temporarily deliver wireless services to inaccessible areas upon emergencies (e.g., in disaster reliefs). Additionally, empowered by wireless power transmission technologies, UAVs can be exploited as wireless chargers to conveniently power numerous energy-efficient UEs in use-case scenarios such as dense sensor networks. With such potential, UAVs are considered as an essential component of the future 6G infrastructure.

For marine, oceanic, as well as wild terrestrial areas, which are impossible or economically challenging to be covered by terrestrial cellular networks, satellites have been since long the most common communication solution. Nevertheless, as two seperate systems, it has been always expensive and inefficient to exchange data between the satellite networks and the terrestrial networks. Looking forward to a future global interconnection that is available anywhere and anytime, it is also strongly suggested to integrate satellite networks into the 6G network as part of it. To pave the way towards a future ISTN, specific approaches and architectural design will be needed for the seamless integration of satellites and UAVs into 6G.

4.10 Machine Learning and Artificial Intelligence

Dense deployment of ML and AI technologies at the edge of RAN to enable edge intelligence has been considered as the most characterizing feature of future 6G networks. On the one hand, most of the aforementioned 6G enabling technologies require a lot of computation to give full play to their performance. For example, fully virtualized and sliced networks will need ML-driven solutions of slice admission control, orchestration, and radio resource management, to maximize the resource utilization rate and reduce congestion [27]. In another instance, ML has been widely considered for user clustering in CoMPs and NOMA applications [13, 19, 47]. In the radio domain, AI is identified as the answer to highly complex non-convex problems such as the environment awareness of IRS [17], and beamforming in new spectrum [18].

The main challenge to the ultra-dense and ubiquitous deployment of ML/AI solutions in RAN is the trade-off between the gain in optimization performance, and the costs in implementation, latency, and security. More specifically, a centralized network intelligence fed by global information aggregated in the central cloud will be able to maximize the performance gain, since it has the knowledge about network and user behavior to the most extent. However, the data exchange between the centralized ML/AI server in the core network and the RAN will generate a significant delay, which will probably violate the latency constraint in some 6G use-case scenarios. Moreover, it may lead to concerns about data privacy regarding the assembly and exploitation the user data. A straightforward solution to address these issues is to migrate most ML/AI algorithms from the central server to the edge cloud, known as the edge intelligence [55], which is discussed further in this section. This can be well supported by the highly cloudified NG-RAN architecture and the RU-DU splitting. However, due to the stochastic behavior of data traffic, the demand for computation can be highly unbalanced in different cells and varying very dynamically. The infrastructural cost to implement edge intelligence will be huge if the algorithms are deployed in a hierarchical and non-cooperative fashion. Distributed/federated ML is known to be effective in smoothing the computation load over space and time, and therefore reduce the cost, but this calls for a deeper decentralization of RAN management in the architecture since it requires more information exchange between the DU and CU with minimal overhead while guaranteeing the data privacy.

4.11 Edge Intelligence

Edge intelligence (EI), which refers to the integration of edge computing and AI, is envisioned to be a key enabling factor for the future 6G technologies. Indeed, there exists a strong demand for EI due to the increasing number of smart devices, terminals, and the internet of intelligent things (IoIT) and the proliferation of intelligent services, for example, self-driving cars, drones, and auto-robots, which are calling for low-latency and reliable ML at the network edge [55]. To this end, this subsection sheds light on the role of EI in the 6G.

Towards addressing mission-critical applications in different vertical industries, EI is a critical key player. Concrete examples of EI for massive and critical MTC in 6G include fast and localized data analysis, and (semi-) centralized resource allocation [43]. The former allows fast and localized data processing, analysis, and content caching at the edge, whereas a more in-depth and holistic data analysis over a larger time scale can be carried out at the core. While in the latter, incorporating MEC servers at the network edge will render (semi-) centralized allocation of communication and computing resources on a fast basis practically feasible.

Furthermore, to enhance system efficiency, EI provides a better solution to constrained optimization problems in edge computing with the help of effective AI technologies. Some examples of AI methods to optimize telecom infrastructure and manage the life-cycle of edge networks are learning-driven communication, intelligent task allocation, the predictive quality of service, and energy management [68].

Besides, driven by the breakthroughs in deep learning and the hardware upgrade on the edge, EI provides a framework for running the training and inference of AI models. This trend is accelerated by the participation of more corporations in the design of chip architectures to support the edge computation paradigm, on the one hand. On the other hand, The recent groundswell interest in decentralized ML training and inference processes fits into this direction. In a word, EI provides AI with a heterogeneous platform full of rich capabilities [16].

To summarize, EI will provide novel business opportunities and technological solutions for various application fields, including but not limited to, personal computing, urban computing, and manufacturing easing their efficient, safe, secure, robust, and resilient wireless networking. In the future 6G era, EI will be a crucial part of everyday computation and smart technologies.

4.12 Context-Awareness

Context-awareness was originally investigated in the area of pervasive computing, where relevant information from the user and his environment, such as location, user identity, etc., were acquired and exploited by the computing system to adapt itself. Several context-aware schemes have been proposed for optimizing cellular network operations such as radio resource management, mobility management, and congestion control. Current RAN architectures are generally designed for a specific RAT, which often fails to fully utilize existing heterogeneous resources. Furthermore, such architecture fails to provide consistent QoE for users moving among different RANs. Context-awareness becomes important in such scenarios, for making more accurate and intelligent management.

The 6G network is expected to be completely context-aware. The network needs to be continuously aware of information concerning all the devices in its environment and capabilities of those devices. There are several open challenges in establishing complete context-awareness and efficiently managing such a system. Some key issues in this aspect are—handling conflicting changes in a system triggered by different applications or network functions based on the same context, monitoring and collecting context information without overwhelming the operation at the core network, maintaining the recentness of acquired context by aligning temporal scales of monitored information.

4.13 Software-Defined Networks

SDN has been a critical enabler technology in 5G networks, combined with the capabilities of NFV, they offer network management flexibility and service modularity. Subsequently, SDN will be indispensable for the evolution of future network generations beyond 5G and will continue to play an integral role within their architectures [36]. In practice, however, SDN technology still faces numerous challenges that thwart full exploitation of its potentials. Some of the challenges that continue to confront SDN in the wake of 6G include: maintaining an end-to-end and current global view of the dynamic network topology and its links states [34], resolving the optimal placement of the SDN controller in the network [2, 35], traffic engineering and forwarding while guaranteeing stringent QoS requirements [58], exploiting AI and ML algorithms for network management automation [8].

5 Conclusions

Present-day cellular networks are facing challenges from dramatically increasing industrial applications, IoT, massive end-user density, novel requirements of extremely low latency, high data rate connectivity, etc. The 5G technology deployment so far has been mainly focusing on tackling the high data rate requirements of the eMBB service type, whereas the advanced and complicated challenges in URLLC and mMTC scenarios still remain open. In this premise, 6G has started to make a deep rethinking on the application of cutting edge key enabling technologies, such as ML, AI, virtualization, etc., in its RAN architecture. Motivated by this vision, first we have studied the legacy RAN architectures related to the 5G mobile networks covering, C-RAN, H-CRAN, V-CRAN, and F-RAN. The architecture of each RAN type was reviewed in detail along with the subtypes and classifications within. Major benefits and key challenges in each RAN architecture were discussed. Furthermore, several differences and similarities were drawn among different RAN architectures, outlining the evolution of RAN architectures from one another within the 5G framework. To this end, we have explored the recently introduced NG-RAN architecture by 3GPP. Last but not least, we highlighted and discussed briefly several key enabling technologies, such as virtualization, use-case-specific-slicing, AI, ML, etc., that are expected to play a significant role in the 6G RAN, and further identified several other future research directions.

References

1. 3GPP, NR; Overall description; Stage-2. Technical Specification TS-38.300, 3GPP, v16.1.0 (2020)
2. V.M. Alevizaki, M. Anastasopoulos, A. Tzanakaki, D. Simeonidou, Joint fronthaul optimization and SDN controller placement in dynamic 5G networks, in *Optical Network Design and Modeling*, ed. by A. Tzanakaki, M. Varvarigos, R. Muñoz, R. Nejabati, N. Yoshikane, M. Anastasopoulos, J. Marquez-Barja (Springer, Cham, 2020), pp. 181–192
3. M.S. Ali, E. Hossain, A. Al-Dweik, D.I. Kim, Downlink power allocation for CoMP-NOMA in multi-cell networks. IEEE Trans. Commun. **66**(9), 3982–3998 (2018)
4. M.S. Ali, E. Hossain, D.I. Kim, Coordinated multipoint transmission in downlink multi-cell NOMA systems: models and spectral efficiency performance. IEEE Wireless Commun. **25**(2), 24–31 (2018)
5. A. Anpalagan, M. Bennis, R. Vannithamby, *Design and Deployment of Small Cell Networks* (Cambridge University Press, Cambridge, 2015). https://doi.org/10.1017/CBO9781107297333
6. E. Calvanese Strinati, S. Barbarossa, J.L. Gonzalez-Jimenez, D. Ktenas, N. Cassiau, L. Maret, C. Dehos, 6G: The next frontier: from holographic messaging to artificial intelligence using subterahertz and visible light communication. IEEE Veh. Technol. Mag. **14**(3), 42–50 (2019)
7. V. Chandrasekhar, J.G. Andrews, A. Gatherer , Femtocell networks: a survey. IEEE Commun. Mag. **46**(9), 59–67 (2008)
8. P. Chemouil, P. Hui, W. Kellerer, Y. Li, R. Stadler, D. Tao, Y. Wen, Y. Zhang, Special issue on artificial intelligence and machine learning for networking and communications. IEEE J. Sel. Areas Commun. **37**(6), 1185–1191 (2019)
9. H. Claussen, L.T.W. Ho, L.G. Samuel, Financial analysis of a pico-cellular home network deployment, in *2007 IEEE International Conference on Communications* (2007), pp 5604–5609
10. H. Claussen, D. Lopez-Perez, L. Ho, R. Razavi, S. Kucera, *Front Matter* (Wiley-IEEE Press, 2018), pp. i–xix
11. CMRI, C-RAN the road towards green Ran. White Paper, CMRI (2011)

12. X. Costa-Perez, J. Swetina, T. Guo, R. Mahindra, S. Rangarajan, Radio access network virtualization for future mobile carrier networks. IEEE Commun. Mag. **51**(7), 27–35 (2013)

13. J. Cui, Z. Ding, P. Fan, N. Al-Dhahir, Unsupervised machine learning-based user clustering in millimeter-wave-NOMA systems. IEEE Trans. Wireless Commun. **17**(11), 7425–7440 (2018)

14. A. Damnjanovic, J. Montojo, Y. Wei, T. Ji, T. Luo, M. Vajapeyam, T. Yoo, O. Song, D. Malladi, A survey on 3GPP heterogeneous networks. IEEE Wireless Commun. **18**(3), 10–21 (2011)

15. S. Dang, O. Amin, B. Shihada, M.S. Alouini, What should 6G be? Nat. Electron. **3**(1), 20–29 (2020)

16. S. Deng, H. Zhao, W. Fang, J. Yin, S. Dustdar, A.Y. Zomaya, Edge intelligence: the confluence of edge computing and artificial intelligence. IEEE Internet Things J. **7**, 7457–7469 (2020). https://doi.org/10.1109/JIOT.2020.2984887

17. M. Di Renzo, A. Zappone, M. Debbah, M.S. Alouini, C. Yuen, J. de Rosny, S. Tretyakov, Smart radio environments empowered by reconfigurable intelligent surfaces: how it works, state of research, and road ahead (2020, preprint). arXiv:200409352

18. A.M. Elbir, CNN-based precoder and combiner design in mmWave MIMO systems. IEEE Commun. Lett. **23**(7), 1240–1243 (2019)

19. M. Elkourdi, A. Mazin, R.D. Gitlin, Optimization of 5G virtual-cell based coordinated multipoint networks using deep machine learning. Int. J. Wireless Mobile Netw. **10**, 1–8 (2018)

20. F.H. Fitzek, P. Seeling, Why we should not talk about 6G (preprint, 2020). arXiv:200302079

21. X. Ge, S. Tu, G. Mao, C. Wang, T. Han, 5G ultra-dense cellular networks. IEEE Wireless Commun. **23**(1), 72–79 (2016)

22. A. Ghosh, R. Ratasuk, B. Mondal, N. Mangalvedhe, T. Thomas, LTE-advanced: next-generation wireless broadband technology [invited paper]. IEEE Wireless Commun. **17**(3), 10–22 (2010)

23. M. Gramaglia, P. Serrano, A. Banchs, G. Garcia-Aviles, A. Garcia-Saavedra, R. Perez, The case for serverless mobile networking, in *The Case for Serverless Mobile Networking* (2020)

24. M. Habibi, M. Nasimi, B. Han, H. Schotten, The structure of service level agreement of slice-based 5G network (2018). arXiv:1806.10426

25. M. A. Habibi, M. Nasimi, B. Han, H.D. Schotten, A comprehensive survey of RAN architectures toward 5G mobile communication system. IEEE Access **7**, 70371–70421 (2019)

26. C. Han, Y. Chen, Propagation modeling for wireless communications in the terahertz band. IEEE Commun. Mag. **56**(6), 96–101 (2018)

27. B. Han, H.D. Schotten, Machine learning for network slicing resource management: a comprehensive survey. ZTE Commun. **4**, 27–32 (2020)

28. T. Huang, W. Yang, J. Wu, J. Ma, X. Zhang, D. Zhang, A survey on green 6G network: architecture and technologies. IEEE Access **7**, 175758–175768 (2019)

29. X. Huang, J.A. Zhang, R.P. Liu, Y.J. Guo, L. Hanzo, Airplane-aided integrated networking for 6G wireless: will it work? IEEE Veh. Technol. Mag. **14**(3), 84–91 (2019)

30. IDC, Worldwide Internet of Things Forecast Update 2015–2019. IDC Report US40983216, IDC (2016)

31. G. Kardaras, C. Lanzani, Advanced multimode radio for wireless mobile broadband communication, in *2009 European Wireless Technology Conference* (2009), pp. 132–135

32. H.R. Karimi, L.T.W. Ho, H. Claussen, L.G. Samuel, Evolution towards dynamic spectrum sharing in mobile communications, in *2006 IEEE 17th International Symposium on Personal, Indoor and Mobile Radio Communications* (2006), pp. 1–5

33. M. Katz, I. Ahmed, Opportunities and challenges for visible light communications in 6G, in *2020 2nd 6G Wireless Summit (6G SUMMIT)*, (2020), pp 1–5

34. S. Khan, A. Gani, A.W. Abdul Wahab, M. Guizani, M.K. Khan, Topology discovery in software defined networks: threats, taxonomy, and state-of-the-art. IEEE Commun. Surv. Tuts. **19**(1), 303–324 (2017)

35. A. Ksentini, M. Bagaa, T. Taleb, On using SDN in 5G: the controller placement problem, in *2016 IEEE Global Communications Conference (GLOBECOM)*, (2016), pp 1–6

36. K.B. Letaief, W. Chen, Y. Shi, J. Zhang, Y.J.A. Zhang, The roadmap to 6G: AI empowered wireless networks. IEEE Commun. Mag. **57**(8), 84–90 (2019)

37. Y. Li, T. Jiang, K. Luo, S. Mao, Green heterogeneous cloud radio access networks: potential techniques, performance trade-offs, and challenges. IEEE Commun. Mag. **55**(11), 33–39 (2017)
38. K. Liang, L. Zhao, X. Zhao, Y. Wang, S. Ou, Joint resource allocation and coordinated computation offloading for fog radio access networks. China Commun. **13**(suppl 2), 131–139 (2016)
39. C. Liaskos, S. Nie, A. Tsioliaridou, A. Pitsillides, S. Ioannidis, I. Akyildiz, A new wireless communication paradigm through software-controlled metasurfaces. IEEE Commun. Mag. **56**(9), 162–169 (2018)
40. Y. Lin, L. Shao, Z. Zhu, Q. Wang, R.K. Sabhikhi, Wireless network cloud: architecture and system requirements. IBM J. Res. Develop. **54**(1), 4:1–4:12 (2010)
41. Y. Liu, Z. Qin, M. Elkashlan, Z. Ding, A. Nallanathan, L. Hanzo, Nonorthogonal multiple access for 5G and beyond. Proc. IEEE **105**(12), 2347–2381 (2017)
42. G.R. MacCartney, T.S. Rappaport, Millimeter-wave base station diversity for 5G coordinated multipoint (CoMP) applications. IEEE Trans. Wireless Commun. **18**(7), 3395–3410 (2019)
43. N.H. Mahmood, H. Alves, O.A. López, M. Shehab, D.P.M. Osorio, M. Latva-Aho, Six key features of machine type communication in 6G, in *2020 2nd 6G Wireless Summit (6G SUMMIT)* (2020), pp. 1–5
44. H.A. Mahmoud, I. Güvenc, A comparative study of different deployment modes for femtocell networks, in *2009 IEEE 20th International Symposium on Personal, Indoor and Mobile Radio Communications* (2009), pp. 1–5
45. P. Marsch, G.P. Fettweis, *Coordinated Multi-Point in Mobile Communications: From Theory to Practice* (Cambridge University Press, Cambridge, 2011)
46. H. Marshoud, V.M. Kapinas, G.K. Karagiannidis, S. Muhaidat, Non-orthogonal multiple access for visible light communications. IEEE Photon. Technol. Lett. **28**(1), 51–54 (2016)
47. F.B. Mismar, B.L. Evans, Machine learning in downlink coordinated multipoint in heterogeneous networks (2016, preprint). arXiv:160808306
48. C. Mouradian, D. Naboulsi, S. Yangui, R.H. Glitho, M.J. Morrow, P.A. Polakos, A comprehensive survey on fog computing: state-of-the-art and research challenges. IEEE Commun. Surv. Tuts. **20**(1), 416–464 (2018)
49. M. Mukherjee, L. Shu, D. Wang, Survey of fog computing: fundamental, network applications, and research challenges. IEEE Commun. Surv. Tuts. **20**(3), 1826–1857 (2018)
50. R.K. Naha, S. Garg, D. Georgakopoulos, P.P. Jayaraman, L. Gao, Y. Xiang, R. Ranjan, Fog computing: survey of trends, architectures, requirements, and research directions. IEEE Access **6**, 47980–48009 (2018)
51. M. Nasimi, F. Hashim, C.K. Ng, Characterizing energy efficiency for heterogeneous cellular networks, in 2012 IEEE Student Conference on Research and Development (SCOReD) (2012), pp. 198–202
52. T. Nihtilä, V. Haikola, HSDPA performance with dual stream MIMO in a combined macrofemto cell network, in *2010 IEEE 71st Vehicular Technology Conference* (2020), pp. 1–5
53. OpenFog Consortium Architecture Working Group, OpenFog reference architecture for fog computing. Technical Report OPFRA001.02081, Open Fog Consortium (2017)
54. C. Patel, M. Yavuz, S. Nanda, Femtocells [industry perspectives]. IEEE Wireless Commun. **17**(5), 6–7 (2010)
55. E. Peltonen, M. Bennis, M. Capobianco, M. Debbah, A. Ding, F. Gil-Castiñeira, M. Jurmu, T. Karvonen, M. Kelanti, A. Kliks, et al., 6G white paper on edge intelligence (2020, preprint). arXiv:200414850
56. M. Peng, S. Yan, K. Zhang, C. Wang, Fog-computing-based radio access networks: issues and challenges. Netw. Mag. Global Internet. **30**(4), 46–53 (2016). https://doi.org/10.1109/MNET.2016.7513863
57. F. Tang, Y. Kawamoto, N. Kato, J. Liu, Future intelligent and secure vehicular network toward 6G: machine-learning approaches. Proc. IEEE **108**(2), 292–307 (2019)
58. S. Tomovic, I. Radusinovic, Toward a scalable, robust, and QoS-aware virtual-link provisioning in SDN-based ISP networks. IEEE Trans. Netw. Service Manag. **16**(3), 1032–1045 (2019)

59. X. Wang, C. Cavdar, L. Wang, M. Tornatore, H.S. Chung, H.H. Lee, S.M. Park, B. Mukherjee, Virtualized cloud radio access network for 5G transport. IEEE Commun. Mag. **55**(9), 202–209 (2017)
60. Q. Wu, R. Zhang, Towards smart and reconfigurable environment: intelligent reflecting surface aided wireless network. IEEE Commun. Mag. **58**(1), 106–112 (2020)
61. J. Wu, Z. Zhang, Y. Hong, Y. Wen, Cloud radio access network (C-RAN): a primer. IEEE Netw. **29**(1), 35–41 (2015)
62. E. Yaacoub, M. Alouini, A key 6G challenge and opportunity—connecting the base of the pyramid: a survey on rural connectivity. Proc. IEEE **108**(4), 533–582 (2020)
63. P. Yang, Y. Xiao, M. Xiao, S. Li, 6G wireless communications: vision and potential techniques. IEEE Netw. **33**(4), 70–75 (2019)
64. M. Yavuz, F. Meshkati, S. Nanda, A. Pokhariyal, N. Johnson, B. Raghothaman, A. Richardson, Interference management and performance analysis of UMTS/HSPA+ femtocells. IEEE Commun. Mag. **47**(9), 102–109 (2009)
65. X. Zhang, C. Han, X. Wang, Joint beamforming-power-bandwidth allocation in terahertz NOMA networks, in *2019 16th Annual IEEE International Conference on Sensing, Communication, and Networking (SECON)* (2019), pp 1–9
66. Z. Zhang, Y. Xiao, Z. Ma, M. Xiao, Z. Ding, X. Lei, G.K. Karagiannidis, P. Fan, 6G wireless networks: vision, requirements, architecture, and key technologies. IEEE Veh. Technol. Mag. **14**(3), 28–41 (2019)
67. J. Zhao, Y. Liu, A survey of intelligent reflecting surfaces (IRSs): towards 6G wireless communication networks (2019, preprint). arXiv:190704789
68. G. Zhu, D. Liu, Y. Du, C. You, J. Zhang, K. Huang, Toward an intelligent edge: wireless communication meets machine learning. IEEE Commun. Mag. **58**(1), 19–25 (2020)
69. L. Zhu, Z. Xiao, X. Xia, D. Oliver Wu, Millimeter-wave communications with non-orthogonal multiple access for B5G/6G. IEEE Access **7**, 116123–116132 (2019)

Chapter 16
Federated Learning in 6G Mobile Wireless Networks

Zhaohui Yang, Mingzhe Chen, Walid Saad, Mohammad Shikh-Bahaei,
H. Vincent Poor, and Shuguang Cui

Abstract In this chapter, the problem of joint transmission and computation
resource allocation for federated learning (FL) over sixth generation (6G) mobile
wireless networks is investigated. In the considered model, each user exploits
limited local computational resources to train a local FL model with its collected
data and, then, sends the trained FL model parameters to a base station (BS) which
aggregates the local FL models and broadcasts the aggregated FL model back to all
the users. Since FL involves learning model exchanges between the users and the
BS, both computation and communication latencies are determined by the required
learning accuracy level, which affects the convergence rate of the FL algorithm. This
joint learning and communication problem is formulated as a delay minimization
problem, where it is proved that the objective function is a convex function of the
learning accuracy. Then, a bisection search algorithm is proposed to obtain the

Z. Yang (✉) · M. Shikh-Bahaei
Centre for Telecommunications Research, Department of Engineering, King's College London,
London, UK
e-mail: yang.zhaohui@kcl.ac.uk; m.sbahaei@kcl.ac.uk

M. Chen
Department of Electrical and Computer Engineering, Princeton University, Princeton, NJ, USA
e-mail: mingzhec@princeton.edu

W. Saad
Wireless@VT, Bradley Department of Electrical and Computer Engineering, Virginia Tech,
Blacksburg, VA, USA
e-mail: walids@vt.edu

H. V. Poor
Department of Electrical and Computer Engineering, Princeton University, Princeton, NJ, USA
e-mail: poor@princeton.edu

S. Cui
Shenzhen Research Institute of Big Data and School of Science and Engineering, The Chinese
University of Hong Kong, Shenzhen, China

© The Author(s), under exclusive license to Springer Nature Switzerland AG 2021
Y. Wu et al. (eds.), *6G Mobile Wireless Networks*, Computer Communications
and Networks, https://doi.org/10.1007/978-3-030-72777-2_16

optimal solution. Simulation results show that the proposed algorithm can reduce delay by up to 27.3% compared to conventional FL methods.

1 Introduction

Due to the explosive growth in data traffic, machine learning and data driven approaches have recently received much attention as a key enabler for future sixth generation (6G) wireless networks [1]. In this regard, standard machine learning approaches require centralizing the training data on a single data center or cloud which cannot preserve privacy and can lead to high communication overhead. However, low-latency and privacy requirements are important in the emerging application scenarios, such as unmanned aerial vehicles, extended reality (XR) services, autonomous driving, which make the standard centralized machine learning approaches inapplicable. Moreover, due to the limited communication resources for data transmission, it is impractical for all wireless devices that are engaged in learning to transmit all of their collected data to a data center or a cloud that can subsequently use a centralized learning algorithm for data analytic or network self-organization. Therefore, it becomes increasingly attractive to process data locally at edge devices.

To this end, *federated learning* (FL) frameworks are needed [2] to meet the demands for applications in 6G mobile wireless networks. FL refers to a type of distributed machine learning algorithms in which multiple devices collaboratively train a global learning model while keeping the training data locally. In particular, wireless devices train their local machine learning models using local data and share the trained local machine learning model parameters instead of the data set. Since the data center cannot access the local data sets at users, FL can help protect data privacy of users.

For wireless communications, FL has the following important roles: (1) exchanging limited local machine learning model parameters instead of the massive training data is energy saving and consumes less wireless resource; (2) training machine learning model parameters locally can effectively reduce the wireless transmission latency; (3) preserving data privacy since in the FL algorithms, the tanning data remains at each device and only the local machine learning model parameters are uploaded; (4) using different learning processes to train several classifiers from distributed data sets increases the possibility of achieving higher accuracy especially on a large-size domain; (5) inherently scalable since the growing amount of data may be offset by increasing the number of computers or processors, and providing a natural solution for large-scale learning where algorithm complexity and memory limitation are always the main obstacles. Since both learning and wireless transmission are considered, there are interactions between FL and wireless communication networks. In particular, FL is important for wireless communications in the following two directions, as shown in Fig. 16.1.

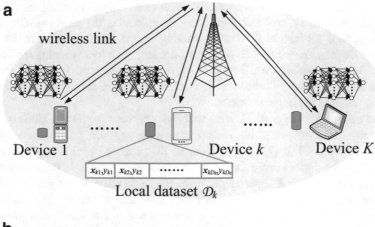

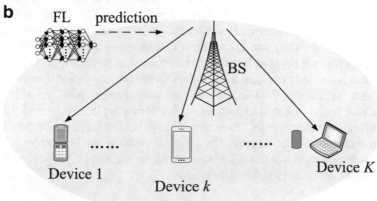

Fig. 16.1 Directions of FL and wireless communication

For FL, the process of decentralized learning needs the wireless devices to transmit over wireless links. To improve the overall performance (such as delay and energy) of the FL process, it is of importance to properly allocate the wireless resources (such as bandwidth, transmit power, and transmit time) for computation and transmission.

For wireless communication, FL have many use cases. For example, federated reinforcement learning (RL) can be used to solve complex convex and nonconvex optimization problems that arise in various use cases such as network control, user clustering, resource management, and interference alignment. Besides, FL enables users to collaboratively learn a shared prediction model while remaining their collected data on their devices for user behaviour predictions, user identifications, and wireless environment analysis. Based on the predicted results, the base station (BS) can efficiently allocate the wireless resources for the devices.

In FL, the wireless devices individually establish local learning models and cooperatively build a global learning model by uploading the local learning model parameters to a BS instead of sharing training data[3–5]. To implement FL over wireless networks, the wireless devices must transmit their local training results over wireless links [6], which can affect the FL performance, because both local training and wireless transmission introduce delay. Hence, it is necessary to optimize the delay for wireless FL implementation.

Some of the challenges of FL over wireless networks have been studied in [7–12]. To minimize latency, a broadband analog aggregation multi-access scheme for FL was designed in [7]. The authors in [8] proposed an FL implementation scheme between devices and access point over Gaussian multiple-access channels. To improve the statistical learning performance for on-device distributed training, the authors in [9] developed a sparse and low-rank modeling approach. The work in in [10] proposed an energy-efficient strategy for bandwidth allocation with the goal of reducing devices' sum energy consumption while meeting the required learning performance. However, the prior works [2, 7–10] focused on the delay/energy consumption for wireless consumption without considering the delay/energy tradeoff between learning and transmission. Recently, in [11] and [12], the authors considered both local learning and wireless transmission energy. In [11], the authors investigated the FL loss function minimization problem with taking into account packet errors over wireless links. However, this prior work ignored the computation delay of local FL model. The authors in [12] considered the sum learning and transmission energy minimization problem for FL, where all users transmit learning results to the BS. However, the solution in [12] requires all users to upload their learning model synchronously.

2 Preliminaries of FL

2.1 Basic Concept and Features of FL

FL is defined to enable the edge devices to collaboratively build a shared learning model and train their collected data locally. Through computation, FL can fully exploit large scale computation and spatial distribution of computing resources. FL can be implemented using one of two architectures: data split and model split. In data split architectures, the data samples are stored in multiple devices with identical machine learning model. In contrast, all data samples are shared by each device, while each device has a split machine learning structure to learn a subset of learning model parameters.

2.2 Flavors of FL

For FL, there are three main types.

2.2.1   Federated Multi-Task Learning

The aim of federated multi-task learning is to build the multi-task learning model by iteratively optimizing the local learning model parameters and the multi-task relationship matrix.

1. Every wireless device computes the local learning model by using its local data set and multi-task relationship matrix.
2. All devices upload the local learning model to the BS via wireless links in the uplink.
3. The BS updates the multi-task relationship matrix, which is broadcast to all devices in the downlink.

The aim of multi-task learning is to learn models for multiple related tasks simultaneously, while the goal in FL is to learn a model over data that has been generated via distributed wireless devices [13]. Both local learning models and multi-task relationship matrix are optimized in federated multi-task learning in a distributed manner.

Federated multi-task learning can be used in many applications for future 6G. For example, consider FL for the activities of wireless devices in a cellular network via their individual text, image, or video data. Since each wireless device generates data in a distinct distribution, it is natural for each wireless device to learn a local model based on their local dataset. However, connections between the local data models may exist (e.g., people have similar habit to use their cell phones), and multi-task learning can be used to improve the performance for each wireless device.

2.2.2   Federated RL

The goal of federated RL is to enable wireless devices to remember what they have learned and what other wireless devices have learned [14]. Federated RL is used in the case where multiple wireless devices making decisions in different environments. In federated RL, each wireless device builds a learning network with the help of other wireless devices.

1. Initially, one edge device obtains its private strategy model learning network through RL in its own environment and upload it to the BS as the shared model.
2. After a while, wireless devices desire to learn navigation by RL in new environments. The wireless devices download the shared model in the BS as the initial actor model in reinforcement learning. Then they wireless devices get their own private learning networks through RL in new environments. After completing the training, wireless devices upload their private learning networks to the BS.
3. In the BS, private learning networks will be fused into shared model, and then a new shared model will be generated. The new shared model can be used by other wireless devices. Other wireless devices will also upload their private learning networks to the BS to promote the evolution of the shared model.

2.2.3   General FL

General FL stands for the most common FL technique. The general FL technique builds a uniform learning model through iteratively updating information between wireless devices and the BS. The FL procedure contains three steps in each iteration: local computation at each wireless device, local FL model parameters transmission for each wireless device, and the result aggregation and broadcast at the BS.

1. Every wireless device needs to compute the result by using its local data set.
2. All wireless devices upload the local prediction parameters to the BS via wireless links in the uplink.
3. The BS aggregates the global prediction model parameters and broadcasts the global prediction model parameters to all wireless devices in the downlink.

According to the preliminaries of FL, both communication and computation resource needs to be considered for FL over wireless communication networks. Consequently, it is of importance to optimize the performance of FL via jointly considering communication and computation resource. In the next Section, we provide the communication and computation model of FL over wireless communication networks.

3   System Model of FL Over Wireless Communication Networks

In this section, we consider the resource allocation problem for FL over wireless communication systems, i.e., 6G wireless communication networks. Consider a cellular network that consists of one BS serving a set $\mathcal{K}$ of K users, as shown in Fig. 16.2. Each user k has a local dataset $\mathcal{D}_k$ with D_k data samples. For each dataset $\mathcal{D}_k = \{\boldsymbol{x}_{kl}, y_{kl}\}_{l=1}^{D_k}, \boldsymbol{x}_{kl} \in \mathbb{R}^d$ is an input vector of user k and y_{kl} is its corresponding output.[1]

3.1   FL Model

For FL, we define a vector $\boldsymbol{w}$ to capture the parameters related to the global FL model that is trained by all datasets. Hereinafter, the FL model that is trained by all users' data set is called *global FL model*, while the FL model that is trained by each user's dataset is called *local FL model*. We introduce the loss function $f(\boldsymbol{w}, \boldsymbol{x}_{kl}, y_{kl})$, that captures the FL performance over input vector $\boldsymbol{x}_{kl}$ and output

---

[1]For simplicity, this chapter only considers an FL algorithm with a single output. Our approach can be extended to the case with multiple outputs [2].

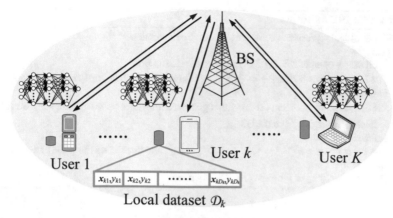

Fig. 16.2 FL over wireless communication networks

y_{kl}. For different learning tasks, the loss function will be different. Since the dataset of user k is $\mathcal{D}_k$, the total loss function of user k will be:

$$F_k(\boldsymbol{w}) = \frac{1}{D_k} \sum_{l=1}^{D_k} f(\boldsymbol{w}, \boldsymbol{x}_{kl}, y_{kl}). \qquad (16.1)$$

In order to deploy FL, it is necessary to train the underlying model. Training is done in order to compute the global FL model for all users without sharing their local datasets due to privacy and communication issue. The FL training problem can be formulated as follows [15]:

$$\min_{\boldsymbol{w}} F(\boldsymbol{w}) = \sum_{k=1}^{K} \frac{D_k}{D} F_k(\boldsymbol{w}) = \frac{1}{D} \sum_{k=1}^{K} \sum_{l=1}^{D_k} f(\boldsymbol{w}, \boldsymbol{x}_{kl}, y_{kl}), \qquad (16.2)$$

where $D = \sum_{k=1}^{K} D_k$ is the total data samples of all users.

To solve problem (16.2), we adopt the FL algorithm in [2], which is summarized in Algorithm 1. In Algorithm 1, at each iteration of the FL algorithm, each user downloads the global FL model parameters from the BS for local computing, while the BS periodically gathers the local FL model parameters from all users and sends the updated global FL model parameters back to all users. We define $\boldsymbol{w}^{(n)}$ as the global FL parameter at a given iteration n. Each user computes the local FL problem:

$$\min_{\boldsymbol{h}_k \in \mathbb{R}^d} \quad G_k(\boldsymbol{w}^{(n)}, \boldsymbol{h}_k) \triangleq F_k(\boldsymbol{w}^{(n)} + \boldsymbol{h}_k)$$

$$- (\nabla F_k(\boldsymbol{w}^{(n)}) - \xi \nabla F(\boldsymbol{w}^{(n)}))^T \boldsymbol{h}_k, \qquad (16.3)$$

Algorithm 1 FL algorithm

1: Initialize global regression vector $\boldsymbol{w}^0$ and iteration number $n = 0$.
2: **repeat**
3:     Each user k computes $\nabla F_k(\boldsymbol{w}^{(n)})$ and sends it to the BS.
4:     The BS computes $\nabla F(\boldsymbol{w}^{(n)}) = \frac{1}{K} \sum_{k=1}^{K} \nabla F_k(\boldsymbol{w}^{(n)})$, which is broadcast to all users.
5:     **parallel for** user $k \in \mathcal{K}$
6:         Solve local FL problem (16.3) with a given learning accuracy η and the solution is $\boldsymbol{h}_k^{(n)}$.
7:         Each user sends $\boldsymbol{h}_k^{(n)}$ to the BS.
8:     **end for**
9:     The BS computes $\boldsymbol{w}^{(n+1)} = \boldsymbol{w}^{(n)} + \frac{1}{K} \sum_{k=1}^{K} \boldsymbol{h}_k^{(n)}$ and broadcasts the value to all users.
10:    Set $n = n + 1$.
11: **until** the accuracy ϵ_0 of problem (16.2) is obtained.

---

by using the gradient method with a given accuracy. In problem (16.3), ξ is a constant value. The solution $\boldsymbol{h}_k$ in problem (16.3) means the updated value of local FL parameter for user k in each iteration, i.e., $\boldsymbol{w}^{(n)} + \boldsymbol{h}_k$ denotes user k' local FL parameter at the n-th iteration. Since it is hard to obtain the actual optimal solution of problem (16.3), we obtain a solution of (16.3) with some accuracy. The solution $\boldsymbol{h}_k^{(n)}$ of problem (16.3) at the n-th iteration with accuracy η means that

$$
\begin{aligned}
G_k(\boldsymbol{w}^{(n)}, \boldsymbol{h}_k^{(n)}) &- G_k(\boldsymbol{w}^{(n)}, \boldsymbol{h}_k^{(n)*}) \\
&\leq \eta(G_k(\boldsymbol{w}^{(n)}, \boldsymbol{0}) - G_k(\boldsymbol{w}^{(n)}, \boldsymbol{h}_k^{(n)*})),
\end{aligned}
\tag{16.4}
$$

where $\boldsymbol{h}_k^{(n)*}$ is the actual optimal solution of problem (16.3).

In Algorithm 1, the iterative method involves a number of global iterations (i.e., the value of n in Algorithm 1) to achieve a global accuracy ϵ_0 of global FL model. The solution $\boldsymbol{w}^{(n)}$ of problem (16.2) with accuracy ϵ_0 means that

$$
F(\boldsymbol{w}^{(n)}) - F(\boldsymbol{w}^*) \leq \epsilon_0(F(\boldsymbol{w}^{(0)}) - F(\boldsymbol{w}^*)),
\tag{16.5}
$$

where $\boldsymbol{w}^*$ is the actual optimal solution of problem (16.2).

To analyze the convergence of Algorithm 1, we assume that $F_k(\boldsymbol{w})$ is L-Lipschitz continuous and γ-strongly convex, i.e.,

$$
\gamma \boldsymbol{I} \preceq \nabla^2 F_k(\boldsymbol{w}) \preceq L\boldsymbol{I}, \quad \forall k \in \mathcal{K}.
\tag{16.6}
$$

Under assumption (16.6), we provide the following lemma about convergence rate of Algorithm 1.

Lemma 16.1 *If we run Algorithm 1 with $0 < \xi \leq \frac{\gamma}{L}$ for*

$$
n \geq \frac{a}{1 - \eta} \triangleq I_0,
\tag{16.7}
$$

iterations with $a = \frac{2L^2}{\gamma^2 \xi} \ln \frac{1}{\epsilon_0}$, we have $F(\boldsymbol{w}^{(n)}) - F(\boldsymbol{w}^*) \leq \epsilon_0(F(\boldsymbol{w}^{(0)}) - F(\boldsymbol{w}^*))$.

The proof of Lemma 1 can be found in [16]. From Lemma 16.1, we can find that the number of global iterations n increases with the local accuracy. This is because more iterations are needed if the local computation has a low accuracy.

3.2 Computation and Transmission Model

The FL procedure between the users and their serving BS consists of three steps in each iteration: Local computation at each user (using several local iterations), local FL parameter transmission for each user, and result aggregation and broadcast at the BS. During the local computation step, each user calculates its local FL parameters by using its local dataset and the received global FL parameters.

3.2.1 Local Computation

We solve the local learning problem (16.3) by using the gradient method. In particular, the gradient procedure in the $(i + 1)$-th iteration is given by:

$$\boldsymbol{h}_k^{(n),(i+1)} = \boldsymbol{h}_k^{(n),(i)} - \delta \nabla G_k(\boldsymbol{w}^{(n)}, \boldsymbol{h}_k^{(n),(i)}), \qquad (16.8)$$

where δ is the step size, $\boldsymbol{h}_k^{(n),(i)}$ is the value of $\boldsymbol{h}_k$ at the i-th local iteration with given vector $\boldsymbol{w}^{(n)}$, and $\nabla G_k(\boldsymbol{w}^{(n)}, \boldsymbol{h}_k^{(n),(i)})$ is the gradient of function $G_k(\boldsymbol{w}^{(n)}, \boldsymbol{h}_k)$ at point $\boldsymbol{h}_k = \boldsymbol{h}_k^{(n),(i)}$. We set the initial solution $\boldsymbol{h}_k^{(n),(0)} = \boldsymbol{0}$.

Next, we provide the number of local iterations needed to achieve a local accuracy η in (16.4). We set $v = \frac{2}{(2-L\delta)\delta\gamma}$.

Lemma 16.2 *If we set step $\delta < \frac{2}{L}$ and run the gradient method for $i \geq v \log_2(1/\eta)$ iterations at each user, we can solve local FL problem (16.3) with an accuracy η.*

The proof of Lemma 2 can be found in paper [16]. Let f_k be the computation capacity of user k, which is measured by the number of CPU cycles per second. The computation time at user k needed for data processing is:

$$\tau_k = \frac{vC_kD_k \log_2(1/\eta)}{f_k} = \frac{A_k \log_2(1/\eta)}{f_k}, \quad \forall k \in \mathcal{K}, \qquad (16.9)$$

where C_k (cycles/bit) is the number of CPU cycles required for computing one sample data at user k, $v \log_2(1/\eta)$ is the number of local iterations for each user as given by Lemma 16.2, and $A_k = vC_kD_k$.

3.2.2 Wireless Transmission

After local computation, all users upload their local FL parameters to the BS via frequency domain multiple access (FDMA). The achievable rate of user k can be given by:

$$r_k = b_k \log_2 \left(1 + \frac{g_k p_k}{N_0 b_k} \right), \quad \forall k \in \mathcal{K}, \tag{16.10}$$

where b_k is the bandwidth allocated to user k, p_k is the transmit power of user k, g_k is the channel gain between user k and the BS, and N_0 is the power spectral density of the Gaussian noise. Due to the limited bandwidth, we have $\sum_{k=1}^{K} b_k \leq B$, where B is the total bandwidth.

In this step, user k needs to upload the local FL parameters to the BS. Since the dimensions of the vector $\boldsymbol{h}_k^{(n)}$ are fixed for all users, the data size that each user needs to upload is constant, and can be denoted by s. To upload data of size s within transmit time t_k, we must have:

$$t_k r_k \geq s. \tag{16.11}$$

3.2.3 Information Broadcast

In this step, the BS aggregates the global prediction model parameters. The BS broadcasts the global prediction model parameters to all users in the downlink. Due to the high power of the BS and large downlink bandwidth, we ignore the downlink time. Note that the local data $\mathcal{D}_k$ is not accessed by the BS, so as to protect the privacy of users, as is required by FL. The delay of each user includes the local computation time and transmit time. Based on (16.7) and (16.9), the delay T_k of user k will be:

$$T_k = I_0(\tau_k + t_k) = \frac{a}{1 - \eta} \left(\frac{A_k \log_2(1/\eta)}{f_k} + t_k \right). \tag{16.12}$$

We define $T = \max_{k \in \mathcal{K}} T_k$ as the delay for training the whole FL algorithm.

3.3 Problem Formulation

We now pose the delay minimization problem:

$$\min_{T, t, b, f, p, \eta} \quad T \tag{16.13}$$

$$\text{s.t.} \quad \frac{a}{1-\eta}\left(\frac{A_k \log_2(1/\eta)}{f_k} + t_k\right) \leq T, \quad \forall k \in \mathcal{K}, \tag{16.13a}$$

$$t_k b_k \log_2\left(1 + \frac{g_k p_k}{N_0 b_k}\right) \geq s, \quad \forall k \in \mathcal{K}, \tag{16.13b}$$

$$\sum_{k=1}^{K} b_k \leq B, \tag{16.13c}$$

$$0 \leq f_k \leq f_k^{\max}, 0 \leq p_k \leq p_k^{\max}, \forall k \in \mathcal{K}, \tag{16.13d}$$

$$0 \leq \eta \leq 1, \tag{16.13e}$$

$$t_k \geq 0, b_k \geq 0, \quad \forall k \in \mathcal{K}, \tag{16.13f}$$

where $t = [t_1, \cdots, t_K]^T$, $b = [b_1, \cdots, b_K]^T$, $f = [f_1, \cdots, f_K]^T$, and $p = [p_1, \cdots, p_K]^T$. $f_k^{\max}$ and $p_k^{\max}$ are, respectively, the maximum local computation capacity and maximum transmit power of user k. (16.13a) indicates that the execution time of the local tasks and the transmit time for all users should not exceed the delay of the whole FL algorithm. The data transmission constraint is given by (16.13b), while the bandwidth constraint is given by (16.13c). Equation (16.13d) represents the maximum local computation capacity and transmit power limits of all users. The accuracy constraint is given by (16.13e).

4   Optimal Resource Allocation

Although the delay minimization problem (16.13) is nonconvex due to constraints (16.13a)–(16.13b), the globally optimal solution is shown to be obtained by using the bisection method.

4.1   Optimal Resource Allocation

Let $(T^*, t^*, b^*, f^*, p^*, \eta^*)$ be the optimal solution of problem (16.13). We provide the following lemma about the feasibility conditions of problem (16.13).

Lemma 16.3 *Problem (16.13) with fixed $T < T^*$ is always feasible, while problem (16.13) with fixed $T > T^*$ is infeasible.*

Proof Assume that $(\bar{T}, \bar{t}, \bar{b}, \bar{f}, \bar{p}, \bar{\eta})$ is a feasible solution of problem (16.13) with $T = \bar{T} < T^*$. Then, solution $(\bar{T}, \bar{t}, \bar{b}, \bar{f}, \bar{p}, \bar{\eta})$ is feasible with lower value of the objective function than solution $(T^*, t^*, b^*, f^*, p^*, \eta^*)$, which contradicts the fact that $(T^*, t^*, b^*, f^*, p^*, \eta^*)$ is the optimal solution. For problem (16.13) with $T = \bar{T} > T^*$, we can always construct a feasible solution $(\bar{T}, t^*, b^*, f^*, p^*, \eta^*)$ to problem (16.13) by checking all constraints. $\qquad\square$

According to Lemma 16.3, we can use the bisection method to obtain the optimal solution of problem (16.13). Denote

$$T_{\min} = 0, \ T_{\max} = \max_{k \in \mathcal{K}} \frac{2a A_k}{f_k^{\max}} + \frac{2a K s}{B \log_2 \left(1 + \frac{g_k p_k^{\max} K}{N_0 B}\right)}. \tag{16.14}$$

If $T > T_{\max}$, problem (16.13) is always feasible by setting $f_k = f_k^{\max}$, $p_k = p_k^{\max}$, $b_k = \frac{B}{K}$, $\eta = \frac{1}{2}$, and

$$t_k = \frac{K s}{B \log_2 \left(1 + \frac{g_k p_k^{\max} K}{N_0 B}\right)}. \tag{16.15}$$

Hence, the optimal T^* of problem (16.13) must lie in the interval $(T_{\min}, T_{\max})$. At each step, the bisection method divides the interval in two by computing the midpoint $T_{\mathrm{mid}} = (T_{\min} + T_{\max})/2$. There are now only two possibilities: 1) if problem (16.13) with $T = T_{\mathrm{mid}}$ is feasible, we have $T^* \in (T_{\min}, T_{\mathrm{mid}}]$ and 2) if problem (16.13) with $T = T_{\mathrm{mid}}$ is infeasible, we have $T^* \in (T_{\mathrm{mid}}, T_{\max})$. The bisection method selects the subinterval that is guaranteed to be a bracket as the new interval to be used in the next step. As such an interval that contains the optimal T^* is reduced in width by 50% at each step. The process continues until the interval is sufficiently small.

With a fixed T, we still need to check whether there exists a feasible solution satisfying constraints (16.13a)–(16.13g). From constraints (16.13a) and (16.13c), we can see that it is always efficient to utilize the maximum computation capacity, i.e., $f_k^* = f_k^{\max}$, $\forall k \in \mathcal{K}$. In addition, from (16.13b) and (16.13d), we can see that minimizing the delay can be done by having: $p_k^* = p_k^{\max}$, $\forall k \in \mathcal{K}$. Substituting the maximum computation capacity and maximum transmission power into (16.13), delay minimization problem becomes:

$$\min_{T, t, b, \eta} \quad T \tag{16.16}$$

$$\text{s.t.} \quad t_k \leq \frac{(1 - \eta)T}{a} + \frac{A_k \log_2 \eta}{f_k^{\max}}, \quad \forall k \in \mathcal{K}, \tag{16.16a}$$

$$\frac{s}{t_k} < b_k \log_2 \left(1 + \frac{g_k p_k^{\max}}{N_0 b_k}\right), \quad \forall k \in \mathcal{K}, \tag{16.16b}$$

$$\sum_{k=1}^{K} b_k \leq B, \tag{16.16c}$$

$$0 \leq \eta \leq 1, \tag{16.16d}$$

$$t_k \geq 0, b_k \geq 0, \quad \forall k \subset \mathcal{K}. \tag{16.16e}$$

We provide the sufficient and necessary condition for the feasibility of set (16.16a)–(16.16e) using the following lemma.

Lemma 16.4 *With a fixed T, set (16.16a)–(16.16e) is nonempty if an only if*

$$B \geq \min_{0 \leq \eta \leq 1} \sum_{k=1}^{K} u_k(v_k(\eta)), \tag{16.17}$$

where

$$u_k(\eta) = -\frac{(\ln 2)\eta}{W\left(-\frac{(\ln 2)N_0\eta}{g_k P_k^{\max}} e^{-\frac{(\ln 2)N_0\eta}{g_k P_k^{\max}}}\right) + \frac{(\ln 2)N_0\eta}{g_k P_k^{\max}}}, \tag{16.18}$$

and

$$v_k(\eta) = \frac{s}{\frac{(1-\eta)T}{a} + \frac{A_k \log_2 \eta}{f_k^{\max}}}. \tag{16.19}$$

Proof To prove this, we first define a function $y = x \ln\left(1 + \frac{1}{x}\right)$ with $x > 0$. Then, we have

$$y' = \ln\left(1 + \frac{1}{x}\right) - \frac{1}{x+1}, \ y'' = -\frac{1}{x(x+1)^2} < 0. \tag{16.20}$$

According to (16.20), y' is a decreasing function. Since $\lim_{t_i \to +\infty} y' = 0$, we have $y' > 0$ for all $0 < x < +\infty$. Hence, y is an increasing function, i.e., the right hand side of (16.16b) is an increasing function of bandwidth b_k. To ensure that the maximum bandwidth constraint (16.16c) can be satisfied, the left hand side of (16.16b) should be as small as possible, i.e., t_k should be as long as possible. Based on (16.16a), the optimal time allocation should be:

$$t_k^* = \frac{(1-\eta)T}{a} + \frac{A_k \log_2 \eta}{f_k^{\max}}, \quad \forall k \in \mathcal{K}. \tag{16.21}$$

Substituting (16.21) into (16.16b), we can construct the following problem:

$$\min_{b,\eta} \sum_{k=1}^{K} b_k \tag{16.22}$$

$$\text{s.t. } v_k(\eta) \leq b_k \log_2\left(1 + \frac{g_k P_k^{\max}}{N_0 b_k}\right), \quad \forall k \in \mathcal{K}, \tag{16.22a}$$

$$0 \leq \eta \leq 1, \tag{16.22b}$$

$$b_k \geq 0, \quad \forall k \in \mathcal{K}, \tag{16.22c}$$

where $v_k(\eta)$ is defined in (16.19). We can observe that set (16.16a)–(16.16e) is nonempty if an only if the optimal objective value of (16.22) is less than B. Since the right hand side of (16.16b) is an increasing function, (16.16b) should hold with equality for the optimal solution of problem (16.22). Setting (16.16b) with equality, problem (16.22) reduces to (16.17). □

To effectively solve (16.17) in Lemma 16.4, we provide the following lemma.

Lemma 16.5 *In* (16.18), $u_k(v_k(\eta))$ *is a convex function.*

Proof We first prove that $v_k(\eta)$ is a convex function. To show this, we define:

$$\phi(\eta) = \frac{s}{\eta}, \quad 0 \le \eta \le 1, \tag{16.23}$$

and

$$\varphi_k(\eta) = \frac{(1-\eta)T}{a} + \frac{A_k \log_2 \eta}{f_k^{\max}}, \quad 0 \le \eta \le 1. \tag{16.24}$$

According to (16.19), we have: $v_k(\eta) = \phi(\varphi_k(\eta))$. Then, the second-order derivative of $v_k(\eta)$ can be given by:

$$v_k''(\eta) = \phi''(\varphi_k(\eta))(\varphi_k'(\eta))^2 + \phi'(\varphi_k(\eta))\varphi_k''(\eta). \tag{16.25}$$

According to (16.23) and (16.24), we have:

$$\phi'(\eta) = -\frac{s}{\eta^2} \le 0, \quad \phi''(\eta) = \frac{2s}{\eta^3} \ge 0, \tag{16.26}$$

and

$$\varphi_k''(\eta) = -\frac{A_k}{(\ln 2) f_k^{\max} \eta^2} \le 0. \tag{16.27}$$

Combining (16.25)–(16.27), we can find that $v_k''(\eta) \ge 0$, i.e., $v_k(\eta)$ is a convex function.

Then, we can show that $u_k(\eta)$ is an increasing and convex function. According to the proof of Lemma 16.4, $u_k(\eta)$ is the inverse function of the right hand side of (16.16b). If we further define function:

$$z_k(\eta) = \eta \log_2 \left(1 + \frac{g_k P_k^{\max}}{N_0 \eta} \right), \quad \eta \ge 0, \tag{16.28}$$

$u_k(\eta)$ is the inverse function of $z_k(\eta)$, which gives $u_k(z_k(\eta)) = \eta$.

According to (16.20), function $z_k(\eta)$ is an increasing and concave function, i.e., $z_k'(\eta) \geq 0$ and $z_k''(\eta) \leq 0$. Since $z_k(\eta)$ is an increasing function, its inverse function $u_k(\eta)$ is also an increasing function.

Based on the definition of concave function, for any $\eta_1 \geq 0$, $\eta_2 \geq 0$ and $0 \leq \theta \leq 1$, we have:

$$z_k(\theta \eta_1 + (1-\theta)\eta_2) \geq \theta z_k(\eta_1) + (1-\theta)z_k(\eta_2). \tag{16.29}$$

Applying the increasing function $u_k(\eta)$ on both sides of (16.29) yields:

$$\theta \eta_1 + (1-\theta)\eta_2 \geq u_k(\theta z_k(\eta_1) + (1-\theta)z_k(\eta_2)). \tag{16.30}$$

Denote $\bar{\eta}_1 = z_k(\eta_1)$ and $\bar{\eta}_2 = z_k(\eta_2)$, i.e., we have $\eta_1 = u_k(\bar{\eta}_1)$ and $\eta_2 = u_k(\bar{\eta}_2)$. Thus, (16.30) can be rewritten as:

$$\theta u_k(\bar{\eta}_1) + (1-\theta)u_k(\bar{\eta}_1) \geq u_k(\theta \bar{\eta}_1 + (1-\theta)\bar{\eta}_2), \tag{16.31}$$

which indicates that $u_k(\eta)$ is a convex function. As a result, we have proven that $u_k(\eta)$ is an increasing and convex function, which shows:

$$u_k'(\eta) \geq 0, \quad u_k''(\eta) \geq 0. \tag{16.32}$$

To show the convexity of $u_k(v_k(\eta))$, we have:

$$u_k''(v_k(\eta)) = u_k''(v_k(\eta))(v_k'(\eta))^2 + u_k'(v_k(\eta))v_k''(\eta) \geq 0,$$

according to $v_k''(\eta) \geq 0$ and (16.32). As a result, $u_k(v_k(\eta))$ is a convex function. □

Lemma 16.5 implies that the optimization problem in (16.17) is a convex problem, which can be effectively solved. By finding the optimal solution of (16.17), the sufficient and necessary condition for the feasibility of set (16.16a)–(16.16e) can be simplified using the following theorem.

Theorem 16.1 *With a fixed T, set (16.16a)–(16.16e) is nonempty if and only if*

$$B \geq \sum_{k=1}^{K} u_k(v_k(\eta^*)), \tag{16.33}$$

where η^* is the solution to $\sum_{k=1}^{K} u_k'(v_k(\eta^*))v_k'(\eta^*) = 0$.

Theorem 16.1 directly follows from Lemmas 16.4 and 16.5. Due to the convexity of function $u_k(v_k(\eta))$, $\sum_{k=1}^{K} u_k'(v_k(\eta^*))v_k'(\eta^*)$ is an increasing function of η^*. As a result, the unique solution of η^* to $\sum_{k=1}^{K} u_k'(v_k(\eta^*))v_k'(\eta^*) = 0$ can be effectively solved via the bisection method. Based on Theorem 16.1, the algorithm for obtaining the minimal delay is summarized in Algorithm 2.

Algorithm 2 Delay minimization

1: Initialize $T_{\min}$, $T_{\max}$, and the tolerance ϵ_0.
2: **repeat**
3:    Set $T = \frac{T_{\min}+T_{\max}}{2}$.
4:    Check the feasibility condition (16.33).
5:    If set (16.16a)–(16.16e) has a feasible solution, set $T_{\max} = T$. Otherwise, set $T_{\min} = T$.
6: **until** $(T_{\max} - T_{\min})/T_{\max} \leq \epsilon_0$.

5 Simulation Results

For our simulations, we deploy $K = 50$ users uniformly in a square area of size
500 m $\times$ 500 m with the BS located at its center. The path loss model is $128.1 +$
$37.6 \log_{10} d$ (d is in km) and the standard deviation of shadow fading is 8 dB [17].
In addition, the noise power spectral density is $N_0 = -174$ dBm/Hz. We use the
real open blog feedback dataset in [18]. This dataset with a total number of 60,021
data samples originates from blog posts and the dimensional of each data sample is
281. The prediction task associated with the data is the prediction of the number of
comments in the upcoming 24 h. Parameter C_k is uniformly distributed in $[1, 3] \times$
10^4 cycles/sample. The effective switched capacitance in local computation is $\kappa =$
10^{-28}. In Algorithm 1, we set $\xi = 1/10$, $\delta = 1/10$, and $\epsilon_0 = 10^{-3}$. Unless specified
otherwise, we choose an equal maximum average transmit power $p_1^{\max} = \cdots =$
$p_K^{\max} = p^{\max} = 10$ dBm, an equal maximum computation capacity $f_1^{\max} = \cdots =$
$f_K^{\max} = f^{\max} = 2$ GHz, a transmit data size $s = 28.1$ kbits, and a bandwidth
$B = 20$ MHz. Each user has $D_k = 500$ data samples, which are randomly selected
from the dataset with equal probability. All statistical results are averaged over 1000
independent runs.

In Fig. 16.3, we show the value of the loss function as the number of iterations
varies for convex and nonconvex loss functions. For this feedback prediction prob-
lem, we consider two different loss functions: convex loss function $f_1(\boldsymbol{w}, \boldsymbol{x}, y) =$
$\frac{1}{2}(\boldsymbol{x}^T \boldsymbol{w} - y)^2$, and nonconvex loss function $f_2(\boldsymbol{w}, \boldsymbol{x}, y) = \frac{1}{2}(\max\{\boldsymbol{x}^T \boldsymbol{w}, 0\} - y)^2$.
From this figure, we can see that, as the number of iterations increases, the value
of the loss function first decreases rapidly and then decreases slowly for both
convex and nonconvex loss functions. According to Fig. 16.3, the initial value of
the loss function is $F(\boldsymbol{w}^{(0)}) = 10^6$ and the value of the loss function decreases to
$F(\boldsymbol{w}^{(500)}) = 1$ for convex loss function after 500 iterations. For our prediction
problem, the optimal model $\boldsymbol{w}^*$ is the one that predicts the output without any
error, i.e., the value of the loss function value should be $F(\boldsymbol{w}^*) = 0$. Thus, the
actual accuracy of the proposed algorithm is $\frac{F(\boldsymbol{w}^{(500)})-F(\boldsymbol{w}^*)}{F(\boldsymbol{w}^{(0)})-F(\boldsymbol{w}^*)} = 10^{-6}$ after 500
iterations. Meanwhile, Fig. 16.3 clearly shows that the FL algorithm with a convex
loss function can converge faster than that the one having a nonconvex loss function.
According to Fig. 16.3, the loss function monotonically decreases as the number of
iterations varies for even nonconvex loss function, which indicates that the proposed
FL scheme can also be applied to the nonconvex loss function.

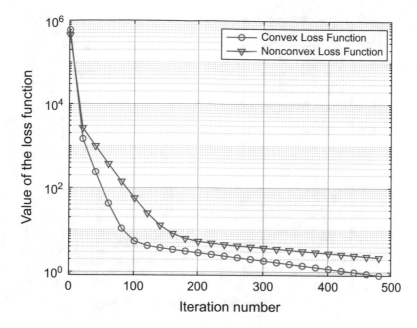

Fig. 16.3 Value of the loss function as the number of iterations varies for convex and nonconvex loss functions

We compare the proposed FL scheme with the FL FDMA scheme with equal bandwidth $b_1 = \cdots = b_K$ (labelled as 'EB-FDMA'), the FL FDMA scheme with fixed local accuracy $\eta = 1/2$ (labelled as 'FE-FDMA'), and the FL time division multiple access (TDMA) scheme in [12] (labelled as 'TDMA'). Figure 16.4 shows how the delay changes as the maximum average transmit power of each user varies. We can see that the delay of all schemes decreases with the maximum average transmit power of each user. This is because a large maximum average transmit power can decrease the transmission time between users and the BS. We can clearly see that the proposed FL scheme achieves the best performance among all schemes. This is because the proposed approach jointly optimizes bandwidth and local accuracy η, while the bandwidth is fixed in EB-FDMA and η is not optimized in FE-FDMA. Compared to TDMA, the proposed approach can reduce the delay by up to 27.3%.

Figure 16.5 shows the delay versus η. From this figure, it is found that the delay is always a convex function with respect to the local accuracy η. It is also found that the optimal η increases with the increase of maximum average transmit power. This is because small η leads to samll number of global iterations, which can decrease the transmission time especially for small maximum average transmit power.

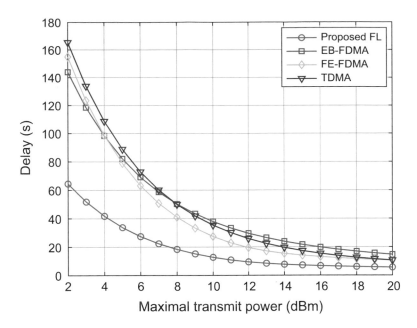

Fig. 16.4 Delay versus maximum average transmit power of each user

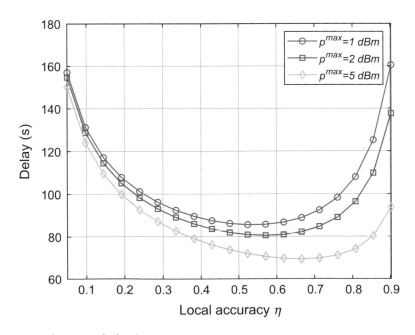

Fig. 16.5 Delay versus the local accuracy η

6 Conclusions

In this chapter, we have addressed the key challenges of combining FL techniques with wireless networks. Moreover, we have investigated the delay minimization problem of FL over wireless communication networks. The tradeoff between computation delay and transmission delay is determined by the learning accuracy. To solve this problem, we first proved that the total delay is a convex function of the learning accuracy. Then, we have obtained the optimal solution by using the bisection method. Simulation results have shown the various properties of the proposed solution.

References

1. W. Saad, M. Bennis, M. Chen, A vision of 6G wireless systems: Applications, trends, technologies, and open research problems. IEEE Netw. **34**, 134–142 (2020)
2. J. Konečný, H.B. McMahan, D. Ramage, P. Richtárik, Federated optimization: Distributed machine learning for on-device intelligence (2016). Preprint arXiv:1610.02527
3. H.B. McMahan, E. Moore, D. Ramage, S. Hampson, B.A.y. Arcas, Communication-efficient learning of deep networks from decentralized data (2016). Preprint arXiv:1602.05629
4. H.H. Yang, Z. Liu, T.Q.S. Quek, H.V. Poor, Scheduling policies for federated learning in wireless networks. IEEE Trans. Commun. **68**, 317–333 (2020)
5. S. Wang, T. Tuor, T. Salonidis, K.K. Leung, C. Makaya, T. He, K. Chan, Adaptive federated learning in resource constrained edge computing systems. IEEE J. Sel. Areas Commun. **37**(6), 1205–1221 (2019)
6. G. Zhu, D. Liu, Y. Du, C. You, J. Zhang, K. Huang, Towards an intelligent edge: Wireless communication meets machine learning (2018). Preprint arXiv:1809.00343
7. G. Zhu, Y. Wang, K. Huang, Low-latency broadband analog aggregation for federated edge learning (2018). Preprint arXiv:1812.11494
8. J.-H. Ahn, O. Simeone, J. Kang, Wireless federated distillation for distributed edge learning with heterogeneous data (2019). Preprint arXiv:1907.02745
9. K. Yang, T. Jiang, Y. Shi, Z. Ding, Federated learning via over-the-air computation (2018). Preprint arXiv:1812.11750
10. Q. Zeng, Y. Du, K.K. Leung, K. Huang, Energy-efficient radio resource allocation for federated edge learning (2019). Preprint arXiv:1907.06040
11. M. Chen, Z. Yang, W. Saad, C. Yin, H.V. Poor, S. Cui, A joint learning and communications framework for federated learning over wireless networks (2019). Preprint arXiv:1909.07972
12. N.H. Tran, W. Bao, A. Zomaya, C.S. Hong, Federated learning over wireless networks: Optimization model design and analysis, in *Proceedings of the IEEE Conference on Computer Communications*. Paris (2019), pp. 1387–1395
13. V. Smith, C.-K. Chiang, M. Sanjabi, A.S. Talwalkar, Federated multi-task learning, in *Advances in Neural Information Processing Systems* (Curran Associates, Inc., Red Hook, 2017), pp. 4424–4434
14. B. Liu, L. Wang, M. Liu, C. Xu, Lifelong federated reinforcement learning: a learning architecture for navigation in cloud robotic systems (2019). Preprint arXiv:1901.06455
15. S. Wang, T. Tuor, T. Salonidis, K.K. Leung, C. Makaya, T. He, K. Chan, When edge meets learning: Adaptive control for resource-constrained distributed machine learning, in *IEEE International Conference on Computer Communications*, Honolulu, HI (2018), pp. 63–71

16. Z. Yang, M. Chen, W. Saad, C.S. Hong, M. Shikh-Bahaei, Energy efficient federated learning over wireless communication networks (2019). Preprint arXiv:1911.02417 (2019)
17. Z. Yang, M. Chen, W. Saad, W. Xu, M. Shikh-Bahaei, H.V. Poor, S. Cui, Energy-efficient wireless communications with distributed reconfigurable intelligent surfaces (2020). Preprint arXiv:2005.00269 (2020)
18. K. Buza, Feedback prediction for blogs, in *Data Analysis, Machine Learning and Knowledge Discovery* (Springer, Berlin, 2014), pp. 145–152

Chapter 17
Role of Open-Source in 6G Wireless Networks

Avinash Bhat, Naman Gupta, Joseph Thaliath, Rahul Banerji, Vivek Sapru, and Sukhdeep Singh

Abstract Telecom domain faces lot of challenges with respect to improvement of flexibility and agility of network deployments. Telecom Service Providers (TSPs) are constantly looking for innovative solutions to reduce network deployment. On the other hand, the requirements for 6G networks will be more stringent when compared to 5G networks. It aims to support terabytes/sec throughput and latency of 0.1 ms or in microseconds. Innovative and collaborative solutions will be required to achieve the requirements of 6G networks. Open source projects like OpenDaylight, OPNFV, OpenStack, M-CORD, ONAP will be play a key role. There are several other multiple initiatives taken by the operators like disaggregation of radio network nodes, introduction of artificial intelligence in the telecom networks, etc. These initiatives provide collaborative environment to develop open source platforms that in turn provides optimized solution in terms of cost and innovation. During the 5G standardization, concepts like Service based Architecture (SBA) were defined. Keeping that as a base, programmable secure plug and play open architecture will play a crucial role during 6G standardization. In this chapter we explore the backbone of 6G wireless i.e. (1) Intelligence and (2) Automation and provide the insights on how open source will further optimize the above two avenues.

1 Introduction

5G is seen as a key tenet of 4th Industrial Revolution and is designed to connect humans and things on a common platform. The Telecommunication providers (TSPs) are key enablers of this transformation. While adopting 5G and implementing the various services across verticals, the TSPs today are grappling with various challenges such as shrinking revenues, traditional processes, long service deployment cycle etc. and are demanding innovation in all areas of their operations.

A. Bhat · N. Gupta · J. Thaliath · R. Banerji · V. Sapru · S. Singh (✉)
Samsung R&D Institute India-Bangalore, Bangalore, Karnataka, India
e-mail: sukh.sandhu@samsung.com

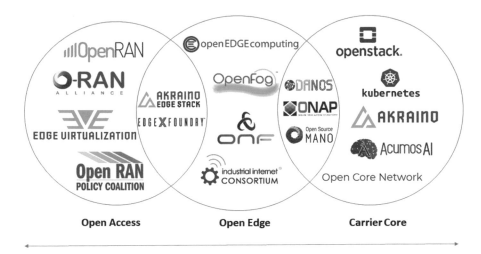

Fig. 17.1 Today's typical open source based 5G network

Unlike the previous generations, this marks a new trend in 5G and even standard bodies have started focusing on other areas apart from the core technology aspect and supporting software based Network functions so as to empower the TSPs. The softwarization of Network function has also brought in a whole new layer of software orchestration into the TSPs network building a foundation for automation. The softwarization of Network functions, has pushed the telecommunication world closer to IT and concepts such as cloud native based solutions are considered key enablers for TSPs transformation. A lot of the cloud native and virtualization concept have been traditionally driven by open source platform and tools and this is now leading to a symbiotic relationship between Open Source and Telecommunication Networks. This trend is irreversible and only going to intervene further in 6G, where we aspire to build a network centred on things.

Telecom communication networks has its own unique challenges that are essential to reduce deployment cost, improve flexibility and agility to network & service deployment models. TSPs are now looking at open source as a key tool to address a few of these challenges. Open source projects like OpenDaylight, OPNFV, OpenStack, M-CORD, ONAP, and others have enabled more rapid innovation across network components. Today's typical open source based 5G network is depicted in Fig. 17.1.[1]

Historically the TSPs Radio Network forms majority of the operator's CAPEX Investment. Further RAN has been based on proprietary solutions, right from the hardware all way up to the application layer. That makes it incredibly difficult to innovate at the same pace as the rest of the network. TSPs are hence pushing the

---

[1]Linux Foundation: https://www.linuxfoundation.jp/full-stack-slide-2/.

O-RAN Alliance initiative to drive new levels of openness in the RAN of next generation wireless systems. O-RAN is going to make RAN smarter and more open than previous generations by developing reference designs that contain more open, inter-operable and standardized interfaces. Crucial to this effort is whitebox network (physical) elements and software applications [1]. Operators driving the disaggregation of Networks (most on Radio side) and introduction of machine learning & artificial intelligence for improved operations based on open source is a clear indication that they are looking for more agility and innovation than currently defined in 5G standards. Further initiatives like Open RAN Policy Coalition are formed to advance the adoption of open and interoperable solutions in the Radio Access Network (RAN) as a means to create innovation, spur competition and expand the supply chain for next generation wireless networks [2]. In US, DARPA has drafted an Open, Programmable, and Secure 5G (OPS-5G) that is targeted to lead the development of a portable standards-compliant network stack for 5G mobile that is open source with a special emphasis on Security [3].

When 5G was designed, the concepts like Network Function Virtualization (NFV) and Software-Defined Network (SDN) principles were well established and hence 3GPP defined a Service-Based Architecture (SBA) along with Control & User Plane Separation (CUPs) to enable a modular and flexible deployment option to TSPs. Similarly it can be expected that Open Programmable Secure deployment architecture will be the centre to the 6G Network standards. 6G networks are going to pose very stringent requirements on network functions to support terabytes/sec throughput and latency of 0.1 ms or in microseconds. This calls for open-source platforms that enable low-level algorithmic development, much deeper into the technologies than any open-source software or hardware has done until now [4]. It is envisioned that Air Interface can be designed on the fly, making the need to include the Artificial Intelligence in Network stack.

The complexity both in Physical layer and in baseband processing will need to increase substantially to achieve the 6G vision. A lot of plug and play flavours of Artificial intelligence is required at the Network to design optimal air interfaces. ML-driven smart surfaces in mobile environments may require continuous retraining, in which the access to sufficient training data, high computational capabilities, and guaranteed low training convergence are needed. At an application level, Holographic radio could be made possible with 6G, where there is a possible integration of imaging and wireless communication. In 5G the basic framework of Automation has been integrated into the network, however meeting the 6G requirements has to happen automatically without any human intervention. DevOps is likely to be a defacto 6G requirement. Additionally, Autonomous Operations (AO) eases the burden on IT Operations by combining humans and machines to dramatically decrease time to resolve an incident. Overall, Open Source will be part of the basic construct of 6G based on the best practices in 5G and further 6G itself offers many opportunities where Open Source based platform focus on Plug and Play Artificial intelligence and Automation have abundant possibilities. In this chapter, we shall discuss how backbone of 6G wireless i.e. Intelligence and Automation be benefited from Open Source.

2   Intelligence

Intelligence is considered to be one of the key drivers of 6G wireless. In this section we introspect the major enablers of Intelligence in 6G: (1) Environment AI, (2) Data Market, (3) Low level algorithmic development and (4) semantic communications and how open source will enrich the capabilities of aforesaid enablers.

2.1   Environment AI

Current/Upcoming 5G based Wireless networks posit that only transmitters and receivers, i.e. the endpoints of communication links will be optimized which will lead to better network performance. This leaves out the propagation environment out of the loop and it has been the same since dawn of modern era of communication. With the emergence of powerful cloud based resources which makes complex AI easier to implement, as well as latest research on controlling the environment to better optimize the radio resources; it is time that 6G focuses on customizing the radio environment and having an AI capable of optimizing the resulting network.

Reconfigurable Meta Surfaces (RMS) is one of the key innovations, which facilitates the management of radio environments. RMSs are envisioned to be configurable, divert/alter multiple waves across line of sight, simultaneously and have low energy footprint. RMSs make the environment a freely configurable and tailored, rather than be an unpredictable entity, as seen today. Renzo et al. [5] and Gacanin and Di Renzo [6] is a great source for further reading on RMS and it's ecosystem.

With multiple network components spread across the landscape, it becomes increasingly complex to deploy and operate them. This makes AI based entities essential to 6G, which involves interacting with the environment, deciding and reacting to incoming data, which optimizes the network in its entirety, and manipulates the path of communication accordingly.

The AI can be based on a reinforced training model, deep learning model, transfer learning based model (there are many more) to get the required result. There are trade-offs for each of the mentioned models so a careful research and planning is needed before deploying the same. Yang et al. [7] and Shafin et al. [8] go into great details on the possible scenarios and current status of each approach.

A distributed and open sourced framework is required which can house some of the key components and have them communicate efficiently. Some of the important components which would be needed includes (i) data collection and storage, (ii) data analytics to decode the environment condition and/or pre-process the data, (iii) a centralized or distributed AI which always tries to adapt and optimize the network, (iv) adaptive policies created by the AI which can be propagated throughout the network and (v) the receivers and transmitters which includes Antennas, RMSs and UE besides many more. There can be many more added components further augment the network.

The reason for push to open source based framework rather than a proprietary solution is to have multiple vendors collaborate on maintaining this platform. It is imperative that all of these components have open interfaces. Having an open, commonly agreed interface, helps decouple all of the components from each other, so that an Operator/Customer can choose any vendor implementation to plug into his framework. This will help drive costs down and the operator can choose the best set of components for his own customized and specialized deployment. Thus having an open interface and an open source based framework for efficient intercommunication among components, will help vendors specialise their components for specific tasks and above all promotes interoperability. This approach is far better than building a proprietary end to end stack which will not be economically viable in a time when 6G's precursor itself is considered very expensive. This framework coupled with open interfaces will enable the network to be self-driving and ever evolving. It will even enable smaller vendors to introduce their own services easily and enable a more varied and vibrant ecosystem with lots of room for innovation from both academic and industrial fields.

Hence the end result we will achieve is a fully automated network which controls both the radio environment (to some extent), as well as all resources in it, and it always maintains a near optimal state at any point of time. This is a dream as well as a challenge for many engineers, and 6G is the perfect opportunity to tackle this head-on.

2.2   Data Market

As the 5G technology is commercialised it has started maturing and proving to be a stepping stone for networks. With 6G we will enter a new era of completely digitised society and hyper connectivity among machines and humans, thus producing tremendously huge amount of data which can be used for intelligently managing and improving the network. The data can be collected from IoT (Internet Of Things) sensors, Autonomous cars, drones or mobile network, and can be used for targeting marketing and sales, improve operations or can be used by operator to improve and manage the network [9]. For instance, by mixing an IP address database with the logs from a website, one can understand where the customers are coming from, then adding demographic data to the mix, one can get some idea of the socio-economic bracket and spending ability of the customers. Such insights aren't limited to analytic use only, we can use it to provide value back to a customer. It provides data monetisation opportunities to operators, using existing infrastructure to offer real-time processed and analysed data.

Given the large amount of business sensitive and personal data generated in 6G security and privacy of data will be of utmost importance. Security at all levels of future systems will be much more critical in the future and 6G needs a network with embedded trust. Some of the data generated by 6G devices and elements in both public and private networks has value for many societal functions and possibly to other private corporations than the one that collects the data.

Creating a open community that transforms how data is collected, prioritized, and shared can create strong drivers for future value, but may also lead to serious privacy and ethical concerns over the location and use of data. In profiting from 6G innovation, modularity and complementarity of technological solutions are of importance. This requires openness, transparency and collaboration. So there is a need for an Open-Source platform to manage sharing of the data and creating contracts to disable compromising the privacy and security of the sensitive data. The openness of the platform created will give data life in the same way that code hosting projects in open source give life to code.

The driving goal is to connect every public and commercially available database to a common platform. The proximity of other data makes all the data more valuable, because of the ease with which it can be found and combined. For example a web page view can be given a geographical location by joining up the IP address of the page request with that from the IP address in an IP intelligence database [10]. With Market places being Open-Source, data can be used across by different domain people to make the most out of the data and everyone can benefit from it (Fig. 17.2).

The Open-Source platform will provide transparency in terms of data privacy and also help in collaboration among different operators to share data for the benefit of the network industry. This not only helps to improve the network but also provide a revenue from the monetisation of the data. Apart from all the above

Fig. 17.2 Data market and open source summary

benefits of open source data market place, using technologies like block chain, big data etc. available in open source solves the problem of standardising the price of the data and maintaining the integrity of the data transactions by providing smart contracts. One more aspect of using open source is that there is a need for a community or a body which will define the fairness to the marketplace by providing the privacy regulations for the data. Access to data and data ownership are increasingly major factors in value creation, and limiting such access is a means of control. Furthermore, how the data itself can be used becomes a key question. The contractual rights and obligations of the different members of a communications ecosystem may describe how the information and data may be used [4]. The challenge, however, will be the mapping of these rights and obligations to the data collected and used by highly adaptive autonomous systems, or smart devices, to create the services of the future. Thus, we believe that to make 6G acceptable to society, the protection of private information will be a key enabler to realize its full potential, and open source will be the driving force for the management of the data marketplaces.

2.3 Low-Level Algorithmic Development

There will be trillions of devices to be served under the umbrella of 6G with diverse capabilities serving diverse use cases which will be more complex than 5G. Under a single 6G New Radio (NR), there will be millions of machines, processing diverse information and having different operational requirements. For instance, from scheduling point of view, there will be different scheduling algorithms for different kinds of devices and use cases. Many use cases operating under a single base station, calls for a plug and play low algorithmic development, which can understand the scheduling and resource requirement deeply per use case, which in turn will help in efficient utilization of resources. To decide importance of use cases per area and its demand, different machine learning techniques will have to be amalgamated with scheduler so that scheduler can take decisions regarding optimal scheduling algorithm per use case well before advance. Low level algorithmic design is the key for choosing appropriate scheduling and machine learning algorithms which can offer flexibility to service providers and the operators to choose the best amongst many. Open source will play a very crucial role to offer the flexibility where in different service providers or developers can write different scheduling algorithms for different use cases and plug it on a common scheduling platform serving different use cases. Also, if there is a requirement for running multiple use cases in a single device at a particular time instance, low level algorithmic developers need to make their algorithm open and inter operable. From physical layer point of view, where there will be huge amount of antennas, RF transceivers, signal processing units, etc. low level algorithmic development becomes very crucial. For such purposes, open-source platforms will again play a very crucial role.

2.4 Semantic Communications

Shannon's classical information theory (CIT) laid down the foundation of old and modern day communications. Modern day communications have already met the predicted limits of Shannon's theory. Therefore, semantic communications is expected to take a leap forward for 6G networks [11]. One of the prominent example is holographic communications wherein a speaker can share the multiple camera video feeds remotely and can create its own hologram [12] which in turn requires huge video data transmissions. Due to uncertainty in millimetre wave or sub-THz link, some packets might get lost which need to be re-transmitted in the conventional system. This may lead to distorted holographic video or image. Hence, transmitter and the receiver should have the capability to predict and construct the hologram in case of packet loss. This can be possible if the sender and the transmitter have access to common shared knowledge about the speaker. In such cases semantic inference can efficiently recover the content based on the shared knowledge. In B5G and 6G networks, the transmitter and receivers highly likely will be embedded with semantic intelligence but the algorithm to create video or an image can be different in both. So, it is very essential to have one common open platform and language where in the semantic applications or algorithms can be developed and plugged in from various developers for various communication task or ultra-modern communication use cases (for e.g. design of optimal air interfaces on the fly for a given environment and set of specific requirements). In a nutshell, open platform for semantic communication can improve communication efficiency and can help in attaining the basic requirement of shared knowledge which in turn can enable optimized semantic inference.

3 Automation

B5G and 6G wireless are expected to be fully automated wherein, right from design and instatiation, to orchestration and healing procedures of Network equipments are performed automatically without intervention of the operators or service providers. DevOps and Autonomous Operations (AO) play a major role in automation of telecom services but lacks interoprability amongst multivendor support. Open source plays a major role in overcoming the aforesaid challenges and is detailed below.

3.1 DevOps

Telecom software is moving towards microservice based architecture and DevOps will play a key role in the telecom industry similar to other industries in the IT domain. We are already seeing telecom vendors mentioning about implementing

DevOps in 5G [13]. In 6G, DevOps will play a critical role in delivering high quality and complex software rapidly. Most of the DevOps concepts and tools were being developed for several years and are now well established in the IT industry. There are many open source tools used for DevOps, which are widely used by non-telecom domains. The open source tools available for DevOps are also frequently updated in the community with new features. These advanced open source tools can be directly reused in the telecom domain instead of reinventing them. In 6G, there are many new critical requirements like ultra-low latency and very high throughput when compared to 5G. In 6G, there will be drastic increase of complexity in terms of hardware, software and number of vertical industries. This will put enormous stress on the telecom domain experts to cater all the requirements of 6G. Utilizing open source software for DevOps will accelerate time to market for many of the new 6G features and will reduce redundant development effort. The focus for telecom industry experts will be to develop software that are microservices concept based and fully virtualized. In 6G, the main challenge of applying DevOps in telecom domain especially in Radio Access Network is the dynamic upgrades of time critical software, which will directly affect user experience even if there is downtime of the order of microseconds. New approaches and research is required to perform dynamic upgrades of 6G software components without causing any impact to service level KPIs. The telecom domain experts need to focus on telecom software design for developing novel solutions related to dynamic upgrades while reusing well established open source tools for creating the end to end DevOps pipeline. Reducing the effort required for developing DevOps tools and reusing open source tools from other domain will increase the speed of innovation in the telecom domain. Automated verification at different stages of the DevOps pipeline, will include testing and providing feedback, which in turn is a crucial aspect. Automated verification can be done using open source implementation of all the components of the system apart from the components under test. The use of open source software and tools for automated verification will ensure substantial improvement of the software quality and significantly reduce the various interoperability challenges faced during deployment. During deployment of 6G networks, operators will prefer to avoid vendor lock in by utilizing software and hardware components from different vendors. Operators will try to maximize the use of open source tools for DevOps and use standardized interfaces to enable effortless integration of multiple vendor software components. The sheer complexity involved in the deployment of 6G networks will also drive operators to use open source tools for the DevOps pipeline to remain cost competitive (Fig. 17.3).

3.2 Autonomous Operations (AO)

Today's operations and incident management systems are mostly manual and governed by static rules that provide partial automation and is supported by dis-aggregated teams (L1,L2,L3) which require lot of co-ordination, manual incidents

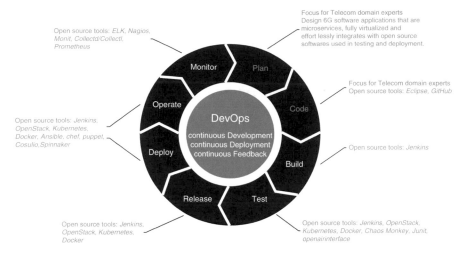

Fig. 17.3 DevOps and open-source

assessment, classification and assignment. With the 5G revolution, micro-services and virtualization of the edge takes centre stage, and more critical workloads like networks elements are being deployed on virtual infrastructures that requires quicker Time to Resolve(T2R) incidents. Also as services become more complex, IT operations need to monitor mobile applications, mIoT devices, across fragmented hybrid cloud deployments and operators working along with other micro-operators to provide services to end customer. All this complexity would push IT operations to become more autonomous by the end of 2030. Autonomous Operations is the next generation of IT automation that would move enterprise IT from the legacy and rules-based solutions of the past to a future where operations are autonomous. Autonomous Operations (AO) eases the burden on IT Operations by combining humans and machines to dramatically decrease the time to resolve an incident. AO increases service availability while driving operational costs down (Fig. 17.4).

By the end of 2030, the centralized and proprietary Autonomous IT operation would not be enough to support the services that emerge from the 6G revolution. These next generation services would be more complex, interdependent, intelligent and autonomous. This would lead to the next evolution in IT Operations, which will be fully autonomous, open and integrated, and will be deployed on the edge.

Services that emerge beyond 5G would be driven by Quantum machine learning and will be completely autonomous. Autonomous mobility [14] is identified as a key theme that would shape our society beyond 2030. As more manual services like field force, delivery vehicles(drones), construction based heavy vehicles, etc. become completely autonomous and rely on edge hosted cloud compute and application, incidents not resolved in microseconds might lead to serious consequences. The requirements for incidents to be resolved in fractions of a second would require operations to be fully autonomous, open and partially deployed at the edge. Fully

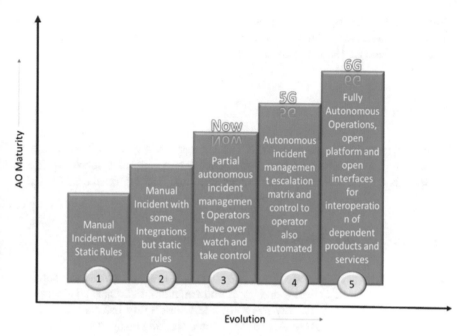

Fig. 17.4 AO and open-source

autonomous operations would be able to detect, prioritize, investigate and remediate incidents completely without human intervention.

Forecasts suggest that by 2030 around 50 billion IoT devices will be in use around the world, creating a massive web of interconnected devices spanning everything from wearables to kitchen appliances [15]. With the advent of 6G, as more devices join the network, an eco-system of highly complex interdependent services would evolve. These services would be offered by different vendors however the service would be highly dependent on each other. This would require operation platform to adopt open interfaces and open platforms where intelligent applications from different vendors can reside and collaborate to find and optimize the performance of all services based on criticality.

By 2030, most of the network will be virtualized, as boundary between network operations and IT operations fade, it would become complex to handle and resolve problems and assure SLAs across these layers. To save OPEX and CAPEX cost, new emerging concepts like micro-operators, ORAN and spectrum sharing would take center stage in 6G. These would trigger operational challenges where multiple telecommunication operators,cloud providers, over the top service provides e.t.c. need to inter-operate to solve/resolve operational issues.

Open-source adoption would not only reduce CAPEX and OPEX costs(which is expected to be one of the biggest challenges for 6G adoption) but also greatly reduce time to resolve network and service related problems. Open-source software and standards would enable telecom service providers, over the top service providers, cloud/infrastructure providers and other new/innovative service provider to work and operate together as an ecosystem rather than working in silos, which would make microseconds level SLA for 6G a reality.

4 Open Issues

We now discuss some open issues that can emerge while realising open source in next generation wireless. The key challenges for adopting open-source discussed are Security and Privacy, Support and Operations and Lacking Business Models.

4.1 Security and Privacy

There has always been concerns around open-source security. Open source software is available for anyone to deploy and try to find vulnerabilities. Once a vulnerability is found it quickly finds its way to the online vulnerability database where it is available for anyone to misuse. Also as the people working on open source software are not governed by any organization it is easier for non-ethical developers to become part of such community and to build backdoors that can be utilized later for attacks. These security and privacy issues would be much tougher to handle, as the scale at which open source software is written and adopted increases drastically for 6G [16].

4.2 Support and Operation Complexity

Proprietary software comes with vendor support contracts. These contracts act as insurance for getting any issues or defects in the software resolved within the agreed SLA as defined in these contracts.Its generally very tough to get quick resolutions of defects from open source communities, hence the enterprise should have the competence to support and resolve any defects in the open-source software on its own. As more of propritary network hardware gets virtualized and the scale and penetration of open source software increases with the 6G revolution, the variety of skills and costs required to support such software becomes huge and poses a big challenge for open source adoption in 6G.

4.3   Business Model

More telecom service providers are adopting open source. There has been a huge surge in new open source projects especially in the telecom space. As more service providers adopt open source, more and more software and network equipment vendors are showing interest in these communities. With proprietary platforms becoming open, these software and network vendors lose their competitive advantage and their profit margins and are trapped unless they understand how to leverage their platforms for commercial gains. This would make it unsustainable for vendors(both software and hardware) to support open-source software communities in the longer run, unless new business models around open source emerge.

5   Concluding Remarks

In this chapter, we introduced the need of open source in the 6G wireless. Subsequently, we put forward discuss two major backbone of 6G i.e. Intelligence and Automation. We discuss the role of open source in providing the swiftness and interoperablity to the key enablers of Intelligence and Automation. Finally, we investigated the open issues and challenges that can be faced while realizing the openness.

References

1. How O-RAN SC completes the open source networking telecommunications stack. https://www.linuxfoundation.org/blog/2019/04/how-o-ran-sc-completes-the-open-source-networking-telecommunications-stack/
2. Open RAN policy coalition. https://www.openranpolicy.org/about-us/
3. Open, programmable, secure 5G (OPS-5G). https://www.darpa.mil/program/open-programmable-secure-5g
4. M. Latva-aho, K. Leppänen, et al., 6G research visions 1. Key drivers and research challenges for 6G ubiquitous wireless intelligence, 6G Flagship, University of Oulu (2019)
5. M.D. Renzo, M. Debbah, D. Phan-Huy, et al., Smart radio environments empowered by reconfigurable AI meta-surfaces: an idea whose time has come. J. Wireless Commun. Netw. **2019**, 129 (2019). https://doi.org/10.1186/s13638-019-1438-9
6. H. Gacanin, M. Di Renzo, Wireless 2.0: towards an intelligent radio environment empowered by reconfigurable meta-surfaces and artificial intelligence (2020). arXiv:2002.11040
7. H. Yang, A. Alphones, Z. Xiong, D. Niyato, J. Zhao, K. Wu, Artificial intelligence-enabled intelligent 6G networks (2019). arXiv:1912.05744
8. R. Shafin, L. Liu, V. Chandrasekhar, H. Chen, J. Reed, J.C. Zhang, Artificial intelligence-enabled cellular networks: a critical path to beyond-5G and 6G. IEEE Wireless Commun. **27**(2), 212–217 (2020). https://doi.org/10.1109/MWC.001.1900323
9. White Paper on Business of 6G "6G Research Visions", No. 3 (2020)
10. Data markets compared. http://radar.oreilly.com/2012/03/data-markets-survey.html

11. J. Bao et al., Towards a theory of semantic communication, in *2011 IEEE Network Science Workshop, West Point* (2011), pp. 110–117. https://doi.org/10.1109/NSW.2011.6004632
12. E. Calvanese Strinati, et al., 6G: the next frontier (2019). https://arxiv.org/pdf/1901.03239.pdf
13. DevOps: fueling the evolution toward 5G networks. https://www.ericsson.com/en/reports-and-papers/ericsson-technology-review/articles/devops-fueling-the-evolution-toward-5g-networks
14. White Paper on Critical and Massive Machine Type Communication Towards 6G. arXiv:2004.14146v2
15. Number of internet of things (IoT) connected devices worldwide in 2018, 2025 and 2030. https://www.statista.com/statistics/802690/worldwide-connected-devices-by-access-technology/
16. G. Gui, M. Liu, F. Tang, N. Kato, F. Adachi, 6G: opening new horizons for integration of comfort, security and intelligence. IEEE Wireless Commun. (2020). https://doi.org/10.1109/MWC.001.1900516

Chapter 18
The Intersection of Blockchain and 6G Technologies

Tri Nguyen, Lauri Lovén, Juha Partala, and Susanna Pirttikangas

Abstract The fifth generation (5G) wireless networks are on the way to be deployed around the world. The 5G technologies target to support diverse vertical applications by connecting heterogeneous devices and machines with drastic improvements in terms of high quality of service, increased network capacity and enhanced system throughput. However, 5G systems still remain a number of security challenges that have been mentioned by researchers and organizations, including decentralization, transparency, risks of data interoperability, and network privacy vulnerabilities. Furthermore, the conventional techniques may not be sufficient to deal with the security requirements of 5G. As 5G is generally deployed in heterogeneous networks with massive ubiquitous devices, it is quite necessary to provide secure and decentralized solutions. Motivated from these facts, in this paper we provide a state-of-the-art survey on the integration of blockchain with 5G networks and beyond. In this detailed survey, our primary focus is on the extensive discussions on the potential of blockchain for enabling key 5G technologies, including cloud computing, edge computing, Network Function Virtualization, Network Slicing, and D2D communications. We then explore and analyse the opportunities that blockchain potentially empowers important 5G services, ranging from spectrum management, data sharing, network virtualization, resource management to interference management, federated learning, privacy and security provision. The recent advances in the applications of blockchain in 5G Internet of Things are also surveyed in a wide range of popular use-case domains, such as smart healthcare, smart city, smart transportation, smart grid and UAVs. The main findings derived from the comprehensive survey on the cooperated blockchain-5G networks and services are then summarized, and possible research challenges with open issues

---

These authors "Tri Nguyen and Lauri Lovén" contributed equally.

T. Nguyen (✉) · L. Lovén · J. Partala · S. Pirttikangas
Faculty of Information Technology and Electrical Engineering, University of Oulu, Oulu, Finland
e-mail: Tri.Nguyen@oulu.fi; Lauri.Loven@oulu.fi; Juha.Partala@oulu.fi;
Susanna.Pirttikangas@oulu.fi

© The Author(s), under exclusive license to Springer Nature Switzerland AG 2021
Y. Wu et al. (eds.), *6G Mobile Wireless Networks*, Computer Communications
and Networks, https://doi.org/10.1007/978-3-030-72777-2_18

are also identified. Lastly, we complete this survey by shedding new light on future directions of research on this newly emerging area.

Keywords 5G networks · Blockchain · 5G Internet of Things · 5G services · Machine learning · Security and privacy

1 Introduction

Satoshi Nakamoto, Feb. 2009

"The central bank must be trusted not to debase the currency, but the history of fiat currencies is full of breaches of that trust. Banks must be trusted to hold our money and transfer it electronically, but they lend it out in waves of credit bubbles with barely a fraction in reserve."

Bitcoin [1] cryptocurrency first appeared in 2008 in relative obscurity. Yet, over the last decade Bitcoin has attracted enormous attention due to the skyrocketing value of its assets, multiplying ten-thousandfold. While a number of competing cryptocurrencies have since emerged, Bitcoin's head start ensured its popularity and, consequently, the highest market capitalization.

Behind Bitcoins success is its technology, which voids the need for trusted intermediaries in monetary transactions. In more detail, instead of trusted intermediaries, Bitcoin's transactions are distributed among a large crowd of anonymous participants, utilizing a core foundation called *blockchain*. Blockchain is, in essence, a public database distributed among a network of participants. It relies on a unique chronological chain of immutable data blocks, shared among all participants, and constantly growing as a result of mounting transactions.

As a potential key component for decentralized systems in general, blockchain technology has been intensively studied for applications besides monetary transactions. Common among these applications is the requirement for trust, that is, an assumption that the actions taken and data provided by some participants are not fraudulent or false. Such applications include, for example, tracking of parcels in a logistic network [2, 3], keeping records of land-owners [4–6], or billing systems in multi-party environments [7, 8]. These applications have, historically, relied on a central authority to provide the trust and manage the book-keeping. Instead, blockchain aims at a *trustless* system, distributing trust among the participants using a consensus mechanism, potentially voiding the need for central authorities.

On the other hand, 6G mobile networks promise to improve latency, data rates, spectrum efficiency, user mobility, connectivity density, network energy efficiency and configurability, and area traffic capacity. Building upon these improvements, 6G will offer an environment for the growth of interactive services and technologies [9–14].

In particular, multi-functionalization, artificial intelligence (AI), and Internet of Things (IoT) stand to benefit from 6G. In more detail, the billions of IoT devices expected to be deployed in the 6G era and communicating with each other as well as with AI agents residing in the cloud or on the edge will require unprecedented performance from the underlying network [9, 15, 16].

Blockchain is also expected to flourish in the 6G era, with three drivers in particular promoting the interplay of blockchain and 6G technologies. First, blockchain requires heavy communication among participants to guarantee the consistency and integrity of the ledger. 6G promises to have the capacity to support the resulting burden on the network. For example, moving from 5G to 6G, the development from ultra-reliable low-latency communication (URLLC) to massive URLLC will improve blockchain latency, reliability and traffic capacity.

Second, decentralizing services and reducing the need for trusted parties, blockchain technology is considered a state-of-the-art solution for next-generation connectivity [9, 13, 17–19]. Indeed, the 6G network can benefit from blockchain-based solutions in functionalities such as spectrum sharing, device-to-device content caching, and resources management.

Third, in addition to blockchain, a whole ecosystem of technological enablers is expected to expand further in the 6G era [9, 20]. The interplay of these enablers with blockchain will further benefit some platforms and services. For example, IoT stands to benefit both from the performance of the 6G networks as well as the decentralization and transparency provided by blockchain. Further, many applications, platforms and verticals, such as vehicular edge networks, smart cities, sharing economy, and social compliance, are expected to benefit from the combination [19, 21].

In this chapter, we will study two of the above drivers, namely, the potential benefits blockchain has to offer for 6G networks and services, and the interplay of blockchain with key enabling technologies for 6G. Further, we provide an overview of the blockchain technology and related concepts, and look into blockchain-based applications leveraging mobile networks. Finally, we look at the challenges related to blockchain-based applications.

The overview of blockchain technology is presented in Sect. 2, and blockchain applications described in Sect. 3. The combination of 6G and blockchain is presented in Sect. 4, challenges presented in Sect. 5, and Sect. 6 concludes the chapter.

2   Blockchain

While distributed databases have been around since at least the 70s [22], blockchain components started to come together in 1990 when Haber and Stornetta proposed the secure time-stamping of digital files to prevent modifications [23]. Further, Bayer et al. [24] proposed Merkle trees [25], that is, complete binary trees built based on a one-way function, to enhance the efficiency and reliability of the time-stamping. Block data integrity was verified by Mazieres and Shasha [26]

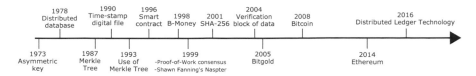

Fig. 18.1 Timeline of the technologies contributing to blockchain

and Li et al. [27], while Bitcoin's Proof-of-Work (PoW) consensus mechanism is partly based on work on the consensus mechanisms of earlier digital currency such as b-money [28], hashcash (with reusable PoW) [29], and especially bitgold.[1] Furthermore, Bitcoin relies on technologies such as asymmetric cryptography (1973) [30], peer-to-peer (P2P) networks (e.g. Shawn Fanning's Napster, 1999) and cryptographic hash functions such as SHA-2 (2001) [31].

The growth of blockchain has ushered in new applications such as smart contracts. The principle of smart contracts was introduced by Szabo [32] in 1996, who defined a smart contract as a protocol that transfers and automatically executes transactions after satisfying specific conditions. In 2014, the concept was successfully realized on the Ethereum blockchain platform [33]. The availability of smart contracts further broadens the scope of blockchain-based interactive applications.

Inspired by the popularity of blockchain, a range of distributed ledger technologies (DLT) have emerged since 2016. A more general concept than blockchain, a DLT still aims to distribute a database among participants, but it is more flexible in terms of structure, design, and network architecture. As such, blockchain is one type of a distributed ledger, while distributed ledgers are instances of distributed databases.

The timeline of technologies contributing to the growth of blockchain is depicted in Fig. 18.1.

2.1   Architecture

Blockchain architecture can be divided in two components: network architecture and data architecture. Blockchain network architecture refers to the means of communication for node discovery and maintenance, the routing protocols used, and the data encryption schemes used in transmission. Blockchain data architecture, as illustrated in Fig. 18.2, refers to the relationships between its data structures as well as the cryptographic schemes used to ensure data integrity and immutability.

In more detail, a blockchain can be decomposed in three parts, namely, the transactions, the data blocks, and the chain of blocks. Transactions use the hash function

---

[1] https://unenumerated.blogspot.com/2005/12/bit-gold.html.

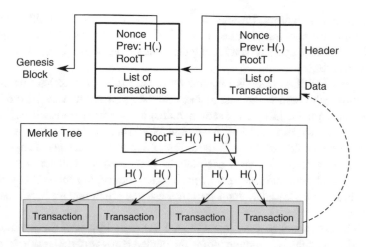

Fig. 18.2 Blockchain comprises a linear chain of linked blocks, each of which consists of a block header part and a block data part. The header stores metadata, including a PoW result (nonce), a hash value of the previous block (Prev), and a root value of Merkle tree transactions (RootT). The RootT value of each block is the final value of a Merkle Tree [25] containing the transactions and their hash values. Two hash values of two transactions are concatenated to form a higher tier value until reaching the root of the tree (RootT)

and asymmetric cryptography to preserve integrity and authenticity, respectively, while data blocks use a hash pointer and a Merkle tree [25] to guarantee both the integrity and the order of transactions. Finally, each block refers to the preceding one with a hash pointer, calculated from the entire part of that preceding block, ensuring the integrity of the whole chain. The first block of the chain is called the genesis block.

2.2   Data Models

According to Belotti et al. [34], blockchains have four types of data models used for tracking transactions: the Unspent Transaction Output (UTXO) model, the account-based model, UTXO$^+$, and the key-value model.

Popularized by Bitcoin, an UTXO represents a digital asset, defined by its chain of ownership listing all its previous owners. To calculate the balance of a Bitcoin participant, the chain of ownership of all received UTXOs is followed before summing their values. A transaction, that is, a transfer of assets between two participants, requires a verification of the UTXOs that are used as inputs for generating the new transaction. Each transaction on a new block, in turn, consists of a set of new outputs from preceding transactions.

As a consequence, the aggregate of all UTXO states is the blockchain state. The UTXO model makes verification of this state expensive, as it requires tracking the full history all transactions. Further, data is duplicated, as output of old transactions become input for new transactions. The UTXO model should thus be applied only when there are few operations affecting the state of the whole blockchain, as with cryptocurrencies.

While the UTXO model is adopted by a number of blockchain-based cryptocurrencies, Ethereum is the first blockchain using an account-based model. The account-based model has each account managing its own transactions, with all accounts in the system forming the global blockchain state. Further, the model supports smart contracts by way of different types of accounts. Ethereum, for example, uses externally owned accounts to keep track of balance, and contract accounts to hold executed code and internal states. Hence, the account-based model provides blockchain-based systems intuitiveness and efficiency for blockchain state operations.

The UTXO$^+$ model aims augment the UTXO model towards the account-based model, without actually implementing accounts. However, the result is seen as complex and unintuitive to use [34]. Corda [35] is an example of a blockchain implementing the UTXO$^+$ model.

The key-value model (table-data model) is currently actively studied, as it can support both the transaction-based blockchain state or the account-based state, depending on the application, thus providing for a wide range of use cases. For example, the Hyperledger Fabric blockchain [36] utilizes the key-value model to represent a collection of key-value pairs for digital currency as well as for asset exchanges. Fabcoin [36], a cryptocurrency based on the Hyperledger Fabric, implements UTXO on the key-value model.

2.3   Consensus Mechanism

Consensus protocol is the most important method for facilitating trust in the decentralized blockchain network. A high abstraction level network communication protocol, the consensus protocol has an important role in forming the blockchain and ensuring the integrity of its data, shared among a number of untrusted participants. In particular, consensus protocol decides which transactions form blocks, and which blocks are chained together.

Blockchain consensus protocol consists of elements such as proposal, propagation, validation, and finalization, and the incentive strategy [37]. While the exact details vary between different blockchain implementations, the Bitcoin blockchain first has a participant, finding the nonce and finishing her PoW, proposing a collection of transactions for a block. The proposed block is propagated to all other participants and then validated to avoid conflicts. Finally, the block is accepted as the next extension to the blockchain, if it fulfills certain conditions such as adding to the longest branch, as exemplified by Bitcoin. Without any intermediaries, the incentive strategy of blockchain consensus protocol is to encourage the participants to obey the protocol by means of a reward, consisting of digital assets newly created in the system.

Bitcoin's remarkable success has brought also consensus mechanisms in the spotlight, greatly increasing research interest on the subject. Yet, consensus mechanisms were first introduced decades ago with, for example, the Byzantine General Problem (BGP) [38] proposed already in 1982.

Indeed, before blockchain, BGP was considered as the prime candidate for a consensus mechanism. In more detail, BGP describes a group of generals attempting to attack a city with a uniform strategy. To decide upon the strategy, the generals have to communicate amongst each other and find consensus on which plan to follow—only some of the generals are actually enemy spies, seeking to thwart consensus or twist it to favor an inferior strategy.

In 1992 and 1999, the PoW [39, 40] and the Practical Byzantine Fault Tolerance (PBFT) [41] protocols were introduced as potential solutions for reaching consensus. Their ideas have been taken into use by the two main types of the blockchain, public and private (see Sect. 2.4), with PoW-based Nakamoto consensus implemented for example by the public Bitcoin blockchain, and the PBFT-based by the private YAC blockchain [42].

2.4   Access

Based on access to transaction data, blockchains implementations can be divided into three categories: public, private, and consortium chains [43–45]:

- A public (open-access, permissionless) blockchain allows all participants full control, including access, contribute, or maintain blockchain's data.
- A private (permissioned) blockchain gives access to transaction data for specific pre-defined participants only.
- A consortium blockchain allows authorized participants to manage and contribute to the blockchain. Participants are authorized by default nodes at the beginning, or by requirements specific to the particular blockchain model.

The three alternatives embody a trade-off between Decentralization and performance. In more detail, a decentralized blockchain where participants have full control on transaction data, verifying and managing blocks, requires an enormous number of messages sent between the participants to build consensus and ensure chain integrity. As a result, performance suffers, with the cost of computation rising and transaction latency and throughput getting worse. On the other hand, a highly centralized system can be made very efficient, with high throughput and low latency.

The concept of centralization is intimately linked with trust. A highly decentralized system, a public blockchain is *trustless*. It needs no trusted parties, that is, some specific, pre-defined nodes, to manage transactions, instead relying on the consensus protocol to ensure integrity. Conversely, private, centralized blockchains place trust on a central authority to manage the chain. Aiming for the middle-ground, consortium blockchains attempt to mitigate the strict division between public and private blockchains by setting up a trusted group of participants to maintain transactions.

Alternatively, blockchain-based systems can be characterized by their decentralization, immutability, integrity, and auditability [34]. As described above, decentralization refers to how blockchain data and the management of the data is distributed among the participants. *Integrity* of the blockchain data must be guaranteed to build trust among participants. Failing integrity, blockchain data can become conflicted or compromised by malicious participants. Moreover, visibility of blockchain data to different parties at different parts of the network contributes to the *auditability* and the *traceability* of data for blockchain-based applications.

2.5 Vulnerabilities

Blockchain presents a number of attack surfaces, subjecting the dependent services and applications to numerous vulnerabilities. Blockchain's vulnerabilities can be roughly divided into protocol vulnerabilities, encompassing issues with blockchain's architecture and protocols, and smart contract vulnerabilities, with issues related to the smart contract programming language and the virtual machine executing the contracts.

2.5.1 Protocol Vulnerabilities

Blockhain aims to distribute an immutable chain of data blocks among the participants. This purpose may be compromised by malicious participants, seeking to break the integrity of the data blocks for their own benefit. While some blockchain-specific attacks are introduced below, blockchain may also be vulnerable to attacks on the network layers. Such attacks include, for example, DNS attacks and distributed denial of services [46].

Sybil Attack, Selfish Mining A *fork* is a situation where two groups of participants have different blocks in their chain, creating two competing *branches* of the chain. Forks are classified into pre-determined *hard forks*, and *soft forks* which appear without warning. Rare events, hard forks may be created legitimately, with a group of participants creating a new branch for an application requiring a new blockchain network. For example, in 2017, the *Bitcoin cash* cryptocurrency [47] was established by a group of Bitcoin developers as a hard fork from classic Bitcoin to improve the size of blocks.

A soft fork, however, is the result of a protocol error. It may be unintentional, or the result of malicious nodes trying to compromise the integrity of the blockchain. Examples include a Sybil attack [48], where a malicious entity employs multiple system identities to unfairly promote a dishonest block, and selfish mining [49], where a group of colluding participants forces the remaining honest participants into wasting their computing resources, thus creating an incentive for the honest participants to join the dishonest group.

Majority Attack A majority attack (or 51% attack) aims to break the consensus for the benefit of a dishonest group, comprising more than 50% of the votes of the participants. In more detail, a majority attack targets the consensus process, based on majority voting. The attack aims, for example, to reverse a transaction by injecting a faulty block into the blockchain. The dishonest group could, for example, first sell bitcoins and then reverse the sales transaction, keeping both the Bitcoins and the sales proceeds [50].

2.5.2 Smart Contract Vulnerabilities

The two main types of vulnerabilities for blockchain-based smart contracts are those in the programming language, and those in the virtual machines, running the code for the contracts [51]. In more detail, smart contracts are computer programs, executed on the virtual machines running on the participant computing nodes. As such, smart contracts are subject to bugs and expose new attack surfaces. Below, we introduce some examples of such vulnerabilities in the Ethereum blockchain and its programming language Solidity, along with sources for more information.

Call to the Unknown Failing to look up a function with a given signature in a target contract, a remote function call will revert to a fallback function. Cleverly setting up the fallback, a malicious contract may inject code to be run by the target, making the program flow of the target contract unpredictable [51].

Re-entrancy Attack A re-entrancy attack [51] also exploits a fallback function, with a malicious contract recursively draining another smart contract of its assets. In more detail, a smart contract exposing an interface for credit withdrawal should update their balance before sending credits to the caller. Otherwise, the malicious contract may set up a special fallback function, triggered upon the arriving credits, to make another withdrawal, which again triggers the fallback, and so on. The DAO attacks[2] in mid-2016 exploited this vulnerability to steal ca. $60M in ether.

Over- and Underflow Attacks The variable types of the smart contract programming language may suffer from over- and underflow attacks [46, 52]. In Solidity, for example, a uint256 type variable is represented by eight bits, with a minimum value of 0 and a maximum value of $2^{256} - 1$. If the value of an uint256 type variable exceeds the maximum by one, it rolls over back to zero, and vice versa. An underflow attack exploits this behavior, employing a transfer which subtracts the attackers balance beyond the minimum, ending up with an extremely high number of credits.[3]

---

[2]https://www.coindesk.com/understanding-dao-hack-journalists.
[3]https://nvd.nist.gov/vuln/detail/CVE-2018-10299.

Short Address Attack The Ethereum Virtual Machine (EVM) has a bug related to the ERC20 standard[4] [46, 52]. The EVM pads a user address with extra 0's in the end if the address is shorter than a certain length. If there is a user address with a genuine 0 in the end, an attacker can impersonate that user by using her address, with the trailing 0 removed, as the EVM will append the missing zero. Moreover, this issue amplifies the number of tokens in the case of transfer.

Immutable Bugs To maintain user trust, smart contracts cannot be changed once published on the blockchain, acting independently according to their program code. Thus, if a smart contract has bugs or vulnerabilities, blockchain offers no direct methods for fixing those bugs. As a result, the resources accumulated by the contract may be at a risk [51].

Generating Randomness Smart contracts sometimes employ pseudo-random numbers, generated with a seed value from an external source. The contracts, however, are distributed, deployed simultaneously at all the participant nodes. A contract function running at different nodes thus should, in most cases, return the same results, or risk inconsistency between nodes.

As a solution, smart contracts often use the hash values of future blocks as their random seed. Identical to all nodes once available, such a seed guarantees identical pseudo-random numbers generated with that seed. A hash value of a future block is, however, unknown at the time of contract deployment, seemingly ensuring randomness. However, given enough resources, a malicious group of participants may influence the formation of the chain of blocks such that the future block and its hash value can, to some extent, be known at contract deployment time [51].

3 Applications

Blockchain-based applications have gone through three major evolutionary steps. To begin with, the success of Bitcoin attracted attention in cryptocurrencies and the transfer and exchange of digital monetary assets. However, the community using digital currencies remains small. Many issues prevent cryptocurrencies from reaching a wider audience, including a limited capacity for uses beyond the low-latency transfer of digital assets between two users.

To support a wider range of use cases, attention moved towards blockchain-based smart contracts. The first blockchain with smart contracts was Ethereum, whose Ethereum Virtual Machine (EVM) is a Turing-complete machine supporting arbitrary operations. With the EVM, Ethereum provides a platform not only for smart contracts with digital enforcement of their conditions, but also for decentralized applications (Dapp) comprising a number of contracts and serving financial and semi-financial use cases. Even further, a number of Dapps could be combined into

---

[4]https://eips.ethereum.org/EIPS/eip-20.

a decentralized autonomous organization (DAO), a virtual entity whose entire life-cycle is executed on the blockchain.

> **Vitalik Buterin, May. 2014**
> A smart contract is the simplest form of decentralized automation, and is most easily and accurately defined as follows: a smart contract is a mechanism involving digital assets and two or more parties, where some or all of the parties put assets in and assets are automatically redistributed among those parties according to a formula based on certain data that is not known at the time the contract is initiated.

Finally, DLTs widened the scope from blockchain technology to more general distributed ledger implementations which adjust and modify the technology to satisfy application-specific requirements. Due to their flexibility, DLTs are expected to become popular especially in IoT applications where potentially billions of devices converse, requiring high performance and adaptability from the ledger.

The best-known example of a DLT for IoT is the IOTA [53], a ledger whose transaction data is stored in *tangle*, a directed acyclic graph (DAG). In more detail, instead of a linear chain of linked data blocks as in a blockchain, tangle is a graph of transactions, each of which refers to two previous ones. As a result, tangle promises the ledger a definitive performance boost in comparison to the linear blockchain.

3.1   IoT

Current centralized IoT systems incur high maintenance cost for the vendor while violating the *security through transparency* principle [54]. Blockchains and DLTs could mitigate these issues by providing a platform for smart contracts encapsulating the required functionality and with operations verified by the customers, thus both decentralizing the IoT system and increasing its transparency.

Smart cities provide a potential application vertical for blockchain and DLT based IoT applications, improving transparency, democracy, decentralization and security [55]. Other important blockchain application verticals include supply chain management, health care, and transportation.

Use of blockchain for supply chain management was first considered in 2016 [56]. Blockchain-based supply chain management provides a trusted environment for data exchange and tracking, promising to ensure the continued quality of products in transit while reducing the probability of mislays. Further, smart contracts support transparent and automatic execution and verification of the supply chain management processes. Improving on traceability, a blockchain-based supply chain provides product tracking traces, helping to forecast demand and reduce the risk of fraud and counterfeit [57].

For healthcare, blockchain could help scaling up and decentralizing the communication and management of current systems, comprising many independent applications for parties such as hospitals, patients, and pharmacies. Moreover, the immutability of blockchain records and support for data provenance make data auditing easier, improving the traceability of stored data as well as the management of critical digital assets such as insurance transactions and patient consent records. Further, blockchain-based healthcare applications improve fault tolerance and security by means of the their decentralized architecture as well as the inbuilt consensus mechanisms and cryptographic schemes [58]. Current blockchain-based healthcare applications include medical record management, insurance claim process, data sharing for telemedicine, opioid prescription tracking, and storage of health care data [59, 60]. However, if sensitive data is stored on the blockchain, it needs to be protected so that its validity can be also verified. Simple encryption may not be sufficient, but zero-knowledge argument schemes need to be applied instead.

For transportation, first blockchain use cases involve vehicle communication and identity management [61, 62], where the need for trust in a heterogeneous network of vehicles can be reduced. Further, blockchain can be applied for mobility-as-a-service [63], where a number of mobility service providers such as taxicabs, public transportation operators and private ride share providers can be brought together behind a single interface, changing focus from a provider-centric view to a user based one. Transparent and verifiable, blockchain-based mobility-as-a-service reduces the need of the service providers to trust each other by way of the trustless blockchain protocol.

3.2 Security

Blockchain can be employed in a number of applications in the security vertical, providing decentralization, trust, integrity, and immutability.

For example, conventional reputation systems are centralized, suffering from single points of failure. Blockchain can help decentralize the systems [64, 65]. Schaub et al. [66] used blockchain for a reputation system for e-commerce, employing blockchain as a public database for feedback from users to product/service providers. Encouraged by the success of blockchain-based reputation system, a number of application areas such as crowdsensing [64, 67], vehicular ad hoc network [68], robotics [69], and education [70] take advantave of blockchain to obtain trust, privacy and data integrity [71].

Further, traditional Public Key Infrastructure (PKI) and Domain Name System (DNS) are based on a centralized architecture, which requires the trust of users and suffers from malicious certificates, man in the middle attacks, denial of service attacks, DNS cache poisoning, and DNS spoofing [72, 73]. With a decentralized, tamper-proof, and public database, blockchain promises to help with the problems. Indeed, a number of studies propose blockchain-based solutions for DNS [73–75], PKI [73, 76–79] and border gateway protocol [75, 80]. Moreover, blockchain has been proposed to replace certification authorities such as those behind Domain Name System Security Extensions or Secure Sockets Layer with a trustless reputation system [81].

Event logs are essential as rich resources of information for forensic investigations [82], providing evidence of problems, bugs and incidents in the systems [83]. As a result, the first target of experienced attackers is the elimination of the log trace on the attack history [84].

To prevent tampering, event logs need to be decentralized, and excel on integrity and trustworthiness [83]. Blockchain can provide a decentralized and tamper-proof logging system, guaranteeing the immutability and integrity of the log data [85–87].

For malware detection, blockchain can be used as a shared database of malware signatures [88–90], enhancing malware detection accuracy and reducing their spread.

4   Blockchain and 6G

Blockchain may benefit 6G networks in numerous ways. In particular, blockchain offers decentralization, trust management, data integrity, as well as self-organization and self-sustainability.

First, based upon a decentralized P2P network, blockchain can improve on the reliability and availability of 6G services by removing some of their single points of failure. Furthermore, blockchain promotes a trustless environment where 6G stakeholders do not need to rely on individual authorities to ensure integrity of the services [91]. Such an environment could be a benefit in, for example, heterogeneous networks, where multiple operators are employed for connections. Using blockchain, a set of services, perhaps implemented as smart contracts, may be universally available for stakeholders, without the need for operators to arbitrate or hand over responsibilities.

Second, the integrity of blockchain data is guaranteed. As such, 6G service events could be stored on blockchain, ensuring auditability and traceability of those services by third parties.

Third, blockchain-based smart contracts are autonomous entities, capable of supporting self-organization and self-sustainability. In more detail, self-organization in a network aims to simplify its management and optimization. A smart contract can observe the operating environment of a 6G service, define trigger conditions, and launch operations based on those triggers to adjust the operating environment. Moreover, the contracts can cross stakeholder borders, ensuring fairness with trustless operation and aiming for global optimization instead of local.

For example, employing spectrum sharing or resource management as self-organizing contracts, the network can react efficiently to system demands. Further, a smart contract may manage access control, keeping track of access conditions and tracing access history for critical assets such as data assets stored on the blockchain.

The impact of blockchain on select enabling technologies for 6G as well as anticipated 6G services are further detailed in the subsections below.

4.1 Enabling Technologies

Network Function Virtualization Network function virtualization (NFV) refers to replacing fixed network components such as load balancers or firewalls with virtualized components (*virtualized network functions*, VNF) which are easy to deploy, migrate and chain together. NFV thus reduces the need for custom hardware and simplifies network management, lowering both capital and operative expenses.

However, the orchestration of VNFs in distributed systems presents several vulnerabilities. To identify and track potential security events, NFV requires auditability and traceability of communication traffic and VNF update history. Such information could be managed by a blockchain-based application [92]. Further, blockchain can provide NFV with authentication, access control, permission management, resource management, and a trustless environment for the heterogeneous NFV stakeholders [93].

Cloud Computing Cloud computing refers to a computing model where computing resources such as networks, servers, storage, applications, and related services are centralized in ubiquitous, convenient and on-demand resource pools, universally accessible via the Internet and offered with minimal management effort for those requiring such resources [94]. For mobile networks, cloud computing offers a platform for NFV as well as, for example, for cloud-based radio access networks (C-RAN), where certain base station functions are centralized as a pool of base station resources [95].

Blockchain can provide cloud-based services in 6G with security, traceability and provenance by, for example, verifying the identity of stakeholders and controlling data access [96] and storing metadata such as access logs [97]. Further, Yang et al. [98, 99] propose a blockchain-based architecture for trusted cloud radio over optical network, based on a decentralized tripartite agreement among vendors, network operators, and network users.

Another application to enhance trust, Yang et al. [100] study using smart contracts on a permissioned blockchain to construct a secure and reliable environment for IoT devices to trade their assets, safe from DDoS attacks. Further, Ma et al. [101] propose blockchain-based distributed key management architecture for IoT access control, encompassing a cloud system and a set of fog systems. Finally, Malomo et al. [102] propose smart contracts for minimizing the breach detection gap in a federated cloud environment comprising several cloud service providers.

Edge Computing Edge computing promises to reduce the communication delay between devices and cloud-based applications while reducing data rates and enhancing privacy. The promises are realized by distributing the cloud-based applications on a continuum between the devices, nearby computing servers at network hubs or base stations, and the cloud [15, 16, 103].

Blockchain is a potential solution for improving many aspects of edge computing [104]. For example, Guo et al. [105] propose an authentication scheme based on a consortium blockchain running on edge servers. The scheme ensures provenance and traceability by storing authentication data and event logs on the chain, while smart contracts provide functions for the authentication service. Further, Wang et al. [106] study an anonymous authentication and key agreement protocol for smart grid, where smart contracts keep record of public keys and provide autonomous triggers for key operations such as updates.

Moreover, a number of studies propose blockchain for edge resource orchestration and brokerage [107–109], with smart contracts providing a trusted marketplace for resource sales and allocation as well keeping a record of the transactions. Yang et al. [110] propose a trusted, cross-domain routing scheme maintaining topology privacy in a heterogeneous multi-access edge computing (MEC) system, while Rahman et al. [111] study a smart contract application providing metadata extraction, storage, analysis, and access control.

Federated Learning Federated learning is a distributed machine learning architecture. A network of nodes each train a neural network model based on local data, assumed independent and identically distributed across the nodes. A central node combines the local models into a global one, maintaining user privacy, and distributes the global model back to the local nodes [112].

As such, federated learning proposes a centralized architecture, requiring the coordination of a central node with storage for a global model. Blockchain could decentralize the architecture [113–115], voiding the need for explicit coordination and single points of failure. Moreover, tracking the lifecycle of the models as well as access to those models on the blockchain enhances trasparency and trust [116].

Further, blockchain can incentivize local nodes to participate in model sharing while discouraging freeriding [117], and manage a reputation system for nodes to ensure high-quality local models [117]. Finally, blockchain could provide a distribution architecture for local model training [118] as well as distributed storage for a federated learning application shifting high-quality data from a distributed large data pool [119].

4.2   6G Services

Spectrum Sharing and Management Data-intensive services and applications such as big data processing, multimedia streaming and AR/VR/XR require high performance data transmissions. However, physical constraints may present a barrier for 5G network operators aiming to support those services [120]. Further, spectrum fragmentation and the current, fixed spectrum allocation policy reduce the availability of spectrum resources [121]. Blockchain can mitigate spectrum management with, for example, two particular on-chain applications: a secure database, and a self-organized spectrum market [122].

Firstly, blockchain can provide a public, secure database for spectrum management. Since blockchain guarantees the integrity of the contained data, it can record information on, say, TV white spaces and other spectrum bands [123]. Further, blockchain may store access history of unlicensed spectrum bands, promoting fairness among users. Moreover, spectrum auction results and transactions between primary users (PUs) and secondary users (SUs) could be stored on-chain to prevent frauds by PUs, guarantee the non-repudiation of auction payments, and prevent unauthorized access by secondary users SUs [124, 125].

Secondly, decentralized and sensing-based dynamic spectrum access can employ blockchain, storing spectrum sensing data to support SUs on selecting spectrum bands with a low utilization rate [126]. In such a solution, the SUs are not only sensing nodes, but also fully-fledged blockchain participants, contributing to the consensus and verifying blocks. An incentive strategy like blockchain-based cryptocurrency reward can encourage participation.

Further, implemented as a blockchain-based smart contract, spectrum management can self-organize. A self-organized spectrum is intricately linked to the implementation of services such as spectrum sensing service or trading of transmission capacity [127, 128]. In more detail, the SUs determine requirements and policies through smart contracts, while the sensing devices later agree and join on those contracts. Hence, mobile network operators can purchase spectrum sensing services from user devices, reducing capital expenses in the deployment of spectrum sensors.

Finally, a self-organized spectrum manager can provide also identity and credibility management services for the spectrum market [129]. Service seekers are registered on-chain where their access credentials can be verified across operator borders.

Information Sharing The rising data rates call for securely and efficiently sharing the data among users. Blockchain, providing a transparent, immutable, trustless and decentralized storage, can mitigate that sharing [130]. While storing the actual content in blockchain is likely not feasible due to the high overhead costs, blockchain can manage data access and key management in particular.

As examples, Zhang and Chen [131] and Wang et al. [132] propose solutions where access to sensitive data is managed with smart contracts, while Bhaskaran et al. [133] study smart contracts on a permissioned blockchain to manage user consent.

Resource Management Decentralized, heterogeneous networks in 5G and 6G require the distribution of computation, capacity, and bandwidth, calling for efficient resource management. However, current resource management solutions are centralized, introducing single points of failure.

As an example, blockchain radio access network [134, 135] balances spectrum usage in a mobile network. A smart contract controls access to the network based on cost, demand, and service time.

Interference Management Devices operating on the same spectrum in a small cell may interfere with each other's connections, reducing overall quality of service. El Gamal and El Gamal aim to reduce interference with a blockchain based incentivization scheme [136]. Further, blockchain can provide a distributed database for cross-tier interference and control access to that data for the benefit of user devices [137].

D2D Communication Device-to-Device (D2D) communication is expected to rise significantly in volume in the coming years [138], especially due to the fast growth in the deployment of IoT devices. Blockchain can provide D2D communication with a trustless environment, authenticating and authorizing the parties, caching data, and ensuring the integrity of the communication [139, 140].

Network Slicing Network slicing is a method for creating virtual communication channels over the physical mobile network with predictable key performance indicators, targeted for customers and use cases requiring guaranteed performance such as critical communication networks [141]. As a use case for NFV, network slicing provides a unified view of VNFs and virtual networks [142].

A number of studies have proposed using blockchain for network slicing brokering and resource management in a secure, automatic, and scalable manner [143–145]. Further, blockchain can provide a trustless environment for stakeholders [146].

5 Challenges

While providing numerous opportunities, the integration of blockchain with 6G mobile networks is not without its challenges, especially in terms of privacy, security and performance. Indeed, while the inherent transparency of blockchain builds trust and promotes verifiability and traceability, it may also impact the privacy of the participants.

Further, the security of blockchain systems is not solely based on cryptographic algorithms, but also on the safety and liveness of the consensus. Safety guarantees the agreement of the network in the presence of multiple faulty participants. Liveness ensures that the network always decides for the acceptance or the rejection of a message. These characteristics are related to properties of the consensus such as validity, integrity/agreement, and termination [147, 148]. By the consensus mechanism we achieve decentralization, but lose performance.

Indeed, the effect on performance is one of the main research questions regarding the application of blockchain in 6G. Blockchain is not optimal for applications that require very low latency. In particular, the consensus mechanism requires significant local computations with verifications, proposals, consensus computations, and related cryptographic tasks. Moreover, without centralized authorities, blockchain protocols incur a heavy overhead cost for bandwidth due to the transmissions required to build consensus. For example, blockchain transactions are transmitted twice to all participants, first at the transaction and then in the confirmed block.

Further, while cryptographic protocols such as zero-knowledge argument schemes can mitigate the privacy issues, their application will further affect the performance of the system.

Finally, forks also hurt blockchain performance, as multiple blocks may need to be collected before data on the chain can be considered valid [149]. For example, Bitcoin's fork solution is a collection of seven consecutive blocks before the decision on a valid block [150]. Due to the awaiting period of block-sequence, the system can decide on the branch gaining the most approval of participants. For example, in Bitcoin, the longest branch gathers the attention of the most powerful computation of the system.

Nevertheless, a PoW consensus similar to Bitcoin will not be optimal for the applications envisioned for 6G due to the massive computational overhead required. A more efficient consensus mechanism with less latency on the block confirmation is needed. There are alternatives such as proof-of-stake, but research is needed to determine their efficiency for the services presented in this chapter.

6   Conclusion

Blockchain is a potential technology for the next generation of mobile networks. 6G expands the scope and applications of blockchain with unprecedented speed, capacity, latency, and connectivity. In return, blockchain offers 6G services with decentralization, trustlessness, transparency, integrity of data, and self-organization.

We introduced the main aspects of blockchain technology, including architecture, data models, and vulnerabilities. We highlighted a number of applications related to mobile networks, and considered the intersection of 6G and blockchain in relation to enabling technologies and 6G services. Finally, we discussed the challenges inherent in blockchain-based applications, especially in terms of privacy, security and performance.

Acknowledgments This work is supported by TrustedMaaS and B-TEA projects by the Infotech institute of the University of Oulu, and the Academy of Finland 6Genesis Flagship (grant 318927).

References

1. S. Nakamoto, *Bitcoin: A Peer-to-Peer Electronic Cash System.* Manubot (2008)
2. G. Perboli, S. Musso, M. Rosano, Blockchain in logistics and supply chain: a lean approach for designing real-world use cases. IEEE Access **6**, 62018–62028 (2018)
3. E. Tijan, S. Aksentijević, K. Ivanić, M. Jardas, Blockchain technology implementation in logistics. Sustainability **11**(4), 1185 (2019)
4. M. Barbieri, D. Gassen, Blockchain can this new technology really revolutionize the land registry system?, in *Responsible Land Governance: Towards an Evidence Based Approach: Proceedings of the Annual World Bank Conference on Land and Poverty* (2017), pp. 1–13

5. V. Thakur, M. Doja, Y.K. Dwivedi, T. Ahmad, G. Khadanga, Land records on blockchain for implementation of land titling in India. Int. J. Inf. Manag. **52**, 101940 (2020)
6. D. Daniel, C. Ifejika Speranza, The role of blockchain in documenting land users' rights: the canonical case of farmers in the vernacular land market. Front. Blockch. **3**, 19 (2020)
7. S. Jeong, N.N. Dao, Y. Lee, C. Lee, S. Cho, Blockchain based billing system for electric vehicle and charging station, in *2018 Tenth International Conference on Ubiquitous and Future Networks (ICUFN)*(IEEE, Piscataway, 2018), pp. 308–310
8. H. Zhang, E. Deng, H. Zhu, Z. Cao, Smart contract for secure billing in ride-hailing service via blockchain. Peer-to-Peer Netw. Appl. **12**(5), 1346–1357 (2019)
9. B. Aazhang, P. Ahokangas, L. Lovén, et al., Key Drivers and Research Challenges for 6G Ubiquitous Wireless Intelligence (White Paper), 1st edn. 6G Flagship, (University of Oulu, Oulu, 2019)
10. Z. Zhang, Y. Xiao, Z. Ma, M. Xiao, Z. Ding, X. Lei, G.K. Karagiannidis, P. Fan, 6G wireless networks: vision, requirements, architecture, and key technologies. IEEE Vehic. Technol. Mag. **14**(3), 28–41 (2019)
11. P. Yang, Y. Xiao, M. Xiao, S. Li, 6G wireless communications: vision and potential techniques. IEEE Netw. **33**(4), 70–75 (2019)
12. E.C. Strinati, S. Barbarossa, J.L. Gonzalez-Jimenez, D. Kténas, N. Cassiau, C. Dehos, 6G: The next frontier (2019). Preprint arXiv:1901.03239
13. W. Saad, M. Bennis, M. Chen, A vision of 6G wireless systems: Applications, trends, technologies, and open research problems (2019). Preprint arXiv:1902.10265
14. N. DOCOMO, White paper 5G evolution and 6G. Accessed, vol. 1 (2020)
15. L. Lovén, T. Leppänen, E. Peltonen, J. Partala, E. Harjula, P. Porambage, M. Ylianttila, J. Riekki, Edgeai: A vision for distributed, edge-native artificial intelligence in future 6G networks, in *The 1st 6G Wireless Summit* Levi, Finland (2019), pp. 1–2
16. E. Peltonen, M. Bennis, M. Capobianco, M. Debbah, A. Ding, F. Gil-Castiñeira, M. Jurmu, T. Karvonen, M. Kelanti, A. Kliks, et al., 6G white paper on edge intelligence (2020). Preprint arXiv:2004.14850
17. Y. Dai, D. Xu, S. Maharjan, Z. Chen, Q. He, Y. Zhang, Blockchain and deep reinforcement learning empowered intelligent 5G beyond. IEEE Netw. **33**(3), 10–17 (2019)
18. I. Ahmad, T. Kumar, M. Liyanage, J. Okwuibe, M. Ylianttila, A. Gurtov, Overview of 5G security challenges and solutions. IEEE Commun. Stand. Mag. **2**(1), 36–43 (2018)
19. T. Nguyen, N. Tran, L. Loven, J. Partala, M.T. Kechadi, S. Pirttikangas, Privacy-aware blockchain innovation for 6G: Challenges and opportunities, in *2020 2nd 6G Wireless Summit (6G SUMMIT)* (IEEE, Piscataway, 2020), pp. 1–5
20. F. Burkhardt, C. Patachia, L. Lovén, et al., *6G White Paper on Validation and Trials for Verticals Towards 2030's*. 6G Flagship (University of Oulu, Oulu, 2020)
21. Y. Lu, X. Zheng, 6G: a survey on technologies, scenarios, challenges, and the related issues. J. Ind. Inf. Integr. **19**, 100158 (2020)
22. R. Davenport, Distributed database technology—a survey. Comput. Netw. (1976) **2**(3), 155–167 (1978)
23. S. Haber, W.S. Stornetta, How to time-stamp a digital document, in *Conference on the Theory and Application of Cryptography* (Springer, Berlin, 1990), pp. 437–455
24. D. Bayer, S. Haber, W.S. Stornetta, Improving the efficiency and reliability of digital time-stamping, in *Sequences II* (Springer, Berlin, 1993), pp. 329–334
25. R.C. Merkle, A digital signature based on a conventional encryption function, in *Conference on the Theory and Application of Cryptographic Techniques* (Springer, Berlin, 1987), pp. 369–378
26. D. Mazieres, D. Shasha, Building secure file systems out of byzantine storage, in *Proceedings of the Twenty-First annual Symposium on Principles of Distributed Computing* (2002), pp. 108–117
27. J. Li, M.N. Krohn, D. Mazieres, D.E. Shasha, Secure untrusted data repository (SUNDR), in *OSDI*, vol. 4 (2004), pp. 9–9
28. W. Dai, B-money. Consulted **1**, 2012 (1998)

29. H. Finney, Rpow-reusable proofs of work. Internet https://cryptome.org/rpow.htm (2004)
30. C.C. Cocks, A note on non-secret encryption. *CESG Memo* (1973)
31. K. Brown, Announcing approval of federal information processing standard (fips) 197, advanced encryption standard (aes). *National Institute of Standards and Technology, Commerce* (2002)
32. N. Szabo, The idea of smart contracts. Nick Szabo's Papers Concise Tutorials, vol. 6 (1997)
33. G. Wood, et al., Ethereum: A secure decentralised generalised transaction ledger. Ethereum Project Yellow Paper **151**(2014), 1–32 (2014)
34. M. Belotti, N. Božić, G. Pujolle, S. Secci, A vademecum on blockchain technologies: when, which, and how. IEEE Commun. Surveys Tutor. **21**(4), 3796–3838 (2019)
35. R.G. Brown, J. Carlyle, I. Grigg, M. Hearn, Corda: an introduction. R3 CEV **1**, 15 (2016)
36. E. Androulaki, A. Barger, V. Bortnikov, C. Cachin, K. Christidis, A. De Caro, D. Enyeart, C. Ferris, G. Laventman, Y. Manevich, et al., Hyperledger fabric: A distributed operating system for permissioned blockchains, in *Proceedings of the Thirteenth EuroSys Conference* (2018), pp. 1–15
37. Y. Xiao, N. Zhang, W. Lou, Y.T. Hou, A survey of distributed consensus protocols for blockchain networks. IEEE Commun. Surveys Tutor. **22**(2), 1432–1465 (2020)
38. L. Lamport, R. Shostak, M. Pease, The byzantine generals problem. ACM Trans. Program. Lang. Syst. **4**(3), 382–401 (1982)
39. C. Dwork, M. Naor, Pricing via processing or combatting junk mail, in *Annual International Cryptology Conference* (Springer, Berlin, 1992), pp. 139–147
40. M. Jakobsson, A. Juels, Proofs of work and bread pudding protocols, in *Secure Information Networks* (Springer, Berlin, 1999), pp. 258–272
41. M. Castro, B. Liskov, et al., Practical byzantine fault tolerance, in *e Proceedings of the Third Symposium on Operating Systems Design and Implementation*, vol. 99 (1999), pp. 173–186
42. F. Muratov, A. Lebedev, N. Iushkevich, B. Nasrulin, M. Takemiya, YAC: BFT consensus algorithm for blockchain (2018). Preprint arXiv:1809.00554
43. I.C. Lin, T.C. Liao, A survey of blockchain security issues and challenges. IJ Netw. Security **19**(5), 653–659 (2017)
44. Z. Zheng, S. Xie, H.N. Dai, H. Wang, Blockchain challenges and opportunities: A survey. International Journal of Web and Grid Services **14**(4), 352–375 (Inderscience Publishers (IEL), 2018)
45. W. Wang, D.T. Hoang, P. Hu, Z. Xiong, D. Niyato, P. Wang, Y. Wen, D.I. Kim, A survey on consensus mechanisms and mining strategy management in blockchain networks. IEEE Access **7**, 22328–22370 (2019)
46. M. Saad, J. Spaulding, L. Njilla, C. Kamhoua, S. Shetty, D.H. Nyang, D. Mohaisen, Exploring the attack surface of blockchain: a comprehensive survey. IEEE Commun. Surveys Tutor. **22**, 1977–2008 (2020)
47. M.A. Javarone, C.S. Wright, From bitcoin to bitcoin cash: A network analysis, in *Proceedings of the 1st Workshop on Cryptocurrencies and Blockchains for Distributed Systems* (2018), pp. 77–81
48. J.R. Douceur, The sybil attack, in *International Workshop on Peer-to-Peer Systems* (Springer, Berlin, 2002), pp. 251–260
49. I. Eyal, E.G. Sirer, Majority is not enough: bitcoin mining is vulnerable. Commun. ACM **61**(7), 95–102 (2018)
50. X. Li, P. Jiang, T. Chen, X. Luo, Q. Wen, A survey on the security of blockchain systems. Future Gener. Comput. Syst. **107**, 841–853 (2017)
51. N. Atzei, M. Bartoletti, T. Cimoli, A survey of attacks on ethereum smart contracts (SOK), in *International Conference on Principles of Security and Trust* (Springer, Berlin, 2017), pp. 164–186
52. H. Chen, M. Pendleton, L. Njilla, S. Xu, A survey on ethereum systems security: vulnerabilities, attacks, and defenses. ACM Comput. Surveys **53**(3), 1–43 (2020)
53. S. Popov, O. Saa, P. Finardi, Equilibria in the Tangle. Computers & Industrial Engineering **136**, 160–172 (Elsevier, 2019)

54. K. Christidis, M. Devetsikiotis, Blockchains and smart contracts for the internet of things. IEEE Access **4**, 2292–2303 (2016)
55. J. Xie, H. Tang, T. Huang, F.R. Yu, R. Xie, J. Liu, Y. Liu, A survey of blockchain technology applied to smart cities: research issues and challenges. IEEE Commun. Surveys Tutor. **21**(3), 2794–2830 (2019)
56. F. Tian, An agri-food supply chain traceability system for china based on RFID & blockchain technology, in *2016 13th International Conference on Service Systems and Service Management (ICSSSM)* (IEEE, Piscataway, 2016), pp. 1–6
57. D. Dujak, D. Sajter, Blockchain applications in supply chain, in *SMART Supply Network* (Springer, Berlin, 2019), pp. 21–46
58. M. Hölbl, M. Kompara, A. Kamišalić, L. Nemec Zlatolas, A systematic review of the use of blockchain in healthcare. Symmetry **10**(10), 470 (2018)
59. T.T. Kuo, H.E. Kim, L. Ohno-Machado, Blockchain distributed ledger technologies for biomedical and health care applications. J. Am. Med. Inf. Assoc. **24**(6), 1211–1220 (2017)
60. P. Zhang, D.C. Schmidt, J. White, G. Lenz, Blockchain technology use cases in healthcare, in *Advances in Computers*, vol. 111 (Elsevier, Amsterdam, 2018), pp. 1–41
61. A. Lei, H. Cruickshank, Y. Cao, P. Asuquo, C.P.A. Ogah, Z. Sun, Blockchain-based dynamic key management for heterogeneous intelligent transportation systems. IEEE Int. Things J. **4**(6), 1832–1843 (2017)
62. L. Li, J. Liu, L. Cheng, S. Qiu, W. Wang, X. Zhang, Z. Zhang, Creditcoin: a privacy-preserving blockchain-based incentive announcement network for communications of smart vehicles. IEEE Trans. Intell. Transport. Syst. **19**(7), 2204–2220 (2018)
63. T.H. Nguyen, J. Partala, S. Pirttikangas, Blockchain-based mobility-as-a-service, in *2019 28th International Conference on Computer Communication and Networks (ICCCN)* (IEEE, Piscataway, 2019), pp. 1–6
64. K. Zhao, S. Tang, B. Zhao, Y. Wu, Dynamic and privacy-preserving reputation management for blockchain-based mobile crowdsensing. IEEE Access **7**, 74694–74710 (2019)
65. Y. Lee, K.M. Lee, S.H. Lee, Blockchain-based reputation management for custom manufacturing service in the peer-to-peer networking environment. Peer-to-Peer Netw. Appl. **13**(2), 671–683 (2020)
66. A. Schaub, R. Bazin, O. Hasan, L. Brunie, A trustless privacy-preserving reputation system, in *IFIP International Conference on ICT Systems Security and Privacy Protection* (Springer, Berlin, 2016), pp. 398–411
67. M. Li, J. Weng, A. Yang, W. Lu, Y. Zhang, L. Hou, J.N. Liu, Y. Xiang, R.H. Deng, Crowdbc: A blockchain-based decentralized framework for crowdsourcing. IEEE Trans. Parallel Distrib. Syst. **30**(6), 1251–1266 (2018)
68. Z. Yang, K. Yang, L. Lei, K. Zheng, V.C. Leung, Blockchain-based decentralized trust management in vehicular networks. IEEE Int. Things J. **6**(2), 1495–1505 (2018)
69. M. Dorigo, et al., Blockchain technology for robot swarms: A shared knowledge and reputation management system for collective estimation, in *Swarm Intelligence: 11th International Conference, ANTS 2018*, Rome, Italy, October 29–31, 2018. Proceedings, vol. 11172 (Springer, Berlin, 2018), p. 425
70. M. Sharples, J. Domingue, The blockchain and kudos: A distributed system for educational record, reputation and reward, in *European Conference on Technology Enhanced Learning* (Springer, Berlin, 2016), pp. 490–496
71. E. Bellini, Y. Iraqi, E. Damiani, Blockchain-based distributed trust and reputation management systems: a survey. IEEE Access **8**, 21127–21151 (2020)
72. J. Yu, M. Ryan, Evaluating web pkis, in *Software Architecture for Big Data and the Cloud* (Elsevier, Amsterdam, 2017), pp. 105–126
73. E. Karaarslan, E. Adiguzel, Blockchain based DNS and PKI solutions. IEEE Commun. Stand. Mag. **2**(3), 52–57 (2018)
74. C. Fromknecht, D. Velicanu, S. Yakoubov, A decentralized public key infrastructure with identity retention. IACR Cryptol. ePrint Arch. **2014**, 803 (2014)

75. A. Hari, T. Lakshman, The internet blockchain: A distributed, tamper-resistant transaction framework for the internet, in *Proceedings of the 15th ACM Workshop on Hot Topics in Networks* (2016), pp. 204–210

76. L. Axon, M. Goldsmith, PB-PKI: A privacy-aware blockchain-based PKI. Proceedings of the 14th International Joint Conference on e-Business and Telecommunications - SECRYPT, (ICETE 2017) (SciTePress, 2017). pp. 311–318. https://doi.org/10.5220/0006419203110318

77. N. Alexopoulos, J. Daubert, M. Mühlhäuser, S.M. Habib, Beyond the hype: On using blockchains in trust management for authentication, in *2017 IEEE Trust-com/BigDataSE/ICESS* (IEEE, Piscataway, 2017), pp. 546–553

78. R. Longo, F. Pintore, G. Rinaldo, M. Sala, On the security of the blockchain bix protocol and certificates, in *2017 9th International Conference on Cyber Conflict (CyCon)* (IEEE, Piscataway, 2017), pp. 1–16

79. H. Orman, Blockchain: The emperors new PKI? IEEE Int. Comput. **22**(2), 23–28 (2018)

80. M. Saad, A. Anwar, A. Ahmad, H. Alasmary, M. Yuksel, A. Mohaisen, Routechain: Towards blockchain-based secure and efficient BGP routing, in *2019 IEEE International Conference on Blockchain and Cryptocurrency (ICBC)* (IEEE, Piscataway, 2019), pp. 210–218

81. M. Vyshegorodtsev, D. Miyamoto, Y. Wakahara, Reputation scoring system using an economic trust model: A distributed approach to evaluate trusted third parties on the internet, in *2013 27th International Conference on Advanced Information Networking and Applications Workshops* (IEEE, Piscataway, 2013), pp. 730–737

82. D. Reilly, C. Wren, T. Berry, Cloud computing: Forensic challenges for law enforcement, in *2010 International Conference for Internet Technology and Secured Transactions* (IEEE, Piscataway, 2010), pp. 1–7

83. S. Zawoad, A. Dutta, R. Hasan, Towards building forensics enabled cloud through secure logging-as-a-service. IEEE Trans. Depend. Secure Comput. **13**(1), 1–1 (2016)

84. M. Bellare, B. Yee, Forward-security in private-key cryptography, in *Cryptographers' Track at the RSA Conference* (Springer, 2003), pp. 1–18

85. J. Cucurull, J. Puiggalí, Distributed immutabilization of secure logs, in *International Workshop on Security and Trust Management* (Springer, Berlin, 2016), pp. 122–137

86. W. Pourmajidi, A. Miranskyy, Logchain: blockchain-assisted log storage, in *2018 IEEE 11th International Conference on Cloud Computing (CLOUD)* (IEEE, Piscataway, 2018), pp. 978–982

87. A. Sutton, R. Samavi, Blockchain enabled privacy audit logs, in *International Semantic Web Conference* (Springer, Berlin, 2017), pp. 645–660

88. C.Noyes, Bitav: Fast anti-malware by distributed blockchain consensus and feedforward scanning (2016). Preprint arXiv:1601.01405

89. J. Gu, B. Sun, X. Du, J. Wang, Y. Zhuang, Z. Wang, Consortium blockchain-based malware detection in mobile devices. IEEE Access **6**, 12118–12128 (2018)

90. R. Fuji, S. Usuzaki, K. Aburada, H. Yamaba, T. Katayama, M. Park, N. Shiratori, N. Okazaki, Investigation on sharing signatures of suspected malware files using blockchain technology, in *International Multi Conference of Engineers and Computer Scientists (IMECS)* (2019), pp. 94–99

91. M. Ylianttila, R. Kantola, A. Gurtov, L. Mucchi, I. Oppermann, Z. Yan, T.H. Nguyen, F. Liu, T. Hewa, M. Liyanage, et al., 6G white paper: Research challenges for trust, security and privacy (2020). Preprint arXiv:2004.11665

92. G.A.F. Rebello, I.D. Alvarenga, I.J. Sanz, O.C.M. Duarte, Bsec-nfvo: A blockchain-based security for network function virtualization orchestration, in *ICC 2019–2019 IEEE International Conference on Communications (ICC)* (IEEE, Piscataway, 2019), pp. 1–6

93. I.D. Alvarenga, G.A. Rebello, O.C.M. Duarte, Securing configuration management and migration of virtual network functions using blockchain, in *NOMS 2018–2018 IEEE/IFIP Network Operations and Management Symposium* (IEEE, Piscataway, 2018), pp. 1–9

94. P. Mell, T. Grance, *The NIST definition of cloud computing: Recommendations of the National Institute of Standards and Technology* (Computer Security Resource Center, 2012)

95. J. Wu, Z. Zhang, Y. Hong, Y. Wen, Cloud radio access network (C-RAN): a primer. IEEE Netw. **29**(1), 35–41 (2015)
96. Y. Zhang, D. He, K.K.R. Choo, Bads: Blockchain-based architecture for data sharing with ABS and CP-ABE in IoT. Wirel. Commun. Mobile Comput. **2018**, 2783658 (2018), p. 9. https://doi.org/10.1155/2018/2783658
97. S. Ali, G. Wang, M.Z.A. Bhuiyan, H. Jiang, Secure data provenance in cloud-centric internet of things via blockchain smart contracts, in *2018 IEEE SmartWorld, Ubiquitous Intelligence & Computing, Advanced & Trusted Computing, Scalable Computing & Communications, Cloud & Big Data Computing, Internet of People and Smart City Innovation (Smart-World/SCALCOM/UIC/ATC/CBDCom/IOP/SCI)* (IEEE, Piscataway, 2018), pp. 991–998
98. H. Yang, H. Zheng, J. Zhang, Y. Wu, Y. Lee, Y. Ji, Blockchain-based trusted authentication in cloud radio over fiber network for 5G. in *2017 16th International Conference on Optical Communications and Networks (ICOCN)* (IEEE, Piscataway, 2017), pp. 1–3
99. H. Yang, Y. Wu, J. Zhang, H. Zheng, Y. Ji, Y. Lee, Blockonet: Blockchain-based trusted cloud radio over optical fiber network for 5G fronthaul. in *Optical Fiber Communication Conference* (Optical Society of America, Washington, 2018), pp. W2A–25
100. H. Yang, J. Yuan, H. Yao, Q. Yao, A. Yu, J. Zhang, Blockchain-based hierarchical trust networking for jointcloud. IEEE Int. Things J. **7**(3), 1667–1677 (2019)
101. M. Ma, G. Shi, F. Li, Privacy-oriented blockchain-based distributed key management architecture for hierarchical access control in the IoT scenario. IEEE Access **7**, 34045–34059 (2019)
102. O.O. Malomo, D.B. Rawat, M. Garuba, Next-generation cybersecurity through a blockchain-enabled federated cloud framework. J. Supercomput. **74**(10), 5099–5126 (2018)
103. J. Haavisto, M. Arif, L. Lovén, T. Leppänen, J. Riekki, Open-source RANs in practice: an over-the-air deployment for 5G MEC (2019). Preprint arXiv:1905.03883
104. R. Yang, F.R. Yu, P. Si, Z. Yang, Y. Zhang, Integrated blockchain and edge computing systems: a survey, some research issues and challenges. IEEE Commun. Surveys Tutor. **21**(2), 1508–1532 (2019)
105. S. Guo, X. Hu, S. Guo, X. Qiu, F. Qi, Blockchain meets edge computing: a distributed and trusted authentication system. IEEE Trans. Ind. Inf. **16**(3), 1972–1983 (2019)
106. J. Wang, L. Wu, K.K.R. Choo, D. He, Blockchain-based anonymous authentication with key management for smart grid edge computing infrastructure. IEEE Trans. Ind. Inf. **16**(3), 1984–1992 (2019)
107. Y. Liu, F.R. Yu, X. Li, H. Ji, V.C. Leung, Resource allocation for video transcoding and delivery based on mobile edge computing and blockchain, in *2018 IEEE Global Communications Conference (GLOBECOM)* (IEEE, Piscataway, 2018), pp. 1–6
108. C. Xia, H. Chen, X. Liu, J. Wu, L. Chen, Etra: Efficient three-stage resource allocation auction for mobile blockchain in edge computing, in *2018 IEEE 24th International Conference on Parallel and Distributed Systems (ICPADS)* (IEEE, Piscataway, 2018), pp. 701–705
109. Y. Liu, F.R. Yu, X. Li, H. Ji, V.C. Leung, Decentralized resource allocation for video transcoding and delivery in blockchain-based system with mobile edge computing. IEEE Trans. Vehic. Technol. **68**(11), 11169–11185 (2019)
110. H. Yang, Y. Liang, J. Yuan, Q. Yao, A. Yu, J. Zhang, Distributed blockchain-based trusted multi-domain collaboration for mobile edge computing in 5G and beyond. IEEE Trans. Ind. Inf. **16**, 7094–7104 (2020)
111. M.A. Rahman, M.M. Rashid, M.S. Hossain, E. Hassanain, M.F. Alhamid, M. Guizani, Blockchain and iot-based cognitive edge framework for sharing economy services in a smart city. IEEE Access **7**, 18611–18621 (2019)
112. Q. Yang, Y. Liu, T. Chen, Y. Tong, Federated machine learning: concept and applications. ACM Trans. Intell. Syst. Technol. **10**(2), 1–19 (2019)
113. D. Preuveneers, V. Rimmer, I. Tsingenopoulos, J. Spooren, W. Joosen, E. Ilie-Zudor, Chained anomaly detection models for federated learning: an intrusion detection case study. Appl. Sci. **8**(12), 2663 (2018)

114. H. Kim, J. Park, M. Bennis, S.L. Kim, Blockchained on-device federated learning. IEEE Commun. Lett. **24**, 1279–1283 (2019)
115. J. Weng, J. Weng, J. Zhang, M. Li, Y. Zhang, W. Luo, Deepchain: Auditable and privacy-preserving deep learning with blockchain-based incentive. IEEE Trans. Depend. Secure Comput. (2019). p. 1. https://doi.org/10.1109/TDSC.2019.2952332
116. Y. Lu, X. Huang, Y. Dai, S. Maharjan, Y. Zhang, Blockchain and federated learning for privacy-preserved data sharing in industrial IoT. IEEE Trans. Ind. Inf. **16**(6), 4177–4186 (2019)
117. J. Kang, Z. Xiong, D. Niyato, S. Xie, J. Zhang, Incentive mechanism for reliable federated learning: a joint optimization approach to combining reputation and contract theory. IEEE Int. Things J. **6**(6), 10700–10714 (2019)
118. Z. Shae, J. Tsai, Transform blockchain into distributed parallel computing architecture for precision medicine, in *2018 IEEE 38th International Conference on Distributed Computing Systems (ICDCS)* (IEEE, Piscataway, 2018), pp. 1290–1299
119. R. Doku, D.B. Rawat, C. Liu, Towards federated learning approach to determine data relevance in big data, in *2019 IEEE 20th International Conference on Information Reuse and Integration for Data Science (IRI)* (IEEE, Piscataway, 2019), pp. 184–192
120. D.C. Nguyen, P.N. Pathirana, M. Ding, A. Seneviratne, Blockchain for 5G and beyond networks: a state of the art survey. J. Netw. Comput. Appl. **166**, 102693 (2020). https://doi.org/10.1016/j.jnca.2020.102693
121. S.K. Sharma, T.E. Bogale, L.B. Le, S. Chatzinotas, X. Wang, B. Ottersten, Dynamic spectrum sharing in 5G wireless networks with full-duplex technology: Recent advances and research challenges. IEEE Commun. Surveys Tutor. **20**(1), 674–707 (2017)
122. Y.C. Liang, Blockchain for dynamic spectrum management, in *Dynamic Spectrum Management* (Springer, Berlin, 2020), pp. 121–146
123. M.B. Weiss, K. Werbach, D.C. Sicker, C.E.C. Bastidas, On the application of blockchains to spectrum management. IEEE Trans. Cognitive Commun. Netw. **5**(2), 193–205 (2019)
124. K. Kotobi, S.G. Bilén, Blockchain-enabled spectrum access in cognitive radio networks, in *2017 Wireless Telecommunications Symposium (WTS)* (IEEE, Piscataway, 2017), pp. 1–6
125. K. Kotobi, S.G. Bilen, Secure blockchains for dynamic spectrum access: a decentralized database in moving cognitive radio networks enhances security and user access. IEEE Vehic. Technol. Mag. **13**(1), 32–39 (2018)
126. Y. Pei, S. Hu, F. Zhong, D. Niyato, Y.C. Liang, Blockchain-enabled dynamic spectrum access: Cooperative spectrum sensing, access and mining, in *2019 IEEE Global Communications Conference (GLOBECOM)* (IEEE, Piscataway, 2019), pp. 1–6
127. S. Bayhan, A. Zubow, A. Wolisz, Spass: Spectrum sensing as a service via smart contracts, in *2018 IEEE International Symposium on Dynamic Spectrum Access Networks (DySPAN)* (IEEE, Piscataway, 2018), pp. 1–10
128. S. Bayhan, A. Zubow, P. Gawłowicz, A. Wolisz, Smart contracts for spectrum sensing as a service. IEEE Trans. Cognit. Commun. Netw. **5**(3), 648–660 (2019)
129. S. Raju, S. Boddepalli, S. Gampa, Q. Yan, J.S. Deogun, Identity management using blockchain for cognitive cellular networks, in *2017 IEEE International Conference on Communications (ICC)* (IEEE, Piscataway, 2017), pp. 1–6
130. K. Fan, Y. Ren, Y. Wang, H. Li, Y. Yang, Blockchain-based efficient privacy preserving and data sharing scheme of content-centric network in 5G. IET Commun. **12**(5), 527–532 (2017)
131. X. Zhang, X. Chen, Data security sharing and storage based on a consortium blockchain in a vehicular ad-hoc network. IEEE Access **7**, 58241–58254 (2019)
132. S. Wang, Y. Zhang, Y. Zhang, A blockchain-based framework for data sharing with fine-grained access control in decentralized storage systems. IEEE Access **6**, 38437–38450 (2018)
133. K. Bhaskaran, P. Ilfrich, D. Liffman, C. Vecchiola, P. Jayachandran, A. Kumar, F. Lim, K. Nandakumar, Z. Qin, V. Ramakrishna, et al., Double-blind consent-driven data sharing on blockchain, in *2018 IEEE International Conference on Cloud Engineering (IC2E)* (IEEE, Piscataway, 2018), pp. 385–391

134. Y. Le, X. Ling, J. Wang, Z. Ding, Prototype design and test of blockchain radio access network, in *2019 IEEE International Conference on Communications Workshops (ICC Workshops)* (IEEE, Piscataway, 2019), pp. 1–6
135. X. Ling, J. Wang, T. Bouchoucha, B.C. Levy, Z. Ding, Blockchain radio access network (B-RAN): towards decentralized secure radio access paradigm. IEEE Access **7**, 9714–9723 (2019)
136. A. El Gamal,H. El Gamal, A single coin monetary mechanism for distributed cooperative interference management. IEEE Wirel. Commun. Lett. **8**(3), 757–760 (2019)
137. D. Lin, Y. Tang, Blockchain consensus based user access strategies in D2D networks for data-intensive applications. IEEE Access **6**, 72683–72690 (2018)
138. F. Jameel, Z. Hamid, F. Jabeen, S. Zeadally, M.A. Javed, A survey of device-to-device communications: research issues and challenges. IEEE Commun. Surveys Tutor. **20**(3), 2133–2168 (2018)
139. L. Jiang, S. Xie, S. Maharjan, Y. Zhang, Joint transaction relaying and block verification optimization for blockchain empowered D2D communication. IEEE Trans. Vehic. Technol. **69**(1), 828–841 (2019)
140. R. Zhang, F.R. Yu, J. Liu, T. Huang, Y. Liu, Deep reinforcement learning (DRL)-based device-to-device (D2D) caching with blockchain and mobile edge computing. IEEE Trans. Wirel. Commun. **19**, 6469–6485 (2020)
141. M. Höyhtyä, K. Lähetkangas, et al., Critical communications over mobile operators' networks: 5G use cases enabled by licensed spectrum sharing, network slicing and QoS control. IEEE Access **6**, 73572–73582 (2018). https://doi.org/10.1109/ACCESS.2018.2883787
142. I. Afolabi, T. Taleb, K. Samdanis, A. Ksentini, H. Flinck, Network slicing and softwarization: a survey on principles, enabling technologies, and solutions. IEEE Commun. Surveys Tutor. **20**(3), 2429–2453 (2018)
143. L. Zanzi, A. Albanese, V. Sciancalepore, X. Costa-Pérez, Nsbchain: A secure blockchain framework for network slicing brokerage (2020). Preprint arXiv:2003.07748
144. A. Adhikari, D.B. Rawat, M. Song, Wireless network virtualization by leveraging blockchain technology and machine learning, in *Proceedings of the ACM Workshop on Wireless Security and Machine Learning* (2019), pp. 61–66
145. D.B. Rawat, A. Alshaikhi, Leveraging distributed blockchain-based scheme for wireless network virtualization with security and qos constraints, in *2018 International Conference on Computing, Networking and Communications (ICNC)* (IEEE, Piscataway, 2018), pp. 332–336
146. B. Nour, A. Ksentini, N. Herbaut, P.A. Frangoudis, H. Moungla, A blockchain-based network slice broker for 5G services. IEEE Netw. Lett. **1**(3), 99–102 (2019)
147. G.F. Coulouris, J. Dollimore, T. Kindberg, *Distributed Systems: Concepts and Design.* (Pearson education, London, 2005)
148. M. Raynal, *Fault-Tolerant Message-Passing Distributed Systems: An Algorithmic Approach* (Springer, Berlin, 2018)
149. L. Wan, D. Eyers, H. Zhang, Evaluating the impact of network latency on the safety of blockchain transactions, in *2019 IEEE International Conference on Blockchain (Blockchain)* (IEEE, Piscataway, 2019), pp. 194–201
150. M. Conti, E.S. Kumar, C. Lal, S. Ruj, A survey on security and privacy issues of bitcoin. IEEE Commun. Surveys Tutor. **20**(4), 3416–3452 (2018)

Chapter 19
Role of Quantum Technology in 6G

Konchady Gautam Shenoy

Abstract The past few years have seen major advances towards practical quantum systems. These include demonstration of quantum entanglement over a gap of over 1200 km, to significantly speeding up the solution of certain combinatorial problems. It stands to reason that eventually, quantum communication techniques may be practically viable in modern communication systems. Certain aspects of this technology are being considered to augment existing technologies, such as secure key distribution. However, this technology hasn't matured to the point where it can take a forefront. In this chapter, we will explore various tenets of quantum communication, describe the challenges and role it will hopefully play in 6G and beyond communications.

1 Introduction

The jump from 4th generation mobile communications (4G) to 5G was fuelled by an unprecedented increase in connected users, bandwidth requirements and data intensive services. In particular, ultra reliable low latency communications (URLLC) have stringent real time constraints as well. With an eye towards proposed 6G communication frameworks, several technologies have been proposed to address the issues. One of those is providing a machine learning framework for end to end communications. The other is quantum assisted and pure quantum communication framework. In this chapter, we will focus on the latter aspect, providing the avenues that would benefit from this technology.

Classical communication and computing relied on using bits, which are the smallest unit of memory comprising of a logical 0 and logical 1. This simple concept was the backbone of all communications and computing as we know it. Quantum

K. G. Shenoy (✉)
Indian Institute of Science, Bangalore, India

computation and communication, in contrast work with units called qubits (quantum bits), which function on quantum phenomena and may be considered a generalization. What sets apart quantum from classical are primarily two phenomena, namely

1. superposition of states (Schrödinger's Cat),
2. entanglement (spooky action at a distance).

These valuable resources are what set quantum technology apart from classical ones. We will go into them in some detail for the remaining sections.

1.1 Superposition

Consider the two dimensional space $\mathbb{C}^2$. We denote the orthonormal basis vectors $|0\rangle = \begin{bmatrix} 1 \\ 0 \end{bmatrix}$ and $|1\rangle = \begin{bmatrix} 0 \\ 1 \end{bmatrix}$ as the *computational* basis of $\mathbb{C}^2$. Thus any vector in $\mathbb{C}^2$ can be written as a complex linear combination of the basis vectors, say $a\,|0\rangle + b\,|1\rangle$ and if $|a|^2 + |b|^2 = 1$, the vector is referred to as a qubit. Now in the classical case, we would only have access to $|0\rangle$ or $|1\rangle$ but now, it is possible to put the qubit in a superposed state $|\psi\rangle = a\,|0\rangle + b\,|1\rangle$, which in and of itself, will be used to boost computing power as will be shown in a later section.

An operator is a map between vectors to vectors. We will be focused on linear operators. More specifically, operators that preserve length of the vector i.e. isometry operators and even more specifically, unitary operators. For qubits, there are a class of 4 unitary operators which are a basis for any linear operator in $\mathbb{C}^2$. These are known as the *Pauli operators* and are given by

1. **Identity operator** $I = \begin{bmatrix} 1 & 0 \\ 0 & 1 \end{bmatrix}$. Action by this Pauli operator doesn't change the vector in any way.

2. **Bit-Flip operator** $X = \begin{bmatrix} 0 & 1 \\ 1 & 0 \end{bmatrix}$. Action by this flips $|0\rangle$ to $|1\rangle$ and vice versa.

3. **Phase-Flip operator** $Z = \begin{bmatrix} 1 & 0 \\ 0 & -1 \end{bmatrix}$. Flips the sign of $|1\rangle$. Leaves $|0\rangle$ unchanged.

4. **Y operator** $Y = \begin{bmatrix} 0 & -i \\ i & 0 \end{bmatrix}$, where $i = \sqrt{-1}$. Essentially a combination of the bit-flip and phase-flip operators. It is equivalent to iXZ.

Note that all Pauli operators are unitary and self inverting. We will use these operators in the applications to be discussed.

When it comes to setting up superpositions, an important operator is the **Hadamard operator** (or Hadamard Gate) given by $H := \frac{1}{\sqrt{2}} \begin{bmatrix} 1 & 1 \\ 1 & -1 \end{bmatrix}$. This is because operating the Hadamard gate on the computational basis yields not only superposed states but a new orthonormal basis given by

$$|+\rangle = H |0\rangle = \frac{1}{\sqrt{2}}(|0\rangle + |1\rangle) \qquad (19.1)$$

$$|-\rangle = H |1\rangle = \frac{1}{\sqrt{2}}(|0\rangle - |1\rangle). \qquad (19.2)$$

Once again, the Hadamard gate is self inverting and unitary. Next we discuss the interpretation and measurement of a superposed qubit. Note that we will use the word qubit and state interchangeably (a state is more general with the qubit dimensions not necessarily 2).

1.2 State Measurements

A *positive operator valued measure (POVM)* is a collection of non-negative operators $\{M_m = E_m^* E_m\}$ with $\sum_m M_m = I$. According to the quantum axioms, when a state, say $|\psi\rangle$, is measured with respect to a POVM, it means that the qubit will evolve to $\frac{E_m|\psi\rangle}{\sqrt{\langle\psi|M_m|\psi\rangle}}$ with probability $\langle\psi| M_m |\psi\rangle$, and output measurement value m where $\langle\psi|$ is the conjugate transpose of $|\psi\rangle$.

When measured in the computational basis i.e. POVM is $M_0 = \begin{bmatrix} 1 & 0 \\ 0 & 0 \end{bmatrix}$ and $M_1 = \begin{bmatrix} 0 & 0 \\ 0 & 1 \end{bmatrix}$, the state $|\psi\rangle = a |0\rangle + b |1\rangle$ has an interesting interpretation. Before measurement, we cannot say that the state is in either $|0\rangle$ or $|1\rangle$ but rather in a hybrid of the two (refer to Schrödinger's cat). However once measured with above POVM, it will collapse to $|0\rangle$ with probability $|a|^2$ and $|1\rangle$ with $|b|^2$. Hence we should take care not to interpret it as "qubit is $|0\rangle$ with probability $|a|^2$, and $|1\rangle$ with $|b|^2$" before measurement as the two notions are different. The latter notion, known as a mixed state is also studied but we will not be covering it here.

1.3 Entanglement

The second aspect of quantum theory that makes it appealing is the concept of entanglement. To understand this, we need to work with at least a two qubit system. The computational basis is now extended to a two qubit (and in general n) by tensoring the two respective spaces. In the two particle case, we have $|00\rangle :=$ $|0\rangle \otimes |0\rangle$, $|01\rangle$, $|10\rangle$ and $|11\rangle$ as the computational basis. So does that mean we can get all states of the tensor system simply by taking a tensor product between spaces?

Unfortunately no as we have to close the space under linearity. Thus there could exist states that are not simple tensor products. These states are called entangled states. We discuss them with a helpful example.

Define the following states, known as **Bell pairs** or **Bell states**:

$$|\Phi^+\rangle = \frac{|00\rangle + |11\rangle}{\sqrt{2}}, \quad |\Phi^-\rangle = \frac{|00\rangle - |11\rangle}{\sqrt{2}}, \tag{19.3}$$

$$|\Psi^+\rangle = \frac{|01\rangle + |10\rangle}{\sqrt{2}}, \quad |\Psi^-\rangle = \frac{|01\rangle - |10\rangle}{\sqrt{2}}, \tag{19.4}$$

Note that not only are these orthogonal to each other but that they are a basis for the tensor space. Moreover, once can switch between one Bell state to the other by merely applying Pauli operators locally. To give an additional insight into entanglement, assume that Alice and Bob have an entangled state $|\Phi^+\rangle$ with Alice possessing first qubit and Bob the second. Suppose Alice performs a local measurement in the computational basis on her qubit and finds that it is 0. Then even though Bob never touched his qubit, the very act of measurement by Alice forced Bob's qubit to $|0\rangle$. Hence the spooky action at a distance moniker.

Two qubits can be entangled using a **CNOT** (Controlled NOT) gate. This is a gate that takes two input qubits, say $|ab\rangle$, where $|a\rangle$ is the control input and $|b\rangle$ is the true input. The output is $|a\rangle\,|a \oplus b\rangle$, where $\oplus$ is modulo 2 addition. To generate the Bell state $|\Phi^+\rangle$, we just need to pass $|+\rangle$ as the control input and $|0\rangle$ as true input. Thus we can entangle two qubits.

2 Deutsch-Jozsa Algorithm

Quantum computing methods are often touted as being superior to classical computing methods. They achieve this by trying a *superposition* of all possible inputs to the system in question. To illustrate the gains, it helps to consider the following problem known as the Deutsch problem [1].

Let $\mathcal{B} := \{0, 1\}$ and $f : \mathcal{B}^n \to \mathcal{B}$ be a function that maps an n bit binary vector to a binary symbol. We say f is *constant* if $f(x) = 0$ (or 1) for all inputs. If $f(x)$ is 0 for half the inputs and 1 otherwise, we call it *balanced*. Suppose we have a black box that implements this function, known as the *oracle* for f. Assume that we do not know the exact description of f but we can evaluate it for any input we give (known as *querying*). The task is to check whether f is balanced or not by querying the oracle. We seek to minimize the number of queries.

This is a problem that, on a classical computer, has a worst case exponential complexity. To see this, noting that we know nothing about f, all we can hope is to keep querying different n length binary vectors until the output changes (from 0 to 1 or 1 to 0) or $2^{n-1} + 1$ queries have resulted in the same result. If probabilistic strategies are allowed, we can do it with far less queries.

The interesting thing is that there is a quantum algorithm called the Deutsch Jozsa algorithm [1] that

- solves this with polynomial number of queries ..., **(Good)**
- where the polynomial is a constant ..., **(Excellent)**
- the constant is 1. **(What!!)**

To understand the scale of improvement, assume that the function takes 1 nano seconds (ns) to evaluate a query. Then for $n = 100$, for a classical system, in the worst case we will have to check $2^{99} + 1$ inputs which, assuming no overheads, would take about **20 trillion years!!** With Deutsch Jozsa, we can get the answer in a matter of **nanoseconds!!**. Note that for $n = 50$, the trillions of years reduces to about a week, but is still computationally intensive. Even if parallel computing was allowed the complexity would still be exponential even though significantly lesser than without. More specifically, the worst case complexity would reduce to $2^{n-M} + 1$, where M is the number of machines used.

The Deutsch-Jozsa algorithm requires that the quantum oracle exist. It can be shown that there is an efficient way to implement oracles for any arbitrary function $f : \mathcal{B}^n \to \mathcal{B}$. The algorithm is as follows.

Algorithm 1 Deutsch-Jozsa algorithm

1: Prepare qubits $|0\rangle^{\otimes n} |1\rangle$.
2: Apply the Hadamard gate to all qubits. Thus the state gets transformed to $\frac{1}{2^{n/2}} \sum_{x \in \mathcal{B}^n} |x\rangle |-\rangle$. Here $|x\rangle$ is understood as the binary expansion (in n bits) of x in qubit form. E.g. for $n = 4$, $|9\rangle = |1001\rangle$. Note that the input is now a uniform superposition of all possible n-length binary inputs.
3: Apply the quantum oracle to the above qubit. This yields, after some simple manipulations,

$$\frac{1}{2^{n/2}} \sum_{x \in \mathcal{B}^n} (-1)^{f(x)} |x\rangle |-\rangle \tag{19.5}$$

4: Discard the last qubit. Apply the Hadamard gate to all qubits. The state is now

$$\frac{1}{2^n} \sum_{x \in \mathcal{B}^n} \sum_{y \in \mathcal{B}^n} (-1)^{f(x)+x^T y} |y\rangle \tag{19.6}$$

where $x^T y = \sum_{i=1}^{n} x_i y_i \mod 2$.
5: Measure the $|0\rangle^{\otimes n}$ state. If f is constant, then with probability 1, we will get all zeros from the measurement. If f is balanced, this probability is 0. Hence by looking at the measurement, we can distinguish the two classes.

For all its power, the Deutsch problem unfortunately has no practical use. It was essentially a toy problem to showcase the power of quantum computing. For algorithms with practical use, Shor's algorithm, for example, can be used to factor large composite numbers. The reader is directed to the excellent text by Nielsen and Chuang [2] for further details on quantum computing algorithms.

In 2019, Google was able to experimentally demonstrate [3] the speed-up advantage by solving a similar toy problem using a 54 qubit Sycamore quantum processor. While the problem they solved did not have any practical utility, it did validate and demonstrate the power and advantages of quantum computing.

3 Superdense Coding

While the previous section illustrated the power of *superposition* in quantum theory, here we consider the other power, namely *entanglement*. In classical communications, over an ideal binary communication channel, the maximum information that can be reliably communicated is one bit per channel use. However, in the quantum paradigm, with the aid of entanglement, it is possible to *double* the maximum information! This method is referred to as *superdense* coding [4].

The setup involves the transmitter (Alice) and receiver (Bob), sharing a Bell pair which is an entangled state of Alice and Bob. Alice has access to her qubit of the shared pair and an ideal quantum channel to Bob. Bob has access to his qubit and the output of the ideal quantum channel. Alice seeks to communicate two bits of classical information, say $b_0 b_1$, where $b_0, b_1 \in \{0, 1\}$. Alice is restricted to what are called *local* operations and measurements, i.e. she can only directly interact with the qubits in her possession. The protocol is as follows.

Algorithm 2 Superdense coding

1: Alice and Bob share a Bell pair that is currently in the state $|\Phi^+\rangle$.
2: Let bits 00 correspond to Bell pair $|\Phi^+\rangle$, 01 to $|\Psi^+\rangle$, 10 to $|\Phi^-\rangle$ and 11 to $|\Psi^-\rangle$. Note that each of these pairs can be got from the others via local unitary operations known as Pauli operators.
3: Depending on which pair of bits Alice wants to communicate to Bob, she applies the appropriate Pauli operators locally to her qubit and converts $|\Phi^+\rangle$ to that qubit. If 00 is to be transmitted, no conversion is required.
4: Alice uses the ideal channel and transmits her qubit to Bob. Now Bob has access to the whole state.
5: Bob measures both qubits in the Bell basis with respect to the encoding above and recovers the two bits.

Communication with the aid of shared entanglement of states is termed as *entanglement assisted* (EA) communications. The protocol above shows that one can send one qubit in a *noiseless quantum* channel for communicating two classical bits. The converse is also true, where we can communicate two bits over a *noiseless classical* channel to transfer one qubit. This is known as quantum teleportation, which we do not discuss here. Further details on these protocols may be found in [2, 4].

Note, however, that the shared entangled state is destroyed at the end of the protocol. Setting up the shared state requires entanglement generation which has its own overheads. This has to be done before every channel use and so, it may be impractical if setting up entanglement itself is challenging. A landmark achievement in this field is the work by Yin et al. [5] which demonstrated entanglement of photons via satellites over a terrestrial distance of 1200 km. However, the two stations were located on mountain tops (the satellite was to beam entangled photons) and it was noted that approximately one in 6 million photons were successfully recovered. Hence, it is still a far cry from practically realizing entanglement but it is a step in the right direction.

4 Error Correction

Error correcting codes [6] are a fundamental aspect of practical communications. These are, in simple words, techniques of adding redundancy to transmitted symbols so as to protect them against errors that may be introduced due to the channel. A simple example of a channel that introduces errors is the binary symmetric channel (BSC) which takes binary inputs and outputs the correct symbol with probability $1 - \alpha$ and bit-flips it with probability α, where $0 < \alpha \leq 1/2$. In the theory of classical error correction, the symbols are usually drawn from a finite field and so, there are only finitely many errors possible. This theory includes block codes (e.g. Hamming code, LDPC codes), polynomial codes (e.g. Reed Solomon code) and convolutional codes (e.g. Turbo codes). Codes such as Turbo and LDPC codes are currently in use in 4G and 5G communications.

In quantum error correction [2], we work with keeping the reliability of qubits as opposed to bits. Unlike the previous case, qubits can be placed in an arbitrary superposition. As a consequence, there are now a potentially uncountable collection of errors. Hence we need to specify a class of errors that can be detected or corrected by the error correction scheme. For qubits, we can consider errors as a consequence of bit-flip ($|0\rangle$ becoming $|1\rangle$ and vice versa) and phase-flip errors (flipping the sign of qubit if it is $|1\rangle$). We could extend classical error correcting codes to quantum in an analogous fashion. However we have to be careful as unlike classical case, it is not possible to create multiple copies of an arbitrary quantum state (no cloning theorem).

4.1 Detecting and Correcting Bit-Flip Errors

A bit-flip is essentially a Pauli X gate applied to that qubit. Fortunately the Pauli gate is self inverting and so, a bit-flip error can be corrected by applying the X gate on the erroneous qubit. To detect a single bit-flip, we describe a classical technique using repetition code. That is $0 \rightarrow 000$ and $1 \rightarrow 111$. Then if an error occurs in any of the three bits, we can both detect and correct it via majority rule.

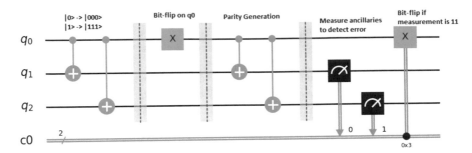

Fig. 19.1 Implementation of bit-flip correction using IBM Qiskit. Above, qubits q_1 and q_2 are initialized to $|0\rangle$ initially. In this circuit, a bit-flip error on q_0 is assumed

To implement this in a quantum code, we prepare $|000\rangle$ for $|0\rangle$ and $|111\rangle$ for $|1\rangle$. Assume the qubit is in $|q_0q_1q_2\rangle$, where q_0 is the qubit to be protected. At the receiver end, assuming at most one bit flip has occured, using CNOT gates, we update $|q_1\rangle \leftarrow |q_0 \oplus q_1\rangle$ and $|q_2\rangle \leftarrow |q_0 \oplus q_2\rangle$. Now we measure q_1 and q_2. If this measures 11, we flip $|q_0\rangle$ by applying an X gate. Otherwise the error is in the ancillary qubits which have served their purpose post measurement. Figure 19.1 shows the implementation using CNOT gates and measurements. However, what makes this scheme powerful is that it even works when q_0 is in an arbitrary superposition!! It is not a difficult exercise to verify this. Note however, that for $|\psi\rangle = a\,|0\rangle + b\,|1\rangle$, the first part will map it to $a\,|000\rangle + b\,|111\rangle$ and not $|\psi\psi\psi\rangle$ (The latter is not possible anyway due to no-cloning theorem unless we know what a and b are).

4.2 Detecting and Correcting Phase-Flip Errors

Suppose instead, we needed to correct phase-flip errors, we note that $X = HZH$, where H is the Hadamard gate and Z the Phase-flip operator. Thus the bit-flip error correcting code as above will work except that we would have to apply a Hadamard gate to the input qubits before and after the phase-flip. However, note that neither scheme can correct both types of errors. Next, we discuss a scheme by Shor that can do both.

4.3 Shor's 9 Qubit Code

In classical coding theory, there is a technique of combining two or more different coding techniques to develop a hybrid code that has, in a sense, the salient features of the constituent codes. This technique, known as code *concatenation* [7], is a widely used technique to derive new codes (usually of longer length) from old ones (of shorter length), e.g. Justesen codes[8].

Shor's 9 qubit code is a concatenation of the bit-flip and phase-flip code. This will result in a code that can correct any one error due to any Pauli matrix (I, X, Y, Z). The steps are provided below. Using the ideas mentioned before, we can show that this can protect against any one error caused by a Pauli gate.

Algorithm 3 Shor's 9 Qubit code

1: Given qubit $|\psi\rangle = a\,|0\rangle + b\,|1\rangle$.
2: Use the phase-flip code to map $|0\rangle \rightarrow |+++\rangle$ and $|1\rangle \rightarrow |---\rangle$.
3: Map $|+\rangle \rightarrow \frac{|000\rangle+|111\rangle}{\sqrt{2}}$ and $|-\rangle \rightarrow \frac{|000\rangle-|111\rangle}{\sqrt{2}}$.
4: Resultant is a 9 qubit code.

An interesting aspect of this code is that not only does it protect against errors of Pauli nature as described but also against arbitrary errors as long as only one qubit is affected [2]. This is an unusual feature of the Shor code that sets it apart from classical error correcting codes. However, as can be seen, protecting one qubit this way required 8 additional qubits. Compare this with a Hamming code which protects 4 bits from one error using only 3 additional bits. Thus the cost of *protecting* a qubit is exceedingly costly.

4.4   Other Quantum Error Correcting Codes

Considering only Pauli errors, it may be interesting to find out what is the minimum number of additional qubits required to protect a qubit. According to the quantum Hamming bound, a minimum of 4 additional qubits are required. What's interesting is that this bound is actually achievable!! The reader is directed to the description of Calderbank-Shor-Steane codes (CSS) code [9, 10] construction of which the Steane's code (7 qubit code) is a popular scheme. The Laflamme 5 qubit code [11], using stabilizer theory is the one that achieves the Hamming bound.

However, compared to classical bits, reliable qubits are costly. If it takes 5 qubits to represent one logical qubit, it puts a high price on the future of reliable quantum computing. Moreover, unless the environment is controlled suitably, qubits tend to interact with the environment and decohere. Thus the qubits inherently have a reliability issue and therefore some sort of fault tolerant computing is the need of the hour. Unfortunately, this puts an even higher price on the number of qubits.

5   Practical Considerations and Future

Each of the prior sections started out on a high and interesting note but concluded on a bleak note. This is the unfortunate reality of quantum computing and communications at present. A lot of research is currently being undertaken to bring it to the *industry* level. The best we can currently hope for are quantum assisted communications where only certain aspects, like quantum key distribution, are implemented.

The qubits themselves are of generally of two types. Note that these names are colloquial and not official.

5.1   Costly but Accurate Qubits (Coolbits)

Currently the implementation of practical quantum computers involves supercooled qubits. In particular, this includes ion trap, nuclear magnetic resonance (nmr) and recent transmission line shunted plasma oscillation (TRANSMON) qubits. The qubits are generally supercooled to the orders of a few tens milli-Kelvin so as to slow down decoherence and interaction with the environment. With the aid of lasers, states may be set as required with an appropriate measuring mechanism. The cooling is carried out by a complicated apparatus known as a dilution refrigerator, which actually takes up majority of the space required of the quantum processor.

The key advantage of these qubits are that they are accurate and suitable for quantum computing needs. However they are quite expensive to design and maintain and not very useful from a communication aspect. In fact, the Sycamore chip, the quantum processor which Google used to claim quantum supremacy [3], was made of TRANSMON qubits and had just 54 qubits, which was an achievement at the time. Also the IBM quantum experience [12] project allows programmers to use their package Qiskit on Python, to write code for quantum processors that can be implemented on their own physical processors.

5.2   Quick but Dirty Qubits (Hotbits)

While the coolbits were more physical in nature, the hotbits are optical. The photon itself can be considered to be a qubit, with various but discrete energy states. To generate single photons, one can use coherent lasers and an electromagnetic cavity [2]. Using optical devices such as beamsplitters, one can generate superposed states. With the aid of Kerr nonlinear devices, one can even obtain entangled states.

The advantage here is that single photon generation is cheap and doesn't require super-cooling. However, these photons are hard to force meaningful interactions on, with Kerr devices permitting weak couplings. Additionally, fiber optic cables that are used for communications, tend to cause photon absorption which adds another layer of undesirability.

5.3   Concluding Remarks

While several key developments have occurred recently, it will still take a long time (at least 20 years) before we can take quantum technology to the realm of engineers. Just as how classical computers, that were huge boxes that were initially the domain of research and defense, evolved to revolutionize and change the lives of average consumers, the same can be said about quantum. It is in the initial stages and we hope for further breakthroughs as one by one of these problems are tackled. For now quantum-assisted technologies [13] including quantum machine learning are being studied for incorporation into beyond 5G and 6G communications. This involves using quantum computing to speed up the operations inherent in neural network implementations (e.g. autoencoders [14]). While autoencoders are being considered for replacing end to end communication models in 5G and beyond 5G communications, the quantum assisted versions should deliver the necessary speedup to support future communications. To conclude, it is still very optimistic to consider implementation in 6G at the time of writing this, save for quantum assisted communications.

References

1. D. Deutsch, R. Jozsa, Rapid solution of problems by quantum computation. Proc. R. Soc. London Ser. A Math. Phys. Sci. **439**(1907), 553–558 (1992)
2. M. Nielsen, I. Chuang, *Quantum Computation and Quantum Information* (Cambridge University Press, Cambridge, 2002)
3. F. Arute, K. Arya, R. Babbush, D. Bacon, J.C. Bardin, R. Barends, R. Biswas, S. Boixo, F.G. Brandao, D.A. Buell et al., Quantum supremacy using a programmable superconducting processor. Nature **574**(7779), 505–510 (2019)
4. J. Watrous, *The Theory of Quantum Information* (Cambridge University Press, Cambridge, 2018)
5. J. Yin, Y. Cao, Y.-H. Li, S.-K. Liao, L. Zhang, J.-G. Ren, W.-Q. Cai, W.-Y. Liu, B. Li, H. Dai et al., Satellite-based entanglement distribution over 1200 kilometers. Science **356**(6343), 1140–1144 (2017)
6. S. Lin, D.J. Costello, *Error Control Coding*, vol. 2 (Prentice Hall, Upper Saddle River, 2001)
7. G.D. Forney, Concatenated codes, 1965
8. J. Justesen, Class of constructive asymptotically good algebraic codes. IEEE Trans. Inf. Theor. **18**(5), 652–656 (1972)
9. A.R. Calderbank, P.W. Shor, Good quantum error-correcting codes exist. Phys. Rev. A **54**(2), 1098 (1996)
10. A. Steane, Multiple-particle interference and quantum error correction. Proc. R. Soc. Lond. Ser. A Math. Phys. Eng. Sci. **452**(1954), 2551–2577 (1996)
11. R. Laflamme, C. Miquel, J.P. Paz, W.H. Zurek, Perfect quantum error correcting code. Phys. Rev. Lett. **77**(1), 198 (1996)
12. IBM quantum experience. https://quantumcomputing.ibm.com/
13. S.J. Nawaz, S.K. Sharma, S. Wyne, M.N. Patwary, M. Asaduzzaman, Quantum machine learning for 6G communication networks: state-of-the-art and vision for the future. IEEE Access **7**, 46317–46350 (2019)
14. T. O'Shea, J. Hoydis, An introduction to deep learning for the physical layer. IEEE Trans. Cognit. Commun. Netw. **3**(4), 563–575 (2017)

Chapter 20
Post-quantum Cryptography in 6G

Juha Partala

Abstract The quantum computing paradigm is fundamentally different from the classical one. There are computational problems we are not able to solve on a contemporary computer, but which we can efficiently solve on a quantum one. One of these problems is the discrete logarithm problem (DLP) which is the basis of modern asymmetric cryptography. Once large-scale quantum computing becomes a reality, these cryptographic primitives need to be replaced with quantum-secure ones. While we are still in the early stages of quantum computing, steps have been taken to prepare for the shift to cryptography that is secure in the post-quantum world. According to the current knowledge, contemporary symmetric cryptography remains secure for the most part even after the advent of quantum computing. Asymmetric primitives based on integer factorization and the DLP need to be replaced. In this chapter, we take a look at the post-quantum secure alternatives for key establishment, public-key encryption and digital signatures. We also discuss their properties and the effect on the performance of the future 6G networks.

1 Introduction

Quantum computing has drawn a lot of interested in the last years. Development has been rapid and commercial devices have been predicted to be available already in the near future [7]. Quantum computing will have a dramatic effect on modern cryptographic algorithms. The security of these algorithms is based on the hardness of specific computational problems. However, it seems that certain problem types are much easier to solve on a quantum computer than on a contemporary one. While we are still in the early stages of quantum computing and large-scale quantum computing can be expected to take a decade or more, steps have been taken to prepare for the shift to cryptography that is secure when quantum computing is a reality.

J. Partala (✉)
Center for Machine Vision and Signal Analysis, University of Oulu, Oulu, Finland
e-mail: Juha.Partala@oulu.fi

431

The sixth generation of communication networks are envisioned to provide nearly unlimited wireless connectivity, ultra-high reliability and ultra-low latency, as well as to be dependable even for critical applications such as eHealth [10]. Computations will be performed on the edges of the network to reduce latency requiring complex trust mechanisms. These properties will place complicated requirements for the network security architecture. In order that the envisioned properties of 6G are realizable in practice, several research questions relating to network security need to be solved [16]. It is evident that the complex security mechanisms of 6G will be based on both symmetric and asymmetric cryptography and quantum computing will affect those mechanisms. Even though security can be achieved against quantum computing, there will be a penalty to efficiency. It will not be straight-forward to satisfy currently envisioned requirements of 6G with quantum secure algorithms.

In this chapter, we take a look at the current status in quantum secure cryptography. We briefly introduce basic concepts of cryptography, such as symmetric and asymmetric primitives and the computational problems underlying contemporary asymmetric cryptography. We also take a look at the recent successes in quantum computing and its predicted development in the future. In the United States, the National Institute of Standards and Technology (NIST) is currently hosting a selection process called NIST PQC for a post-quantum cryptography standard. These primitives will provide post-quantum secure key exchange, public-key encryption and digital signatures in a standardized way. We describe the cryptographic primitives in the current stage of the competition and discuss their properties and efficiency in regard to 6G.

The chapter is organized as follows. First, we introduce the basic concepts of cryptography and quantum computing in Sect. 2. Section 3 is devoted to recent developments in quantum computer engineering. The development in the application of cryptography towards 6G is laid out in Sect. 4. Section 5 describes the state-of-the-art alternatives for post-quantum secure key establishment, public-key encryption and digital signatures. Finally, Sects. 6 and 7 provide the discussion and conclusion.

2 Cryptography and Quantum Computing

2.1 Cryptography

Cryptography studies techniques for securing transactions, information and computations. It encompasses fundamental techniques for modern communications such as confidentiality, message integrity, key exchange and digital signatures. The security of cryptographic schemes is based on the assumed infeasibility of specific computational problems. First, a rigorous and precise definition of security is formulated. Based on the security definition, a cryptographic algorithm is proven to satisfy it by reducing an infeasible problem to the problem of breaking the algorithm. That is, if an adversary is able to violate the security of that algorithm, he/she is able to also solve the hard computational problem. Therefore, a cryptographic algorithm can be at most as secure as the underlying problem and a lot of research is devoted into the study of those problems.

Cryptography can be roughly divided into two classes:

1. Symmetric cryptography, where communication participants share a secret key. This class contains classical cryptographic primitives such as stream and block ciphers, cryptographic hash functions and message authentication codes. Symmetric cryptography always requires a trusted channel to establish the shared secret key.
2. Asymmetric cryptography, also called public-key cryptography, does not require a shared secret key. Instead, a private and public key pair is used. This class contains, for example, key exchange schemes to establish shared keys for symmetric primitives, public key encryption and digital signatures. Of the key pair, only the private key is kept secret. The public key can be used, for example, to encrypt a message intended to the owner of that key or to verify digital signatures. The private key is needed, for example, to decrypt or to generate signatures.

Asymmetric cryptography is more susceptible to quantum computing and we will mostly concentrate on it. Regarding the solvability of computational problems, polynomial time computation is typically viewed as feasible computation. If an algorithm finishes in polynomial time with respect to the length of its input, then it can be also executed in practice. A problem that cannot be solved in polynomial time is generally considered to be infeasible.

Every non-prime integer n can be factored into a product of smaller integers. The problem of factoring arbitrary integers has proved out to be hard and no polynomial-time classical algorithm is known. Integer factorization is the underlying computational problem of several asymmetric cryptographic schemes. These schemes include, most notably, the RSA public-key encryption scheme [14]. Another well-known computational problem is the discrete logarithm problem (DLP) in finite cyclic groups. The DLP asks to find an integer x given g^x, where g is a generator of the group. It underlies another widely used class of asymmetric schemes: those based on the Diffie-Hellman key exchange scheme [3]. The original scheme applies modular arithmetic and no polynomial-time classical algorithm is known for the DLP on such groups. ElGamal public-key encryption and the Digital Signature Algorithm (DSA) also apply the DLP. The problem seems to be even harder on cyclic groups based on elliptic curves and the so called elliptic curve DLP (ECDLP). Currently, the elliptic curve based Diffie-Hellman scheme is the most efficient key exchange scheme available when considering both key length and performance.

2.2   Quantum Computing

Quantum computing is based on quantum bits, *qubits*. Unlike ordinary bits that can be either in state 0 or 1, qubits can be in a superposition (a linear combination with complex coefficients) of these two states. Furthermore, multiple qubits can be entangled meaning that the quantum state of a single qubit cannot be described independently of the others. Similarly to ordinary computers, operations can be carried out on the qubits. In the quantum circuit model, reversible transformations are applied to construct quantum logic gates. Such gates can then be connected into a circuit performing an arbitrary computation thus implementing a quantum computer. Other equivalent models of computing, such as adiabatic or measurement based quantum computing, also exist. However, our discussion will be based on the quantum circuit model.

Computational problems that are solvable on a contemporary computer can be also solved on a quantum one. The converse is also true. Any problem solvable on a quantum computer is also solvable on an ordinary one. However, the superposition of entangled qubits facilitates an exponential speedup on solving specific problem types compared to the classical model of computation. This means that while any computation can be performed on both types of computers, in practice certain problems seem to be much easier in the quantum computing model. However, the number of entangled qubits is the enabling factor of such algorithms. The algorithm can be executed and finished fast if and only if enough qubits are available. Therefore, the number of qubits that we can entangle and keep coherent determines whether we can consider a specific problem feasible in practice.

Quantum computing affects both classes of cryptography but not in equal measure. Symmetric cryptography survives for the most part. However, asymmetric primitives in wide use need to be updated. There are two important quantum algorithms that affect cryptography: Grover's algorithm and Shor's algorithm. The former affects both symmetric and asymmetric schemes, while the latter targets asymmetric schemes.

2.2.1   Grover's Algorithm

Let f be an injective function with N different inputs and suppose that we are only able observe the input and output behavior of f. That is, f is viewed as a black box. In order to find a secret input to f, we need to make $O(N)$ queries into the black box in the worst case, since our last try could be the right value. In the quantum computing paradigm, the secret value can be found faster. Grover's algorithm finds the secret value on a quantum computer in time $O(\sqrt{N})$ [6], which also seems to be optimal [1]. This may seem only a quadratic improvement, but it is significant in practice.

Grover's algorithm affects both symmetric and asymmetric schemes. For example, due to the quadratic speedup, symmetric 128-bit encryption can be broken in approximately 2^{64} steps and 256-bit encryption in 2^{128} steps. In general, this means that the key length of any cryptographic scheme needs to be at least doubled to maintain the current level of security in the quantum computing model. The lowest security options, such as AES-128, will be rendered obsolete in the post-quantum era. However, the higher security versions, such as AES-196 and AES-256 will remain secure with a lower security level. The same is true for any other symmetric primitive applying a secret key such as message authentication schemes.

Regarding cryptographic hash functions, the Grover's algorithm applies to the problem of finding preimages. In the quantum model, preimage resistance is halved. However, due to the birthday attack and its effect on the collision-resistance of hash functions, the digest length is already on an adequate level. The birthday attack is a classical algorithm that finds a collision (m_1, m_2) such that $H(m_1) = H(m_2)$, where H is a cryptographic hash function, in $O(\sqrt{N})$ steps. Due to the birthday attack, message digests computed using cryptographic hash functions need to have a length that is at least double of the security parameter. That is, to have a security level of 128 bits against collisions, the digest has to be 256 bits. In the quantum computing model, the birthday attack can be theoretically mounted in $O(\sqrt[3]{N})$ steps [2]. However, such an attack would also require $O(\sqrt[3]{N})$ qubits making it infeasible in practice. Quantum algorithms for collision-finding targeting specific hash function constructions also exist [9]. However, to the best of current knowledge, hash function security remains largely unaffected in the quantum computing model.

2.3 Shor's Algorithm

Shor's algorithm is a quantum algorithm for factoring integers in polynomial time [15]. On a classical computer, the fastest algorithm for integer factorization is sub-exponential time. Therefore, integer factorization represents a practical separation of the traditional and quantum computing models. Due to the polynomial time solvability of factoring, RSA is not secure in the quantum circuit model. The DLP can be also solved in polynomial time using Shor's algorithm for both the original modular aritmetic based groups and the elliptic curve groups.

The practical solvability of the factoring problem and the DLP on a quantum computer depends on the security parameter. For example, when solving the DLP on the group of multiplication modulo a prime p, $\mathbb{Z}_p^*$, where p is an n-bit prime, Shor's algorithm needs at least $2n + 3$ qubits. The same is true for factorization. For the ECDLP, there exists an algorithm using at most $5n + 8\sqrt{n} + 4\lceil \log_2 n \rceil + 10$ qubits [13]. Therefore, the deciding factor in the future security of the DLP and factoring-based schemes is the development of the number of qubits. NIST recommendations for the binary length of the composite n for factoring based schemes or for the prime p for the DLP and ECDLP based schemes together with the minimum number of required qubits have been collected into Tables 20.1 and 20.2. Note that factoring and modular multiplication based cryptography is more secure than elliptic curve cryptography with these parameter choices in the quantum computing model.

Table 20.1 The required bit length of the composite integer n for factoring based schemes and that of the prime modulus p for DLP based schemes using a specific security level. On the right, the minimum number of qubits required to solve the corresponding instance in polynomial time on a quantum computer

Security parameter [bits]	Factoring and DLP [bits]	Number of qubits
80	1024	2051
112	2048	4099
128	3072	6147
196	7680	15,363
256	15,360	30,723

Table 20.2 The required bit length of the prime p of the underlying finite field for the elliptic curve DLP. On the right, the minimum number of qubits required to solve the corresponding ECDLP in polynomial time on a quantum computer

Security parameter [bits]	ECDLP [bits]	Number of qubits
80	160	944
112	224	1282
128	256	1450
196	384	2124
256	512	2788

3 Development of Quantum Computing

In order to build a physical quantum computer using the quantum circuit model, five requirements need to be met [4]:

1. We need a scalable physical system, where the qubits are instantiated.
2. We have to be able to initialize and reinitialize the system into an initial state, where the qubits are entangled with each other.
3. The entangled qubits need to stay in coherence long enough in order to perform meaningful computations. Coherence means that the qubits remain entangled with each other and do not get interfered or entangled with, for example, the environment thus ruining the computation.
4. There has to be a universal set of quantum gates that can be implemented in practice to be able to carry out arbitrary computations.
5. We have to be able to measure the qubits to read out the result of the computation.

Solving these engineering problems is not easy in practice. For example, decoherence of the qubits is easily caused by interaction with the outside environment. In fact, decoherence and noise are the limiting factors of contemporary quantum computation. Novel methods and development in quantum error correction among other things are needed before large-scale quantum computing is possible.

Despite the numerous engineering challenges, quantum computing has progressed rapidly during the last decade and the development in the number of qubits has been especially rapid in the last 5 years. In 2020, state-of-the-art quantum

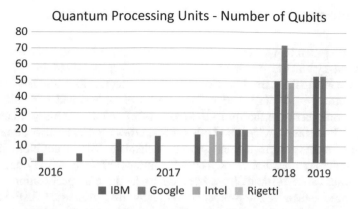

Fig. 20.1 Development in the number of qubits in published quantum processing units. However, it should be noted that development in coherence and other factors influencing computability are not reflected in this chart

computing processors can practically operate on 53 qubits. Such processors have been published by IBM and Google. In 2019 Google announced that it has achieved "quantum supremacy": solving a problem that would be infeasible on a classical computer. Notable published quantum processing units from IBM, Google, Intel and Rigetti Computing and their respective qubit lengths have been depicted in Fig. 20.1. However, it should be noted that the number of logical qubits might be lower than that of physical ones due to errors and error correction. In addition, the coherence times are not the same for all of these computers. These issues are not reflected in the chart. In fact, the number of qubits does not give the full picture. IBM measures the development using a "quantum volume" metric. It is based on the width and depth of a circuit that a quantum computer can successfully compute. In addition to the number of qubits, quantum volume attempts to capture progress in all the five requirements by incorporating, for example, the fidelity of the quantum gates.

Precise predictions on the development of the quantum volume or the number of qubits are obviously impossible to make. However, there are predictions that are based on the rate of development observed during the last decade. At IBM, the current hope is that quantum volume will double annually. There are also other optimistic estimates. The "Neven's Law", which can be considered a quantum analogue of Moore's Law, predicts that quantum computing power experiences a double exponential growth relatively to classical computing. Naturally, it is impossible to be certain whether these predictions will hold or whether a more modest development will realize. However, based on the recent developments in quantum computing, it makes sense to prepare for the possibility that large-scale quantum computing is a reality in the lifetime of the future 6G networks.

4 Cryptography and the Development Towards 6G

Up to the 4th generation, as well as in the currently implemented 5G, authentication is based on symmetric cryptography. A shared key is stored in a Subscriber Identity Module (SIM) card. Functionality of the authentication and key agreement (AKA) protocol in 4G is based on symmetric primitives and authentication is performed only between the user equipment (UE) and the mobility management entity. In 5G, service providers and other third parties need to be also authenticated requiring a more flexible authentication mechanism [5]. In addition, the SIM card methodology does not work well with Internet-of-Things (IoT) devices and therefore alternative methods have been devised. The 5G specification currently includes three authentication protocols: 5G AKA, EAP-AKA and EAP-TLS. The first two are based on symmetric cryptography. However, the third one is based on asymmetric cryptography and has been included, especially, to support IoT environments. The Extensible Authentication Protocol (EAP) supports multiple authentication mechanisms directly on the data link layer. Transport Layer Security (TLS) is a set of cryptographic protocols designed to provide authentication, confidentiality and message integrity and is widely used on the internet.

The current 5G standard does not address the issue of quantum computing. As of August 2020, the standard specifies three ciphers for symmetric encryption and integrity: SNOW 3G, the Advanced Encryption Standard (AES) and ZUC. SNOW 3G and ZUC are stream ciphers, while AES is a block cipher that is used in the counter mode for encryption. Message integrity is provided using the same three algorithms with AES applied in the CMAC mode. The standard defines the key lengths to be 128 bits for each of these algorithms for both encryption and message integrity. However, as we have observed before such a length is not sufficient in the quantum computing model. Authentication in TLS is based on certificates and a public-key infrastucture (PKI). The inclusion of EAP-TLS into 5G makes the internet and its PKI an integral part of the wireless communication architecture and will be inherited into 6G. In fact, the core network of 5G will be implemented as a set of microservices communicating over the internet. The development towards internet technologies is expected to continue in 6G and future generations.

It is envisioned that 6G will constitute the main boundary that connects the digital world to the physical world. Critical applications, such as remote health monitoring, are envisioned to be built based on the connectivity of the 6G network. The development towards cloud and edge native infrastructures that started in 5G is expected to continue in 6G. Virtualization and software-defined functionality will increase. Computations will be performed both in the cloud and on the edge of the network to reduce latency. The importance of the robustness of the trust mechanisms will increase as cyber attacks could even endanger the physical safety of the individuals. The number of Internet-of-Things (IoT) devices will drastically increase. The security of the core network of 6G will also depend heavily on the security of the internet and its security mechanisms such at TLS.

TLS and the PKI have been built using asymmetric cryptography that does not satisfy post-quantum security. However, there are efforts towards standardized post-quantum secure cryptography. These efforts are heavily influenced by the importance of TLS. In fact, the suitability of a particular scheme as a replacement for elliptic curve primitives in TLS will be one of the most important deciding factor which post-quantum secure scheme will eventually be standardized. We shall take a look at the post-quantum alternatives and their characteristics in the next section.

5   Post-quantum Secure Asymmetric Cryptography for 6G

There are public-key primitives that are generally considered to be quantum-safe. These include classic methods such as the McEliece cryptosystem [11] and NTRU [8]. These schemes have survived decades of attacks both in the classical and quantum models and can thus be considered post-quantum secure. However, compared to ECDLP based schemes, efficiency may be poor or the key sizes big. In the recent years, a lot of research has been devoted into the design of efficient post-quantum secure cryptography and standardization efforts are ongoing. The NIST PQC competition for a post-quantum cryptography standard is expected to provide post-quantum secure key exchange and public-key encryption, as well as to augment the Digital Signature Standard (DSS).

In 2020, the competition is in its third round and is expected to yield a standardized set of quantum-secure public-key primitives at the latest after the fourth and final round in 2024. Currently, there are four key exchange schemes and three digital signature schemes considered as finalists in round three and will be considered for standardization already after the third round is over. In addition, eight algorithms have been chosen as alternatives and may be included into the standard after the fourth round. Although there are other standardization efforts, the attention of the cryptographic community is currently fixed on the NIST competition. Therefore, current state-of-the-art results are reflected in its outcomes. It can be argued that the schemes chosen to advance into the third round will offer the best security-performance trade-off and thus will be the main contenders for post-quantum secure 6G. In the following, we briefly describe and analyze these schemes regarding their security assurance, efficiency in key generation, encryption/signing and decryption/verification performance. In addition, we take a look at the private and public key lengths, as well as at the length of the ciphertexts. These parameters will affect the performance of the security protocols and 6G communications.

5.1   Key Establishment and Public-Key Encryption

Post-quantum secure key establishment methods (KEMs) and public key encryption algorithms attempt to replace the Diffie-Hellman key exchange scheme and the RSA cryptosystem. In the general case, public-key encryption schemes are

Table 20.3 Security and the private and public key lengths, as well as the ciphertext length of the third round finalists of the NIST PQC for post-quantum key establishment and public-key encryption (lowest security level)

Scheme	Security	Private key [bytes]	Public-key [bytes]	Ciphertext [bytes]
Classic McEliece	+++++	6452	261,120	128
CRYSTALS-Kyber	+++	1632	800	736
NTRU	++++	935	699	699
SABER	+	992	672	736

Table 20.4 Performance of the third round finalists of the NIST PQC for post-quantum key establishment and public-key encryption (lowest security level, Intel Haswell architecture)

Scheme	Key generation [cycles]	Encryption [cycles]	Decryption [cycles]
Classic McEliece	93,309,536	44,576	132,452
CRYSTALS-Kyber	118,044	161,440	190,206
NTRU	12,506,668	761,236	1,940,870
Saber	98,000	139,000	151,000

required to satisfy a well-established security definition of indistinguishability under an adaptive chosen message attack (IND-CCA2). For the ephemeral use cases, indistinguishability under the chosen plaintext attack (IND-CPA) is sufficient. In the following, we describe the NIST PQC third round finalists and the alternative schemes for KEMs and public key encryption. The security assurance, as well as the key and ciphertext lengths have been collected into Table 20.3. Here, the security is evaluated as the relative assurance on the security of the underlying problem, as well as on the supplied security proofs. It should be noted that some problems have been under attack for decades, while others are more recent suggestions. The performance of the reference implementations on Intel architecture can be found in Table 20.4. The numbers in these tables have been collected from the official webpages and documents of the NIST submissions.

5.1.1 Classic McEliece

The McEliece scheme is based on the original McEliece cryptosystem from 1978 [11]. Its security is based on the NP-hardness of decoding a random linear code. A private key includes the generator matrix of a Goppa error-correcting code and a public-key is derived by scrambling the generator matrix. Due to its long history, the McEliece cryptosystem and its underlying problem has been extensively studied and can thus be considered a conservative, secure choice for post-quantum security. The private and public keys are very large; in several millions of bits for secure parameters in the quantum setting. The ciphertext size is very small and encryption and decryption are efficient, making McEliece a good choice for scenarios where public keys do not need to be generated and exchanged often.

5.1.2   CRYSTALS-Kyber

CRYSTALS is a cryptographic suite that contains the key-encapsulation mechanism Kyber, as well as the digital signature scheme Dilithium. CRYSTALS is based on module lattices and the Module Learning With Errors (MLWE) problem thought to be hard even on a quantum computer. The Learning With Errors (LWE) is a well-studied problem and can be considered secure. The MLWE problem is younger and not as well-studied. However, no algorithms attacking the MLWE that would not apply to the LWE have been found.

Kyber has a simple specification and enables relatively easy adjustment of the security parameter even for optimized implementations. In the current specification, key lengths for 128-bit security (Kyber-768) are 2400 bytes for the private key, 1184 bytes for the public-key and 1088 bytes for the ciphertext. According to NIST, performance is good for most applications. CRYSTALS-Kyber is one of the structured lattice schemes in the third round of the competition and at most one of those is planned to be selected for the standard.

5.1.3   NTRU

NTRU is another structured lattice scheme originally suggested already in the 1990s [8]. Its security is based on the problem of factoring polynomials and depends on the shortest vector problem (SVP) in a lattice. Contrary to the other structured lattice schemes remaining in the competition, NTRU is not based on the LWE problem. Although previously patented, NTRU is currently in the public domain. Due to its age, NTRU is a well-established scheme and has been already standardized by IEEE (IEEE P1363.1), as well as the American National Standards Institute (ANSI) for financial services (X9.98). It has been studied for over 20 years and, therefore, its security can be considered to be on a stronger foundation compared to the other lattice schemes. For the lowest parameter set `ntruhps2048509` designed to offer 128 bits of security, the private key is 935 bytes and the public key and the ciphertext are both 699 bytes. Encryption is relatively fast. However, performance is not on the level of Kyber or Saber. Key generation and decryption are costly compared to the other lattice schemes.

5.1.4   Saber

Saber is a structured lattice scheme based on the the Module Learning With Rounding (MLWR) problem. It is a variant of the MLWE problem, where errors have been replaced by rounding. The algorithm has been designed to be simple, flexible and efficient. Only power-of-two integer moduli are used making secure software and hardware implementation easier. The MLWR problem is relatively new and the security of Saber cannot be currently reduced to the MLWE problem, which NIST sees as a mild concern. For the lowest security level, private keys are

992 bytes, public keys are 672 bytes and the ciphertext length is 736 bytes. Saber offers the best performance among the lattice-based finalists of round three.

5.1.5 Alternate Candidates

In addition to the three finalists, five schemes were chosen to advance to the third round as alternatives. The schemes may be standardized later. The alternative schemes are the following.

1. **BIKE** is a code-based scheme similar to McEliece. Specially structured codes are applied to offer a more balanced performance that approaches that of the lattice-based schemes. However, due to the added structure, the security assurance is lower than for McEliece.
2. **FrodoKEM** is a lattice based scheme that applies the original LWE problem. The LWE problem is the most-studied computational primitive regarding lattice cryptography. Therefore, FrodoKEM can be considered to offer better security guarantees than the structured lattice schemes. However, performance is worse compared to the structured schemes.
3. **HQC** is a code-based scheme with security based on the quasi-cyclic syndrome decoding with parity problem. It has a good security assurance, but the public key and ciphertext lengths are bigger than those of BIKE.
4. **NTRU Prime** consists of two lattice-based schemes. One of these is based on the assumptions of the original NTRU, while the other is inspired by the RLWE problem.
5. **SIKE** follows a completely different approach than the previous schemes. Its security is based on the hardness of computing isogenies of elliptic curves, the supersingular isogeny Diffie-Hellman (SIDH) problem. It has the smallest public keys and ciphertexts of the described schemes. However, the performance is worse than most of the other schemes and the SIDH problem is still less-studied than the other problems.

5.2 Digital Signatures

Post-quantum digital signature schemes are designed to replace or augment the Digital Signature Standard (DSS). Methods in the competition are required to satisfy existential unforgeability under an adaptive chosen message attack (EUF-CMA). In the following, we describe the three third round finalists for post-quantum digital signatures in the NIST PQC competition, as well as the alternative schemes that were also advanced to the third round. The security assurance, as well as the public key and signature lengths have been collected into Table 20.5. As with the key establishment schemes, the security is evaluated as the relative assurance on the security of the underlying problem, as well as on the supplied security proofs.

Table 20.5 Security assurance and public key and signature lengths of the third round finalists for post-quantum digital signatures (lowest security level)

Scheme	Security	Public-key [bytes]	Signature [bytes]
CRYSTALS-Dilithium	+++	1184	2044
Falcon	+++	897	657.38
Rainbow	+	148,500	64

Table 20.6 Performance of the third round finalists for post-quantum digital signatures (lowest security level, Intel architecture)

Scheme	Key generation [cycles]	Signing [cycles]	Verification [cycles]
CRYSTALS-Dilithium[a]	242,532	1,058,483	272,800
Falcon[b]	26,136,000	814,464	158,040
Rainbow[c]	1,302,000	601,000	350,000

[a] Haswell architecture
[b] Skylake architecture using the native floating point hardware (SSE2)
[c] Skylake architecture

The performance of the reference implementations can be found in Table 20.6. The numbers in these tables have been collected based on the documents of the submissions to the NIST PQC.

5.2.1   CRYSTALS-Dilithium

Dilithium is a lattice based digital signature scheme based on the same cryptographic suite CRYSTALS underlying the Kyber key establishment scheme. The security is based on the MLWE problem. The scheme has been designed to be easy to implement securely and efficiently by applying the same parameter set for all security levels. For the same reason, contrary to many other lattice based digital signatures, randomness is generated solely using the uniform distribution which is easy to implement regardless of the platform. Dilithium has good performance regarding key generation, signing and verification and performs well in real world situations. The key and signature lengths are also relative small for a post-quantum scheme; for the lowest security level, the public key is 1184 bytes and a signature 2044 bytes.

It should be noted that only one of the lattice based digital signature schemes will be included into the standard.

5.2.2   Falcon

Falcon is another lattice based digital signature scheme and therefore a contender with Dilithium to be standardized. The security of Falcon is based on the Short Integer Solution (SIS) problem over the same lattice structure to NTRU. Compared to Dilithium, Falcon is more complex to implement securely due to floating point operations and Gaussian sampling. Due to the application of NTRU lattices, signatures are shorter than for other lattice schemes. For the lowest security level, public keys are 897 bytes and the signatures 657.38 bytes. Signing and verification are also efficient and scale well when security is increased. However, similar to NTRU, key generation is less efficient. Table 20.6 lists the performance of the Falcon reference implementation on the Intel Skylake architecture with the hardware floating point operations provided by the SSE2 instruction set [12].

5.2.3 Rainbow

Rainbow follows a different approach compared to the other third round finalists. It is a digital signature scheme based on multivariate polynomials. The underlying construction is a so called multi-layered unbalanced Oil-and-Vinegar scheme. The construction is not as well-studied as the underlying problems of the lattice schemes. Rainbow signatures are small, down to 64 bytes (512 bits) for the lowest security level. However, public keys are large with 148,500 bytes for the same level. Table 20.6 lists the performance of the Rainbow reference implementation on the Intel Skylake architecture without special instructions. However, if the AVX2 vector instructions are available, performance can be increased by over 80% resulting in very fast signing and verification. However, key generation is costly and does not scale well with the security parameter. Due to the large keys, NIST does not see Rainbow suitable as a general purpose digital signature scheme such as those standardized in FIPS 186-4. Rainbow is suitable for scenarios, where vector instructions are available, keys do not have to be generated and exchanged often and small signatures are required.

5.3 Alternate Candidates

1. **GeMSS** follows the multivariate polynomial approach similar to Rainbow, but is based on a different computational assumption that is better-studied. It has even bigger public keys and slower signing than Rainbow. However, signatures are even shorter.
2. **Picnic** is a signature scheme based on a non-interactive zero-knowledge proof of knowledge. It is designed to be highly modular. Its building blocks can be easily exchanged for alternative ones. The public keys are small, but signatures are large and both signing and verification are slow.
3. **SPHINCS+** is constructed entirely of a cryptographic hash function. Therefore, its post-quantum security guarantees can be considered among the strongest of the submissions. Public keys are very small. However, signing is very slow and the signatures are large meaning that the performance would be poor if contemporary digital signatures were replaced with SPHINCS+.

6   Discussion

Quantum computer engineering has progressed rapidly during the last five years. According to the predictions of IBM, quantum volume will double annually in the near future. Provided that such a development is realized, contemporary cryptographic schemes would come under quantum attacks during the lifetime of 6G networks. At the same time, 6G is envisioned to provide connectivity for applications such as remote health monitoring. Steps need to be taken to secure wireless connectivity against quantum attacks in order to provide reliability for these critical applications.

In the 5G specification, symmetric algorithms such as SNOW 3G, AES and ZUC are applied with 128-bit keys for both encryption and message integrity. These key lengths are not sufficient in the post-quantum world. Due to the Grover's algorithm, the key lengths of symmetric primitives need to be doubled. In the future 6G networks, symmetric cryptography has to be implemented with at least 256-bit keys in order to maintain the current security level against quantum attacks. Fortunately, block ciphers such as AES remain secure and applicable. The same is true for contemporary cryptographic hash functions such as SHA-2 and SHA-3. According to the current knowledge, quantum computing does not pose significant challenges to the protocols implemented solely using symmetric primitives provided that the key lengths are adjusted. However, these key lengths need to be maintained also if the communication shifts from 6G to older generation networks. That is, changes need to made to pre-6G network standards and equipment to assure post-quantum security. It should be also noted that the increase to 256 bits in key length will incur a penalty to performance.

In 5G, the core network functionality has become dependent on the internet. Such a development will continue even further in 6G. It will be impossible to separate the security of 6G from the security of the internet. However, the current lack of security on the internet and the amount of security and privacy threats will pose significant challenges to the envisioned dependability of the 6G networks especially regarding critical applications such as remote health monitoring. The security architecture will become highly complex and hard to implement securely. The security mechanisms will be based on those developed for the internet such as TLS, IPSec and DNSSEC and their cryptographic primitives. However, these protocols have been designed for typical internet applications and may not be optimal for the envisioned applications and latency requirements of the 6G networks.

Currently, internet security protocols such as TLS are not secure against quantum attacks due to factoring and DLP based public key cryptography. Fortunately, standardization for post-quantum secure replacements are ongoing. Finalists of the third round in the NIST PQC offer adequate performance and key sizes for typical applications of TLS. However, the applicability of those schemes regarding 6G may differ from their applicability regarding the standard use of TLS on the internet. For example, the strict latency requirements envisioned for 6G place strict requirements on the size of cryptographic keys and the efficiency of the algorithms.

Regarding key establishment and public key encryption, its use cases in 6G need to be carefully evaluated and specified. If public keys do not need to be generated often or exchanged, then Classic McEliece will offer the most succinct ciphertexts with efficient encryption and decryption. However, its public keys are very large and key generation is very slow. If new keys are frequently generated and exchanged, which is the case when ephemeral keys are applied, then Kyber and Saber offer the best performance. NTRU offers faster key generation than McEliece and key lengths similar to Kyber and Saber, but its encryption and decryption are considerably slower.

Digital signatures will be heavily applied through the PKI and the EAP-TLS protocol and its successors in 6G. Falcon offers the smallest public-keys of the three NIST PQC third round finalists. Its signatures are also smaller than those of Dilithium and it offers reasonable signing and verification performance. However, key generation is very slow and its secure implementation is harder than for Dilithium. For a use case, where new keys are needed, Dilithium offers significantly faster key generation than the other two candidates. It also has the benefit of sharing the same design base with Kyber if both key establishment and digital signatures are needed. Rainbow is an interesting candidate. Its signatures are significantly smaller than those of the other schemes. Its performance is also good on hardware that supports vector instructions. However, such hardware might be rare for resource constrained IoT devices. In addition, the public keys are very large making the scheme unsuitable for scenarios where the public keys are not pre-stored on the device.

There are no post-quantum secure cryptographic algorithms that simultaneously offer very small keys and ciphertexts/signatures and have efficient key generation, encryption and decryption or signing and verification. Trade-offs need to made when contemporary asymmetric primitives are replaced with post-quantum secure ones. Such a replacement necessarily incurs costs either in the communication or operational efficiency of the network. Research is needed to identify the correct application of post-quantum secure cryptography in order to satisfy the envisioned performance and functionality of the 6G architecture.

7   Conclusion

Practical quantum computing is expected to be a reality during the lifetime of 6G networks. Therefore, the security architecture of the future 6G needs to provide security against quantum attacks. While symmetric cryptography will remain secure by updating the key length, contemporary public-key cryptography will not. We note that symmetric primitives will need key lengths of at least 256 bits for current level of security. Due to the development towards internet and cloud based architecture, 6G will be dependent on the public key infrastructure of the internet and its security mechanisms. We review the state-of-the-art post-quantum secure public-key primitives for key establishment, encryption and digital signatures selected into the third round of the NIST PQC competition for post-quantum secure cryptography standardization. These schemes have significant differences in their operational characteristics such as key or signature lengths and algorithmic performance. Careful consideration is needed for their optimal application in the 6G security architecture.

Acknowledgments This work is supported by the TrustedMaaS project by the Infotech institute of the University of Oulu, and the Academy of Finland 6Genesis Flagship (grant318927).

References

1. C.H. Bennett, E. Bernstein, G. Brassard, U. Vazirani, Strengths and weaknesses of quantum computing. SIAM J. Comput. **26**(5), 1510–1523 (1997). https://doi.org/10.1137/S0097539796300933
2. G. Brassard, P. HØyer, A. Tapp, Quantum cryptanalysis of hash and claw-free functions, in *LATIN'98: Theoretical Informatics*, ed. by C.L. Lucchesi, A.V. Moura (Springer, Berlin, Heidelberg, 1998), pp. 163–169
3. W. Diffie, M. Hellman, New directions in cryptography. IEEE Trans. Inf. Theory **22**(6), 644–654 (1976)
4. D.P. DiVincenzo, The physical implementation of quantum computation. Fortschritte der Physik **48**(9–11), 771–783 (2000). https://doi.org/10.1002/1521-3978(200009)48:9/11<771::AID-PROP771>3.0.CO;2-E
5. D. Fang, Y. Qian, R.Q. Hu, Security for 5G mobile wireless networks. IEEE Access **6**, 4850–4874 (2018)
6. L.K. Grover, A fast quantum mechanical algorithm for database search, in *Proceedings of the Twenty-Eighth Annual ACM Symposium on Theory of Computing, STOC '96* (Association for Computing Machinery, New York, NY, 1996), pp. 212–219. https://doi.org/10.1145/237814.237866
7. L. Gyongyosi, S. Imre, A survey on quantum computing technology. Comput. Sci. Rev. **31**, 51–71 (2019). https://doi.org/10.1016/j.cosrev.2018.11.002. http://www.sciencedirect.com/science/article/pii/S1574013718301709
8. J. Hoffstein, J. Pipher, J.H. Silverman, NTRU: a ring-based public key cryptosystem, in *Algorithmic Number Theory*, ed. by J.P. Buhler (Springer, Berlin, Heidelberg, 1998), pp. 267–288
9. A. Hosoyamada, Y. Sasaki, Finding hash collisions with quantum computers by using differential trails with smaller probability than birthday bound. in *Advances in Cryptology – EUROCRYPT 2020*, ed. by A. Canteaut, Y. Ishai (Springer International Publishing, Cham, 2020), pp. 249–279
10. M. Latva-Aho, K. Leppänen, Key drivers and research challenges for 6G ubiquitous wireless intelligence. Tech. rep., 6G Flagship, University of Oulu, Finland (2019). http://urn.fi/urn:isbn:9789526223544
11. R.J. McEliece, A public-key cryptosystem based on algebraic coding theory. DSN Progr. Rep. **44**, 114–116 (1978)
12. T. Pornin, New efficient, constant-time implementations of Falcon (2020). https://falcon-sign.info/falcon-impl-20190918.pdf. Accessed 14 Aug 2020
13. J. Proos, C. Zalka, Shor's discrete logarithm quantum algorithm for elliptic curves. Quantum Inf. Comput. **3**(4), 317–344 (2003)

14. R.L. Rivest, A. Shamir, L. Adleman, A method for obtaining digital signatures and public-key cryptosystems. Commun. ACM **21**(2), 120–126 (1978). https://doi.org/10.1145/359340. 359342
15. P.W. Shor, Algorithms for quantum computation: discrete logarithms and factoring, in *Proceedings 35th Annual Symposium on Foundations of Computer Science* (1994), pp. 124–134
16. M. Ylianttila, R. Kantola, A. Gurtov, L. Mucchi, I. Oppermann, Z. Yan, T.H. Nguyen, F. Liu, T. Hewa, M. Liyanage, A. Ijaz, J. Partala, R. Abbas, A. Hecker, S. Jayousi, A. Martinelli, S. Caputo, J. Bechtold, I. Morales, A. Stoica, G. Abreu, S. Shahabuddin, E. Panayirci, H. Haas, T. Kumar, B.O. Ozparlak, J. Röning, 6G white paper: research challenges for trust, security and privacy. Tech. rep., arXiv eprint 2004.11665 (2020). https://arxiv.org/abs/2004.11665

Chapter 21
6G: Open Issues and Concluding Remarks

Aloknath De, Harpreet S. Dhillon, Madhan Raj Kanagarathinam, and Abhishek Roy

Through generation of communication technologies, the human society has become connected more and more intensely. And communication is no more tethered, wireless communication has become all-pervasive. As 5G systems are getting deployed in major parts of the world, academia and industry have initiated research activities towards conceptualizing a 6G communication system. Beyond enhancing certain features of 5G system, 6G is expected to satisfy unprecedented requirements that 5G possibly can't meet. These requirements are in consonance with mega trends that are driving technology evolution. In this book, we have delineated many of the underlying requirements pertinent to architecture, performance and integrity of such an advanced communication system.

Users are experiencing a plethora of rich-media services. It is not just voice, but data, image, animation, video, multimedia have become an integral part of our communication content. Also, users are inclined to edit, augment and share impromptu. Being able to interact with these media instantaneously implies fast computing and high-rate transmission, resulting in low latency. Since the advent of internet, humans are becoming more and more connected, and more so wirelessly in recent

A. De (✉)
CTO, Samsung R&D India, Bangalore, Karnataka, India
e-mail: aloknath.de@samsung.com

H. S. Dhillon
Bradley Department of ECE, Virginia Tech, Blacksburg, VA, USA

M. R. Kanagarathinam
Mobile Communication R&D Department, Samsung R&D Institute India-Bangalore, Bangalore, India

Department of Computer Science and Engineering, Indian Institute of Technology Madras, Chennai, India

A. Roy
Wireless Standards Team, MediaTek USA Inc., Irvine, CA, USA

© The Author(s), under exclusive license to Springer Nature Switzerland AG 2021
Y. Wu et al. (eds.), *6G Mobile Wireless Networks*, Computer Communications
and Networks, https://doi.org/10.1007/978-3-030-72777-2_21

449

time. With realization of internet-of-things, it is also going beyond people-to-people communication. A paradigm shift is in progress—people get considered as end-points and machines gain their stature in human-machine multimedia interactions.

Human-to-machine and machine-to-machine communication are on the growth trajectory. On another landscape, physical and digital world are getting intertwined slowly but steadily to build integrated systems. These new systems are often referred to as Phygital Systems or Cyber-Physical Systems (CPS). It is envisaged that 6G communication network will be the backbone for such manifestation. When it is realized fully, it will serve as a social infrastructure that enriches people's daily lives. The trend reveals that each generation of wireless system has taken approximately a decade to research, develop, standardize, productize and deploy. 6G system is no exception. With its ambitious goal, it may take up to 2030 to deploy such a system widely.

In an overall sense, 6G vision [1–3] is to make the world hyper-connected: a world that enables faster sharing of content, a world that presents rich-media content. Novel 6G services will emerge due to advances in communications alongside associated technologies such as sensing, imaging and displaying. Exponential growth of cutting-edge technologies such as AI, IoT, Robotics are driving directional shift in the industry. The cross-play of technologies and the definition of new services will shape 6G system incarnation. Many research clusters in the world are actively studying plausible technologies that bolster 6G vision. In this book, the authors from different corners of the globe have presented their perspectives and the solution approaches on 6G.

We have covered here seven dimensions of 6G research:

1. Base layer design for Tera-hertz communication.
2. High-precision network, dynamic topology and open source.
3. Machine-communication in IoT with thrust on energy efficiency.
4. All-pervasive AI with learnability and explainability.
5. Edge and fog computing with split processing.
6. Security, privacy and trustworthiness of system.
7. Rich-media use cases and services.

1   Tera-Hertz Communication

Wireless data traffic volume pertaining to connected things is expected to grow due to sizeable increase in the device connections in a given cubic meter area. Additionally, rich-media services such as sending holographic videos will require a bandwidth that is currently unavailable in the mm-wave spectrum. Hence, a broader RF spectrum bandwidth has become a necessity and can only be found at the sub-THz and THz bands. FCC in US has opened the spectrum between 95 GHz and 3000 GHz for experimental use and unlicensed applications to encourage the development of new wireless communication technologies. Following this trend, THz band (0.1–10 THz) is likely to be used in 6G communication system. The absorption lines for oxygen and water are mostly located in the THz band. This requires studying tractable yet accurate THz multipath channel models for both indoor and outdoor environments. This poses serious challenge as there is lack of existing efficient devices that can generate and detect signal at THz frequencies. As severe path-loss and atmospheric absorption is quite evident in THz, it has to be overcome by utilizing very large antenna arrays at base stations. Sophisticated techniques for beamforming have to be developed for energy efficiency. Multiple-input, multiple-output (MIMO) is now going to shape up as ultra-MIMO for an overall system performance.

Communication links at THz frequency band will depend on line-of-sight as well as focused-reflected paths, not on scattering and diffracting paths. The question arises as to whether OFDM will remain waveform of choice or some other waveform will become more suitable to support GHz-wide channels and yet of low-complexity for THz operation. Intelligent Reflecting Surfaces (IRSs) that tune the wireless propagation environment with an array of IRS units are expected to play a critical role in 6G system design. Semiconductor technologies based on InP, GaAs, SiGe and even CMOS are generating power in the range of mW with reasonable efficiency, but further research in this direction is a must to make it a commonplace technology in 6G. Energy-efficiency will further demand associated system improvement in IC and RF technologies from ADC-DAC to low-loss antennas. Early chapters of the book have covered the design of Physical and Access control layers and suggested techniques to overcome some of these challenges.

2 High-Precision Network Configuration

Instead of fixed topological configuration, flexible network deployment helps to accommodate an increase in data traffic, or to fill holes in populace coverage, or to form an ad-hoc network for provisioning critical services. In 6G, mesh-type network can be a major topology towards deploying flexible and adaptive network. An automated addition, configuration and optimization of nodes will, in turn, reduce significantly the effort for network planning. In order to realize a high-performing network, the multi-pathing, multi-homing and dynamic mobility are to be considered. Therefore, high-precision and dynamic network configuration will entail additional work to maintain service continuity.

As 6G communication system strives to have coverage of the entire globe, the design grows in complexity as well as interactions among constituent software modules. For realizing complex network functions, open source modules are being conceptualized for core networks and base stations. This lowers barriers to market entry, promote interoperability and reduce development cycles. Open and Intelligent Radio Access Network (O-RAN), Open Network Automation Platform (ONAP) are two established open source platforms that are making revolution in the network design. Architecturally, 6G system is likely to include 'open source'-based intelligent software modules. However, the openness of communication will increase system vulnerabilities and attack surfaces which need to be counteracted through other means.

The very concept of cellular structure is evolving and it will become quite a cell-less system by 6G deployment time. On another front, the group mobility is becoming a commonplace. Communication systems are being designed to provide an efficient support of mobile devices that are moving as a group on a bus, a train or an aeroplane. In 6G, for globe-wide coverage, non-terrestrial network components are also embraced. (Nano)-satellites, high-altitude platform station (HAPS) are becoming part and parcel of integrated communication system. They are different in nature than the terrestrial networks, in terms of supporting large cell sizes, moving cells, long propagation delays, large path-loss and large Doppler shifts. Three chapters have highlighted network-linked challenges and plausible solution approaches.

3 Machine-Type Communication

By 2030, smart cities, smart manufacturing and other smart entities would be shaping up significantly [4]. Ubiquitous connectivity anytime and anywhere will facilitate new possibilities at home, in entertainment, in work, and in providing citizen services. The recently-introduced 5G New radio is natively designed for supporting machine-type communication (MTC) in order to improve the overall efficiency of different verticals, from healthcare to logistics. Yet, as far as machine communication is considered, this is the first-generation of technology. Many demanding requirements of MTC can't fully be supported by 5G network. And these requirements are only expected to grow leaps and bounds in the next decade.

Machine-type devices (MTDs) are of various categories, forms and size. MTDs, at the extreme end of the IoT ecosystems, are mostly low-power sensors and they are powered by micro-batteries or energy harvesters. The fundamental question that arises as to what design enablement has to happen for such low-cost, low-power receivers. How do we realize highly-efficient sleep modes to support ultra-low power MTC devices? How can new service classes characterizing mission-critical and dependable MTCs be supported through multifaceted connectivity? During this decade, it is expected that UAV-based swarms will be deployed in shop floors, connected logistics and emergency responses. They will collectively perform a set of tasks in a distributed fashion for a common goal. We will need robust connectivity solutions for such complex but localized MTC networks.

With variety of heterogeneous MTDs and service requirements, MTC service classes for 6G will have to be enlarged. The current literature suggests that there will be five classes under two broad categories of critical MTC (cMTC) and massive MTC (mMTC): (a) dependable cMTC supporting extreme ultra-reliability and low latency along with security, positioning information, e.g., autonomous driving, (b) broadband cMTC supporting mobile broadband data with high reliability and low latency, e.g., robotic-aided surgery, (c) scalable cMTC supporting massive connectivity with high reliability and low latency, e.g., factory monitoring, (d) globally-scalable mMTC supporting ultra-wide network coverage throughout all space dimensions, e.g., underwater communication and (e) zero-energy mMTC supporting massive deployment of devices with long battery life and network lifetime, e.g., precision agriculture. Two of the chapters have described how 6G will facilitate the growth of MTC in the context of IoT.

4 Augmented Intelligence

6G will need to enable greater levels of autonomy, improve human-machine interaction and achieve resilient connectivity in diverse environments. In 6G, the system will have to be designed so as to include AI at different layers of communication stack. Knowing possibilities with AI upfront will make 6G system more profound. In that case, the suitable intelligence would be incorporated during the design phase itself instead of a posteriori application in 5G [5, 6]. AI may be used for local decision such as source compression or channel coding. It may be used for joint parametric optimization such as source-channel coding. Or even AI could be used in end-to-end analysis such as system error performance with possible reasons of error.

There is no denial that spectrum is a scarce resource. Therefore, even if an operator owns license for certain spectrum, in future, an underutilization will call for spectrum sharing with other operators who have a dire need for services. With the help of AI algorithms, allowing opportunistic use of the underutilized spectrum by others will make the best use of the limited spectrum resources. The key challenge in this dynamic spectrum sharing process is avoiding collision of spectrum usage among different entities. Theoretically, to prevent such collisions, network operators could exchange all relevant spectrum access information. But, that would impose a significant communication overhead. Therefore application of AI, with a limited amount of information exchange, would be helpful for predicting spectrum usage by other entities. Like the spectrum sharing issue, AI could be applied judiciously in resolving many focused issues in 6G system.

On the other hand, as we move from conventional algorithms to deep learning-based methods, we lose a certain degree of trust in the system as the algorithmic results are often not explainable. We can't understand the impact of poor data, biased data or malicious data in such solutions. From the legal standpoint, Explainable AI (XAI) will become a necessity in communication system. Neural Network model pruning brings the lower complexity and, in turn, more explainability. But, the challenge is to keep high performance of algorithms, yet increase in explainability of results. In this regard, the progress in research is quite evident in the PHY and MAC layers, but they will have to be extended beyond. As 6G manages a wide range of safety-critical tasks and mission-critical services, the need for including a framework with increased XAI is a must for enabling trust in future wireless systems.

Another strong application of AI is in the context of MTC and IoT. The digital replica of a physical entity, including people, devices, objects, systems and even places, in a virtual world is called a digital twin. Without having a temporal or spatial constraint, the users will be able to analyze and monitor the reality in a virtual world. Interestingly, even action in physical world can be triggered by interacting with its digital twin by defining event, capturing device-state and modifying as appropriate. In fact, the introduction of AI could complete any action automatically without human intervention. However, these would mean tera-bytes per second data transmission for monitoring a cubic meter area. Can 6G system design incorporate explainable AI-based digital twin whereby real system provides the necessary performance whereas the XAI twin will offer explanations? In this book, two chapters have addressed the role of AI in 6G.

5 Edge Computing with Computational Split

Smartphone has been the most prominent smart device, but now intelligent devices include smart watch to sound buds, virtual reality (VR) headsets to augmented reality (AR) glasses, smart sensors to intelligent displays, home appliances to assistive robots, delivery drones to autonomous vehicles and so on. In future, intelligent devices will utilize higher sensory resolution and leverage 6G communication infrastructure to deliver performance superior than that is possible by human being. In fact, it will become a norm that intelligent devices leverage, process and present information to human being in a perceptually meaningful way. However, given that the devices will tend to be become thinner and lighter, not always processing can be done on-device. For example, AR glasses should be as light, thin and small as regular glasses to meet the user's expectations.

The aspiration in service and software requirement for 6G is so high that the future hardware evolution may not be able to cope up at the same pace. 5G brings in edge computing, 6G will usher in communications and computing convergence. Devices would do local computation and execution for processing fast as well as preserving privacy. However, when the algorithmic complexity is high, the processing could be split and a part of the computation could be offloaded to the other adjacent devices. It will be prudent to evolve an open source framework or a flexible open standard than any proprietary definition of boundary for computation offload. For example, it could give choice to a device to do complete operation; or to do at select pre-specified level and hand over to other processing entities. This will also require synchronization of a large amount of data, context and the program itself among network entities.

A broad vision for 6G Edge Intelligence, encompassing transition from IoT to 'Internet of Intelligent Things', is brewing. For example, sensors and edge processing systems embedded within vehicles will utilize real-time wireless connectivity of an Intelligent Transport System (ITS). In such intelligent systems, the processing is done sometimes with few powerful fog servers nearer to the core network. At times, the computation is done with many edge servers closer to the user devices. Over two chapters, we have covered the edge and fog computing alongside federated learning deployed across intelligent devices.

6 Security, Privacy and Trustworthiness

In a connected world, accrued benefits are enormous; but so are security concerns. User devices can be hacked until and unless these devices provide a sufficiently secure trusted environment. Conventional authentication, authorization and accounting processes are neither cost-effective nor scalable if directly applied to a large set of connected MTDs. To secure the data in MTC in the age of quantum computing, lightweight and flexible quantum computer-resistant (or post-quantum) encryption and authentication schemes need to be considered. Quantum key distribution (QKD) is another direction towards ensuring the long term security of data. The security of QKD is based on the quantum effects requiring an optical channel, whereas the post-quantum cryptography aims at circumventing the threat posed by quantum computers.

Future transaction systems are getting built on the premise of 'zero trust'. Distributed ledger technology (DLT) allows value transactions between parties through a decentralized trust process. By 2030, MTC networks will expand DLT's application horizons as the need increases to transfer valuable, authenticated sensor data, services or micro-payments among IoT and intelligent devices. Once data privacy agreement is respected, user information could be analyzed for betterment of services. In fact, even anonymized rendition of personal identifiable information (PII) could be utilized to enhance quality-of-experience for the services provided by network operators. 6G envisages far more immersive engagement with the network. The diversity and volume of novel IoT devices and their control systems will continue to pose significant security and privacy risks; and additional threat vectors will emerge as we move towards 6G system.

The challenges in creating a trustworthy 6G are quite multidisciplinary in nature; they span from technology, regulation, techno-economics, politics to ethics. As 6G moves toward THz spectrum with much higher bandwidth, more densification and cloudification for a hyper-connected world is being pursued utilizing the concepts of security function softwarization and virtualization. The development of AI blurs the line between reality and fake content and helps to create ever more intelligent attacks. Therefore, trust modeling, trust policies and trust mechanisms need to be defined comprehensively. Now, with the wide deployment of AI, even the neural network models have to be made attack-proof by developing defense mechanisms that can recognize targeted attacks against Deep Learning and XAI engines. Three chapters of the book have discussed the roles of trust, security and privacy in next generation networks; unique and interwoven dimensions are analyzed. In particular, they have also captured the role of quantum technology in 6G and the emergence of cryptographic technology post-quantum era.

7   Rich-Media Services

Many new services would be created through hyper-connectivity realized through 6G, namely truly immersive extended reality (XR), high-fidelity mobile hologram and digital replica. XR implementation requires advanced device form-factors, such as hand-held components to support mobile and active software content. Yet, the progress in hardware performance, especially mobile computing power and battery capacity can't fully keep pace with the boom of XR requires. These would have to be overcome by offloading computation to more powerful devices or servers around. Whether it is interactive tactile internet, or a fast-moving game, system latency has to be low. Performance target would include air latency less than 100 micro-seconds, end-to-end latency less than 1 ms, and extremely low jitter in the order of microseconds.

Hologram is a next-generation media technology that can present gestures and facial expressions by means of a holographic display. Real-time capturing, transmission and 3D rendering techniques will ensure that mobile devices are able to render media for 3D hologram displays. For such real-time services, hundred of times greater than the current 5G system data transmission rate would be required. A detailed calculation [1] establishes that hologram display over a mobile device would require at least half Tbps vis-à-vis the typical peak data rate of 20 Gbps in present 5G systems.

The exponential growth in the number of connected machines will require 6G to support about 10^7 devices per square kilometre. This is ten times larger than the connection density requirement of 5G. The surging need for the ultra-dense deployment of industrial IoT devices will require 3D connectivity. Unmanned aerial vehicles (UAV)-based swarms have started finding application in a myriad of use cases. Industry 4.0 has brought in dominant data collection and analytics. Moving forward, Industry 5.0 will involve much more interactivity, both local and remote. AI-infused digital twin and cyber-physical systems would demand the power of 6G system. Our societies will become digitized and data-driven; and many IoT verticals will reap the benefits of emerging technologies. The first few chapters of this book have outlined meaningful use cases leading to new business models.

To sum all up, United Nations' 2030 Sustainable development Goals (SDG) are towards building an inclusive, trustworthy and self-sustainable society. A powerful communication infrastructure could address many of these goals directly or indirectly. Many countries around the globe are gearing up for serious research to lead this revolution. For 6G, it is expected that ITU-R will start working to define a 6G vision soon. Commercialization of 6G systems is expected to happen as early as 2028 while broader deployment may occur around 2030.

References

1. Samsung Research, 6G: The Next Hyper-Connected Experience for All, 2020
2. 6gchannel.com, Series of Whitepapers on 6G (Key Drivers, Networking, Machine learning, Edge Intelligence, Trust and other themes), 6G Flagship Project, University of Oulu, Finland, 2019 and 2020
3. NTT Docomo, '5G Evolution and 6G', Whitepaper, January 2020
4. N.H. Mahmood et al., Critical and Massive Machine Type Communication towards 6G, Whitepaper in arXiv, April 2020
5. S. Ali et al., Machine Learning in Wireless Communication Networks, 6G Whitepaper, April 2020
6. W. Guo, Explainable Artificial Intelligence (XAI) for 6G: Improving Trust Between Human and Machine, November 2019

Correction to: 6G Mobile Wireless Networks

Yulei Wu, Sukhdeep Singh, Tarik Taleb, Abhishek Roy, Harpreet S. Dhillon, Madhan Raj Kanagarathinam, and Aloknath De

Correction to:
Y. Wu et al. (eds.), *6G Mobile Wireless Networks*,
Computer Communications and Networks
https://doi.org/10.1007/978-3-030-72777-2

Author Osama Zwaid Alsulami name is updated to Osama Zwaid Aletri throughout the book.

The updated version of these chapters can be found at
https://doi.org/10.1007/978-3-030-72777-2
https://doi.org/10.1007/978-3-030-72777-2_10
https://doi.org/10.1007/978-3-030-72777-2_14

Index

© The Author(s), under exclusive license to Springer Nature Switzerland AG 2021   459
Y. Wu et al. (eds.), *6G Mobile Wireless Networks*, Computer Communications and Networks, https://doi.org/10.1007/978-3-030-72777-2

Printed in the United States
by Baker & Taylor Publisher Services